MW01518294

The Handbook of Environmental Chemistry

Founded by Otto Hutzinger

Editors-in-Chief: Damià Barceló · Andrey G. Kostianoy

Volume 20

The Handbook of Environmental Chemistry
Recently Published and Forthcoming Volumes

Emerging Organic Contaminants and Human Health

Volume Editor: Damià Barceló

With contributions by

A. Alastuey · N. Ali · B. Artíñano · P.J. Babin · C. Balducci ·
C. Barata · D. Barceló · M. Casado · A. Cecinato · A. Covaci ·
M.L. de Alda · M.S. Díaz-Cruz · A.C. Dirtu · M. Faria ·
M. Farré · M.A. Fernández · A. Galletti · M.J. García Galán ·
T. Geens · A. Ginebreda · B. Gómara · M.J. González ·
A.C. Ionas · A. Jelić · M. Köck-Schulmeyer · C.M. Lino ·
M. Llorca · M.J. López de Alda · P. López-Mahía ·
C.M. Manaia · G. Malarvannan · L. Meisel · J.M. Navas ·
O.C. Nunes · A. Olivares · E. Oliveira · S. Pelayo · A. Pena ·
F. Pérez · M. Petrović · B. Piña · C. Postigo · X. Querol ·
D. Raldúa · S.D. Richardson · L. Roosens · L.J.G. Silva ·
B. Thienpont · N. Van den Eede · I. Vaz-Moreira ·
P. Verlicchi · M. Viana

Springer

Editor
Prof. Dr. Damià Barceló
Department of Environmental Chemistry
IDAEA-CSIC
Barcelona, Spain
and
Catalan Institute for Water Research (ICRA)
Scientific and Technological Park of the
University of Girona
Girona, Spain

The Handbook of Environmental Chemistry
ISSN 1867-979X   ISSN 1616-864X (electronic)
ISBN 978-3-642-28131-0   ISBN 978-3-642-28132-7 (eBook)
DOI 10.1007/978-3-642-28132-7
Springer Heidelberg New York Dordrecht London

Library of Congress Control Number: 2012935241

Printed on acid-free paper

Springer is part of Springer Science+Business Media (www.springer.com)

The Handbook of Environmental Chemistry
Also Available Electronically

The Handbook of Environmental Chemistry is included in Springer's eBook package *Earth and Environmental Science*. If a library does not opt for the whole package, the book series may be bought on a subscription basis.

For all customers who have a standing order to the print version of *The Handbook of Environmental Chemistry,* we offer free access to the electronic volumes of the Series published in the current year via SpringerLink. If you do not have access, you can still view the table of contents of each volume and the abstract of each article on SpringerLink (www.springerlink.com/content/110354/).

You will find information about the

– Editorial Board
– Aims and Scope
– Instructions for Authors
– Sample Contribution

at springer.com (www.springer.com/series/698).

All figures submitted in color are published in full color in the electronic version on SpringerLink.

Aims and Scope

Since 1980, *The Handbook of Environmental Chemistry* has provided sound and solid knowledge about environmental topics from a chemical perspective. Presenting a wide spectrum of viewpoints and approaches, the series now covers topics such as local and global changes of natural environment and climate; anthropogenic impact on the environment; water, air and soil pollution; remediation and waste characterization; environmental contaminants; biogeochemistry; geo-ecology; chemical reactions and processes; chemical and biological transformations as well as physical transport of chemicals in the environment; or environmental modeling. A particular focus of the series lies on methodological advances in environmental analytical chemistry.

Series Preface

With remarkable vision, Prof. Otto Hutzinger initiated *The Handbook of Environmental Chemistry* in 1980 and became the founding Editor-in-Chief. At that time, environmental chemistry was an emerging field, aiming at a complete description of the Earth's environment, encompassing the physical, chemical, biological, and geological transformations of chemical substances occurring on a local as well as a global scale. Environmental chemistry was intended to provide an account of the impact of man's activities on the natural environment by describing observed changes.

While a considerable amount of knowledge has been accumulated over the last three decades, as reflected in the more than 70 volumes of *The Handbook of Environmental Chemistry,* there are still many scientific and policy challenges ahead due to the complexity and interdisciplinary nature of the field. The series will therefore continue to provide compilations of current knowledge. Contributions are written by leading experts with practical experience in their fields. *The Handbook of Environmental Chemistry* grows with the increases in our scientific understanding, and provides a valuable source not only for scientists but also for environmental managers and decision-makers. Today, the series covers a broad range of environmental topics from a chemical perspective, including methodological advances in environmental analytical chemistry.

In recent years, there has been a growing tendency to include subject matter of societal relevance in the broad view of environmental chemistry. Topics include life cycle analysis, environmental management, sustainable development, and socio-economic, legal and even political problems, among others. While these topics are of great importance for the development and acceptance of *The Handbook of Environmental Chemistry,* the publisher and Editors-in-Chief have decided to keep the handbook essentially a source of information on "hard sciences" with a particular emphasis on chemistry, but also covering biology, geology, hydrology and engineering as applied to environmental sciences.

The volumes of the series are written at an advanced level, addressing the needs of both researchers and graduate students, as well as of people outside the field of "pure" chemistry, including those in industry, business, government, research establishments, and public interest groups. It would be very satisfying to see these volumes used as a basis for graduate courses in environmental chemistry. With its high standards of scientific quality and clarity, *The Handbook of*

Environmental Chemistry provides a solid basis from which scientists can share their knowledge on the different aspects of environmental problems, presenting a wide spectrum of viewpoints and approaches.

The Handbook of Environmental Chemistry is available both in print and online via www.springerlink.com/content/110354/. Articles are published online as soon as they have been approved for publication. Authors, Volume Editors and Editors-in-Chief are rewarded by the broad acceptance of *The Handbook of Environmental Chemistry* by the scientific community, from whom suggestions for new topics to the Editors-in-Chief are always very welcome.

Damià Barceló
Andrey G. Kostianoy
Editors-in-Chief

Volume Preface

Global changes, including socio-demographic and environmental issues, are challenges to our society. Drivers of global change are climate change, population growth, urbanization, industrialization, and rising income, living standards, and water and energy demand. These forces will be confounded by slowing productivity growth, falling investment in irrigation and agriculture worldwide, loss of biodiversity, risk to public health, and water scarcity, among other issues. Future population growth and water scarcity pose significant risks to global food security, as it has been already pointed out by Professor John Beddington, UK Government Chief Scientist, in March 2009, by the so-called "Perfect Storm" of problems by 2030 [1]. This perfect storm involves food shortages, scarce water, and insufficient energy resources that threaten to unleash public unrest, cross-border conflicts, and mass migration as people flee from the worst affected regions.

It is nevertheless remarkable that water, sanitation, and health nexus were among the earliest issues being reported. The connection between human health and well-being and access to sufficient drinking water has long been recognized. Public health and epidemiology started on the concept of water-borne diseases, and the nature of human exposure to bacteria in polluted waters has driven the mandate for sanitation and hygiene, still important throughout the world today. Already in 1514 anonymous maps displayed drainage to improve public health in Italy. Through the London epidemics of 1849 and 1854, Snow [2] verified his hypothesis that contaminated water was the critical variable in cholera transmission, when he plotted cases and the area of water distribution.

But we know that human exposure to different contaminants takes place also via food, air, and dust. The influence of diet on human concentration of persistent organic pollutants or the links between air pollution and adverse health effects has been recognized.

A lot of information already exists on regulated contaminants and human health, but there is less information on the influence of the so-called emerging contaminants and nanomaterials. Due to the fact that most of the emerging organic contaminants are not regulated, a few studies are available in relation to human health issues. For this reason I think that this book is timely due to increased interest in the last years to bridge human health with environmental and food contamination. The establishment of relationships between human health and levels some of these emerging contaminants in body fluids is taking place at global scale, from USA, China, and EU countries. Links between antibiotic resistance due to the use of large amounts of antibiotics for human and veterinary purposes, or the direct relationship between levels of drinking water disinfection by products with bladder cancer, asthma, genotoxicity, and cytotoxicity were established. One of the most recent issues of concern is the use of nanomaterials in food industry via food packaging and their way that these nanomaterials migrate to the food. The European Food Safety Authority (EFSA) has already published a report on that emerging issue.

The book contains 14 chapters that cover several chemical groups of emerging organic contaminants, several of them are persistent, bioaccumulative, and toxic (PBT) while others are associated with other effects such as endocrine disruption, antibiotic resistance, and bioaccumulation in biota. One of the groups with more chapters on this book are the pharmaceuticals with emphasis on antibiotics and on all the problems associated with the increased pharmaceutical products used in hospitals as well as the issue of ecopharmacovigilance that was introduced in 2008. Other emerging contaminants reported are brominated flame retardants, polar pesticides, phthalates, phosphate esters, perfluorinated compounds, personal care products, musks, and illicit drugs among others. The various chapters describe levels in environmental, food, and health matrices with the exception of the two chapters of the book that dealt with toxicological and ecotoxicological issues of the emerging contaminants.

This book is intended for a broad audience, from analytical chemists, environmental chemists and engineers, toxicologists, ecotoxicologists, and epidemiologists working already in this field as well as newcomers including students in their first years of their Ph.D. who want to learn more about this issue. Finally, I would like to thank all the authors for their time and efforts in preparing the corresponding chapters that make this book unique in this HEC series.

Barcelona, Spain   Damià Barceló

References

1. Charles H, Godfray J, Beddington JR et al (2010) Food security: The challenge of feeding 9 billion people. Science 327:812–818
2. Snow J (1855) On the mode of communication of Cholera. John Churchill

Contents

Emerging Organic Contaminants and Nanomaterials in Food

Marinella Farré and Damià Barceló

Abstract Governments all over the world are intensifying their efforts to improve food safety. These efforts come as a response to an increasing number of food safety problems and rising consumer concerns. In addition, the variety of toxicant residues in food is continuously increasing as a consequence of industry development, new agricultural practices, environmental pollution, and climate change. This paper reviews the major groups of emerging contaminants in food, as well as, the levels of concentrations reported and the analytical approaches presented for their detection with special emphasis on more fast and cost-efficient methods of detection.

The four main groups of emerging food contaminants that are discussed here are:

1. Industrial organic pollutants: Perfluorinated compounds (PFCs), polybrominated diphenylethers (PBDEs), new pesticides, and nanomaterials.
2. Pharmaceutical residues: Antibiotics and coccidiostats
3. Biotoxins: Emerging groups of marine biotoxins

Keywords Biotoxins, Coccidiostats, Food contaminants, LC-MS/MS, Nanomaterials antibiotics, Perfluorinated compounds, Pesticides, Polybrominated diphenylethers

M. Farré (✉)
Department of Environmental Chemistry, Institute of Environmental Assessment and Water Studies, IDAEA-CSIC, C/Jordi Girona 18-26, 08034 Barcelona, Spain
e-mail: mfuqam@cid.csic.es

D. Barceló
Department of Environmental Chemistry, Institute of Environmental Assessment and Water Studies, IDAEA-CSIC, C/Jordi Girona 18-26, 08034 Barcelona, Spain

Catalan Institute of Water Research (ICRA), C/Emili Grahit, 101, 17003 Girona, Spain

D. Barceló (ed.), *Emerging Organic Contaminants and Human Health*,
Hdb Env Chem (2012) 20: 1–46, DOI 10.1007/698_2011_137,
© Springer-Verlag Berlin Heidelberg 2012, Published online: 2 March 2012

Contents

1   Introduction

Contaminants are substances that have not been intentionally added to food. These substances may be present in food as a result of the various stages of its production, packaging, transport, or holding. They also might result from environmental contamination. Since contamination generally has a negative impact on the quality of food and may imply a risk to human health, the EU has taken measures to minimize contaminants in foodstuffs.

There are many thousands of chemical substances in food; most of them being of natural origin. A number, however, are man-made and arise from the use of agrochemicals, or due to pollution of water, soil and air, or occur during food preparation/processing. In addition, food may contain biological contaminants. A range of additives may also be added for a variety of purposes (e.g., to enhance the flavor, color, improve stability).

Therefore, while consumers expect the food that they eat to be safe, as a consequence of industrial development, pollution, and climate change, the variety of food contaminants are increased. Currently, one of the great challenges in food safety is controlling the risks associated with mixtures of contaminants, which continuously are changing.

Among the most prominent groups of emerging food contaminants can be considered industrial origin contaminants as perfluorinated compounds (PFCs), polybrominated biphenyls (PBBs), the new generation of pesticides, nanomaterials, and emerging groups of marine biotoxins (such as palytoxins and spirolides). Many of them are of particular concern because they can cause severe damages in human health; for example, some of them are suspected to be cancer promoters. Other selected compounds have been related to endocrine disruptor effects or can be accumulated and biomagnified through the food chain.

In this review, we have been selected the most relevant groups of emerging food contaminants. We also included some groups of pharmaceuticals of special concern such as antibiotics which can create bacterial resistances and which are illegally used as growth promoters. The main sources of the selected groups of contaminants will be discussed together with their toxicological data and concentrations reported during the last few years. The strategies for their analysis including sample preparation, separation, and detection will be presented.

2 Sources of Food Contamination, Properties and Toxicological Properties

2.1 Industrial Origin Compounds

Since the industrial revolution in the nineteenth century, the knowledge on chemistry was developed rapidly, together with several industries. Currently, under the REACH legislation more than 100,000 compounds have been pre-registered for use within the European Union. Chemicals are present in all kinds of industrial applications and consumer products. However, some of these compounds or their degradation products can cause damage to the environment and human health.

Food safety, have to face the possible contamination produced during the whole process, including those from environmental contamination or used directly related to the food production (pesticides, veterinary drugs, contamination associated with cooking, processing, packaging, and conservation, among others).

In addition, some compounds are classified as persistent organic pollutants (POPs), because of their resistance to degradation and can be bioaccumulated, show long-range transportation, and are toxic. Most POPs are lipophilic and their uptake rates in organisms are higher than the rate of depuration. This results in an accumulation in aquatic, terrestrial organisms and in humans. Further transfer-up in the food chain can lead to elevated levels in top predators (biomagnification). Their toxic properties can cause serious health damages such as the development of certain types of cancer, metabolic dysfunctions, and endocrine disruption. Initially, 12 chlorinated compounds were classified as POPs and following the ratification of the Stockholm Convention, parties took action in order to reduce the emissions of the 12 POPs. The production and use of POPs was substantially decreased (such as p,p'-DDT), and almost completely stopped for some compounds in most countries. However, some groups of compounds largely used and produced during the last decades meet the definition of POP. Examples of these new POPs are perfluorinated compounds (PFCs), brominated flame retardants (BFRs), such as polybrominated diphenyl ethers (PBDEs), and hexabromocyclododecane (HBCD). The restriction and replacement of some of these compounds should be carefully evaluated. Also, the possible alternative compounds that have to be taken into account include effectiveness, persistence, bioaccumulative effects, toxicological properties, economic feasibility of their production, and human and environmental risk assessment.

On the other hand, during recent years, nanotechnology has emerged, presenting a great variety of new materials, allowing new applications for all industrial sectors including food industry. Nevertheless, while nanotechnology is successfully introduced in the industry, their possible risk to the environment and human health is not well understood and assessed. In addition, food industry has developed a variety of applications based on engineered NPs and nanomaterials, such as high loadings of vitamins and health benefits active in food, new methods for flavor stabilization, as well as, natural food-coloring dispersions can be developed. But NPs and NMs also appear as a new group of possible food contaminants, whereas their detection and characterization in food is poorly developed, and the potential risks of the application or associated contamination in food need to be understood.

In this section, emerging food contaminants with industrial origin, including perfluorinated compounds, polybrominated compounds, new pesticides, and nano-materials will be discussed.

2.1.1 Perfluorinated Compounds

PFCs comprise a large group of compounds characterized by a fully fluorinated hydrophobic linear carbon chain attached to one or more hydrophilic head. PFCs repel both water and oil, and are therefore ideal chemicals for surface treatments. These compounds have been used for many industrial applications including stain repellents (such as Teflon), textile, paints, waxes, polishes, electronics, adhesives, and food packaging [1].

Usage and disposal of PFCs has led to the widespread distribution of these chemicals in the environment. Furthermore, PFOS and PFOA, as well as other perfluorocarboxylic acids (PFCAs) are stable degradation products and/or meta-bolites of neutral PFCs like fluorotelomers alcohols (PFTOHs), perfluorinated sulfonamides (PFASAs), and perfluorinated sulfonamide ethanols (PFASEs) [2]. PFCs are bioaccumulative attached to proteins and, these compounds have been detected in different water matrices [3, 4], wildlife [5, 6], fish [6], and humans [7]. In addition, PFCs are biomagnified in the food chains [5, 8], leading to increased levels in animal-derived foods. Main sources of human exposure to PFCs have been identified through: drinking water, food, and dust inhalation. Bioaccumulation in fish has been shown to be one of the main influences of PFCs in dietary exposure. Food preparation is another relevant source of food contamination [9], but prelimi-nary data on the influence of domestic cookware on levels of PFCs in the prepara-tion of food indicated no elevated levels for a limited number of experiments [10]. Packaging may also introduce PFCs used in greaseproof packaging for fast foods and special packaging. In these situations, PFCs enter into food via migration from the food package [9].

2.1.2 Polybrominated Flame Retardants

Polybrominated biphenyl (PBBs) and polybrominated diphenyl ethers (PBDEs) are classes of additive congeners that were used in a wide range of products, including paints, plastics, textiles, and electronics to reduce their flammability. Electrical and electronic equipment waste (WEEE) is generated on the order of millions of tons annually, bringing significant risk to human health and the environment [11]. The problem connected with the use of brominated flame retardants PBB and PBDE in polymers is the recycling of wastes containing these chemicals, resistance to biodegradation, and the potential for long-range transport and bioaccumulation in the lipid compartments of biota. During the last decade the production and usage of PBBs were discontinued because of their harmful effects, mainly in disturbing steroid hormone secretion. Due to their persistence and long-range transport, both PBBs and PBDES have been detected in many places and populations on Earth [12]. Photochemical transformation studies conducted by the same authors proved that PBBs have a great ability to debrominate higher brominated isomers, leading to identified and unidentified structures of lower brominated PBBs which are believed to be environmentally relevant since they can be formed by natural sunlight and reach the marine food chain [13]. The most probable route for exposure of the general human population to PBBs and PBDEs is through diet, but very little is known about PBB and PBDEs concentrations in fish, which are at the basic level of the food chain. Comparing both groups of brominated compounds, PBBs showed an even higher biomagnification potential than PBDEs, although PBB concentrations were in general lower than those of PBDEs [14, 15]. In a recent study, the contamination level of PBBs in fish from the North and Baltic Sea, freshwater fish from Poland, and cod liver oils from two different Polish manufacturers were assessed. PBB concentrations were also measured in foodstuff samples from Poland, like pork fat, beef meat, and butter. The main conclusion of this study was that almost all fish classes and fish products were contaminated [16].

2.1.3 Pesticides

The largest groups of food pollutants are pesticides. Pesticide testing in foodstuffs is a challenging application involving the simultaneous trace analysis of a wide range of agrochemicals possessing a wide range of physicochemical properties, effects, and toxicities. Considering the number or registered compounds (currently more than 1,000), the continuous introduction of new compounds in the market, and the possible presence of trace amounts of these compounds and their degradation products in food, pesticide monitoring is one of the highest interest fields in food safety. For this reason, numerous regulations such as the European Union directives have set maximum residue limits (MRLs) for pesticides in food [17].

During the last decades a vast literature has reported new methods for analysis, as well as, the levels encountered in foodstuffs. In parallel, many reviews [18–23] have been devoted to revise analytical methods, effects, and levels of concentration of pesticides in food. Due to the high importance of pesticide residue analysis, such discussion is included here, although this review will just present the last tendencies for pesticides control in food.

2.1.4   Nanomaterials

Finally, the last group to be considered is nanomaterials. Nanoscience is an emerging area of science that has the potential to generate radical new products and processes. Concepts in nanoscience provide a sound framework for developing an understanding of the interactions and assembly behavior of food components into microstructure, which influence food structure, rheology, and functional properties at the macroscopic scale. Advances in processes for producing nanostructured materials coupled with appropriate formulation strategies have made possible the production and stabilization of nanoparticles that have potential applications in the food and related industries. During the last few years with the emerging use of nanotechnological materials in many industrial areas, nano-particles appeared as emerging groups of possible food contaminants, first because of their inclusion in the food chain [24, 25] and second as residues from the new technologies in the food industry [26]. One of the main applications of new nanomaterials in food industry is in food packaging, and the main problems are related to the absence of identification and migration control method resources for nanoparticles in food and also to the emergency of risk evaluation from potentially toxic nanoparticles presented in food [27].

2.2   *Pharmaceutical Residues: Antibiotics and Coccidiostats*

Some groups of pharmaceutical residues are of special concern in food safety to consumers, producers, and regulatory agencies. This is the case for antibiotics, which can instigate consequences in human health by producing allergies or reducing their effectiveness against infections, due to extensive or inappropriate use. The antibacterial resistance caused by this extensive use of pharmaceuticals is a potential problem for human medicine since antibiotic-resistant bacteria can pass through the food chain to people. In veterinary practice, antibiotics are utilized at therapeutic levels primarily to treat diseases and to prevent infections. They are also used at subtherapeutic levels to increase feed efficiency and to promote growth in food-producing animals. The frequent and sometimes illegal use of antibiotics may result in residues being found at different concentration levels in products of animal origin such as milk and meat.

   Another important source of antibiotics in human diet is through the ingestion of farmed fish. Farmed fish and shrimp are produced in crowded facilities with inade-quate or nonexistent regulation of antibiotic use. The detection of chloramphenicol (vital for treating typhoid fever) residues in shrimp has activated public awareness worldwide.

Unfortunately, drug resistance in food-borne pathogens is a fact which cannot be avoided since antibiotics are being used in food animals [28]. This resistance covers both intrinsic (natural), such as vancomycin resistance, a glycopeptide antibiotic, in *E. gallinarum*, and transferable (acquired) resistance, like the ampicillin, tetracyclines, macrolides, aminoglycosides chloramphenicol, quinolones resistance in *E. faecium* [29]. Emergent multiresistant strains of *Salmonella* spp. have been associated with an increasing number of human infections in many countries like Spain, UK, Denmark, and Greece since the mid-1990s [30].

2.3 Biotoxins: Emerging Groups of Marine Biotoxins

Marine biotoxins are produced by naturally occurring marine phytoplankton. Marine algal toxins are responsible for more than 60,000 intoxication/year with an overall mortality of about 1.5%. These substances can accumulate in aquatic animals intended for human consumption like filter-feeding mollusks. The toxins are thermoresistant compounds; therefore, normal cooking, freezing, or smoking cannot destroy them.

Most common groups of marine biotoxins are Diarrheic shellfish poisoning (DSP), Paralytic shellfish poisoning (PSP), Amnesic shellfish poisoning (ASP), Neurologic shellfish poisoning (NSP), Azaspiracid shellfish poisoning (AZP), Ciguatera fish poisoning (CFP), Palytoxins, and Spirolides.

Different studies have reported the relations between pollution, climate change, and toxic algal blooms. Certain microalgae in seas and oceans, including dinoflagellates and diatoms, can form extensive monocultures. The conditions favoring this growth include water temperature, sunlight, competing microorganisms, nutrients (eutrophication), wind, and directions of currents. A perceived Increase of harmful algal blooms has been globally registered with important ecological and economic consequences due to their effects on coastal marine resources. In the last decades, various marine dinoflagellates usually living in tropical and subtropical areas, such as benthic dinoflagellates producing palytoxin have been detected in European marine waters [31].

Under recent research EU FP Programs, a great effort has been carried out in order to develop reliable methods for detection of some of these targets in shellfish and water. Some of these methods have been validated in formal collaborative studies, and therefore reference materials are available. The most studied groups are ASP, DSP, and PSP. However, some groups of marine biotoxins are emerging groups that are yet nonregulated. There is a lack of standards and fully validated reliable methods of detection. Those are CFP (causing a range of gastrointestinal, cardiovascular, and neurological symptoms that occur within 1–6 h of ingesting contaminated fish with the toxin and can last for days, months, or years), Palytoxin (typical symptoms of palytoxin poisoning are angina-like chest pains, breathing difficulties, tachycardia, unstable blood pressure, and hemolysis. The onset of symptoms is rapid, and death usually follows just minutes after) and Spirolides (which is a large group with a wide range of toxic effects). This review presents the most recent advances in detection of these toxins for food safety.

3   Analytical Approaches

3.1   Industrial Origin Compounds

3.1.1   Perfluorinated Compounds

During recent years, a great effort of development has been carried out to detect PFCs in environmental samples and biota and different reviews have been published [32, 33]. However, a limited number of works have reported the concentrations levels of PFCs in food.

One of the major problems encountered in PFCs analysis is the cross contamination during the analytical process [34]. The major source of contamination of PFCs in laboratories is the contact with laboratory materials made of, or containing, fluoropolymers such as polytetrafluoroethylene or perfluoroalkoxy compounds [34]. On the other hand, different causes of losses have been identified associated with the adsorption of sample containers, such as glass [35] or polymeric containers such as polypropylene (PP) and high-density ethylene (HDPE) container surfaces. Biodegradation and biotransformations should be also prevented. Good results were obtained when conservations were conducted in the freezer or using combinations of solvents like acetonitrile, and freezing [36] among others.

Due to their inertia and lack of immunogenicity, PFCs are analyzed by instrumental analytical techniques. Such existing methods have been included in this chapter due to the emerging relevancy of PFCs in food safety (Table 1).

For the extraction of food, procedures based on ion-paired extraction have been widely used. This method uses tetrabutylammonium (TBA) hydroxide solution as ion-pairing agent at pH 10 and ethyl acetate as the extractant and has been widely applied for the extraction of food [72, 83]. However, this method has shown to have some disadvantages such as (1) co-extraction of lipids and other matrix constituents and (2) the wide variety of recoveries observed, which are related to matrix effects mentioned above. Liquid solid extraction (LSE) has been also applied to the analysis of biota and food samples [10]. In this case, target compounds are extracted from food by soaking the sample in methanol and shaking for 30 min. After cleanup using active carbon, the extract is ready for analysis. The method does not suffer from matrix effects, and recoveries were in the 80–110% range.

SFE relies on the use of a gas compressed at a pressure and temperature above the critical point (Pc). It consists of a dense gas state in which the fluid combines hybrid properties of liquid and gas. When this technique was launched in 1978, it

Table 1 Summary of analytical methods for emerging food contaminants

Analytes	Analytes	Analytes	Analytes	Analytes	Analytes	Analytes
PBDEs						
PBDEs	Chicken fat	Homogenization, extraction with methylene chloride, filtration and purification through GPC (triphasic silica column and alumina column).	Gas Chromatography (J&W DB-5MS)	Magnetic sector	<1 µg/kg	[37]
BDE #17, #28, #47, #49, #66, #71, #77, #85, #99, #100, #119, #126, #138, #153, #154 & #183.	Diverse and unspecified solid and liquid food samples	Homogenization, blending with 200 ml hexane and 75 g acid modified silica gel and purification through multilayer column.	Gas Chromatography (J&W DB-5MS)	Magnetic sector	50 ng/kg	[38]
BDE #47, #85, #99, #100, #153 & #154.	Chicken fat, beef fat, fish muscle.	MSPD with florisil and purification through an acidic-silica SPE cartridge (elution with 20 ml of hexane). Concentration to 0.5 ml and SPE through a neutral silica cartridge (elution with n-hexane–dichloromethane 80:20)	Gas Chromatography (Agilent HP-5)	[63]Ni ECD	150 ng/kg	[39]
3 monoBDEs, 7 diBDEs, 8 triBDEs, 6 tetraBDEs, 7 pentaBDEs, 5 hexaBDEs, 3 heptaBDEs and the BDE #209.	Fish (muscle and liver)	PLE (hexane:CH2Cl2 1:1)	Gas Chromatography (HP-5MS & DB-5 ms for the analysis of BDE #209)	NICI-MS	2–19 ng/Kg	[40]

(continued)

Table 1 (continued)

Analytes	Analytes	Analytes	Analytes	Analytes	Analytes	Analytes
BDE #47, #99 & #100	Fish (Muscle tissues of salmon and conger eel and liver tissues of sea bass) green mussel	Homogenization, MSPD with sodium sulfate, microwave-assisted extraction with pentane-dichloromethane (1:1) and purification with GPC	Gas Chromatography (DB-5MS)	Q-MS	<100 ng/Kg	[41]
BDE #17, #28, #47, #66, #85, #99, #100, #153, #154 & #183	Adipose tissues from marine mammals, chicken and trout	Adipose Tissues, chicken and trout: MSPD with silica gel/anhydrous sodium sulfate powder, purification thorug GPC extraction with 400 mL of 1:1 (v/v) acetone/hexane mixture	Gas Chromatography (VF-5MS Factor Four, Varian)	IT-MS	0.07–1.3 pg (instrumental limit of detection)	[42]
BDE #17, #28, #47, #66, #85, #99, #100, #153, #154 & #183	Palm oil	Palm oil: Extraction by dialysis in hexane using a semi-permeable membrane. Purification thorugh multilayer column filled with neutral silica, silica modified with sulfuric acid (44%, w/w), and silica modified with KOH.	Gas Chromatography (VF-5MS Factor Four, Varian)	IT-MS	0.07–1.3 pg (instrumental limit of detection)	[42]
BDE #17, #28, #47, #66, #71, #85, #99, #100, #138, #153, #154, #183 & #190	Butter	Dissolution with hexane, purification by GPC	Gas Chromatography (VF-5MS Factor Four, Varian)	QIST-MS	–	[43]

Analytes	Matrix	Sample preparation	Separation	Detection	Concentration	Ref
1 monoBDE, 2 diBDEs, 2 triBDEs, 5 tetraBDEs, 3 hexaBDEs, 1 heptaBDE and the decaBDE	Fish, crab, mussel and oyster	PLE (acetone: dichloromethane 1:3), purification by GPC and filtration through glass fiber filters.	Gas Chromatography (HP-5MS, Agilent)	Q-MS	–	[44]
BDE #28, #47, #99, #100 & #153	Fish filets	Fish filet: ASE (hexane: dichloromethane 1:1), acidic wash and reconstitudion with 50% MeOH	–	ELISA	0.017 (BDE #47) to 370 (BDE #209)	[45]
Mono- to deca-BDEs, OH-PBDEs, MeO-PBDEs	Fish tissue	PLE (hexane: dichloromethane,1:1, v/v), GPC and florisil clean-up (mono-deca-PBDEs), Derivatization by acetylation (OH-, MeO-PBDEs)	Gas Chromatography (J&W DB5)	Magnetic sector	Dry wt basis: 113 pg/glw (monoBDEs), 3.4–39 pg/glw (di-decaBDEs), Wet wt basis: 3.6–41.4 pg/glw	[46]
Tetracyclines: OTC, TC, CTC	Honey	Extraction with buffer pH 4 (citric acid in sodium phosphate dibasic hepthydrate), SPE: C8, ethyl acetate:methanol (75:25, v/v) elution	HPLC (Zorbax Eclipse XDB-C8), sodium acetate and calcium chloride pH 6.9	Fluorescence	8 µg/kg	[47]
Penicillins: AMP, Pen-G, OXA, CLO	Bovine milk	MM-CPE based on mixed micellar extractants of Triton X-114 and CTAB	HPLC (Vydac C18), Phosphate buffer pH 6.6: MeOH (55:45, v/v)	Photodiode array	2–3 ng/ml	[48]
Fluoroquinolones: CIP, DAN, ENR, SAR	Milk, Pig kidney	Selective SPE (SupelMIP, 3% NH4OH in MeOH:water, 1:1, v/v)	CE (36% methanol in 125 mM Phosphoric acid pH 2.8)	LIF	0.17–0.98 µg/kg (milk) 1.1–10.5 µg/kg (kidney)	[49]
OTC	Fish	Methanol:water (7:3, v/v), 0.1 M EDTA	Indirect competitive ELISA	Photometry, time resolved fluorometry	16 µg/kg (photometry), 8 µg/kg (fluorometry)	[50]

(continued)

Table 1 (continued)

Analytes	Analytes	Analytes	Analytes	Analytes	Analytes	Analytes
100 veterinary drugs (macrolides, penicillins, quinolones, sulfonamides, tetracyclines, ionophores, coccidiostat, amphenicols)	Meat, fish, egg	Deproteinization with acetonitrile:Milli-Q water (6:4, v/v), SPE (StrataX), egg: MeOH:ethyl acetate (1:1, v/v), meat and fish: methanol:acetonitrile (1:1, v/v)	UPLC (BEH C18), 0.1% formic acid in acetonitrile: 0.1% formic acid (9:1, v/v)	microTOF-ESI-MS	Detection capability (CCβ): 1.9–2,119 µg/kg	[51]
Macrolides: ERTMC, JSMC, RXTMC, SRMC, TMCS, TLDMC, TLS	Meat, fish	PLE (MeOH, 80 °C)	HPLC (Kromasil C18), 1% acetic acid in Milli-Q water:acetonitrile	ESI-MS	18–51 µg/kg	[52]
Quinolones: OXO, FLU	Fish, shellfish	Microextraction with decanoic acid, tetrahydrofuran, water	HPLC (Kromasil C18), 0.01 M oxalic acid: acetonitrile-water (75:25, v/v)	Fluorescence	Quantitation limits: 6.5–22 µg/kg (OXO), 50–600 µg/kg (FLU)	[53]
Nitroimidazoles: MET, Macrolides: SPY	Fish muscle	Deproteinization by 0.2% orthophosphoric acid-methanol (6:4, v/v), SPE (C18 Sep-Pak) methanol elution	HPLC (C18), phosphate buffer pH 2.4:acetonitrile	UV	2–5 µg/kg (MET), 5–25 µg/kg (SPY)	[54]
Florfenicol	Liver, meat	Online MIP: elution with MeOH:HAc:SDS (88:10:2)	–	Fluorescence	< 2.5 µg/kg	[55]
Amphenicols: CAP, FF, TAP	Shrimp	SFE (ethyl acetate for supercritical CO2), in situ derivatization (BSTFA with 1% TMCS in ethyl acetate)	NCI-GC (TR-5MS)	MS	0.009–0.017 µg/kg	[56]
Ionophores: LAS, MAD, MON, NAR, SAL, SMD Chemical Coccidiostats: CL, DEC, DCL, HLF, DNC, ROB	Chicken liver	ACN extraction, SepPak Alumina N, SPE: Oasis HLB (MeOH)	HPLC (Luna Phenyl Hexyl), ACN (A), MeOH (B) and 0.01 M ammonium formate buffer pH 4.0 (C)	MS/MS	< 1 µg/kg, 10.9 µg/kg (LAS)	[57]

Analytes	Matrix	Sample preparation	Detection	LOD/Range	Ref.
Penicillins: AMX, AMP, Pen-G, Pen-V, OXA, CLO, DIC, NAF	Porcine tissue, milk	Derivatization (piperidine), SPE: Oasis HLB (MeOH: H2O, 80:20, v/v)	MS/MS	2.7-14 μg/kg	[58]
Penicillins: AMX, AMP, Pen-G, Pen-V, OXA, CLO, DIC, NAF	Porcine tissue, milk	Deproteinization with 0.1 phosphate buffer pH 8 and hexane, SPE (C18), ACN elution	nano-LC (C18 silica particle with polystyrene-based monolith outlet frit), 0.1% FA in water (A), ACN (B)	UV or MS	2.3-4.1 μg/L (UV) 0.01-0.51 μg/L (MS)	[59]
Fluoroquinolones: NOR, CIP, OFL, ENR, RUF	Milk	Deproteinization with 5% acetic acid in ACN, defatting with n-hexane, SPE: C18, 25% 1 M aq. Ammonia in MeOH elution	HPLC (VP-ODS C18), MeOH:10 mM aq. Ammonium acetate (25:75, v/v)	ESI-MS/MS	0.2 μg/kg	[60]
5 tetracyclines, 7 macrolides, 3 aminoglycosides, 8 β-lactams, 2 amphenicols, 17 sulfonamides	Honey	1. 1 ml ACN extraction, centrifugation and evaporation to dryness. 2) Add 0.5 ml 10% TCA and 1 ml of ACN. Ultrasonic bath, centrifugation and collected supernatant in a neutral pH. Evaporated to dryness.3) 40 μl NFPA and 1 ml ACN; ultrasonic bath. Supernatant was evaporated to dryness. 4) Fridge few min (4 °C), pH neutralized. 1 ml ACN and ultrasonic bath. Supernatant evaporated to dryness. 2. The 4 dried extracts, resuspended with 100 μl of MeOH/Water (20/80) and filtered through 0.45 μm nylon filter.	HPLC (Zorbax SB-C18 reverse phase column). Water 1 mM NFPA and 0.5% HFo : ACN/Met (50/50)	MS/MS	27.1-80.9 μg/Kg	[61]

(continued)

Table 1 (continued)

Analytes	Analytes	Analytes	Analytes	Analytes	Analytes	Analytes
46 Tetracyclines, quinolones, macrolides, sulfonamides, lincosamides, diamino-pyrimidine derivatives	Milk	2 g of sample. Deproteinization with 20% TCA in water and pH 4 buffer (0.1 M citric acid hydrate:0.2 M disodium hydrogen phosphate (6:4, v/v) SPE (Oasis HLB), MeOH elution	HPLC (C18 column AQUA), 0.2% formic acid in water (A), 0.2% formic acid in ACN (B)	MS/MS	CCα: 6.6–243 μg/kg	[62]
Aminoglycosides: TOB, NEO, PAR, GEN, APR, HYG, KAN, AMK, STR, DIS, SPEC	Animal tissues, honey, milk	5% TCA and 0.2 M HFBA extraction, SPE (double HLB), ACN:0.2 M HFBA (8:2, v/v) elution	HPLC (C18 UG120), 20mMHFBA in ACN: water, 5:95, v/v (A), 20mMHFBA in ACN: water, 50:50, v/v (B)	ESI-MS/MS	CCα: 9–246 μg/kg (milk), 9–41 μg/kg (honey), 20–5,279 μg/kg (kidney), 22–1,643 μg/kg (liver), 18–594 μg/kg (muscle)	[63]
Tetracyclines: OTC, TC, CTC, DXC	Animal tissues	PLE (Water, 70 °C), SPE (HLB), 10 mM FA in MeOH elution	HPLC (Xterra C18), 10 mM FA in MeOH (A), 10 mM FA in water (B)	ESI-MS/MS	0.1–0.3 μg/kg	[64]
Pesticides						
Pesticides: organophosphates, N-methylcarbamates, benzimidazoles, azoles, ureas, benzoylphenylureas, carboxylic acids, triazines, uracils, chloroacetamides, triazols and ketoenoles	Olive oil	QhEChERS: 3 g homogenized sample mixed with water, addition of ACN, NaCl, an. MgSO4, centrifugation Clean-up: Mix with PSA sorbent, an. MgSO4, C18, graphitized carbon black, 0.45 μm PTFE filtering	LC (C8), HPLC water, 0.1% FA (A), ACN (B)	MS/MS (LINAC)	0.01–6.6 μg/kg	[65]

Analyte	Matrix	Sample preparation	Separation	Detection	LOD/LOQ	Ref.
32 Pesticides	Eggs, milk, avocado.	1. (a) QuEChERS: 15 ml ACN (1%AcH), MgSO$_4$, AcNa; centrifugation; (b) Dispersive SPE C18 and primary and secondary amine sorbents with MgSO$_4$. 2. MSPD: C18, anhydrous Na$_2$SO$_4$/Florisil column/ 2 ml of ACN to elute the sample	GC HPLC	MS MS/MS	< 10 ng/g	[66]
Pesticides	Fruit and vegetable	QhEChERS: 15 g sample, addition of ACN, NaCl, an. MgSO4, centrifugation Clean-up: Mix with PSA sorbent, an. MgSO4, C18, graphitized carbon black, 0.45 μm PTFE filtering	LC (C8), HPLC water, 0.1% FA (A), ACN (B)	TOF MS/MS (QTrap)	1–10 ng/g	[67]
33 Pesticides	Fruit-based soft drinks	SPE: HLB	LC (C18)	ESI-TOFMS	LOQ: 0.02–2 μg/L	[68]
Fenthion & metabolites (fenthion sulfone, fenthion sulfoxide, fenoxon, fenoxon sulfoxide, fenoxon sulfone)	Oranges	Homogenization, mixed with ethyl acetate, an. Na2SO4, evaporate, re-homogenization, filter through an. Na2SO4, rinse with ethyl acetate, evaporate, rinse with MeOH	LC (IT-MS): Luna C18, UPLC (for QqTOF-MS): Acquity C18, methanol (A), 10 mM ammonium formate in water (B)	IT-MS, QqT OF-MS	0.4–5.6 μg/kg (TOF), 3–6 μg/kg (IT)	(Picó 2007) [69]
Perfluorinated compounds						
PFOS, PFHS, PFBS, PFOSA, PFDoDA, PFUnDA, PFDA, PFNA, PFOA, PFHpA, PFHxA	Mussels and oysters	Alkaline digestion/SPE (Oasis WAX)	HPLC	MS/MS	–	[70]

(continued)

Table 1 (continued)

Analytes	Analytes	Analytes	Analytes	Analytes	Analytes	Analytes
PFBS, PFHxS, PFOS, PFOSA, PFHxA, PFHpA, PFOA, PFNA, PFDA, PFUnDA, PFDoDA, PFOcDA	Water	SPE: Oasis WAX	HPLC (Betasil C18), 2 mM ammonium acetate in water/methanol	MS/MS	0.005–0.01 ng/ll	[71]
PFHxS, PFOS, PFHxA, PFHpA, PFOA, PFNA, PFDA, PFUnDA, PFDoDA, PFOSA	Blood and liver from farm and pet animals	Ionic pair extraction: 2 x MTBA-HSO$_4$/MTBE extraction; centrifugation. Filtered by 0.1 μm nylon fiber	HPLC (Betasil C18), 2 mM ammonium acetate in water/methanol	MS/MS	0.01-0.05 ng/ml	[72]
PFPA, PFBS, PFOA, i,p-PFNA, PFNA, PFOS, PFDA, L-PFDS	Fish	Sea sand homogenization, PLE (water), SPE (OasisWAX 3), elution with 0.1% NH4OH in MeOH, ionic-pair extraction (MTBE)	LC (LiChroCART-LiChrospher 100 RP-18), 20 mM ammonium acetate in water (A), 20 mM ammonium acetate in MeOH (B)	QLIT-MS/MS	0.005–0.05 ng/ml	[73]

Nanomaterials

Analytes	Analytes	Analytes	Analytes	Analytes	Analytes	Analytes
Fullerenes C60 and C70 and cuntionalized fullerene N-methylfulleropyrrolidine C60	Fresh Water	Filtration with nylon membrane filters; sonication of the filters with three volumes of toluene; rotavaporation and reconstitution in a 1.5 mL of toluene-methanol (2:1, v/v) mixture	LC: (Purospher Star RP-18 endcapped column; 125 × 2.0 mm, particle size 5 μm), isocratic elution with toluene-methanol (1:1, v/v) mixture, (0.4 mL/min).	QqLIT-MS	75-250 ng/l	[74]

Biotoxines

Analytes	Analytes	Analytes	Analytes	Analytes	Analytes	Analytes
28 biotoxins: okadaic acid, dinophysistoxins, yessotoxins, azaspiracids, pectenotoxins, spirolides, okadaic acid fatty acid esters	Shellfish	Homogenization, MeOH extraction, HT tuffryn 0.2 μm membrane filtering	LC (X-Bridge C18), water with 6.7 mM ammonium hydroxide, pH 11 (A), ACN/water, 90:10, v/v, with 6.7 mM ammonium hydroxide, pH 11 (B)	both negative and positive ESI-MS	0.8–9.1 pg on-column	[75]

Analyte	Matrix	Method	Technique	Value	Ref
Tetrodotoxin	Shellfish	Na$^+$ channel blockers biosensor		86 fg	[76]
Yessotoxin	Shellfish	Microplate fluorescence assay		0.2 µM	[77]
Yessotoxin	Mussels	Fluorescently labeled Phosphodiesterases I using fluorescence polarization		2 µM (spiked sample), 2.66 µg (contaminated sample)	[78]
Saxytoxin	Shellfish	Fluorescence		<38 µg/100 g meat (lowest reported value for contaminated samples)	[79]
Palytoxin	Shellfish	Microplate assay with human neuroblastoma cells		0.2 ng/ml	[80]
Gymnodimine and 13-desmethyl C spirolid	Shellfish	Fluorescence polarization inhibition assay		80–2,000 µg kg^{-1} and 85–700 µg kg^{-1}	[81]
Okadaic acid and dinophysistoxin-2	Shellfish	Homogenization, MeOH extraction, high speed blending and centrifugation	LC (BDS-HypersilC8) water 2 mM ammonium formate and 50 mM formic acid (A) and acetonitrile:water (95:5) containing 2 mM ammonium formate and 50 mM formic acid (B). QLIT-MS/MS	3.7 mg/kg	[82]
Okadaic acid, dinophysistoxins, yessotoxins, azaspiracids, pectenotoxins, spirolides, okadaic acid fatty acid esters	Shellfish	Homogenization, MeOH extraction, HT tuffryn 0.2 µm membrane filtering	LC (X-Bridge C18), water with 6.7 mM ammonium hydroxide, pH 11 (A), ACN/water, 90:10, v/v, with 6.7mMammoni umhydroxide, pH 11 (B). both negative and positive ESI-MS	0.8–9.1 pg on-column	[75]

OXO oxolinic acid, *FLU* flumequine, *MET* metronidazole, *SPY* spiramycin, *MIP* moleculrar imprinted polymers, *SDS* sodium dodecylsulphate, *CAP* chloramphenicol, *FF* florfenicol, *TAP* triamphenicol, *SFE* Supercritical fluid extraction, *BSTFA* N,O-bis(trimethylsilyl) trifluoroacetamide, *TMCS* trimethyl-chlorosilane (in ethyl acetate), *NCI* negative chemical ionization, *LAS* lasalocid, *MAD* maduramycin, *MON* monensin, *NAR* narasin, *SMD* semduramy-cin, *CL* Clazuril, *DEC* decoquinate, *DCL* diclazuril, *HLF* halofuginone, *DNC* dinitrocarbanilide, *ROB* robenidine, *AMX* amoxicillin, *Pen-V* penicillin-V, *DIC* dicloxacillin, *NAF* nafcillin, *DIF* difloxacin, *OFL* ofloxacin, *RUF* rufloxacin, *NIC* nicarbazin was regarded with great expectations; nevertheless, SFE did not ever get the practical acceptance that was anticipated due to known limitations in the extraction of polar compounds. Addition of an organic modifier or selecting a different supercritical fluid is possible but some ideal properties of CO_2 are lost. The appearance of PLE claiming the advantages of both SFE and liquid solid extraction with high extraction efficiency diverted the attention from SFE. PLE have to be considered the technique of choice for the analysis of emerging compounds in both environmental and food samples, because compared to the other extraction techniques, PLE provides good recoveries and saves time and organic solvents. Different examples of application can be found in Table 1. Example Llorca et al. [84], reported the extraction of 18 perfluorinated compounds in sewage sludge by means of PLE extraction with methanol at 70°C and 100 bar of pressure, followed by SPE clean-up and LC separation and analysis in a hybrid quadrupole-linear ion trap (QqLIT) tandem MS. In this work, selected PFCs were detected with method limits of quantification (MLOQ) in the ng/kg range [84].

Several studies have been carried out using KOH digestion followed by solid phase extraction (SPE). The method was initially reported by Yamashita et al. [85] and was later applied in different studies [70]. For liquid samples analysis, protocols based on filtration followed by SPE have been also widely applied. Due to the different polarities of PFCs, different extraction SPE cartridges have been explored. Broadly, good recoveries were reported using Oasis WAX-SPE cartridges for short-chain (C_4–C_6) compounds. For longer chain PFCs, less polar phases (C_{18} and Oasis HLB) may be applied. When an ion-pairing agent is used that decreases the polarity of the ion pair complex, a nonpolar solvent (e.g., MTBE) may be used. Nonionic PFCs may be extracted from the matrix by nonpolar media (C_{18} SPE or hexane). Moderate polar media (Oasis HLB and Oasis WAX SPE, a hexane–acetone mixture or acetonitrile) have also been applied for extraction of nonionic PFCs.

During recent years purification techniques working online with chromatographic systems have been applied to the analysis on different groups of food contaminants. However, in the case of PFCs a number of works have been developed. Gosetti et al. [86] developed a rapid online SPE ultra high performance liquid chromatography tandem mass spectrometry (UPLC-MS/MS) method for the identification and quantitation of 9 PFCs in environmental, biological, and food samples. Limits of detection (LOD) and limits of quantification (LOQ) range from 3 to 15 ng/L and from 10 to 50 ng/L, respectively. Recoveries were greater than 82.9%. Another promising approach for online clean up is the turbulent flow chromatography (TFC), which shows great potential for the rapid, direct analysis of PFCs in complex samples such as food. TFC is achieved by the use of high flow rates and extraction columns using large particle size stationary phases. When coupled with mass spectrometric detection, this technique allows the direct analysis of complex samples with very rapid chromatography and, therefore, extremely high throughout. Currently, this approach has been successfully applied for the analysis of biota.

For quantitative aspects, LC is the separation technique of choice during recent years. LC separation of PFCs has been mainly carried out with C_{18} and C_8 columns. Taniyasu et al. explored the chromatographic properties and separation of short-chain PFAs on RP-C18 and ion exchange columns [87]. The results suggested that RP columns are not suitable for the analysis of short-chain PFAs, as ion-exchange columns, a suitable alternative, have superior retention properties for more hydrophilic substances.

Due to the complexity of the food samples, it is possible that the presence of some compounds in the matrix interferes with analyte determination; even when working in LC–MS/MS, certain compounds present in the sample can affect the initial ionization of the analyte through what is often called ion suppression/enhancement or matrix effects.

ESI operating in the negative ion (NI) mode has been the interface most widely used for the analysis of anionic PFCs. In addition, ESI has also been optimized for the determination of neutral compounds, such as the sulphonamides PFOSA, Et-PFOSA, and t-Bu-PFOS. The use of atmospheric pressure photoionization (APPI) has been explored by Takino and collaborators [88]. The authors found the main advantage of this technology to be the absence of matrix effects, but the limits of detection were considerably higher than those obtained by LC–ESI-MS/MS.

LC-MS/MS performed using triple quadrupole mass spectrometer (QqQ) combined with multiple reaction monitoring (MRM) is one of the more widely applied detector, as well as, to be one of the better suited for quantification of PFCs. Nowadays the performance of ion trap (IT) and time of flight (TOF) have been also considered for trace quantification of PFCs. PFCs contain carboxylic, sulfonic, hydroxy, or sulphonamide group. They have acidic properties and can therefore dissociate. Therefore, electrospray ionization in the negative mode (ESI($-$)) is best suited. LC-(ESI)-MS/MS is the technique of choice for the analysis of PFCs in food samples, and allows limits of detection in the pg–ng/g in food and feed samples. Pseudomolecular ions are formed such as [M–K]$^-$ for PFOS (m/z 499), [M–H]$^-$ for PFOA (m/z 413), and PFOSA (m/z 498, which are generally selected as precursor ions for MS^2 experiments using ion trap and triple quadrupole instruments. Berger et al. presented a comparison between IT, QqQ, and TOF instruments [89]. Tandem mass spectrometry showed excellent specificity, but the matrix background is eliminated by the instrument; thus, it cannot be visualized. Applying TOF-MS gives an estimation of the amount of matrix left in the extract, which could impair the ionization performance and the high mass resolution of the TOF-MS instrument offers excellent specificity for PFCs identification after a crude sample injection.

3.1.2 Polybrominated Flame Retardants

For PBDEs and PBBs, sample treatment procedures have typically been based on protocols previously established for trace classic POPs.

3.1.3 Biological Approaches

A limited number of biological analytical approaches have been published for the determination of polybrominated flame retardants and most of them are based on immunochemical techniques. A sensitive magnetic particle enzyme-linked immu-noassay was developed to analyze PBDEs in food and environmental samples [45]. Using this assay, fifty samples could be analyzed in about 1 h after sample cleanup, with a LOD below 0.1 ppb, approximately the same as instrumental analysis, toward the following brominated diphenyl ether (BDE) congeners: BDE-47, BDE-99, BDE-28, BDE-100, and BDE-153. The congeners most readily recognized in the enzyme-linked immunosorbent assay (ELISA) were BDE-47 and BDE-99 with cross-reactivity of BDE-28, BDE-100, and BDE-153 being less than 15% relative to BDE-47. The clean up methods prior to ELISA were matrix dependent. no pretreatment was needed for environmental water samples, while fish, milk, and soil samples required various degrees of clean up. Recently, the suitability of such methods has been assessed for the analysis of fish and crabs and comparison with GC/ECD-ITMS results demonstrated that ELISA provides a timely and cost-effective screening method [90].

An ELISA to monitor polybrominated diphenyl ether BDE-47. 2,2',4,4'-Tetrabromodiphenyl ether (BDE-47), a dominant PBDE congener of toxicological concern, was developed by Ahn et al. [91]. The optimized competitive indirect ELISA conditions showed a linear detection range of 0.35–8.50 μg/L with an IC_{50} value of 1.75 μg/L. Little or no cross-reactivity (<6%) was observed toward related PBDE congeners containing the BDE-47 moiety and other halogenated compounds. The use of a magnetic particle-based competitive direct ELISA increased the sensitivity by tenfold over the indirect ELISA.

On the other hand, limited advances have been made with regard to the devel-opment of biosensors toward the detection of this group of compounds. Marchesini et al. [92], reported on an SPR biosensor based on the thyroxine (T4) transport disrupting activity to screen chemicals (including PBDEs) that may interfere with the thyroid system [92].

Kim et al. [93], presented a new application of the noncompetitive phage anti-immunocomplex assay (PHAIA) by converting an existing competitive assay to a versatile noncompetitive sandwich-type format using immunocomplex binding phage-borne peptides to detect the brominated flame retardant, brominated diphenyl ether 47 (BDE 47). The resulting PHAIAs showed variable sensitivities, and the most sensitive peptide had a dose–response curve with an SC50 (concen-tration of analyte producing 50% saturation of the signal) of 0.7 ng/ml BDE 47 and a linear range of 0.3–2 ng/ml, which was nearly identical to the best heterologous competitive format (IC50 of 1.8 ng/ml, linear range of 0.4–8.5/ml). However, the PHAIA was 1,400-fold better than homologous competitive assay.

Recently, a novel immunoassay has been developed for the quantitative determi-nation of polybrominated biphenyls using indirect competitive format. The new method was optimized concerning the coating conjugate and antibody concentration, incubation time and temperature, the tolerance to organic solvents and so on. Under optimized conditions, PBB15 can be determined in the concentration range of 0.01–100 μg/L with a detection limit of 0.02 μg/L. The cross-reactivities of the assays were below 8%. While water samples could be analyzed directly [94].

3.1.4 Instrumental Methods

Most existing methods are based on instrumental analysis involving exhaustive sample pretreatment and preconcentration steps, followed by purification and fractionation before final chromatographic separation and detection. For fat and oil samples, dissolving the lipids in an appropriate solvent is usually the first treatment. This has been achieved by melting the fat at 90°C followed by LLE or direct solid liquid extraction (SLE) with an apolar solvent [37], column extraction with a mixture of apolar solvents after drying of the sample with anhydrous Na_2SO_4, Soxhlet extraction and/or sonication with apolar solvents. Typically, sample intake is between 0.5 g and 1 g and quantitative recoveries >60% have been reported.

Column extraction using a multilayer column containing appropriate sorbents for a preliminary purification has been used for biological tissues. Fernandes et al. [38] reported a method using 400 mL of DCM:n-hexane (2:3, v/v) to elute PBDEs from a multilayer column containing 10 g of food. Alternatively, matrix solid-phase dispersion (MSPD) usually results in a more efficient retention of impurities and lower solvent consumption. Martínez et al. [39] reported high recoveries (81–106%) for six tetra- to hexa-BDEs using only 0.5 g naturally contaminated fish tissue (WELL-WMF-01) dispersed on 1.5 g Florisil® and packed on top of acidified silica. Enhanced extraction techniques, such as PLE or Microwave-assisted extraction (MAE) have also been used. Eljarrat et al. [95] developed a selective PLE for the simultaneous analysis of PBDEs and total hexabromocyclo-dodecane (HBCD) stereoisomers in fish tissue. For MAE, purification should be carried out off-line after separation of the solvent from the matrix components, but its high extraction efficiency allowed a significant reduction of the extraction time [41] compared to conventional extraction procedures.

The most applied analytical method for PBDEs is still GC–MS; however, recently new analytical approaches such as ITD-MS for environmental samples [96] or quadrupole ion-storage MS (QTrapMS) for semipermeable membrane devices (SPMDs) samples [97] have been evaluated for the analysis of PBDEs. Gómara et al. [42] have established a method for the determination of 10 PBDE congeners by means of GC–ITD-MS operated in EI. Parameters affecting isolation and fragmenta-tion of precursor ions in the ion-trap were optimized in order to achieve the best robustness and sensitivity for all PBDE congeners investigated. The fragment ions monitored correspond to loss of two Br atoms from the molecular ion. Using GC-QTrapMS for the analysis of 20 PBDEs, differences in isolation and fragmentation patterns for PBDEs congeners with degree of bromination were observed [98]. Recently, different food analysis applications have been reported, for example, for the assessment of PBDEs in Tibetan butter [43], or in fish and crabs [44]. Compre-hensive two-dimensional gas chromatography (GC × GC), a technique that offers excellent separation power [20] has been evaluated also for the analysis of PBDEs. Recently, Lacorte et al. applied comprehensive gas chromatography coupled to high resolution mass spectrometry-based method for the determination of PBDEs and their hydroxylated and methoxylated metabolites in sediment, fish tissue, and milk samples [46]. LC techniques in the analysis of PBDEs have been also investigated [99]. The use of atmospheric pressure photoionization (APPI) may facilitate the analysis of PBDEs and phenolic compounds, such as TBBP-A, in the same run. Negative APPI gave better sensitivity than positive APPI, but due to the low resolving power of LC compared to GC, this technique has not been further optimized for PBDE analysis.

3.1.5 Pesticides

Most of the analytical methods for the analysis of pesticides in food are based on instrumental approaches based on chromatography coupled to mass spectrometry. However, a great effort of development has been paid to develop rapid screening methods based on biological methods, such as, enzyme linked immunosorbent assays (ELISA).

3.1.6 Biological Methods

A huge number of immunoassays for pesticides detection in water and food have been presented during the last decades, and different review articles have been revised it [100–109].

Among recent developments, a fluorescent polarization immunoassay (FPIA) for simultaneous determination of organophosphorus pesticides (OPs) using a broad-specificity monoclonal antibody was developed and has been presented by Xu et al. The effects of tracer structure, tracer concentration, antibody dilution, methanol content, and matrix effect on FPIA performance were studied. The FPIA can detect five OPs simultaneously with a limit of detection below 10 ng ml^{-1}. The time required for the equilibrium of antibody–antigen interaction was less than 10 min. The recovery from spiked vegetable and environmental samples ranged from 71.3% to 126.8%, with the coefficient of variations ranging from 3.5% to 14.5%. The developed FPIA was applied to samples, followed by confirmation with high-performance liquid chromatography-tandem mass spectrometry (HPLC-MS/MS) analysis. The developed FPIA demonstrated good accuracy and reproducibility, and is suitable for rapid and high-throughput screening for OP contamination with high efficiency and low cost [110]. In another example, an ELISA for the determination of Glyphosate in food was developed by Selvi et al. In this method, avian antibodies (IgY) were used to recognize glyphosate. The assay was specific to glyphosate with a limit of detection of 2 ppb. Mean analytical recovery of glyphosate in different food samples was 9.00–134.00% [111].

More advanced biological technologies for rapid detection of pesticides are biosensors. In this field, recent advances in miniaturization of analytical systems and newly emerging technologies offer platforms with greater automation and multiplexing capabilities than traditional biological binding assays. Multiplexed bioanalytical techniques provide control agencies and food industries with new possibilities for improved, more efficient monitoring of food contaminants. In this sense, should be pointed out planar-array and suspension-array technologies for their application in pesticides detection. However, most applications are low implemented and are validated for the analysis of water samples.

3.1.7 Instrumental Analysis

Recent trends in pesticide analysis in food aims for reduced sample pretreatments or simplified methodologies (as QuEChERS approaches), the use of online purification processes, the use of new adsorbents (such as molecular imprinted polymers (MIPs) and nanomaterials) for the extraction and clean-up processes, and focused on the development of large multiresidue methods, most of them based on LC-MS/MS. In spite of the relevant role of LC-MS/MS, GC-MS-based methods still play an important role in pesticide analysis in food. Despite the development achieved in the immunochemical approaches, the need for multi-residue methods has supported the development and use of instrumental techniques.

So far, LSE is the most popular for extracting contaminants in food. However, over the last years LPME in its different application modes (single drop microextraction, dispersive liquid–liquid microextraction and hollow fiber-LPME) has been increasingly applied to food analysis because of its simplicity, effectiveness, rapidity, and low consumption of organic solvents. Different applications have been recently reviewed by Asensio-Ramos et al. [112]

A great acceptance has been gained by the QuEChERS (quick, easy, cheap, effective, rugged, and safe)-based method [113–133]. One of the most recently reviews published by Pareja et al. in 2011 [126] about the determination of pesticide residues in rice by LC-MS indicated that the QuEChERS is the most frequently employed compared with other methodologies. Klein and Alder proposed a simple method based on methanol extraction followed by SPE [134]. The method known as QuEChERS utilizes acetonitrile for extraction and $MgSO_4$:NaCl to induce partitioning of acetonitrile extract from the water of the sample and dispersive SPE to cleanup. The method has been validated for different food matrices including eggs, milk, fruits, and meat [66, 135].

Carbon nanotubes could be also used in a format of disc. A comparison study showed that the double-disk system (comprising two stacked disks with 60 mg of CNTs) exhibited extraction capabilities that were comparable to those of a commercial C_{18} disk with 500 mg sorbent for nonpolar or moderately polar compounds. Moreover, the former system was more powerful than the latter for extracting polar analytes. The triple-layered CNTs disk system showed good extraction efficiency when the sample volume was up to 3,000 mL. Katsumata et al. [136] obtained very high enrichment factor for preconcentration of atrazine and simazine (3,900 and 4,000, respectively, for 200 mL of sample solution when only 30 mg of MWCNTs was used in the format of disk.

Carbon-encapsulated magnetic nanoparticles (CEMNPs) are core–shell materials with similar surface characteristics to carbon nanotubes and this similarity enables to use them as solid sorbents. They are composed of the magnetic core (10–100 nm in diameter), which is tightly coated by a carbon coating built from parallel stacked graphitic layers [137]. Encapsulation approach primarily protects the nanoparticles against the external environment, hampers aggregation, and also provides the ability for surface functionalization. This process may improve also the dispersion stability of core–shell nanomaterials in a wide range of suspending solvents. An attractive property of CEMNPs is that magnetic nanoparticles can readily be isolated from sample solution by the application of an external magnetic field. Compared with nonmagnetic nanoparticles, the proposed sorbent material avoids the time-consuming column passing and filtration operation and shows great analytical potential in preconcentration of large volumes of real water samples.

Similarly to SPE, for the SPME technique CNTs with high porosity and large adsorption area seems to be a good candidate for SPME coating. In addition, the more thermal and physical resistance of carbon nanotubes in comparison with commercial SPME coatings, are the other important characteristics from the practical point of view. However, this technique has been just applied in environmental analysis till now.

Successfully examples of applications have been reported for the analysis of pesticides in food using PLE during the extraction step [138–147]. Research had been conducted to optimize the effects of extraction temperature, number of extraction cycles, and various extraction solvent mixture compositions on the extraction effectiveness and recoveries of pesticides from food. Besides, cleanup sorbent material(s) can also be imbedded in the extraction cells so that cleanup can also be processed simultaneously with extraction. Although it has the advantages of low solvent consumption and short extraction period, the initial cost is high, large amount of unwanted matrix substances are co-extracted and some unstable compounds, such as endrin yielded low recoveries.

Microwave-assisted extraction (MAE) is another recent technique that can also be applied for extracting pesticides from food [142, 148, 149]. By applying microwave energy to the extraction solvent, the highly localized temperature and pressure cause heating of the matrix and migration of target analytes from the sample material to the surrounding solvent rapidly.

For the extraction and clean-up clean of liquid samples, or for clean-up of extract solid-phase extraction has been extensively applied [126, 150–152]. In order to enhanced the selectivity, different immunosorbents have been proposed during last decade [153, 154], as well as, using molecular-imprinted polymers (MIPs) [154, 155]. However, the last approach to enhance extraction rates and selectivity are based on the use of new nanomaterials [156]. High extraction efficiency has been reported using nanomaterials. Carbon nanomaterials exhibit a strong adsorption ability for a wide variety of pesticides. Because of their advantageous characteristics (high adsorption capacity, good thermal stability, wide pH range of application), carbon nanostructures have been employed in SPE and SPME. A considerable number of chromatographic and electrophoretic methods using SPE technique have been described. Mostly, SPE cartridges filled with CNTs were applied for the preconcentration of pesticide and herbicide residues from environmental waters. In several published papers, sorption on CNTs has been examined for different pesticides [157–161].

Since the analytical point of view most of current analytical methods are based on LC-MS/MS, but for some classes of pesticides GC-MS continues being the technique of choice. The use of quadrupole ion trap (QIT) to analyze multiple pesticide residues is limited to several multiclass pesticides in fruit [162], because of the limited number of ions that can be isolated at the same time. For this reason, the use of several time windows is required and this is indeed a strong limitation in practice. The use of hybrid triple quadrupole linear ion trap (QqLIT) mass spectrometer has provided significant contribution to the development of high-sensitive multiresidue analytical methods for pesticide control. An example of application is the method reported by Hernando et al. for the analysis of pesticide residues in olive oil [65].

LC-MS/MS and LC-time-of-flight-MS (LC-TOFMS) are powerful and complementary techniques that can independently cover the majority of the challenges related with pesticide residues food control. The sequential combination of both systems benefits from their complementary advantages and assists to increase the performance and to simplify routine large-scale pesticide multiresidue methods. García-Reyes et al. developed an approach consisting of three stages: (1) automated pesticide screening by LC-TOFMS, (2) identification by LC-TOFMS accurate mass measurements, and (3) confirmation and quantitation by LC-MS/MS of 100 target pesticides in crops [67]. In this work, the final confirmation of the positive findings was carried out using two MRM transitions and accurate quantitation was performed by LC-MS/MS using a hybrid triple QqLIT mass spectrometer. Examples of applications to real samples showed the potential of the proposed approach, including the detection of nonselected a priori compounds as a typical case of retrospective evaluation of banned or misused substances. In addition, the use of this QqLIT instrument also offers additional advantageous scanning modes (enhanced product ion and MS^3 modes) for confirmatory purposes in compounds with poor fragmentation. On the other hand, the main advantages of LC-TOFMS include the high sensitivity available in full-scan acquisitions and high resolution (10 000–12 000) as compared to other MS analyzers, such as triple quadrupole mass spectrometers. LC-TOFMS also reduces the chance of false positives, but it can be less sensitive than QqQ-MS in quantification, when specific MRM transitions are monitored.

Recent studies have focused more on degradation/transformation products of pesticides (e.g., from hydrolysis, oxidation, biodegradation, or photolysis) because those can be present at greater levels in food and the environment than the parent pesticide itself, and sometimes are as toxic or more toxic than the parent pesticide. New pesticides have also become available on the market (such as glyphosate and organophosphorus pesticides), and studies are being conducted to understand their fate and transport in the environment. Therefore, researchers are increasingly using TOFMS and Q-TOFMS to enable the identification of new pesticide and their degradation products. For example, the occurrence of carbamates and their transformation products in vegetables and fruits was assessed [227]. This is an important issue because several of the transformation products are more toxic than the parent pesticide. Soler et al. compared four mass analyzers for determining carbosulfan and its metabolites by LC/MS [162]. Of the analyzers investigated, single quadripole, QqQ, QIT, and Q-TOF, the triple quadrupole-MS provided the highest sensitivity, with detection limits of 0.04–0.4 µg/kg compared to single quadrupole (4–70 µg/kg), QIT (4–25 µg/kg), and Q-TOF (4–23 µg/kg). Other examples are the confirmation of fenthion metabolites using QIT and QTOF [163], and from the same research team the identification of degradation products from the postharvest treatment of pears and apples by ultra-high pressure liquid chromatography (UPLC) coupled to QTOF/MS [164].

Regarding new generation of pesticides the ability to separate enantiomers and produce a single enantiomeric isomer has been exploited by pesticide manufacturer industry recently, creating new markets for their products. The development of enantiomerically enriched pesticides may actually benefit the environment, as less material could potentially be applied to crops, less may be accumulated in the environment, and there may be fewer unintended side effects on nontarget species. However, more research is needed to confirm this possibility. Most research to-date has investigated chiral profiles in surface waters, soil, vegetation, and fish. The most commonly used analytical techniques to separate and measure chiral isomers include the use of chiral columns with GC and LC (often including the use of mass spectrometry). CE and CE/MS are also often used.

3.1.8 Nanomaterials

The growth of the application of nanotechnology products is expected in food industry, and nowadays over 200 companies are conducting R&D into the use of nanotechnology in either agriculture, engineering, processing, packaging or delivery of food, and nutritional supplements [25]. In some countries, nanomaterials are already used in food industry for food packaging, with nanoclays as diffusion barriers and nano-silver as antimicrobial agents [165]. However, the use of nanomaterials in the food sector is a new source of exposure to these materials through, for example, the migration of nanoparticles from food packaging into food, in addition to other expected routes such as through environmental pollution and then via drinking water, food chain, etc.

A range of analytical techniques is available for providing information on concentration and properties, including microscopy approaches, chromatography, centrifugation and filtration, spectroscopy, and related techniques.

The most popular tools for the visualization of engineered nanoparticles are electron and scanning probe microscopes. The visualization, the state of aggregation, dispersion sorption, size, structure, and shape can be observed by means of atomic force microscopy (AFM), scanning electron (SEM), and transmission electron microscopy (TEM). Analytical tools (mostly spectroscopic) can be coupled to electron microscopes for additional elemental composition analysis, generally known as analytical electron microscopy (AEM). For example, energy dispersive X-ray spectroscopy (EDS) can be combined with SEM and TEM permitting a clear determination of the composition of elements heavier than oxygen. Quantitative analysis, however, leads to approximately 20% uncertainty, and electron energy loss spectroscopy (EELS) can be used combined with TEM for quantitative analysis with uncertainties as low as 10% [166]. Main disadvantages are that these are destructive techniques, charging effects caused by accumulation of static electric fields at the specimen, and they have to be operated under vacuum conditions; therefore, sample preparation is required for food or biological samples. Other microscope techniques are scanning transmission electron microscope (STEM) and X-ray microscopy (XRM). To image nanoparticles in their natural state without modification can be achieved using environmental scanning electron microscope (ESEM) and WetSEMTM capsule which can operate under wet conditions. Imaging under fully liquid conditions is also possible using atomic force microscopy (AFM) [167]. An oscillating cantilever is scanning over the specimen surface and electrostatic forces (down to 10^{-12} N) are measured between the tip and the surface. 3D surface profiles are achieved but the main limitation of this technique is the possible overestimation of the lateral dimension of nanoparticles if the geometry of the tip is larger than the particles being probed.

Chromatographic approaches have been also used to separate nanoparticles from samples coupled to different detectors, such as ICP-MS, MS, DLS. The best known technique for size separation is size exclusion chromatography (SEC). A size exclusion column is packed with porous beads, as the stationary phase, which retain particles, depending on their size and shape. This method has been applied to the size characterization of quantum dots, single-walled carbon nanotubes, and polystyrene nanoparticles [168, 169]. Another approach is hydrodynamic chromatography (HDC), which separates particles based on their hydrodynamic radius. HDC has been connected to the most common UV–Vis detector for the size characterization of nanoparticles, colloidal suspensions, and biomolecules [170–172].

A wide range of spectroscopic methods are available for nanoparticle analysis and characterization. Scattering techniques useful for nanoparticle characterization include light scattering methods, such as static (SLS) and dynamic light-scattering (DLS), or neutron scattering, such as small-angle neutron scattering (SANS). Laser-induced breakdown detection (LIBD) is a laser-based technique featuring extremely low detection limits, which is capable of analyzing the size and concentration of colloids, depending on the measured breakdown probability (BP). LIBD is, therefore, a highly promising tool for nanoparticle characterization, although it cannot distinguish between different types of particles and requires particle-specific size calibration [173].

Mass spectrometry is also used for nanoparticles investigations. Two ionization techniques often used with liquid and solid biological samples include electrospray ionization (ESI) and matrix-assisted laser desorption/ionization (MALDI). Inductively coupled plasma (ICP) sources are mainly used for metal analysis. In general, these techniques have been applied coupled to liquid chromatography. LC-MS, for example, have been applied for the investigation of carbon-based nanoparticles in aquatic and biological media [74, 174, 175].

However, the analysis of nanoparticles in food is in its initial phase of development. In addition, when measuring nanoparticles in different media, apart from data on concentrations, information on the size distribution and properties of the particles is also required. Therefore, no single technique can provide all this information, so a range of analytical techniques is required.

Many analytical tools are theoretically suitable for the characterization of nanoparticles, but requirements for analysis of engineered nanoparticles in natural and food-related samples will differ greatly from their analysis in pure media. At the moment, analytical methods are required to reliably detect and characterize nanoparticles in complex matrices, such as foodstuff.

3.2   Pharmaceutical Residues: Antibiotics and Coccidiostats

Antibiotics are utilized in veterinary practice, at therapeutic levels, primarily to treat diseases and to prevent infections as well as in animal production, at subtherapeutic levels, to increase feed efficiency, and to promote growth in food-producing animals [176]. Residues have been found in products of animal origin, such as milk and meat and this can instigate serious consequences in human health. Allergies to sensitive individuals as well as reduced effectiveness against infections can be the result of extensive or inappropriate use of antibiotics in food animals. The fact that antibiotic-resistant bacteria can pass through the food chain to people is of special concern in food safety and that is the reason why this group of compounds has been included in this review, even though it does not fall under the category of emerging contaminants in food.

A variety of biological approaches for the rapid detection of antibiotics in food exist, including traditional microbial inhibition tests, immunoassays, and biosensors (Table 1).

Microbial inhibition tests involve incubating a susceptible organism in the presence of the sample. These tests are generally reliable and cost-effective but require incubation for several hours before the result can be visualized. Different types of microbial inhibition tests are commercially available and several of them have been validated with Bacillus stearothermophilus disc diffusion assay (BsDA) being an official standard test for regulatory use in reference laboratories [177]. The Delvotest® (Gist-brocades BV, The Netherlands) is one of the oldest and most known microbial inhibitor tests, with many different versions in the market, such as, Delvotest® SP, capable of detecting more substances, like sulphonamides, and more sensitivity to tylosin, erythromycin, neomycin, gentamicin, trimethoprim, and other antimicrobials. Other similar tests exist, such as the Charm AIM-96 test, Charm Farm Test-"Vial," and the Charm Farm Test-"Mini Vial" manufactured by

Charm Sciences Inc. (USA). The "Screening Test for Antibiotic Residues" (STAR) five-plate test was evaluated for the screening of 66 antibiotics in milk [178].

Several immunoassays have been reported for the analysis of antibiotics in food [179]. Among them, ELISA assays offer a good alternative option for analysis of antimicrobials at screening level, due to their simple use, high sensitivity, and fast response. However, final confirmatory analysis should be carried out for positive samples in order to achieve specific identification and quantification of target analytes. A recent multianalyte ELISA for the simultaneous determination of the most frequently used antibiotic families in the veterinary field has been developed by combining two individual ELISAs; one for sulphonamide and fluoroquinolone antibiotics and the other for ß-lactam antibiotics and detectability reached in full-fat milk samples was below the European maximum residue limits [180]. FitzGerald et al., reported a competitive ELISA for the rapid measurement of 11 active β-lactams in milk, tissue, urine, and serum with an EC50 of 2 μg/L [181], whereas Pimpitak et al. used monoclonal antibodies to detect furaltadone metabolite in shrimps with an IC50 of 5.33 ng/ml [181].

Electrochemical immunosensors are a powerful tool for the analysis of antibacterials in food and different configurations have been presented during recent years. For example, an amperometric immunosensor was reported by Wu et al. [182], for penicillin quantification in milk, with a linear range from 0.25 to 3 ng/ml and a limit of detection of 0.3 μg/L [182]. Other types of transduction have been also explored, like a label-free impedimetric flow injection immunosensor for the detection of penicillin G.

The Surface Plasmon Resonance (SPR) principle has been especially applied in extensively used optical biosensors. Two SPR immunoassays for detection of β-lactam antibiotics in milk were developed by Gustavsson et al. [183] and applied to real sample analysis by Sternesjö et al. [184]. The two assays showed similar performance with respect to detection limits for penicillin G (1.2 and 1.5 μg/kg, for 2- and 3-peptide assays, respectively), while other β-lactams were detected at or near their respective MRLs. In another example, Raz et al. [185] examined the possibilities of implementing direct competitive immunoassay formats for small and large molecule detection on a microarray, using imaging surface plasmon resonance (iSPR) [185]. Weigel et al. [186] presented a comparison between SPR biosensors for fluoroquinolone antibiotics in poultry muscle, fish, and egg and already established methods (LC-MS/MS and screening assay) [186]. The method based on SPR correctly identified all contaminated samples and demonstrated advantages in sensitivity and analysis time compared to the established ones. Recently, Raz et al. [185] have performed a label-free and multiplex detection of antibiotic residues in milk using iSPR. During recent years, innovative resonance methods have been introduced thanks to the possibilities opened by the new nanotechnological materials. An immunonanogold resonance-scattering spectral method for penicillin G [93, 187] and a waveguide-interrogated optical immuno-sensor (WIOS) for sulfonamides [188] have been reported for milk. Sulfapyridine, used as the reference sulfonamide, was detected with the immunosensor in buffer and in milk with a limit of detection (IC$_{90}$) of 0.2 $\pm$ 0.1 μg/L and 0.5 $\pm$ 0.1 μg/L, respectively. The developed immunosensor presents great potential as a generic sensing device for fast and early detection of food contaminants in-situ by nonskilled users.

Biosensors based on microbial immobilization have also been used for food applications, e.g., the inexpensive and rapid high-throughput bacterial biosensor developed by Virolainen et al. for rapid detection of tetracyclines and their 4-epimer derivatives in poultry meat [188, 189].

In addition to the biological-based approaches, instrumental multiclass methods for antibacterial and coccidiostat detection in food samples are rapidly increasing during the last years since regulations concerning the presence of such chemicals in animal-derived foodstuffs are becoming more stringent (Table 1). The challenges that these types of analyses pose to the analysts mainly have to do with the complexity of the matrix and the different physicochemical properties of the antibacterial families. A purification and preconcentration step is often required prior to analysis in order to minimize matrix effects and reach the desired sensitivities. Protein precipitation, usually done with organic solvents [62, 190], defatting, usually with hexane and acid hydrolysis in the case of honey [190, 191] is the first step employed prior to analysis.

The development of multiclass methods for the detection of antibacterials and coccidiostats in food samples has shown a growing interest during the last years since the regulations concerning the presence of such chemicals in animal-derived foodstuffs is becoming more and more stringent. The challenges that these types of analyses pose to the analysts mainly have to do with the complexity of the matrix and the different physicochemical properties of the antibacterial families. Therefore, very often, a purification and preconcentration step is required prior to analysis in order to minimize matrix effects and reach the desired sensitivities [192, 193].

The most common food matrices analyzed include meat, fish, milk, egg, and honey. The first step usually employed prior to analysis is protein precipitation, which is usually done with organic solvents [59, 60, 62, 194, 195], defatting, usually with hexane [60], and acid hydrolysis in the case of honey [190, 191].

A sample extraction step follows usually done by pressurized liquid extraction in the case of solid samples, like meat and fish [196] and animal tissue [64]. The two studies focused on single groups of antibiotics, macrolides, for which extraction with methanol at 80°C [196] took place and tetracyclines extracted with water at 70°C [64]. Kukusamunde et al. explored mixed micellar cloud point extraction (MM-CPE) for the detection of four hydrophilic penicillins (ampicillin, cloxacillin, penicillin G, and oxacillin) in milk. This was achieved by a mixed micellar extractant consisting of Triton X-114 and the cationic cethyl trimethylammonium bromide (CTAB). The method, using HPLC-photodiode array, achieved LODs in the range of 2–3 ng/ml [197]. In another application also dealing with penicillins in milk, porcine tissue and feed samples, van Holthoon et al. developed a derivatization procedure in order to avoid degradation of the unstable penicillins and conversion to stable piperidine derivatives [58]. The applicability of supercritical fluid (CO_2) extraction (SFE) in situ derivatization was recently explored for amphenicols in shrimp and showed that low LODs (0.009–0.017 μg/kg) can be achieved even without additional clean-up [56].

However, in most cases a clean-up step is necessary in order to remove matrix interferences that reduce the signal for target analytes. This is usually accomplished by solid phase extraction (SPE), performed either *offline* or *online* by the use of reversed-phase cartridges, usually polymeric Oasis HLB [57, 58, 62, 64, 198] or C18 [59, 60]. Tandem SPE has been also explored in food analysis in order to achieve cleaner extracts [199, 200]. Selective SPE has also been explored by the application of molecularly imprinted polymers (MIPs), recently applied for four fluoroquinolones in milk and pig kidney [49] and four tetracyclines in foodstuff samples like lobster, duck, honey, and egg [55]. Online SPE is being increasingly employed during the last years [201, 202] as it significantly reduces manual sample handling, thus providing high-throughput analysis and minimization of inaccuracies.

Separation and detection techniques for antibacterials in food mainly focus on the use of LC coupled to MS or tandem MS. Nevertheless, recent studies have suggested capillary electrophoresis coupled to laser-induced fluorescence (LIF) as a way of improving sensitivity [49], HRLC coupled to microTOF-ESI-MS as a highly selective, sensitive, and quick screening method for 100 veterinary drugs in fish, meat, and egg samples [195], and nanoscale LC coupled to UV or ion trap MS, with LODs in the range 0.01–0.51 μg/L (nanoLC-MS) and the possibility that even lower limits could be achieved by using triple quadrupole MS [59].

Scientific progress over the past years shows the increasing efforts of researchers worldwide toward the sensitive multianalyte determination of antibacterials in foodstuffs by means of fast and simple clean-up procedures and high sample throughput analyses. This trend, we believe, will become more evident as the need for wide monitoring of food contaminants will increase as new information on toxicity and human health effects will come into the light. Following this trend, the technique of choice for efficient extraction of analytes from complex matrices and that will gain more attention in the following years is online SPE. This technique provides high robustness and throughput analysis by reducing analysis time and solvent consumption as well as sample manipulation and therefore sources of errors. Concerning separation and detection, the method of choice is LC coupled to MS or tandem MS, which however makes the need for clean-up or sample dilution inevitable.

3.3 Marine Biotoxins

3.3.1 Immunoassays

Specific antibodies, either monoclonal or polyclonal have been produced against a large variety of biotoxins, and a high number of assays based on different formats have been reported, showing high sensitivities for quantitative or semiquantitative measurements. In the case of marine biotoxins, substantial efforts to develop immunoassays have been undertaken to detect most common groups of toxins in shellfish, DSP, PSP, ASP, NSP, and CFP toxins. Some of them are excellent candidates for the development of a range of functional immunochemical-based detection assays for this group of toxins, as, for example, the assay reported by Steward et al. based on monoclonal antibodies against okadaic acid and dinophysistixins-1,-2 [203]. One of the main problems associated to immunochemical approaches to detect marine biotoxins is the lack of the adequate cross-reactivity to detect all the analogs of each group. For some groups, this creates an uncertainty gap that prevents the antibody approach as a useful tool to detect marine biotoxins on a routine basis. For those chemical groups with a reduced number of analogs, as is the case of Domoic acid, immunochemical methods could be suitable approaches. Despite these limitations, immunoassays seem to be a promising tool for routine analysis of shellfish toxins, due to the high sample throughput and relative low cost. Moreover, they require neither sophisticated and expensive equipment nor skilled personnel.

3.3.2 Bioassays

The more classical approach to assess the presence of marine biotoxins in seafood is the in vivo mouse bioassay. It is based on the administration of suspicious extracted shellfish samples to mice, the evaluation of the lethal dose and the toxicity calculation according to reference dose response curves, established with reference material. It provides an indication about the overall toxicity of the sample, as it is not able to differentiate among individual toxins. This is a laborious and time-consuming procedure; the accuracy is poor, it is nonspecific and generally not acceptably robust. Moreover, the mouse bioassay suffers from ethical implications and it is in conflict with the EU Directive 86/609 on the Protection of Laboratory Animals. Despite the drawbacks, this bioassay is still the method of reference for almost all types of marine toxins, and is the official method for PSP toxins.

The investigation of the presence of marine biotoxins in water, phytoplankton, and food has been achieved by several in vitro assays. However, alternatives to the animal bioassay for marine toxins have not been sufficiently evaluated in interlaboratory studies needed to demonstrate their scientific validity. In addition, these methods continue to be time consuming and expensive for intensive monitoring programs, and present some difficulties for their automation.

3.3.3 Instrumental Techniques

A high number of works have been reported the development of LC-MS and LC-MS/MS for the determination of biotoxins in aquatic environments with limits of detection in the range of ng-pg/L concentrations; however, much less work have been performed to detect emerging groups of marine biotoxins such as

(CFP, Palytoxin, Spirolide). Emerging groups of marine biotoxins comprises those groups less studied and without a specific regulation. The main advantages of instrumental approaches based on LC-MS are high selectivity, specificity, and sensitivity. For example, a method based on LC-QqQ-MS operating in selected ion monitoring (SIM) and MRM mode under positive ionization to detect palytoxin has been developed [31]., and it was decisive for the confirmation of these toxins for first time along the Italian coast. Other examples of works carried out during the last year for the analysis of emerging biotoxins are [75, 82, 204–206]. In this work, main advantages of instrumental applications have been the confirmation of target groups of toxins. For example, González et al. [82], the presence of free okadaic acid (OA) and dinophysistoxin-2 (DTX-2) as well as esters of these toxins. The results also revealed the presence of minor amounts of 13-desmethyl spirolide C (SPX-1) in the analyzed samples, although this toxin has never been reported before in Spain. The combination of different MS modes of operation, just as enhanced MS (EMS) and MS^3 experiments, allowed to confirm the first occurrence of spirolides in Spanish shellfish.

However, the major limitations of instrumental analysis for marine biotoxins are, first sometimes the lack of standards, second the required time of analysis, are expensive techniques to be applied in routine analysis but the main limitation is the lack of information about the possible presence of other nontarget marine biotoxins.

4 Occurrence of Selected Groups in Food Samples

In this section, different studies reporting the concentration levels of selected compounds in food samples. A summary is presented in Table 2.

4.1 Industrial Origin Compounds

4.1.1 PFCs

A relatively low number of studies have been published on the analysis of PFCs in food samples. A recent study looking at wild fish from Northern Germany revealed levels for PFOS ranging from 8.2 to 225 ng/g fresh weight (fw) [215], which were notably lower than values reported for fish from Japan (3–7,900 ng/g fw) [83] and fish from Spain (0.1–50 ng/g fw) [73] but higher than levels found in freshwater fish from Sweden (0.5–23 ng/g fw). The differences observed in the above studies can be probably attributed to variations in the fat content of the samples analyzed. Mussels and oysters have also shown the presence of PFOS in ranges 0.11–0.59 ng/g ww, along with PFHS (0.06–0.51 ng/g ww), PFBS (0.009–0.030 ng/g ww), PFOSA (0.04–2.96 ng/g ww), PFDoDA (0.2 ng/g ww), PFDA (0.13 and 0.12 ng/g) [70].

Table 2 Occurrence of emerging contaminants in the food chain

Analytes	Matrix	Levels	Reference
PBDEs			
PBDEs	Chicken fat	∑PBDEs: 1.76–39.43 ng/g	[37]
BDE #47, #85, #99, #100, #153 & #154.	Dogfish liver sample	BDE #47: 2.4 ng/g; BDEs #85, #99, #100, #153 & #154: <LoD	[207]
3 monoBDEs, 7 diBDEs, 8 triBDEs, 6 tetraBDEs, 7 pentaBDEs, 5 hexaBDEs, 3 heptaBDEs and the decaBDE (209).	Fish muscle and liver	PBDEs: 0.2–436 (muscle) and 0.1–446 ng/g (liver)	[40]
BDE #17, #28, #47, #66, #85, #99, #100, #153, #154 & #183	Adipose tissues from marine mammals, chicken and trout	Whale blubber: 0.7 (BDE#28) to 38 ng/kg (BDE #47) Polar bear fat: 0.0053 (BDE #28) to 6.5 ng/kg (BDE #47) Trout: 17 (BDE #183) to 102,175 ng/kg (BDE #153) Chicken: 0.46 (BDE #28) to 17 ng/kg (BDEs #47 & #99)	[42]
BDE #17, #28, #47, #66, #85, #99, #100, #153, #154 & #183	Palm oil	4.3 (BDE #28) to 32 ng/kg (BDE #47)	[42]
BDE #17, #28, #47, #66, #71, #85, #99, #100, #138, #153, #154, #183 & #190	Butter	∑PBDEs: 18.0–955 ng/kg	[43]
1 monoBDE, 2 diBDEs, 2 triBDEs, 5 tetraBDEs, 3 hexaBDEs, 1 heptaBDE and the decaBDE.	Fish, crab, mussel and oyster	∑PBDEs: 42.6–192 ng/g of lipid	[44]
27 PBDEs congeners	Fish and Crabs	∑PBDEs: 135–518 ng/g	[90]
BDE #28, #47, #66, #71, #85, #99, #100, #138, #153, #154, #183 & #190	Fish	∑PBDEs: 3.5–604	[208]
Pesticides			
OCPs – aldrin, DDE (op′, pp′), DDD (op′,pp′), DDT (op′, pp′), dieldrin, endosulfan (a, b), endrin, HCH (a, b, c), heptachlor, heptachlor epoxide, HCB	Eggs, chicken, meat	5–1,300 ng/g lipid wt. (eggs), 4–130 ng/g lipid wt. (chicken), 5–2,890 ng/g lipid wt. (meat)	[209]
222 pesticides	Processed foods (dumpling, curry, French fries, fried chicken, fried fish)	Chlorpropham was found in a processed food sample at 40 ng/g.	[210]

Table 2 (continued)

Analytes	Matrix	Levels	Reference
47 pesticides	Cooked wheat flour and polished rice	Tricyclazole was found in 12 samples at levels between 0.015 and 0.111 mg/kg. Fenobucarb was found in 3 samples at levels between 0.031 and 0.089 mg/kg.	[211]
236 pesticides	Baby-food	Chlorpropham was detected in sample with a concentration below 10 ng/g.	[212]
Antioxidant pesticides & metabolites used in postharvest treatment: imazalil (IMZ)/ ethoxyquin (EQ) and thiabendazole (TBZ)/ diphenylamine (DPA), Hydroxy-DPA, n-phenyl-4-quinoneimine, methoxy-DPA, demethyl-EQ, demethyldehydro-EQ, EQ-dimer, methyl-EQ, EQ-N-oxyl, 2,2,4,-trimethyl-6-quinolone	Pears, Apples (postharvest treatments)	0.002–0.672 mg/kg (EQ), 0.94–11.86 mg/kg (IMZ), 0.024–0.902 mg/kg (DPA), 0.012–2.59 mg/kg (TBZ)	[213]
Perfluorinated compounds			
PFOS, PFHS, PFBS	Wildlife: birds, fish	Fish: 3–7,900 ng/g (PFOS), 4–19 ng/g (PFHS), 36–151 ng/g (PFBS), Birds: 68–1,200 ng/g (PFOS), 2.6–10 ng/g (PFHS), <45 ng/g (PFBS)	[83]
PFPA, PFBS, PFOA, i,p-PFNA, PFNA, PFOS, PFDA, L-PFDS	Edible Fish	0.23-23.04 ng/g fw	[73]
PFOS, PFHS, PFBS, PFOSA, PFDoDA, PFUnDA, PFDA, PFNA, PFOA, PFHpA, PFHxA	Mussels, oysters	PFOS: 114–586 pg/g (w/w), PFHS: 63–512 pg/g (w/w), PFBS: 9–30 pg/g (w/w), PFOSA: 38–2,957 pg/g (w/w) PFDoDA: 196 pg/g, PFDA: 132 pg/g and 119 pg/g (values for single occurrence)	[70]
PFOS, PFHxS, PFDoDA, PFUnDA, PFDA, PFNA, PFOA	Farm, pet animals	PFOS: 67 ng/g (Chicken), 54 ng/g (pigs), 34 ng/g (cattle)	[72]

(continued)

Table 2 (continued)

Analytes	Matrix	Levels	Reference
4 PFSs, PFOSA, 10 PFCAs	Freshwater fish	PFOS: 0.5–23 ng/g fresh wt., PFHsS: 0.02–0.8 ng/g, PFOSA: <0.3–3.3 ng/g	[214]
PFOS, PFOA	Wild fish	PFOS: 8.2–225 ng/g fw	[215]
PFOS, PFOSA, C7-C14 PFCAs	Macroalgae, bivalves, fish, sea ducks, marine mammals	0.1 and 40 ng/g ww	[5]
Antibiotics			
Fluoroquinolones: NOR, CIP, OFL, ENR, RUF	Milk	85.30–0.34 ng/g	[60]
5 tetracyclines, 7 macrolides, 3 aminoglycosides, 8 β-lactams, 2 amphenicols, 17 sulfonamides	Honey	41–147 µg/kg	[61]
Macrolides: ERTMC, JSMC, RXTMC, SRMC, TMCS, TLDMC, TLS	Meat	58–87 µg/kg	[52]
Tetracycline, chlorotetracycline, oxytetracycline, doxycycline	Meat	24–160 µg/kg	[64]
Florfenicol	Liver, meat	186.2–1423.5 µg/kg	[55]
Amphenicols: CAP, FF, TAP	Shrimp	47–592 µg/kg	[56]
Biotoxins			
Spirolide biotoxins: 13-desMeC SPX, 13,19-didesMeC SPX, 27-OH-13,19-didesMeC SPX, YTX, homoYTX, CarboxyYTX, CarboxyhomoYTX, 45-OHYTX, 45-OhhomoYTX, PTX-2sa	Shellfish	11–7,950 ng/g	[205]
Putative Palytoxin	Plankton	1,350 ng (plankton pellet), 1,950 ng (butanol plankton extract)	[216]
Okadaic acid (OA), dinophysistoxins (DTXs), pectenotoxins (PTXs), azaspiracids (AZAs) and spirolides	Shellfish	9–2,012 ng/g (mainly OA group toxins)	[204]
pectenotoxin-2, spirolide-A and their derivatives	Shellfish	2–585 ng/g	[206]

HBCD hexabromocyclododecane, *HCB* hexachlorobenzene, *p,p′-DDT* Dieldrin, 1,1,1-trichloro-2,2-bis-(4-chlorophenyl)-ethane, *p,p′-DDE* 1,10-dichloro-2,20-bis(4-chlorophenyl) ethylene, *p,p′-DDD* 1,10-dichloro-2,20-bis-(4-chlorophenyl)ethane, *PFOS* perfluorooctane sulfonate, *PFHS* perfluorohexane sulfonate, *PFBS* perfluorobutane sulfonate

Chicken, pig, and cattle samples have revealed PFOS mean values of 67 ng/g, 54 ng/g, and 34 ng/g, respectively [72] .

4.1.2   PBDEs, PCBs, PBBs

On the other hand, there is a plethora of studies monitoring the occurrence of PBDEs, PCBs, and PBBs in food samples, with the majority of results reporting the occurrence of these compounds in fish, shellfish, and crustacean samples. Levels detected are in the range 3.5–604 ng/g lipid weight (lw) [208], 14–60 ng/g lw [217], 43–192 ng/g lw [44] for the sum of 12 and 10 PBDEs in biota, 7 PBDEs in fish, 20 PBDEs in lower-trophic-level coastal marine species, respectively. An Elisa screening on Hawaiian euryhaline fish and crabs showed ranges of 135–518 ng/g lw [90]. A study on HBCD revealed levels between 90 and 4,863 ng/g ww in fish [218]. However, the levels found for PCBs in bivalves, crabs, and fish samples were significantly higher (14–7,340 ng/g) than PBDE levels [44]. Recent monitoring of dairy products has shown PBDE concentrations in the range of 0.31 ng/g and 0.02–0.96 ng/g for 17 PBDEs in milk and 12 PBDEs in butter, respectively. Also in the case of butter, PCB levels were higher than the ones for PBDEs with values ranging from 0.14 to 2.52 ng/g [219].

4.1.3   Pesticides

As has been discussed before, recent developments are focused on large multiresidue methods of analysis. Some examples of recent applications of multiresidue methods in cooked and processed food are those reported recently by Kitacawa et al. [210], and Lee et al. [211], Kitakawa et al., presented a methods for the analysis of 222 pesticides in different processed foods. On the other hand, Lee et al. presented a LC-MS/MS method for the quantitative determination of 44 pesticide residues with hydrolyzable functional group in five types of vegetables. In another example [212]. a simple multiresidue method has been developed for the routine determination of 236 pesticides and degradation products in meat-based baby food. This approach combines a modified QuEChERS sample preparation method using a triple partitioning extraction step with water/ACN/hexane and a system composed of GC with programmable temperature vaporization injector hyphenated to an IT-MS.

Another import field of development is the investigation of pesticides transformation products in food. Some examples are the investigation of the acaricide amitraz and its transformation products, 2,4-dimethylaniline (DMA), 2,4-dimethylformamidine (DMF), and N-2,4-dimethylphenyl-N-methylformamidine (DMPF) in pears [220] Antioxidant pesticides as well as their metabolites used in postharvest treatment have been investigated in pears and apples with concentrations in the ranges 0.002–0.672 ng/g (ethoxyquin), 0.94–11.86 ng/g (imazalil), 0.024–0.902 ng/g (diphenylamine), 0.012–2.59 ng/g (thiabendazole).

The study reveals for the first time the presence of some EQ metabolites in fruits, with levels exceeding several times those of the parent compounds (Picó 2010).

About new emerging pesticides such as chiral pesticides most of the works are focused on environmental occurrence [221, 222].

4.1.4 Nanomaterials

As we mentioned previously, the research on the field of nanomaterials is at a primitive stage and literature mainly focuses on the benefits of using such particles for environmental load reduction, waste treatment, and source pollution control, as well as the toxicological and health issues accompanying the use of such materials. As a consequence, there are still few methods developed for food matrices and even lesser monitoring schemes applied. Currently, no data have been noticed reporting the occurrence on nanomaterial residues in food and just one work has been published till now reporting the occurrence of nanoparticle (fullerenes) in real environmental samples [6].

4.2 Pharmaceutical Residues: Antibiotics and Coccidiostats

Unfortunately, in the case of antibiotics and coccidiostats, considering the sometimes conflicting interests between the pharmaceutical industry, the regulatory agencies and the scientific community, not many results can be found in the literature regarding survey studies for those substances. A small number of publications are therefore available, mostly in an abstract form, which usually do not emphasize values detected. An application of an enzyme immunoassay for monitoring of milk samples revealed 41.3% of samples (151 total samples) containing antimicrobial residues in Brazil, with one sample exceeding the corresponding MRL (200 µg/kg) for streptomycin-dihydrostreptomycin. Also, 4 samples were above the zero tolerance level for chloramphenicol [223]. In a similar study using a semiquantitative ELISA to monitor 60 ultra-heat-treatment milk samples from Turkey, the authors detected high incidence rate of chloramphenicol (28 samples) and tetracycline (40 samples) [228]. Another study on raw cow's milk with an LC-UV method revealed low concentrations of tetracycline residues and 51% of the samples containing oxytetracycline [224]. Dairy products, eggs and meat, poultry, and meat tissue samples from Kuwait were screened by the Charm II test and confirmed by LC-MS/MS for tetracycline residues. The study showed 5% of poultry and 18% of milk samples were above the permitted limits [225]. Finally, honey samples from the Italian market showed a total of 6.3% of all samples containing the antibacterial drugs analyzed with sulphonamides being the most occurring, followed by tetracyclines, streptomycin, tylosin, and chloramphenicol [226].

4.3 Biotoxins: Emerging Group of Marine Biotoxins

Biotoxins, mainly of the spirolide family, have been principally monitored in shellfish samples. Reported toxin levels vary from 2–585 ng/g [206] to 11–7,950 ng/g [205] in shellfish from France and Italy, respectively, whereas reported values for Spanish mussels were in the range of 13–20 ng/g [82]. This study employed the use of different MS modes of operation, enhanced MS (EMS) and MS3 experiments in order to confirm the first occurrence of spirolides in Spanish shellfish. Okadaic acids were the principal toxin contaminants found in shellfish samples from Galicia, Spain, with levels reaching 2,012 ng/g [204]. Putative palytoxin was for the first time detected in Italian waters at levels of 1,350 ng for plankton pellet) and 1,950 ng for butanol extract; thus, it was suggested to be the causative agent responsible for the Genoa 2005 outbreak showing respiratory illness in people exposed to marine aerosols [31].

5 Conclusions

Whereas there is a legislation establishing official methods for regulated food contaminants and food control, laboratories should be accredited by the ISO 17025 standard; for emerging contaminants these measures are not established. Therefore, for new classes of contaminant, alternative approaches to method validation and quality control should be developed. During the current European Commission's Framework 7 these lacks and the need to develop harmonize analytical approaches covering emerging groups of compounds have been recognized and different food safety initiatives have been taken.

During the last few years, new analytical methods turned to multiclass methods in order to face up the large number of compounds that should be screened, and rapid methods of detection.

Instrumental screening methods based on exact mass measures have increased for multiscreening purposes offering adequate uncertainty and possible identification of nontarget compounds. On the other hand, biological approaches offer as well another possibility for rapid and cost-effective alternative.

Acknowledgments This work was supported by the Spanish Ministry of Science and Innovation, through the project CEMAGUA (CGL2007-64551/HID). Waters is acknowledged for providing the SPE cartridges. Lina Kantiani acknowledges the Generalitat de Catalunya (2009FI_B 00660) and the Alexander S. Onassis foundation (F-ZD 09/2007-2008).

References

1. Clara M, Scheffknecht C, Scharf S, Weiss S, Gans O (2008) Emissions of perfluorinated alkylated substances (PFAS) from point sources – identification of relevant branches. Water Sci Technol 58:59–66
2. Carmosini N, Lee LS (2008) Environ Sci Technol 42:6559–6565
3. Furdui VI, Stock NL, Ellis DA, Butt CM, Whittle DM, Crozier PW, Reiner EJ, Muir DCG, Mabury SA (2007) Environ Sci Technol 41:1554–1559
4. Kim SK, Kannan K (2007) Environ Sci Technol 41:8328–8334
5. Kelly BC, Ikonomou MG, Blair JD, Surridge B, Hoover D, Grace R, Gobas FAPC (2009) Environ Sci Technol 43:4037–4043
6. Llorca M, Farré M, Picó Y, Barceló D (2009) J Chromatography A 1216:7195–7204
7. Kärrman A, Domingo JL, Llebaria X, Nadal M, Bigas E, van Bavel B, Lindström G (2009) Environ Sci Pollut Res 35:1–9
8. Kannan K, Tao L, Sinclair E, Pastva SD, Jude DJ, Giesy JP (2005) Arch Environ Contam Toxicol 48:559–566
9. Tittlemier SA, Pepper K, Seymour C, Moisey J, Bronson R, Cao XL, Dabeka RW (2007) J Agric Food Chem 55:3203–3210
10. Powley CR, George SW, Ryan TW, Buck RC (2005) Anal Chem 77:6353–6358
11. REPORT - E-waste, the hidden side of IT equipment's manufacturing and use Schwarzer, S, De Ono A, Peduzzi P, Giuliani G, Kluser S (2005) United Nations Environment Programme Geneva
12. Vetter W, Von Der Recke R, Symons R, Pyecroft S (2008) Rapid Commun Mass Spectrom 22:4165–4170
13. von der Recke R, Vetter W (2007) J Chromatogr A 1167:184–194
14. de Wit CA, Herzke D, Vorkamp K (2010) Sci Total Environ 408:2885–2918
15. Vorkamp K, Dam M, Riget F, Fauser P, Bossi R, Hansen AB (2004) NERI technical report 97
16. Giero J, Grochowalski A, Chrzaszcz R (2010) Chemosphere 78:1272–1278
17. García-Reyes JF, Ferrer C, Gómez-Ramos MJ, Fernández-Alba AR, García-Reyes JF, Molina-Díaz A (2007) TrAC – Trends Anal Chem 26:239–251
18. Picó Y, Blasco C, Font G (2004) Mass Spectrom Rev 23:45–85
19. Soler C, Maoes J, Picó Y (2008) Crit Rev Anal Chem 38:93–117
20. Dalläge J, Beens J, Brinkman UAT (2003) J Chromatogr A 1000:69–108
21. Fernández-Alba AR, García-Reyes JF (2008) TrAC – Trends Anal Chem 27:973–990
22. Crinnion WJ (2009) Altern Med Rev 14:347–359
23. Beltran J, López FJ, Hernández F (2000) J Chromatogr A 885:389–404
24. Ferry JL, Craig P, Hexel C, Sisco P, Frey R, Pennington PL, Fulton MH, Scott IG, Decho AW, Kashiwada S, Murphy CJ, Shaw TJ (2009) Nat Nanotechnol 4:441–444
25. Chaudhry Q, Scotter M, Blackburn J, Ross B, Boxall A, Castle L, Aitken R, Watkins R (2008) Food Addit Contam Part A Chem Anal Control Expo Risk Assess 25:241–258
26. Šimon P, Chaudhry Q, Bakoš D (2008) J Food Nutr Res 47:105–113
27. Nevzorova VV, Gmoshinsky IV, Khotimchenko SA (2009) Vopr Pitan 78:54–60
28. Threlfall EJ, Ward LR, Frost JA, Willshaw GA (2000) Int J Food Microbiol 62:1–5
29. Foulquie Moreno MR, Sarantinopoulos P, Tsakalidou E, De Vuyst L (2006) Int J Food Microbiol 106:1–24
30. Threlfall EJ (2002) FEMS Microbiol Rev 26:141–148
31. Ciminiello P, Dell'Aversano C, Fattorusso E, Forino M, Magno GS, Tartaglione L, Grillo C, Melchiorre N (2006) Anal Chem 78:6153–6159
32. Díaz-Cruz MS, García-Galán MJ, Guerra P, Jelic A, Postigo C, Eljarrat E, Farré M, López de Alda MJ, Petrovic M, Barceló D (2009) TrAC – Trends Anal Chem 28:1263–1275
33. Schulte C (2007) PFT - Nachweise in der Umwelt und Theorien zur Verbreitung, 12 (2007) 67–71

34. Taniyasu S, Kannan K, Man KS, Gulkowska A, Sinclair E, Okazawa T, Yamashita N (2005) J Chromatogr A 1093:89–97
35. Martin JW, Kannan K, Berger U, De Voogt P, Field J, Franklin J, Giesy JP, Harner T, Muir DCG, Scott B, Kaiser M, Jarnberg U, Jones KC, Mabury SA, Schroeder H, Simcik M, Sottani C, Van Bavel B, Kärrman A, Lindström G, Van Leeuwen S (2004) Environ Sci Technol 38:248A–255A
36. Wang N, Szostek B, Buck RC, Folsom PW, Sulecki LM, Capka V, Berti WR, Gannon JT (2005) Environ Sci Technol 39:7516–7528
37. Huwe JK, Lorentzsen M, Thuresson K, Bergman A (2002) Chemosphere 46:635–640
38. Fernandes A, White S, D'Silva K, Rose M (2004) Talanta 63:1147–1155
39. Martínez A, Ramil M, Montes R, Hernandez D, Rubí E, Rodríguez I, Torrijos RC (2005) J Chromatogr A 1072:83–91
40. Eljarrat E, Barceló D (2004) TrAC – Trends Anal Chem 23:727–736
41. Bayen S, Lee HK, Obbard JP (2004) J Chromatogr A 1035:291–294
42. Gomara B, Herrero L, Bordajandi LR, González MJ (2006) Rapid Commun Mass Spectrom 20:69–74
43. Wang Y, Yang R, Wang T, Zhang Q, Li Y, Jiang G (2010) Chemosphere 78:772–777
44. Mizukawa K, Takada H, Takeuchi I, Ikemoto T, Omori K, Tsuchiya K (2009) Mar Pollut Bull 58:1217–1224
45. Shelver WL, Parrotta CD, Slawecki R, Li QX, Ikonomou MG, Barcelo D, Lacorte S, Rubio FM (2008) Chemosphere 73:S18–S23
46. Lacorte S, Ikonomou MG, Fischer M (2010) J Chromatogr A 1217:337–347
47. Peres GT, Rath S, Reyes FGR (2010) Food Control 21:620–625
48. Artigas P, Gadsby DC (2003) Proc Natl Acad Sci USA 100:501–505
49. Lombardo-Agüí M, García-Campaña AM, Gámiz-Gracia L, Cruces Blanco C (2010) J Chromatogr A 1217:2237–2242
50. Chafer-Pericos C, Maquieira A, Puchades R, Miralles J, Moreno A, Pastor-Navarro N, Espinós F (2010) Anal Chim Acta 662:177–185
51. Hallegraeff GM (2003) Manual on harmful marine microalgae. Hallegraeff GM, Anderson DM, Cembella Ad (eds). UNESCO, Saint Berthevin, 25–49
52. McMahon T, Silke J (1996) Harmful Algae News 14:2
53. Costi EM, Sicilia MD, Rubio S (2010) J Chromatogr A 1217:1447–1454
54. Fladmark KE, Serres MH, Larsen NL, Yasumoto T, Aune T, Daskeland SO (1998) Toxicon 36:1101–1114
55. Ge S, Yan M, Cheng X, Zhang C, Yu J, Zhao P, Gao W (2010) J Pharm Biomed Anal 52:615–619
56. Liu W-L, Lee R-J, Lee M-R (2010) Food Chem 121:797–802
57. Olejnik M, Szprengier-Juszkiewicz T, Jedziniak P (2009) J Chromatogr A 1216:8141–8148
58. van Holthoon F, Mulder PPJ, van Bennekom EO, Heskamp H, Zuidema T, van Rhijn HA (2010) 1163 Anal Bioanal Chem Anal Bioanal Chem 396:3027–3040
59. Hsieh SH, Huang HY, Lee S (2009) J Chromatogr A 1216:7186–7194
60. Tang Q, Yang T, Tan X, Luo J (2009) J Agric Food Chem 57:4535–4539
61. Doucette GJ, Logan MM, Ramsdell JS, Van Dolah FM (1997) Toxicon 35:625–636
62. Bohm DA, Stachel CS, Gowik P (2009) J Chromatogr A 1216:8217–8223
63. Adams ME, Olivera BM (1994) Trends Neurosci 17:151–155
64. Blasco C, Corcia AD, Picó Y (2009) Food Chem 116:1005–1012
65. Hernando MD, Ferrer C, Ulaszewska M, García-Reyes JF, Molina-Díaz A, Fernández-Alba AR (2007) Anal Bioanal Chem 389:1815–1831
66. Lehotay SJ, Mastovska K, Yun SJ (2005) J AOAC Int 88:630–638
67. García-Reyes JF, Hernando MD, Ferrer C, Molina-Díaz A, Fernández-Alba AR (2007) Anal Chem 79:7308–7323
68. James KJ, Sierra MD, Lehane M, Braníƒa Magdalena A, Furey A (2003) Toxicon 41:277–283

69. Tubaro A, Florio C, Luxich E, Sosa S, Della Loggia R, Yasumoto T (1996) Toxicon 34:743–752
70. So MK, Taniyasu S, Lam PKS, Zheng GJ, Giesy JP, Yamashita N (2006) Arch Environ Contam Toxicol 50:240–248
71. So MK, Miyake Y, Yeung WY, Ho YM, Taniyasu S, Rostkowski P, Yamashita N, Zhou BS, Shi XJ, Wang JX, Giesy JP, Yu H, Lam PKS (2007) Chemosphere 68:2085–2095
72. Guruge KS, Manage PM, Yamanaka N, Miyazaki S, Taniyasu S, Yamashita N (2008) Chemosphere 73:S210–S215
73. Llorca M, Farré M, Picó Y, Barceló D (2009) J Chromatogr A 1216:7195–7204
74. Farré M, Pérez S, Gajda-Schrantz K, Osorio V, Kantiani L, Ginebreda A, Barceló D (2010) J Hydrol 383:44–51
75. Gerssen A, Mulder PPJ, McElhinney MA, de Boer J (2009) J Chromatogr A 1216:1421–1430
76. Cheun B, Endo H, Hayashi T, Nagashima Y, Watanabe E (1996) Biosens Bioelectron 11:1185–1191
77. Alfonso A, Vieytes MR, Yasumoto T, Botana LM (2004) Anal Biochem 326:93–99
78. Alfonso C, Alfonso A, Pazos MJ, Vieytes MR, Yasumoto T, Milandri A, Poletti R, Botana LM (2007) Anal Biochem 363:228–238
79. Gawley RE, Mao H, Haque MM, Thorne JB, Pharr JS (2007) J Org Chem 72:2187–2191
80. Espiña B, Cagide E, Louzao MC, Fernandez MM, Vieytes MR, Katikou P, Villar A, Jaen D, Maman L, Botana LM (2009) Biosci Rep 29:13–23
81. Fonfría ES, Vilariño N, Espiña B, Louzao MC, Alvarez M, Molgó J, Aroz R, Botana LM (2010) Anal Chim Acta 657:75–82
82. González AV, Rodríguez-Velasco ML, Ben-Gigirey B, Botana LM (2006) Toxicon 48:1068–1074
83. Taniyasu S, Kannan K, Horii Y, Hanari N, Yamashita N (2003) Environ Sci Technol 37:2634–2639
84. Llorca M, Farre M, Pico Y, Barcelo D (2011) J Chromatogr A, 1218: 4840–4846
85. Yamashita N, Kannan K, Taniyasu S, Horii Y, Okazawa T, Petrick G, Gamo T (2004) Environ Sci Technol 38:5522–5528
86. Barceló D, Petrovic M (2006) TrAC – Trends Anal Chem 25:191–193
87. Taniyasu S, Kannan K, Yeung LWY, Kwok KY, Lam PKS, Yamashita N (2008) Anal Chim Acta 619:221–230
88. Takino M, Daishima S, Nakahara T (2003) Rapid Commun Mass Spectrom 17:383–390
89. Berger U, Haukas M (2005) J Chromatogr A 1081:210–217
90. Xu T, Cho IK, Wang D, Rubio FM, Shelver WL, Gasc AME, Li J, Li QX (2009) Suitability of magnetic particle immunoassay for the analysis of PBDEs in Hawaiian euryhaline fish and crabs in comparison with gas chromatography/electron capture detection-ion trap mass spectrometry. Environmental Pollution 157:417–422
91. Ahn KC, Gee SJ, Tsai HJ, Bennett D, Nishioka MG, Blum A, Fishman E, Hammock BD (2009) Environ Sci Technol 43:7784–7790
92. Marchesini GR, Meimaridou A, Haasnoot W, Meulenberg E, Albertus F, Mizuguchi M, Takeuchi M, Irth H, Murk AJ (2008) Toxicol Appl Pharmacol 232:150–160
93. Kim HJ, Rossotti MA, Ahn KC, González-Sapienza GG, Gee SJ, Musker R, Hammock BD (2010) Anal Biochem 401:38–46
94. Chen HY, Zhuang HS (2012) Advanced Materials Research (Volumes 347 – 353) 1537–1541
95. Eljarrat E, De La Cal A, Raldua D, Duran C, Barcelo D (2004) Environ Sci Technol 38:2603–2608
96. Salgado-Petinal C, Garcia-Chao M, Llompart M, Garcia-Jares C, Cela R (2006) Anal Bioanal Chem 385:637–644
97. Yuà V, Pardo O, Pastor A, De La Guardia M (2006) Anal Chim Acta 557:304–313
98. Larrazábal D, Martínez MA, Eljarrat E, Barceló D, Fabrellas B (2004) J Mass Spectrom 39:1168–1175

99. Debrauwer L, Riu A, Jouahri M, Rathahao E, Jouanin I, Antignac JP, Cariou R, Le Bizec B, Zalko D (2005) J Chromatogr A 1082:98–109
100. Buchanan I, Liang HC, Khan W, Liu Z, Singh R, Ikehata K, Chelme-Ayala P (2009) Water Environ Res 81:1731–1816
101. Farré M, Pérez S, Gonzalves C, Alpendurada MF, Barceló D (2010) TrAC – Trends Anal Chem 29:1347–1362
102. Gómiz-Gracia L, García-Campaña AM, Soto-Chinchilla JJ, Huertas-Pérez JF, González-Casado A (2005) TrAC – Trends Anal Chem 24:927–942
103. Jiang X, Li D, Xu X, Ying Y, Li Y, Ye Z, Wang J (2008) Biosens Bioelectron 23:1577–1587
104. Maqbool U, Jamil Qureshi M (2008) Pak J Sci Ind Res 51:281–292
105. Rebe Raz S, Haasnoot W (2011) TrAC – Trends Anal Chem 30:1526–1537
106. Ricci F, Volpe G, Micheli L, Palleschi G (2007) Anal Chim Acta 605:111–129
107. Rodriguez-Mozaz S, Lopez de Alda MJ, Barcelo D (2009) Achievements of the RIANA and AWACSS EU projects: immunosensors for the determination of pesticides, endocrine disrupting chemicals and pharmaceuticals. Hdb Env Chem 5J:33–46
108. Van Emon JM (2010) J AOAC Int 93:1681–1691
109. Xu ZL, Shen YD, Beier RC, Yang JY, Lei HT, Wang H, Sun YM (2009) Anal Chim Acta 647:125–136
110. Xu ZL, Wang Q, Lei HT, Eremin SA, Shen YD, Wang H, Beier RC, Yang JY, Maksimova KA, Sun YM (2011) Anal Chim Acta 708:123–129
111. Selvi AA, Sreenivasa MA, Manonmani HK (2011) Food Agric Immunol 22:217–228
112. Asensio-Ramos M, Ravelo-Perez LM, Gonzaez-Curbelo MA, Hernandez-Borges (2011) 1218:7415–7437
113. Afify AEMMR, Mohamed MA, El-Gammal HA, Attallah ER (2010) J Food Agric Environ 8:602–606
114. Chung SWC, Chan BTP (2010) J Chromatogr A 1217:4815–4824
115. Garrido Frenich A, Martínez Vidal JL, Pastor-Montoro E, Romero-González R (2008) Anal Bioanal Chem 390:947–959
116. Gilbert-López B, García-Reyes JF, Lozano A, Fernández-Alba AR, Molina-Díaz A (2010) J Chromatogr A 1217:6022–6035
117. Grimalt S, Sancho JV, Pozoa OJ, Hernándeza FE (2010) J Mass Spectrom 45:421–436
118. Kmellár B, Abrankó L, Fodora P, Lehotay SJ (2010) Food Addit Contam Part A Chem Anal Control Expo Risk Assess 27:1415–1430
119. Koesukwiwat U, Lehotay SJ, Miao S, Leepipatpiboon N (2010) J Chromatogr A 1217:6692–6703
120. Kovalczuk T, Lacina O, Jech M, Poustka J, Hajšlová J (2008) Food Addit Contam Part A Chem Anal Control Expo Risk Assess 25:444–457
121. Lehotay SJ, Mastovska K, Lightfield AR, Gates RA (2010) J AOAC Int 93:355–367
122. Lehotay SJ, O'Neil M, Tully J, García AV, Contreras M, Mol H, Heinke V, Anspach T, Lach G, Fussell R, Mastovska K, Poulsen ME, Brown A, Hammack W, Cook JM, Alder L, Lindtner K, Vila MG, Hopper M, De Kok A, Hiemstra M, Schenck F, Williams A, Parker A (2007) J AOAC Int 90:485–520
123. Lehotay SJ, Son KA, Kwon H, Koesukwiwat U, Fu W, Mastovska K, Hoh E, Leepipatpiboon N (2010) J Chromatogr A 1217:2548–2560
124. Liu G, Rong L, Guo B, Zhang M, Li S, Wu Q, Chen J, Chen B, Yao S (2011) J Chromatogr A 1218:1429–1436
125. Mastovska K, Dorweiler KJ, Lehotay SJ, Wegscheid JS, Szpylka KA (2010) J Agric Food Chem 58:5959–5972
126. Pareja L, Fernández-Alba AR, Cesio V, Heinzen H (2011) TrAC – Trends Anal Chem 30:270–291
127. Podhorniak LV, Kamel A, Rains DM (2010) J Agric Food Chem 58:5862–5867
128. Riedel M, Speer K, Stuke S, Schmeer K (2010) J AOAC Int 93:1972–1986
129. Romero-González R, Frenich AG, Vidal JLM (2008) Talanta 76:211–225

130. Romero-González R, Garrido Frenich A, Martínez Vidal JL, Prestes OD, Grio SL (2011) J Chromatogr A 1218:1477–1485
131. Schreiber A, Wittrig R (2010) Agro Food Ind Hi Tech 21:18–21
132. Wang J, Chow W, Leung D (2010) Anal Bioanal Chem 396:1513–1538
133. Wong J, Hao C, Zhang K, Yang P, Banerjee K, Hayward D, Iftakhar I, Schreiber A, Tech K, Sack C, Smoker M, Chen X, Utture SC, Oulkar DP (2010) J Agric Food Chem 58:5897–5903
134. Klein J, Alder L (2003) J AOAC Int 86:1015–1037
135. Lehotay SJ, De Kok A, Hiemstra M, Van Bodegraven P (2005) J AOAC Int 88:595–614
136. Katsumata H, Kojima H, Kaneco S, Suzuki T, Ohta K (2010) Microchem J 96:348–351
137. Li L, Xu Y, Pan C, Zhou Z, Jiang S, Liu F (2007) J AOAC Int 90:1387–1394
138. Brutti M, Blasco C, Picó Y (2010) J Sep Sci 33:1–10
139. Chuang JC, Wilson NK (2011) J Environ Sci Health B 46:1532–4109
140. Font G, Ruiz MJ, Fernández M, Picó Y (2008) Electrophoresis 29:2059–2078
141. Schreck E, Geret F, Gontier L, Treilhou M (2008) Talanta 77:298–303
142. Barriada-Pereira M, González-Castro MJ, Muniategui-Lorenzo S, López-Mahña P, Prada-Rodríguez D, Fernández-Fernández E (2007) Talanta 71:1345–1351
143. Blasco C, Font G, Picó Y (2005) J Chromatogr A 1098:37–43
144. Frenich AG, Salvador IM, Vidal JLM, López-López T (2005) Anal Bioanal Chem 383:1106–1118
145. Mendiola JA, Herrero M, Cifuentes A, Ibañez E (2007) J Chromatogr A 1152:234–246
146. Soler C, Hamilton B, Furey A, James KJ, Mañes J, Picó Y (2007) Anal Chem 79:1492–1501
147. Soler C, Mañes J, Picó Y (2006) J Chromatogr A 1109:228–241
148. Fuentes E, Báez ME, Quiñones A (2008) J Chromatogr A 1207:38–45
149. Guillet V, Fave C, Montury M (2009) J Environ Sci Health B 44:415–422
150. Aguilera-Luiz MM, Plaza-Bolaños P, Romero-González R, Vidal JLM, Frenich AG (2011) Anal Bioanal Chem 399:2863–2875
151. Amendola G, Pelosi P, Dommarco R (2010) J Environ Sci Health B 45:789–799
152. Fenik J, Tankiewicz M, Biziuk M (2011) TrAC – Trends Anal Chem 30:814–826
153. Picó Y, Fernández M, Ruiz MJ, Font G (2007) J Biochem Biophys Methods 70:117–131
154. Xie W, Han C, Qian Y, Ding H, Chen X, Xi J (2011) J Chromatogr A 1218:4426–4433
155. Smith RM (2003) J Chromatogr A 1000:3–27
156. Pyrzynska K (2011) Chemosphere 83:1407–1413
157. Al-Degs YS, Al-Ghouti MA, El-Sheikh AH (2009) J Hazard Mater 169:128–135
158. Asensio-Ramos M, Hernandez-Borges J, Ravelo-Pérez LM, Rodriguez-Delgado M (2008) Electrophoresis 29:4412–4421
159. López-Feria S, Cárdenas S, Valcárcel M (2009) J Chromatogr A 1216:7346–7350
160. Li Q, Wang X, Yuan D (2009) J Environ Monit 11:439–444
161. Guan SX, Yu ZG, Yu HN, Song CH, Song ZQ, Qin Z (2011) Chromatographia 73:33–41
162. Soler C, Hamilton B, Furey A, James KJ, Mañes J, Picó Y (2006) Rapid Commun Mass Spectrom 20:2151–2164
163. Picó Y, Farré M, Soler C, Barceló D (2007) Anal Chem 79:9350–9363
164. Picó Y, Farré Ml, Segarra R, Barceló D (2010) Talanta 81:281–293
165. Sanguansri P, Augustin MA (2006) Trends Food Sci Technol 17:547–556
166. Mavrocordatos D, Pronk W, Boller M (2004) Analysis of environmental particles by atomic force microscopy, scanning and transmission electron microscopy. Water Sci Technol 50:9–18
167. Balnois E, Papastavrou G, Wilkinson KJ (2007) Force microscopy and force measurements of environmental colloids. In: Wilkinson KJ, Lead JR, editors. Environmental Colloids and Particles: Behaviour, Structure and Characterization. Chichester: Wiley, pp 405–468
168. Krueger KM, Al-Somali AM, Falkner JC, Colvin VL (2005) Anal Chem 77:3511–3515
169. Huang X, McLean RS, Zheng M (2005) Anal Chem 77:6225–6228
170. Williams A, Varela E, Meehan E, Tribe K (2002) Int J Pharm 242:295–299

171. Chmela E, Tijssen R, Blom MT, Gardeniers HJGE, Van den Berg A (2002) Anal Chem 74:3470–3475
172. Blom MT, Chmela E, Oosterbroek RE, Tijssen R, Van Den Berg A (2003) Anal Chem 75:6761–6768
173. Bundschuh T, Knopp R, Kim JI (2001) Colloids Surf A Physicochem Eng Asp 177:47–55
174. Isaacson CW, Bouchard D (2010) J Chromatogr A 1217:1506–1512
175. Isaacson CW, Usenko CY, Tanguay RL, Field JA (2007) Anal Chem 79:9091–9097
176. Kantiani L, Farré M, Barceló D (2009) TrAC – Trends Anal Chem 28:729–744
177. Charm SE, Chi R (1988) J Assoc Off Anal Chem 71:304–316
178. Gaudin V, Maris P, Fuselier R, Ribouchon JL, Cadieu N, Rault A (2004) Food Addit Contam 21:422–433
179. Loomans EEMG, Van Wiltenburg J, Koets M, Van Amerongen A (2003) J Agric Food Chem 51:587–593
180. Adrian J, Pinacho DG, Granier B, Diserens JM, Sánchez-Baeza F, Marco MP (2008) Anal Bioanal Chem 391:1703–1712
181. Pimpitak U, Putong S, Komolpis K, Petsom A, Palaga T (2009) Food Chem 116:785–791
182. Wu H, Yang W, Ma J (2008) Chemistry Bulletin/Huaxue Tongbao 71:394–397
183. Gustavsson E, Degelaen J, Bjurling P, Sternesjo Ã (2004) J Agric Food Chem 52:2791–2796
184. Sternesjo A, Gustavsson E (2006) J AOAC Int 89:832–837
185. Raz SR, Bremer MGEG, Haasnoot W, Norde W (2009) Anal Chem 81:7743–7749
186. Weigel S, Pikkemaat MG, Elferink JWA, Mulder PPJ, Huet AC, Delahaut P, Schittko S, Flerus R, Nielen M (2009) Food Addit Contam Part A Chem Anal Control Expo Risk Assess 26:441–452
187. Jiang Z, Li Y, Liang A, Qin A (2008) Luminescence 23:157–162
188. Adrian J, Pasche S, Diserens JM, Sánchez-Baeza F, Gao H, Marco MP, Voirin G (2009) Biosens Bioelectron 24:3340–3346
189. Virolainen NE, Pikkemaat MG, Elferink JWA, Karp MT (2008) J Agric Food Chem 56:11065–11070
190. Hammel YA, Mohamed R, Gremaud E, LeBreton MH, Guy PA (2008) J Chromatogr A 1177:58–76
191. Sheridan RS, Meola JR (1999) J AOAC Int 82:982–990
192. Marazuela MD, Bogialli S (2009) Anal Chim Acta 645:5–17
193. Moreno-Bondi MC (2009) Anal Bioanal Chem 395:921–946
194. Maher HM, Youssef RM, Khalil RH, El-Bahr SM (2008) J Chromatogr B Analyt Technol Biomed Life Sci 876:175–181
195. Peters RJB, Bolck YJC, Rutgers P, Stolker AAM, Nielen MWF (2009) J Chromatogr A 1216:8206–8216
196. Berrada H, Borrull F, Font G, Marco RM (2008) J Chromatogr A 1208:83–89
197. Kukusamude C, Santalad A, Boonchiangma S, Burakham R, Srijaranai S, Chailapakul O (2010) Talanta 81:486–492
198. Paschoal JAR, Reyes FGR, Rath S (2009) Anal Bioanal Chem 394:2213–2221
199. Bailón-Pérez MI, García-Campaña AM, del Olmo-Iruela M, Gámiz-Gracia L, Cruces-Blanco C (2009) J Chromatogr A 1216:8355–8361
200. Zhu WX, Yang JZ, Wei W, Liu YF, Zhang SS (2008) J Chromatogr A 1207:29–37
201. Kantiani L, Farré M, Sibum M, Postigo C, De Alda ML, Barceló D (2009) Anal Chem 81:4285–4295
202. Li J, Chen L, Wang X, Jin H, Ding L, Zhang K, Zhang H (2008) Talanta 75:1245–1252
203. Stewart I, Eaglesham GK, Poole S, Graham G, Paulo C, Wickramasinghe W, Sadler R, Shaw GR (2009) Toxicon 256:804–812
204. Villar-González A, Rodríguez-Velasco ML, Ben-Gigirey B, Botana LM (2007) Toxicon 49:1129–1134
205. Ciminiello P, Dell'Aversano C, Fattorusso E, Forino M, Tartaglione L, Boschetti L, Rubini S, Cangini M, Pigozzi S, Poletti R (2010) Toxicon 55:280–288

206. Amzil Z, Sibat M, Royer F, Masson N, Abadie E (2007) Mar Drugs 5:168–179
207. Naar J, Bourdelais A, Tomas C, Kubanek J, Whitney PL, Flewelling L, Karen Steidinger JL, Baden DG (2002) Environ Health Perspect 110:179–185
208. Gao Z, Xu J, Xian Q, Feng J, Chen X, Yu H (2009) Chemosphere 75:1273–1279
209. Flewelling LJ, Naar JP, Abbott JP, Baden DG, Barros NB, Bossart GD, Bottein MYD, Hammond DG, Haubold EM, Heil CA, Henry MS, Jacocks HM, Leighfield TA, Pierce RH, Pitchford TD, Rommel SA, Scott PS, Steidinger KA, Truby EW, Van Dolah FM, Landsberg JH (2005) Nature 435:755–756
210. Kitagawa Y, Okihashi M, Takatori S, Okamoto Y, Fukui N, Murata H, Sumimoto T, Obana H (2009) J Food Hyg Soc Jpn 50:198–207
211. Lee SJ, Park S, Choi JY, Shim JH, Shin EH, Choi JH, Kim ST, El-Aty AMA, Jin JS, Bae DW, Shin SC (2009) Biomed Chromatogr 23:719–731
212. Przybylski C, Segard C (2009) J Sep Sci 32:1858–1867
213. Lotierzo M, Henry OYF, Piletsky S, Tothill I, Cullen D, Kania M, Hock B, Turner APF (2004) Biosens Bioelectron 20:145–152
214. Berger U, Glynn A, Holmström KE, Berglund M, Ankarberg EH, Tarnkvist A (2009) Chemosphere 76:799–804
215. Schuetze A, Heberer T, Effkemann S, Juergensen S (2010) Chemosphere 78:647–652
216. Ciminiello P, Dell'Aversano C, Fattorusso E, Magno S, Tartaglione L, Cangini M, Pompei M, Guerrini F, Boni L, Pistocchi R (2006) Toxicon 47:597–604
217. Szlinder-Richert J, Barska I, Usydus Z, Grabic R (2010) Chemosphere 78:695–700
218. Guerra P, de la Cal A, Marsh G, Raldúa D, Barata C, Eljarrat E, Barceló D (2009) J Hydrol 378:355
219. Adou K, Bontoyan WR, Sweeney PJ (2001) J Agric Food Chem 49:4153–4160
220. Tokman N, Soler C, Farré ML, Picó Y, Barceló D (2009) J Chromatogr A 1216:3138–3146
221. Sekhon BS (2009) J Pestic Sci 34:1–12
222. Pérez-Fernández V, García MA, Marina ML (2010) J Chromatogr A 1217:968–989
223. Machinski M Jr, Bando E, Oliveira RC, Ferreira GMZ (2009) J Food Prot 72:911–914
224. Navratilova P (2008) Czech J Food Sci 26:393–401
225. Al-Mazeedi HM, Abbas AB, Alomirah HF, Al-Jouhar WY, Al-Mufty SA, Ezzelregal MM, Al-Owaish RA (2010) Food Addit Contam Part A Chem Anal Control Expo Risk Assess 27:291–301
226. Baggio A, Gallina A, Benetti C, Mutinelli F (2009) Food Addit Contam Part B Surveill 2:52–58
227. Takino M, Yamaguchi K, Nakahara T (2004) J Agric Food Chem 25:727–735
228. Unusan N (2009) International Journal of Food Sciences and Nutrition 60:359–364

Pharmaceuticals in Drinking Water

Aleksandra Jelić, Mira Petrović, and Damià Barceló

Abstract Pharmaceuticals are a group of emerging contaminants that has received noticeable attention over the past decade. Continual development of the advanced instruments and improved analytical methodologies made possible detection of these microcontaminants in low levels in different environmental matrices. Traces of pharmaceuticals have also been found in groundwater and surface water that are used for drinking water supply. Therefore, concern has been raised over the potential risk to human health from exposure to the pharmaceutical residues via drinking water. Still, there is no evidence that any serious risk could arise from low concentrations of pharmaceuticals found in drinking water. Anyhow, there is much more to be understood about long-term, low-level exposure to a mixture of pharmaceuticals and their metabolites. In the following chapter, we give a brief overview of the technologies commonly applied for drinking water treatment, with reference to pharmaceutical removal, and we review available literature data on the occurrence of pharmaceuticals in finished drinking water.

Keywords Drinking water, Occurrence, Pharmaceuticals, Removal

A. Jelić (✉)
Institute of Environmental Assessment and Water Research (IDAEA-CSIC), C/ Jordi Girona, 18-26, 08034 Barcelona, Spain
e-mail: aljqam@cid.csic.es

M. Petrović
Catalan Institute for Water Research (ICRA), c/Emili Grahit 101, 17003 Girona, Spain

Catalan Institution for Research and Advanced Studies (ICREA), Passeig Lluis Companys 23, 80010 Barcelona, Spain

D. Barceló
Institute of Environmental Assessment and Water Research (IDAEA-CSIC), C/ Jordi Girona, 18-26, 08034 Barcelona, Spain

Catalan Institute for Water Research (ICRA), c/Emili Grahit 101, 17003 Girona, Spain

King Saud University (KSU), P.O. Box 2455, 11451 Riyadh, Saudi Arabia

D. Barceló (ed.), *Emerging Organic Contaminants and Human Health*,
Hdb Env Chem (2012) 20: 47–70, DOI 10.1007/698_2011_133,
© Springer-Verlag Berlin Heidelberg 2012, Published online: 17 January 2012

Contents

Abbreviations

AOP	Advanced oxidation processes
DBP	Disinfection by-products
DWTP	Drinking water treatment plant
ED	Electrodialysis
EDC	Endocrine-disrupting compounds
GAC	Granular activated carbon
NF	Nanofiltration
NOM	Natural organic matter
NSAIDs	Nonsteroidal anti-inflammatory drugs
MF	Microfiltration
RO	Reverse osmosis
PAC	Powdered activated carbon
UF	Ultrafiltration
WWTP	Wastewater treatment plant

1 Introduction

Pharmaceutical products have an important role in the treatment and prevention of disease in both humans and animals. They are designed either to be highly active and interact with receptors in humans and animals or to be toxic for many infectious organisms. Because of the nature, they can also bioaccumulate and have unintended effects on animals and microorganisms in the environment. Although the effects of the pharmaceuticals are investigated through safety and toxicology studies, the potential environmental impacts of their production and use are less understood and have recently become a topic of research interest [1].

Hundreds of tons of pharmaceuticals is dispensed and consumed annually world-wide. The usage and consumption of pharmaceuticals have been increasing consistently due to the discoveries of new drugs, the expanding population, and the inverting age structure in the general population, as well as due to expiration of patents with resulting availability of less-expensive generics [2]. Recent development of the advanced instruments and improved analytical methodologies made possible detection of pharmaceuticals in low levels in different environmental matrixes. Traces of pharmaceuticals have been reported in ng to low μg per liter, in wastewater, surface water, ground water, and drinking water [3]. There have been a few examples of adverse effects of pharmaceuticals to the ecosystem. A widely used anti-inflammatory drug diclofenac caused the decline of Oriental white-backed vulture (*Gyps bengalensis*) in the Indian subcontinent [4]. It was shown that the pharmaceutical, ingested via food, was responsible for renal disease in the scavenging birds. Similar effects of diclofenac on renal function of rainbow trout (*Oncorhynchus mykiss*) were observed after prolonged exposure (28 days) to an environmentally relevant concentration range [5]. Concern has been raised among governmental and nongovernmental water regulators, water suppliers, and the public regarding the potential risks to human health from exposure to the pharmaceutical residues via drinking water. In the following paragraphs, we aim to summarize literature information on the occurrence of pharmaceuticals and their removal during drinking water treatment.

2   Routes of Pharmaceuticals to the Environment

Pharmaceuticals find their way to the environment through many pathways (Fig. 1), but the principal way is through the discharge of raw and treated sewage from residential users or medical facilities [6]. Through the excretion via urine and feces, extensively metabolized drugs are released into the receiving waters. On the other side, the topically applied pharmaceuticals (when washed off) and the expired and unused ones (when disposed directly to trash or sewage), as well as released from drug manufacturing plants [7–9], may pose a direct risk to the environment because they enter sewage in their unmetabolized and powerful form [10]. Besides the wastewater effluents discharged to water bodies or reused for irrigation, biosolids

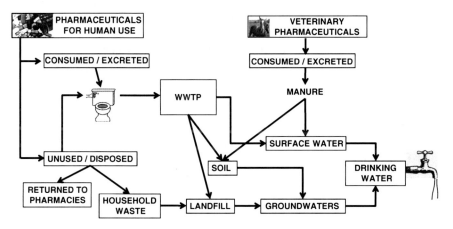

Fig. 1 Main routes of pharmaceuticals to the environment

used in agriculture as soil amendment or disposed to landfill are another significant route of pharmaceuticals to the environment [11] and drinking water supplies.

3 Drinking Water Treatment

Drinking water treatment plants (DWTP) were designed to remove different impurities (e.g., pathogens and organic and inorganic suspended matter) present in raw water, in order to provide clean and safe water that meets the drinking water standards. The type of treatment applied mostly depends on the characteristics (physical, chemical, and bacteriological) of the raw water. Groundwater is the largest body of freshwater in the European Union [12]. If the groundwater is not under the influence of surface water, it is generally less treated than surface water; it is typically subjected to aeration and disinfection processes for microbial pathogens removal. But groundwater can be rich in dissolved solids, especially carbonates and sulfates of calcium and magnesium, as well as chloride and bicarbonate, depending on the strata through the groundwater flowed. Thus, an additional treatment may be required in order to provide pleasant water for drinking and household use. Surface water is exposed to direct wet weather runoff and to the atmosphere and is easily contaminated; thus, there are strict regulations for the treatment of such water when used for drinking purposes. For the groundwater under the influence of surface water, as well, should be applied a multiple-barrier drinking water treatment approach.

A variety of treatment operation and processes are designed and applied to remove contaminants from raw water and improve it for drinking purposes. The process units are arranged in sequential series of treatment processes. The most commonly used processes include coagulation (flocculation and sedimentation), filtration, and disinfection. The particles present in raw water (pathogens, clay, organic material, metals) are usually too small to settle out from the water column. Therefore, alum and iron salts, alone or in combination with other metal salts, are used to promote coagulation of the small particles into larger ones, which can sediment easier. The water is left undisturbed to allow the coagulants to settle out. Still, the water contains small particles as clays and silts, natural organic matter, precipitates from coagulation treatment processes, iron and manganese, and microorganisms, which have to be removed. Thus, it is run through a series of filters which capture and remove particles still remaining in the water column. Typically, beds of sand or charcoal are used for filtration of the water, enhancing the effectiveness of disinfection. If inorganic contaminants cannot be removed by filtration and sedimentation, ion exchange (or water softening) processes are used. Typically, ion exchange is used to treat hard water (Ca, Mg), but it can also be used to remove arsenic, chromium, excess fluoride, radium, and uranium. It cannot remove bacteria, silt and sand, and many other organic and inorganic compounds. All treatment methods have limitations; this is why the water quality conditions require a combination of various treatment processes. Disinfection, the last unit in the drinking water treatment, is commonly done with chlorination, ozone, or ultraviolet radiation. Ozone is a powerful disinfectant. Ultraviolet radiation is an effective disinfectant and treatment for relatively clean source waters. Still, neither of these two is effective in controlling biological contaminants in the distribution pipes. Chlorination is the most used disinfection treatment. Chlorine, chloramines, or chlorine dioxide are most often used because they are very effective disinfectants, not only at the treatment plant but also in the pipes of the distribution system. After the disinfection step, the water is considered to be safe enough to drink and is sent for distribution to homes and businesses.

Still, a number of studies have shown that the conventional water (both wastewater as drinking water) treatments are not effective enough in removing microcontaminants, among which are pharmaceuticals [13–15]. Even though advanced treatment processes are able to achieve higher removal rates, they still do not obtain complete removal of pharmaceuticals [16, 17]. Nevertheless, the toxicological relevance of various contaminants should be considered before investing in additional advanced and very costly treatment technologies for the removal of pharmaceuticals, or other microcontaminants, from drinking water.

4 Removal of Pharmaceuticals During Drinking Water Treatment

Both wastewater and drinking water treatment plants were basically designed to remove pathogens and organic and inorganic suspended and flocculated matter, and neither of these treatment technologies have been designed specifically to remove pharmaceuticals that may be present in sewage water or drinking water sources. Even though the new and expensive treatment technologies have been developed to deal with health and environmental concerns associated with findings of nowadays research, the progress was not as enhanced as the one of the analytical detection capabilities. This is the main reason that we find a variety of microcontaminants detected in different environmental matrices including surface, groundwater, and drinking water.

Even though not complete, treatment plants achieve some level of removal of pharmaceutical residues. The extent to which one compound can be removed during water treatment is influenced not only by chemical and biological properties of the compound but also of the water characteristics, operational conditions, and treatment technology used. Various mechanisms influence the behavior of microcontaminants during different stages of drinking water treatment. Actually, coagulation/flocculation, activated carbon adsorption, and membrane systems eliminate contaminants by physical adsorption or separation, while oxidation processes such as chlorination and ozonation (with or without advanced oxidation processes, AOP) rely on chemical oxidation reactions. Besides sorption and physicochemical transformation, biotransformation and photodegradation may also be responsible for the removal of pharmaceuticals from water. All these mechanisms are limited in some way: (1) because pharmaceuticals have been designed to be biologically stable; (2) the sorption depends not only on the type and properties of the organic contaminant of interest but also on the suspended solids; and (3) even though they are photoactive (because many of them have aromatic rings, heteroatoms, and other functional groups that could be susceptible to photodegradation) or easily oxidized, they may give by-products of environmental concern. Therefore, no treatment unit itself is capable of "cleaning" the water of pharmaceuticals (or other organic microcontaminants). Only a combination of various physicochemical processes, a multistep treatment, can provide better contaminant removal.

4.1 Physical Separation/Adsorption

Coagulation and flocculation are an essential part of drinking water treatment. They aid in removing suspended solids (turbity) and natural organic matter (NOM), and some inorganics too (e.g., copper, mercury, arsenic, and fluoride). But they are almost ineffective in removing organic trace-level contaminants, including pharmaceuticals, from source water [14]. If a compound of concern is hydrophobic enough (Kow > 5), then some removal could be expected due to the binding to the surfaces of coagulates that are purged through subsequent sedimentation and filtration. Still, this "some" is much less than enough. For the anti-inflammatories naproxen, ibuprofen, ketoprofen, and diclofenac, a maximum removal of 20% was achieved during coagulation/flocculation [15, 17–19]. A very persistent pharmaceutical, anticonvulsant carbamazepine, was not affected by coagulation/ flocculation even when present at 1 μg/L concentration [20].

During filtration step, sorption plays an important role for the removal of pharmaceuticals. It depends not only on the compound charge and hydrophobicity but also on the properties of the solid phase [18, 20]. The sorption of pharmaceuticals to suspended solids present in the treated water, or to porous media used for filtration, is very complex and difficult to assess due to the fact that these compounds can sorb through different interactions: hydrophobic, electrostatic, and/or chemical [21]. Two forms of activated carbon: granular (GAC) and powdered activated carbon (PAC) are adsorbents with highly porous surface most commonly used for drinking water treatment. The primary purpose of activated carbons was to adsorb organic molecules that cause taste and odor, mutagenicity, and toxicity, as well as natural organic matter (NOM) that causes color and can also react with chlorine to form disinfection by-products (DBPs). Later on, activated carbons were used for the removal of pesticides (USEPA2001), and in the last decade, it was applied for removal of endocrine-disrupting compounds (EDC) and pharmaceuticals. Hydrophobic interactions are the dominant mechanism in activated carbon adsorption of organic compounds [22]. Therefore, the pharmaceuticals with higher log Dow (pH-dependent n-octanol/water coefficient) are expected to be easily removed by activated carbon. Pore size distribution and structure have a large influence on both adsorption capacity and kinetics.

Their removal efficiency depends on contact time, organic loading, carbon type, and chemical structure and solubility [20, 23, 24]. Removal of trace organic contaminants is also influenced by the concentration of NOM in source water because both the contaminants and NOM compete for available carbon surface adsorption sites [25]. Activated carbon can be very effective for the removal of pharmaceuticals, especially the hydrophobic ones. In general, PAC applications were known for more than 70 years, but the implementation of GAC increased in the 1980s [26, 27]. PAC is normally added seasonally or event specific at many conventional DWTP; thus, it might offer better removal due to the fact that is fed to the system fresh, not recycled. But since pharmaceuticals are released continuously into the environment, GAC adsorbers are more suitable treatment option. GAC is also very effective. The US EPA identified packed-bed GAC as a "best available technology" for treating numerous regulated organic pollutants [17]. In a study carried out by Snyder et al. [23], both PAC and GAC were capable of removing nearly all compounds evaluated by greater than 90% in different strategies of water treatment. In a full-scale plant, with relatively high levels of TOC, using GAC without regular replacement and regeneration provided very little removal. Indeed, regeneration and replacement are critical for excellent removal using GAC. Combination of GAC with nanofiltration, as a kind of pretreatment, was shown to be very effective in removing pharmaceuticals because NF removes all the NOM from the feed water and reduce the concentrations of more hydrophilic compounds as well [28]. Westerhoff et al. [17] found that the addition of 5 mg/L of PAC with a 4-h contact time removed 50% to >98% of volatile compounds and 10% to >95% of polar compounds, among them are pharmaceuticals fluoxetine and trimethoprim. The removal depended on the PAC dosages (higher dosage, higher removal) and was independent on the initial concentration of the compounds. In pilot-scale experiments performed by Ternes et al. [20], after pulverized GAC filtration, diclofenac and bezafibrate were not detected, the concentration of carbamazepine and primidone were reduced by 75%, and the concentration of clofibric acid was reduced by ca. 20%. Besides clofibric acid, ibuprofen, sulfamethoxazole, and diclofenac were some of the compounds found to be most resistant to activated carbon removal [29]. Despite short filter contact times of 1.5–3 min in a full-scale DWTP studied by Stackelberg and colleagues [15], GAC filtration accounted for 53% of the removal of organic contaminants, including pharmaceuticals, from the water phase. Huerta-Fontela et al. [16] studied the removal of 55 pharmaceuticals, hormones, and metabolites during full-scale drinking water treatment that involved GAC filtration after ozonation step. They found that only 3 out of 14 pharmaceuticals, remained after ozonation, were completely removed by GAC, i.e., acebutolol, diazepam, and diltiazem; but more than half were removed with more than 75% efficiency. Six organic compounds, including carbamazepine and ibuprofen, were completely removed during GAC filtration in a DWTP in Seoul [30].

Membrane treatment is a very effective technology used in a broad range of applications. It has been used for desalinization, specific ion removal, nutrients and suspended solids removal, and nowadays plays an important role in removal of trace-level organic contaminants dissolved in water. It is a very advantageous technology since it provides good removal of contaminants without formation of (possibly toxic) by-products. The most important industrial membrane-based technologies are reverse osmosis (RO), electrodialysis (ED), nanofiltration (NF), ultrafiltration (UF), and microfiltration (MF). UF and MF are microporous, and they are basically similar in that the mode of separation is molecular sieving through very fine membrane pores (pore flow mechanism). The mechanism of separation by RO membranes is quite different. In RO membranes, the pores are so small, which fall within the range of thermal motion of the polymer chains, and the mechanism of separation is different. In RO membranes, solute permeates the membrane by dissolving in the membrane material and diffusing down a concentration gradient (solution-diffusion model) [31]. Thus, separation occurs because of the difference in solubilities and mobilities of different solutes in the membrane. NF membranes falls between microporous (e.g., UF) and solution-diffusion membranes (e.g., RO). Besides membrane pore diameter that indicates the way the membranes are used, physicochemical properties (e.g., molecular weight, hydrophobicity, polarity, chemical nature, etc.) of compounds of interest, as well as feed water and the operation conditions, affect the removal efficiency of membranes. Size exclusion, electrostatic repulsion, adsorption, and diffusion mechanisms have been identified as the key factors for the organic compound transport through membranes [32–35]. As in the case of activated carbon filtration, the presence of NOM affects membrane performance [36]. Despite the growing number of publication on the removal of pharmaceuticals from water using membrane technologies, more information is still needed. Most of the studies were performed on a laboratory scale and, in many cases, in deionized water. The results of the lab-scale studies where the compounds of interest were added to deionized water should be taken with caution, given the fact that those were performed in absence of NOM that was proven to alter membrane potentials. There are very few studies on the performance of a full-scale NF and RO membrane treatment in rejecting micropollutants. Yet in these studies, NF and RO membranes were generally employed as tertiary treatment in WWTP or for treating saline groundwater [23, 35, 37]. As most pharmaceuticals fall into range of 150–500 μ in molecular size, it is expected that they could be easily removed by RO and NF membranes and less by UF. Kim et al. [30] found that UF did not affect the concentrations of studied pharmaceuticals and some EDCs in raw water, while RO and NF processes showed excellent removal rates (>95%) in a full-scale WWTP. In lab-scale experiments, Yoon et al. [37] obtained similar results, concluding that retention in these membranes is mainly affected by membrane pore size. They also found that more polar, less volatile, and less hydrophobic compounds had less retention than less polar, more volatile, and more hydrophobic compounds, which indicated that retention by NF and UF was clearly governed by hydrophobic adsorption. The results of a study performed by Snyder et al. [23] showed that RO membranes were capable of removing nearly all compounds investigated to levels lower than method limits of quantification. Additionally, they pointed out the importance of considering the disposal of permeates from NF and RO that contain the rejected compounds. Watkinson et al. [38] studied removal of 28 human and veterinary antibiotics in WWTP and

DWP, and they found that RO membrane reduced the concentration of antibiotics present in the feed (beta-lactams, quinolones, macrolides, and sulphonamides) by approximately 94%. Polar or charged compounds that interact with the membrane surface are expected to be better removed than less polar or neutral compounds [37, 39]. Radjenovic et al. [40] reported very high rejections (>90%) in both NF and RO membranes, applied in a full-scale DWTP, for negatively charged (ketoprofen, diclofenac, and sulfamethoxazole) and positively charged pharmaceuticals (sotalol and metoprolol). Similar results (i.e., rejection of >90%) were reported in a NACWA report [41] where tight NF and RO were applied for a group of negatively charged pharmaceuticals. Kimura et al. [33] demonstrated that charged compounds could be rejected to a greater extent (>90%) using polyamide NF and ultra-low-pressure RO, regardless of physicochemical properties of the tested compounds. The RO exhibited slightly better rejection than NF likely because the RO had a lower molecular weight cutoff (MWCO) than NF. The retention of sulfamethoxazole and ibuprofen was shown to increase greatly as the compounds transform from their neutral to negatively charged species when the solution pH increased above their pKa value [42]. For neutral species, despite the decrease in rejection for both membranes, RO performance was significantly better than NF [43]. Rejection of noncharged compounds was found to be influenced mainly by the size of the compounds. Kimura et al. [44] reported that retention of neutral compounds varied depending on molecular size, polarity, and membrane materials, ranging from 57% to 91%, with better performance on polyamide NF than on cellulose NF. Adams et al. [18] studied the removal of antibiotics using a low-pressure RO system with a cellulose acetate membrane. The rejection rate for the antibiotics was around 90% from distilled and river water, and it was even higher (99% and 99.9%) using with two and three RO units in series. Also, Khiari et al. [29] indicated a double-pass RO system as a very good solution for complete removal of target compounds from water. A multibarrier approach, RO/AOP or double-pass RO, is most successful in the removal of trace contaminants [23].

4.2   Chemical Disinfection

Disinfection is the process for deactivation of pathogenic microbes (e.g., bacteria, algae, spores, and viruses) in water, but it has the potential to remove some trace organic contaminants through oxidation. Chlorine (OCl^- and $HOCl$), chlorine dioxide (ClO_2), chloramines, and ozone (O_3) are the most common chemical disinfectants in use nowadays. Besides, ozone is combined with hydrogen peroxide or UV (pH > 8) (AOPs), thus becoming even more powerful oxidizing agent. Factors that can influence chemical disinfection efficiency are water quality (e.g., types and quantity of organic contaminants, pH, and temperature), disinfectant concentration, and contact time. Free chlorine (OCl^- and $HOCl$) can produce oxidation, hydrolysis, and deamination reactions with a variety of chemical substrates. Chloride dioxide is a strong oxidant, as well, and is very effective over wide range of pH (pH 5–10), and on pH > 8.5 even more effective than chlorine. Free chlorine can react with ammonia or organic amine to produce chloramines, which are much less reactive than free chlorine with pharmaceuticals and EDCs [45]. The strongest oxidizing agent of the here-mentioned chemical disinfectants is ozone. Ozone reacts with organic contaminants through either direct reaction with molecular ozone or through the formation of free radicals, including the hydroxyl radical. Both molecular ozone and hydroxyl radical pathways can lead to transformation of organic compounds. Hydroxyl radicals react nonselectively with high rate constants with organic molecules; thus, they can contribute to the oxidation of ozone-recalcitrant compounds [46, 47]. They can attack unsaturated bonds (forming aldehydes, ketones, or carbonyl compounds) or can participate in electrophilic reactions with aromatic compounds. Amino acids, proteins, and nucleic acids quickly react with ozone [48].

The major issue regarding chemical disinfection is the formation of disinfection by-products (DBPs). It has been shown that ozonation followed by chlorination or chloramination considerably increases the formation of DBPs [49]. The investigated DBPs include iodo-trihalomethanes and iodo-acids, which are found at highest levels after chloramination; halonitromethanes and haloaldehydes, which are enhanced by preozonation; a highly mutagenic MX compound ([3-chloro-4-(dichloromethyl)-5-hydroxy-2(5H)-furanone]), which is enhanced by chlorine dioxide-chlorine-chloramines; and nitrosamines concentrations, which are increased by chloramination. Still, the oxidation of organic matter gives better biodegradable fraction that can be degraded during postbiological treatment [50]. Although it assists the formation of DBPs, ozone was recommended as an alternative disinfectant to reduce the concentrations of DBPs, likely due to its selectivity (AWWA Water Quality and Treatment Handbook).

Although the disinfectants are primarily added in order to deactivate microbes and oxidize organic matter present in untreated water, they can achieve a certain level of pharmaceutical removal as well. This will depend upon the chemical structure of pharmaceuticals and treatment conditions such as pH and oxidant dose [18, 30, 51–54]. Westerhoff et al. [17] reported moderate to high removals of pharmaceuticals during chlorination at pH 5.5 at bench scale, although suggesting that pharmaceuticals were likely transformed to oxidation by-products. When analyzed, untreated and treated water from 31 conventional water treatment plants across Missouri that use chlorine and chloramines as water disinfectants, Wang et al. [55] reported almost complete removal of acetaminophen by both oxidation processes. The "removal" here could refer to "transformation," when it is known that chlorine reacts rapidly with phenolic compounds, as acetaminophen, through the reaction between HOCl and the deprotonated phenolate [15, 17, 45]. Bedner and Maccrehan [56] found 11 degradation products of acetaminophen after chlorination in pure water at pH 7 and in wastewater, of which two TPs (i.e., 1, 4-benzoquinone and N-acetyl-p-benzoquinone) are toxic. Similar was observed for the NSAID naproxen that was transformed into several intermediates and not completely removed in the experiments performed by Boyd et al. [57]. Gibs et al. [58] found that only 22 compounds, out of 98 studied pharmaceuticals and other trace organic compounds, reacted with free chlorine within 24 h. Chlorination was effective for fluoroquinolones, sulfonamides, and analgesics/anti-inflammatories. This is in agreement with other studies that showed that primary or secondary amines, such as diclofenac, sulfamethoxazole, and trimethoprim, were very reactive with chlorine [17]. While chlorination is very effective in removing sulfonamides [17, 55, 59], chloramination treatment does not provide such a good removal at typical dosage concentrations of 3 mg/L [60]. It could be expected that carbamazepine, which contains a urea group and aromatic rings, is easily oxidized by chlorination step [17], but Wang et al. [55] reported low removal of this compound during disinfections step at pH > 7. Westerhoff et al. [17] noted that the removal of pharmaceuticals and EDCs was always equal or higher at pH 5.5 than those at ambient pH. It was explained by the fact that at lower pH, a powerful oxidant hypochlorous acid accounts for nearly 99% of the free chlorine [61].

Chemical oxidation using ozone, alone or combined with AOPs, has been proven to be capable of removing/transforming pharmaceuticals during both wastewater and drinking water treatment [20, 51, 54, 62]. It has been also found that ozonation is more effective than chlorine and UV treatment [63], and its efficacy is often independent of initial pollutant concentration [47, 64, 65]. Ozonation (2.5 mgL_1) was highly effective (>70%) for all the studied pharmaceuticals detected in untreated waters of 20 DWTPs from geographically diverse locations across the United States in a study by Snyder et al. [63]. Westerhoff et al. [17] reported very high removal, i.e., >80%, of all the studied pharmaceuticals (various NSAIDs, antibiotics, etc.) by ozonation during the bench-scale water treatment. Only ibuprofen that has an electron-withdrawing functional group on the aromatic ring was removed with a slightly lower efficiency [17]. Bench-scale experiments conducted with surface waters, spiked with 16 pharmaceuticals and EDCs, provided evidence that ozone is effective (>80%) for removing trace organic contaminants from water with ozone doses typically applied in drinking water treatment [62]. Hollander et al. [46] studied the removal of 220 micropollutants in the full-scale municipal WWTP upgraded with postozonation followed by sand filtration. They found that ozonation contributed 40–50% (for naproxen, benzotriazole, atenolol, clarithromycin), 60–70% (for metoprolol, 5-methylbenzotriazole, sulfamethoxazole), and >80% (for diclofenac, carbamazepine, trimethoprim) to the elimination of the whole wastewater treatment process (relative to the influent concentration). Compounds with high rate constants ($>10^4$ M^{-1} s^{-1}), such as carbamazepine, diclofenac, macrolide antibiotics, and sulfonamide antibiotics, that have aromatic systems, amine moieties, or double bonds, were oxidized to concentrations below limits of quantifications. Similar results concerning the oxidation of carbamazepine were reported in the literature [66, 67]. Dimethylamino functional group in the molecule of macrolide antibiotics is the target of the oxidation by ozone, and it makes these molecules easily eliminated even at low ozone dose [68]. They also found that clarithromycin was oxidized by ozone mainly to the corresponding N-oxides that were no longer biologically active. On the other hand, Radjenovic et al. [69] identified six transformation products of roxithromycin, whereas two products exhibited high refractoriness to ozonation. Furthermore, the intact tertiary amine moiety of roxithromycin in these products suggests that the antimicrobial activity of the parent compound will be preserved.

While electron-donating functional groups facilitate the oxidation of aromatic rings, electron-withdrawing groups make the aromatic ring less reactive with ozone [17]. In order to improve the removal of such compounds (e.g., clofibric acid, ibuprofen, diclofenac), ozonation is conducted in the presence of hydrogen peroxide (O_3/H_2O_2) or under UV irradiation (O_3/UV) [20, 54, 70]. During the advanced ozonation, the oxidation capacity of ozone is increased due to the formation of hydroxyl radicals that are more powerful oxidizing agents and much less selective with organic compounds, thus allowing easier transformation of organic micropollutants [17, 47]. This is the basic principle of AOPs: enhancement of the formation of reactive moieties that attack organic molecules, oxidizing them to less complex intermediates. AOPs include homogeneous and heterogeneous photocatalysis based on near-UV or solar visible irradiation, ozonation, electrolysis, the Fenton's reagents, ultrasound, and wet air oxidation and less conventional processes like ionizing radiation, microwaves, pulsed plasma, and the ferrate reagent. Depending on the untreated water characteristics and the treatment objective itself, AOPs can be employed either alone or coupled with other physicochemical and biological processes [70]. The most popular AOPs applied in combination are H_2O_2/UV, O_3/UV, H_2O_2/O_3, $H_2O_2/O_3/UV$, $UV/TiO_2/H_2O_2$, ultrasound/ Fenton's reagents, UV/Fenton's reagent, wet air oxidation/H_2O_2, and electrolysis/ Fenton's reagent [71]. They have been widely studied for the removal of various organic contaminants, including pharmaceuticals, from aqueous solution, but only a few studies have been focused on drinking water sources, i.e., surface and ground waters, and AOPs effect on toxicity and estrogenic activity. In general, the advanced oxidation technologies have been shown to be very effective in removing pharmaceuticals from water [48]. In comparison with chemical and biological processes, AOPs might be more environmentally friendly because they neither transfer pollutants between phases as in adsorption or chemical precipitation nor produce sludge as in biochemical [72]. In most full-scale water treatment plants, AOPs usually refer to a combination of processes that involve O_3, H_2O_2, and/or UV light. Technology selection and the combination of AOPs will depend upon water characteristics, organic compound properties, and economic costs.

4.3   Physical Disinfection

Physical disinfection of drinking water is mainly carried out through photodegradation by ultraviolet radiation (UV). Because many pharmaceuticals have chromophores that absorb UV wavelength, UV irradiation can lead to some transformation. Comparing to chemical oxidation, such as chlorination, UV disinfection has an advantage of minimizing the formation of DBPs [73]. However, at a typical (low-energy) UV dose of 40 mJ/cm^2 applied for drinking water disinfection, UV was ineffective for the removal of most target compounds [29]. Higher doses of UV are usually required to cause any substantial transformation [29, 74]. The efficiency of UV photolysis can be enhanced in combination with hydrogen peroxide, which dissociates into hydroxyl radicals. Such a reactive moiety as hydroxyl radical induces faster oxidation rates of organic compounds and facilitate the degradation processes. A number of publications reported excellent removal of organic micropollutants by UV/H_2O_2 oxidation in bench-scale experiments [75–80]. Besides the absorbance spectrum of a pharmaceutical, the efficiency of UV/H_2O_2 treatment depends upon the quantum yield of photolysis, the concentration of hydrogen peroxide employed, and the water matrix [70]. The presence of NOM in water may induce radicals scavenging and decrease degradation [73, 79]. Nevertheless, Doll and Frimmel [81] suggested that NOM can act as inner filter, as radical scavenger, and/or precursor of reactive species such as singlet oxygen, solvated electrons, superoxide anion, hydroxyl radicals, etc., which are able to degrade anthropogenic organic compounds.

5 Occurrence of Pharmaceuticals in Finished Drinking Water

Few thousands of prescription and over-the-counter pharmaceutical products are registered and approved for usage nowadays, with around 1,300 unique active ingredients (Orange book, FDA). They differ in mode of action, chemical structure, physicochemical properties, and metabolism. For simplicity, all the registered therapeutic drugs are divided into 14 groups according to the organ or system on which they act and their chemical, pharmacological, and therapeutic properties – Anatomical Therapeutic Chemical Classification System (ATC system) (WHO Collaborating Centre for Drug Statistics Methodology – WHOCC). Due to the volume of prescription, the toxicity, and the evidence for presence in the environment, nonsteroidal anti-inflammatory drugs (NSAIDs), antibiotics, beta-blockers, antiepileptics, blood lipid–lowering agents, antidepressants, hormones, and antihistamines have been the most studied pharmaceutical groups [82]. Although there have been many research publications focused on the occurrence, fate, and effects of pharmaceuticals in the environment, we have data on the occurrence of only 10% of the registered active compounds, and very few information on their effects in the environment. Even less information are available regarding the occurrence and fate of the active and/or nonactive transformation products of pharmaceuticals.

As pharmaceuticals and their metabolites find their way to the environment primarily via the discharge of raw and treated sewage from residential users or medical facilities, most information on their occurrence and fate is related to WWTPs and the receiving waters. Much less is known about their presence in DWTPs and finished (tap) water. This can be attributed not only to the performances of the analytical technologies to detect a diverse range of low levels of pharmaceuticals and their metabolites but also to the high costs of the technologies. The lack of regulations concerning those microcontaminants might be another reason for the scarce information. The majority of the available data on the occurrence of pharmaceuticals in surface and ground water, as sources of drinking water, and tap drinking water come from targeted research surveys performed in universities or research centers in the USA and the EU. Naturally, the occurrence of pharmaceuticals and their metabolites/transformation products will depend on the rate of production, the dosage and frequency of administration and usage, the metabolism, and environmental persistence, as well as on the removal efficiency of WWTPs and DWPs. It is also important to mention that many compounds frequently detected in samples of surface water and ground-water are not detected in samples of finished water, indicating that DWPs achieve a quite good level of treatment reducing their concentration below the LOQs or (a worse case) that the compounds are transformed to intermediates that are not determined by the used methodologies. Therefore, the analysis of unknown and/or unmeasured transformation products is very important in order to understand the fate of pharmaceuticals during DTP, as well as the potential human-health issues associated with chronic exposure to these compounds through drinking water [83].

Table 1 shows the occurrence of the selected and most investigated pharmaceuticals in drinking water as found in the literature (adopted from [13, 103]). NSAIDs and antiepileptics were mostly studied and detected in drinking water. NSAIDs are the most used class of drugs for treatment of acute pain and inflammation. They are administered both orally and topically and available as prescription and over-the-counter (nonpre-scription) drugs. High consumption and way of administration of NSAIDs result in elevated concentration reported in the effluent from WWTPs, receiving waters, which then influence drinking water.

Antiepileptic carbamazepine is one of the most studied and detected pharmaceuticals in the environment. It is heavily or not degraded during wastewater treatment, and many studies have found it ubiquitous in various environment matrices (groundwater, river, soil) [104]. Thus, it is not that surprising that carba-mazepine was frequently detected in drinking water [83–86]. Another antiepileptic, phenytoin, and one psycholeptic, meprobamate, were also detected in tap water. Benotti et al. [84] detected these two compounds in more than half of the samples of treated and tap water, with the maximum concentrations of 16 ng/L and 40 ng/L, respectively. In this survey, the presence of 20 pharmaceuticals, 25 known or potential EDCs, and 6 other wastewater contaminants were studied in source water, treated drinking, and tap water from 19 U.S. DWTPs sampled during 2006–2007. The results showed that 13 of 15 studied distribution systems contained detectable concentrations of at least one target compound, and no pharmaceutical was detected in two plants that withdrew water with no direct input of wastewaters. Phenytoin was also detected in 11 out of 12 analyzed drinking water samples (with eliminations of 96% during drinking water treatment) in a study carried out by Huerta-Fontela et al. [16]. They assessed the presence of 42 pharmaceuticals, including psychiatric drugs, angiotensin agents, antihistaminics, beta-blockers,

Table 1 Pharmaceuticals detected in tap water worldwide

Therapeutic group	Compound	Maximal concentration detected (ng/L)	References
C03 (Cardiovascular system/diuretics)	Hydrochlorothizide	10	[16]
C07(Cardiovascular system/beta-blocking agents)	Atenolol	23	[16]
		0.84	[84]
		2	[85]
	Sotalol	3	[16]
C10 (Cardiovascular system/lipid-modifying agents)	Bezafibrate	1.9	[86]
		2.2	[85]
		27	[87]
	Clofibric acid	50–270	[87–90]
		10–165	[91]
		Low ng/L	[92]
		5.3	[93]
	Gemfibrozil	70	[94]
		0.8	[86]
		1.2	[84]
M01 (Musculoskeletal system/anti-inflammatory and antirheumatic product)	Diclofenac	6–35	[87, 89]
		2.5	[95]
		<10	[92]
M02 (Musculoskeletal system/topical products for joint and muscular pain)		2	[96]
		1	[85]
	Ibuprofen	3	[87]
		0.6	[95]
		8.5	[97]
		1.3	[85]
		1,350	[98]
	Ketoprofen	8	[97]
		7	[85]
		3	[95]
	Salycilic acid	19	[85]
N02 (Nervous system/analgesics)	Phenazone	250–400	[99, 100]
	Propyphenazone	80–240	[89, 99, 100]
	Acetaminophen	45	[85]
		210.1	[95]
	Codein	30	[15]
N03 (Nervous system/antiepileptics)	Phenytoin	16	[84]
		10	[16]
	Carbamazepine	24	[94]
		140	[15]
		43.2	[95]
		9.1	[86]
		258	[83]
		10	[84]
		32	[85]
		60	[89]
	10,11-epoxy-carbamazepine	2	[16]

(continued)

Table 1 (continued)

Therapeutic group	Compound	Maximal concentration detected (ng/L)	References
N05 (Nervous system/psycholeptics)	Diazepam	10	[101]
		23.5	[93]
	Meprobamate	40	[84]
		5.9	[102]

and cardiac agents; 6 EDCs; and 6 main metabolites in a full-scale DWP which included exhaustive treatments such as ozonation and granular activated filtration GAC. Only five pharmaceuticals were found in finished waters at ng/L levels. Besides phenytoin, atenolol and hydrochlorothiazide were detected in almost all the analyzed samples, in low ng/L concentrations.

Other compounds commonly detected in samples of drinking water belong to beta-blocking and lipid-regulating agents. As NSAIDs, beta-blockers are not highly persistent, but they are present in the environment because of their high volume of use. Due to the same mode of action of beta-blockers, it has been found that the mixture of beta-blockers showed concentration addition indicating a mutual specific nontarget effect on algae [105]. These compounds are generally found in aqueous phase because of their low sorption affinity and elevated biodegradability [106]. Atenolol has been frequently identified in wastewaters in concentrations ranging up to 1 μg/L [107]. As a result of the incomplete removal during conventional wastewater treatment, this compound was also found in surface waters in the ng/L to low mg/L range [108]. It was not detected in the finished water from DWTPs employing ozone. However, 18–75% of the atenolol concentrations measured in source water of seven plants employing chlorine (and no O_3) indicated that it may not be efficiently removed by chlorine [84]. Huerta-Fontela et al. [16], as well, reported poor removal of atenolol during chlorine treatment. However, this compound was more persistent towards ozonation, despite its reactive secondary amine moiety, and they found it in 10 out of 12 finished waters in average concentration of 12 ng/L (max 23 ng/L).

Clofibric acid is a pharmacologically active metabolite of lipid-lowering agents clofibrate, etofyllin clofibrate, and etofibrate. Among lipid-lowering drugs and pharmaceuticals, in general, this is one of the most frequently detected in the environment and one of the most persistent drugs, with an estimated persistence in the environment of 21 years [82]. It has been detected in the ng/L range concentrations in influent, without big difference in the concentrations at the effluent. Low ng/L concentrations of clofibric acid, but still detectable, i.e., 3.2–5.3 ng/L, were found in finished water from DWTP in Lodi, Italy [93]. This drug was detected in samples of potable water collected from DWTPs from Berlin area, in concentrations up to 270 ng/L [90]. It was shown that the concentrations of clofibric acid found in tap water samples of the individual DWTPs correlate well with the proportions of ground water recharge used by the particular water works in drinking water production [109].

Among the most frequently prescribed drugs are also antibiotics, used in both human and veterinary medicine. High global consumption of up to 200 thousand tons per year [110] and high percentage of antibiotics that may be excreted without undergoing metabolism (up to 90%) result in their widespread presence in the environment [111]. Sulfonamide, fluoroquinolone, and macrolide antibiotics show the highest persistence and are frequently detected in wastewater and surface waters [111]. A macrolide antibiotic – erythromycin – is one of three pharmaceuticals included in the final US EPA report CCL-3 from September 2009 as priority drinking water contaminant, based on health effects and occurrence in environmental waters [3]. Tylosin, azithromycin, and roxithromycin were detected in finished waters from two DWTPs in the upper Scioto River Basin, Ohio, USA [112]. Tylosin, many times used as a growth promoter for livestock, has been also detected in drinking water from DWTP Lodi, Italy, in the range from 0.6 to 1.7 ng/L [93]. Although a number of antibiotics, belonging to different families, were detected in source waters, they were not detected in finished water or were present in much lower levels in finished waters in North Carolina, USA [113]. Watkinson et al. [114] studied the occurrence of 28 antibiotics in watersheds of South East Queensland, Australia, and they did not find any antibiotic above detection limits in any of the drinking water samples. However, no data are available to confirm whether the antibiotics were removed or just transformed during drinking water treatment.

As shown in the preceding paragraphs, pharmaceutical compounds have been detected in very low ng/L concentrations in finished waters. It is clear that that the concentrations of individual compounds in finished waters are far below doses used in therapy. For example, with a minimal single therapeutic dose of carbamazepine of 100 mg (a dose ranges from 100 to 1,600 mg), a person would need to take around 400 thousand liters of water that contains the maximum reported concentration for carbamazepine in finished water, i.e., 258 ng/L (Table 1), to ingest the therapeutic dose. To date, there is no evidence that exposure to low concentrations of pharmaceuticals, through drinking water or fish consumption, is harmful to human health. Nevertheless, most of the data come from the acute toxicity studies for individual pharmaceutical compounds, and little is known about potential health effects associated with long-term chronic ingestion of low concentrations through drinking water [115, 116]. A major issue is to assess the effects of a mixture of pharmaceuticals, their metabolites, and/or transformation products (potentially) present in freshwater. Additionally, there are also other trace organic compounds, among which is a variety of by-products that may be occurring in water due to anthropogenic activities. Risk assessment studies concerning the human health impact of pharmaceuticals generally have healthy adults as the targeted population but do not take into consideration more sensitive population such as children, pregnant women and fetus, and allergic people [117]. Although the risk to health arising from exposure to pharmaceuticals in drinking water appears to be small, more research is necessary to properly evaluate the issue before drawing final conclusions.

6 Conclusion

Nowadays, research shows that pharmaceutical residues at trace quantities are widespread in the environment. They have also been detected in drinking water, though in very low ng/L concentrations. This indicates that the applied water treatment technologies are not efficient enough to eliminate pharmaceuticals from natural waters used as drinking water sources. The removal of pharmaceutical (and other organic) compounds will depend not only on their physicochemical and biological properties but also on source water characteristics, operational conditions, and treatment technology used. It is very difficult to predict an outcome of such a multifactor system. Even more difficult is to assess human health risks from exposure to innumerable pharmaceuticals and their transformation products that might be present in drinking water. So far, the results of scientific reports point out that it is unlikely that any serious risk can arise from low concentrations of pharmaceuticals found in drinking water. Anyhow, more information on quality, quantity, and toxicity of mixtures of pharmaceuticals and their metabolites are definitely needed before drawing any conclusion on the risk to human health from long-time exposure to low concentrations of these microcontaminants via drinking water. Studying more, it seems like the gaps in our knowledge about pharmaceuticals in the environment and drinking water are becoming deeper.

References

1. Boxall ABA (2004) European Molecular Biology Organization (EMBO) Report 5, 1110
2. Daughton CG (2003) Cradle-to-cradle stewardship of drugs for minimizing their environmental disposition while promoting human health. I. rational for and avenues toward a green pharmacy. Environ Health Perspect 111:757–774
3. Richardson SD (2010) Environmental mass spectrometry: emerging contaminants and current issues. Anal Chem 82:4742–4774
4. Oaks JL, Gilbert M, Virani MZ, Watson RT, Meteyer CU, Rideout BA, Shivaprasad HL, Ahmed S, Iqbal Chaudhry MJ, Arshad M, Mahmood S, Ali A, Ahmed Khan A (2004) Diclofenac residues as the cause of vulture population decline in Pakistan. Nature 427:630–633
5. Schwaiger J, Ferling H, Mallow U, Wintermayr H, Negele RD (2004) Toxic effects of the non-steroidal anti-inflammatory drug diclofenac: part I: histopathological alterations and bioaccumulation in rainbow trout. Aquat Toxicol 68:141–150
6. Rahman MF, Yanful EK, Jasim SY (2009) Endocrine disrupting compounds (EDCs) and pharmaceuticals and personal care products (PPCPs) in the aquatic environment: implications for the drinking water industry and global environmental health. J Water Health 7:224–243
7. Larsson DGJ, de Pedro C, Paxeus N (2007) Effluent from drug manufactures contains extremely high levels of pharmaceuticals. J Hazard Mater 148:751–755
8. Li D, Yang M, Hu J, Ren L, Zhang Y, Li K (2008) Determination and fate of oxytetracycline and related compounds in oxytetracycline production wastewater and the receiving river. Environ Toxicol Chem 27:80–86

9. Li D, Yang M, Hu J, Zhang Y, Chang H, Jin F (2008) Determination of penicillin G and its degradation products in a penicillin production wastewater treatment plant and the receiving river. Water Res 42:307–317

10. Ruhoya ISR, Daughton CG (2008) Beyond the medicine cabinet: an analysis of where and why medications accumulate. Environ Int 34:1157–1169

11. Barron L, Tobin J, Paull B (2008) Multi-residue determination of pharmaceuticals in sludge and sludge enriched soils using pressurized liquid extraction, solid phase extraction and liquid chromatography with tandem mass spectrometry. J Environ Monit 10:353–361

12. Directive (2006) 2006/118/EC of the European Parliament and of the Council of 12 December 2006 on the protection of groundwater against pollution and deterioration

13. Mompelat S, Le Bot B, Thomas O (2009) Occurrence and fate of pharmaceutical products and by-products, from resource to drinking water. Environ Int 35:803–814

14. POSEIDON (2001–2004) Detailed report related to the overall duration. http//www. eu-poseidon.com

15. Stackelberg PE, Gibs J, Furlong ET, Meyer MT, Zaugg SD, Lippincott RL (2007) Efficiency of conventional drinking-water-treatment processes in removal of pharmaceuticals and other organic compounds. Sci Total Environ 377:255–272

16. Huerta-Fontela M, Galceran MT, Ventura F (2011) Occurrence and removal of pharmaceuticals and hormones through drinking water treatment. Water Res 45:1432–1442

17. Westerhoff P, Yoon Y, Snyder S, Wert E (2005) Fate of endocrine-disruptor, pharmaceutical, and personal care product chemicals during simulated drinking water treatment processes. Environ Sci Technol 39:6649–6663

18. Adams C, Asce M, Wang Y, Loftin K, Meyer M (2002) Removal of antibiotics from surface and distilled water in conventional water treatment processes. J Environ Eng 128:253 (258pages)

19. Vieno N, Tuhkanen T, Kronberg L (2006) Removal of pharmaceuticals in drinking water treatment: effect of chemical coagulation. Environ Technol 27:183–192

20. Ternes TA, Meisenheimer M, McDowell D, Sacher F, Brauch H-J, Haist-Gulde B, Preuss G, Wilme U, Zulei-Seibert N (2002) Removal of pharmaceuticals during drinking water treatment. Environ Sci Technol 36:3855–3863

21. Meakins NC, Bubb JM, Lester JN (1994) Fate and behaviour of organic micropollutants during wastewater treatment processes: a review. Int J Environ Pollut 4:27–58

22. Snyder SA, Westerhoff P, Yoon Y, Sedlak DL (2003) Pharmaceuticals, personal care products, and endocrine disruptors in water: implications for the water industry. Environ Eng Sci 20:449–469

23. Snyder SA, Adham S, Redding AM, Cannon FS, DeCarolis J, Oppenheimer J, Wert EC, Yoon Y (2007) Role of membranes and activated carbon in the removal of endocrine disruptors and pharmaceuticals. Desalination 202:156–181

24. Yoon Y, Westerhoff P, Snyder SA, Esparza M (2003) HPLC-fluorescence detection and adsorption of bisphenol A, 17β-estradiol, and 17α-ethynyl estradiol on powdered activated carbon. Water Res 37:3530–3537

25. Li Q, Snoeyink VL, Mariãas BJ, Campos C (2003) Elucidating competitive adsorption mechanisms of atrazine and NOM using model compounds. Water Res 37:773–784

26. Clark RM, Lykins BW Jr (1989) Granular activated carbon: design, operation, and cost. Lewis Publishers, Chelsea, Michigan

27. Sontheimer H, Crittenden JC, Summers RS (1988) Activated carbon for water treatment. 2nd edn, in English. DVGW-Forschungsstelle, Karlsruhe, Germany

28. Heijman SGJ, Verliefde ARD, Cornelissen ER, Amy G, van Dijk JC (2007) Influence of natural organic matter (NOM) fouling on the removal of pharmaceuticals by nanofiltration and activated carbon filtration. Water Sci Technol Water Supply 7:17–23

29. Khiari D (2007) Endocrine disruptors, pharmaceuticals, and personal care products in drinking water: an overview of AwwaRF Research to Date. Awwa Research Foundation, Denver, CO

30. Kim SD, Cho J, Kim IS, Vanderford BJ, Snyder SA (2007) Occurrence and removal of pharmaceuticals and endocrine disruptors in South Korean surface, drinking, and waste waters. Water Res 41:1013–1021
31. Baker RW (2004) Membrane technology and applications, 2nd edn. Wiley, England
32. Kimura K, Amy G, Drewes J, Watanabe Y (2003) Adsorption of hydrophobic compounds onto NF/RO membranes: an artifact leading to overestimation of rejection. J Membr Sci 221:89–101
33. Kimura K, Amy G, Drewes JE, Heberer T, Kim T-U, Watanabe Y (2003) Rejection of organic micropollutants (disinfection by-products, endocrine disrupting compounds, and pharmaceutically active compounds) by NF/RO membranes. J Membr Sci 227:113–121
34. Nghiem LD, Schäfer AI, Elimelech M (2004) Removal of natural hormones by nanofiltration membranes: measurement, modeling, and mechanisms. Environ Sci Technol 38:1888–1896
35. Nghiem LD, Schäfer AI, Elimelech M (2005) Pharmaceutical retention mechanisms by nanofiltration membranes. Environ Sci Technol 39:7698–7705
36. Agenson KO, Oh J-I, Urase T (2003) Retention of a wide variety of organic pollutants by different nanofiltration/reverse osmosis membranes: controlling parameters of process. J Membr Sci 225:91–103
37. Yoon Y, Westerhoff P, Snyder SA, Wert EC (2006) Nanofiltration and ultrafiltration of endocrine disrupting compounds, pharmaceuticals and personal care products. J Membr Sci 270:88–100
38. Watkinson AJ, Murby EJ, Costanzo SD (2007) Removal of antibiotics in conventional and advanced wastewater treatment: implications for environmental discharge and wastewater recycling. Water Res 41:4164–4176
39. Urase T, Sato K (2007) The effect of deterioration of nanofiltration membrane on retention of pharmaceuticals. Desalination 202:385–391
40. Radjenović J, Petrović M, Ventura F, Barceló D (2008) Rejection of pharmaceuticals in nanofiltration and reverse osmosis membrane drinking water treatment. Water Res 42:3601–3610
41. Snyder S, Lue-Hing C, Cotruvo J, Drewes JE, Eaton A, Pleus RC, Schlenk D (2010) Pharmaceuticals in the Water Environment, NACWA Report
42. Nghiem LD, Schäfer AI, Elimelech M (2006) Role of electrostatic interactions in the retention of pharmaceutically active contaminants by a loose nanofiltration membrane. J Membr Sci 286:52–59
43. Xu P, Drewes JE, Bellona C, Amy G, Kim T-U, Adam M, Heberer T (2005) Rejection of emerging organic micropollutants in nanofiltration-reverse osmosis membrane applications. Water Environ Res 77:40–48
44. Kimura K, Toshima S, Amy G, Watanabe Y (2004) Rejection of neutral endocrine disrupting compounds (EDCs) and pharmaceutical active compounds (PhACs) by RO membranes. J Membr Sci 245:71
45. Pinkston KE, Sedlak DL (2004) Transformation of aromatic ether- and amine-containing pharmaceuticals during chlorine disinfection. Environ Sci Technol 38:4019–4025
46. Hollender J, Zimmermann SG, Koepke S, Krauss M, McArdell CS, Ort C, Singer H, von Gunten U, Siegrist H (2009) Elimination of organic micropollutants in a municipal wastewater treatment plant upgraded with a full-scale post-ozonation followed by sand filtration. Environ Sci Technol 43:7862–7869
47. von Gunten U (2003) Ozonation of drinking water: Part I. Oxidation kinetics and product formation. Water Res 37:1443–1467
48. Langlais B, Reckhow DA, Brink DR (1991) Ozone in drinking water treatment: application and engineering. AWWARF and Lewis Publishers, Boca Raton, FL
49. Richardson SD, Plewa MJ, Wagner ED, Schoeny R, DeMarini DM (2007) Occurrence, genotoxicity, and carcinogenicity of regulated and emerging disinfection by-products in drinking water: a review and roadmap for research. Mutat Res/Rev Mutat Res 636:178–242

50. Hammes F, Salhi E, Köster O, Kaiser H-P, Egli T, von Gunten U (2006) Mechanistic and kinetic evaluation of organic disinfection by-product and assimilable organic carbon (AOC) formation during the ozonation of drinking water. Water Res 40:2275–2286

51. Huber MM, Canonica S, Park G-Y, von Gunten U (2003) Oxidation of pharmaceuticals during ozonation and advanced oxidation processes. Environ Sci Technol 37:1016–1024

52. Huber MM, Göbel A, Joss A, Hermann N, Löffler D, McArdell CS, Ried A, Siegrist H, Ternes TA, von Gunten U (2005) Oxidation of pharmaceuticals during ozonation of municipal wastewater effluents: a pilot study. Environ Sci Technol 39:4290–4299

53. Ternes TA, Stüber J, Herrmann N, McDowell D, Ried A, Kampmann M, Teiser B (2003) Ozonation: a tool for removal of pharmaceuticals, contrast media and musk fragrances from wastewater? Water Res 37:1976–1982

54. Zwiener C, Frimmel FH (2000) Oxidative treatment of pharmaceuticals in water. Water Res 34:1881–1885

55. Wang C, Shi H, Adams CD, Gamagedara S, Stayton I, Timmons T, Ma Y (2011) Investigation of pharmaceuticals in Missouri natural and drinking water using high performance liquid chromatography-tandem mass spectrometry. Water Res 45:1818–1828

56. Bedner M, MacCrehan WA (2005) Transformation of acetaminophen by chlorination produces the toxicants 1,4-benzoquinone and N-acetyl-p-benzoquinone imine. Environ Sci Technol 40:516–522

57. Boyd GR, Zhang S, Grimm DA (2005) Naproxen removal from water by chlorination and biofilm processes. Water Res 39:668–676

58. Gibs J, Stackelberg PE, Furlong ET, Meyer M, Zaugg SD, Lippincott RL (2007) Persistence of pharmaceuticals and other organic compounds in chlorinated drinking water as a function of time. Sci Total Environ 373:240–249

59. Dodd MC, Huang C-H (2004) Transformation of the antibacterial agent sulfamethoxazole in reactions with chlorine: kinetics, mechanisms, and pathways. Environ Sci Technol 38:5607–5615

60. Chamberlain E, Adams C (2006) Oxidation of sulfonamides, macrolides, and carbadox with free chlorine and monochloramine. Water Res 40:2517–2526

61. Westerhoff P, Chao P, Mash H (2004) Reactivity of natural organic matter with aqueous chlorine and bromine. Water Res 38:1502–1513

62. Broséus R, Vincent S, Aboulfadl K, Daneshvar A, Sauvé S, Barbeau B, Prévost M (2009) Ozone oxidation of pharmaceuticals, endocrine disruptors and pesticides during drinking water treatment. Water Res 43:4707–4717

63. Snyder SA, Trenholm RA, Bruce GM, Snyder EM, Pleus RC (2008) Toxicological relevance of EDCs and pharmaceuticals in drinking water. Awwa Research Foundation, Denver, USA

64. Alum A, Yoon Y, Westerhoff P, Abbaszadegan M (2004) Oxidation of bisphenol A, 17beta-estradiol, and 17alpha-ethynyl estradiol and byproduct estrogenicity. Environ Toxicol 19:257–264

65. Hu J-y, Aizawa T, Ookubo S (2002) Products of aqueous chlorination of bisphenol A and their estrogenic activity. Environ Sci Technol 36:1980–1987

66. McDowell DC, Huber MM, Wagner M, von Gunten U, Ternes TA (2005) Ozonation of carbamazepine in drinking water: identification and kinetic study of major oxidation products. Environ Sci Technol 39:8014–8022

67. Reungoat J, Macova M, Escher BI, Carswell S, Mueller JF, Keller J (2010) Removal of micropollutants and reduction of biological activity in a full scale reclamation plant using ozonation and activated carbon filtration. Water Res 44:625–637

68. Lange F, Cornelissen S, Kubac D, Sein MM, von Sonntag J, Hannich CB, Golloch A, Heipieper HJ, Möder M, von Sonntag C (2006) Degradation of macrolide antibiotics by ozone: a mechanistic case study with clarithromycin. Chemosphere 65:17–23

69. Radjenovic J, Godehardt M, Petrovic M, Hein A, Farre M, Jekel M, Barcelo D (2009) Evidencing generation of persistent ozonation products of antibiotics roxithromycin and trimethoprim. Environ Sci Technol 43:6808–6815

70. Klavarioti M, Mantzavinos D, Kassinos D (2009) Removal of residual pharmaceuticals from aqueous systems by advanced oxidation processes. Environ Int 35:402–417
71. Fatta-Kassinos D, Hapeshi E, Malato S, Mantzavinos D, Rizzo L, Xekoukoulotakis NP (2010) Removal of xenobiotic compounds from water and wastewater by advanced oxidation processes. In: Fatta-Kassinos D, Bester K, Kümmerer K (eds) Xenobiotics in the urban water cycle. Springer, Netherlands, pp 387–412
72. Ince NH, Apikyan IG (2000) Combination of activated carbon adsorption with light-enhanced chemical oxidation via hydrogen peroxide. Water Res 34:4169–4176
73. Pereira VJ, Weinberg HS, Linden KG, Singer PC (2007) UV degradation kinetics and modeling of pharmaceutical compounds in laboratory grade and surface water via direct and indirect photolysis at 254 nm. Environ Sci Technol 41:1682–1688
74. Rosenfeldt EJ, Linden KG (2004) Degradation of endocrine disrupting chemicals bisphenol A, ethinyl estradiol, and estradiol during UV photolysis and advanced oxidation processes. Environ Sci Technol 38:5476–5483
75. Andreozzi R, Caprio V, Marotta R, Radovnikovic A (2003) Ozonation and H_2O_2/UV treatment of clofibric acid in water: a kinetic investigation. J Hazard Mater 103:233–246
76. Andreozzi R, Caprio V, Marotta R, Vogna D (2003) Paracetamol oxidation from aqueous solutions by means of ozonation and H_2O_2/UV system. Water Res 37:993–1004
77. Kim I, Yamashita N, Tanaka H (2009) Performance of UV and UV/H_2O_2 processes for the removal of pharmaceuticals detected in secondary effluent of a sewage treatment plant in Japan. J Hazard Mater 166:1134–1140
78. Pereira VJ, Linden KG, Weinberg HS (2007) Evaluation of UV irradiation for photolytic and oxidative degradation of pharmaceutical compounds in water. Water Res 41:4413–4423
79. Vogna D, Marotta R, Andreozzi R, Napolitano A, d'Ischia M (2004) Kinetic and chemical assessment of the UV/H_2O_2 treatment of antiepileptic drug carbamazepine. Chemosphere 54:497–505
80. Vogna D, Marotta R, Napolitano A, Andreozzi R, d'Ischia M (2004) Advanced oxidation of the pharmaceutical drug diclofenac with UV/H_2O_2 and ozone. Water Res 38:414–422
81. Doll TE, Frimmel FH (2005) Removal of selected persistent organic pollutants by heterogeneous photocatalysis in water. Catal Today 101:195–202
82. Khetan SK, Collins TJ (2007) Human pharmaceuticals in the aquatic environment: a challenge to green chemistry. Chem Rev 107:2319–2364
83. Stackelberg PE, Furlong ET, Meyer MT, Zaugg SD, Henderson AK, Reissman DB (2004) Persistence of pharmaceutical compounds and other organic wastewater contaminants in a conventional drinking-water-treatment plant. Sci Total Environ 329:99–113
84. Benotti MJ, Trenholm RA, Vanderford BJ, Holady JC, Stanford BD, Snyder SA (2009) Pharmaceuticals and endocrine disrupting compounds in U.S. drinking water. Environ Sci Technol 43:597–603
85. Vulliet E, Cren-Olivé C, Grenier-Loustalot M-F (2011) Occurrence of pharmaceuticals and hormones in drinking water treated from surface waters. Environ Chem Lett 9:103–114
86. Loos R, Wollgast J, Huber T, Hanke G (2007) Polar herbicides, pharmaceutical products, perfluorooctanesulfonate (PFOS), perfluorooctanoate (PFOA), and nonylphenol and its carboxylates and ethoxylates in surface and tap waters around Lake Maggiore in Northern Italy. Anal Bioanal Chem 387:1469–1478
87. Stumpf M, Ternes TA, Haberer K, Seel P, Baumann W (1996) Nachweis von arzneimittelrückständen in kläranlagen und fließgewässern. Vom Wasser 86:291–303
88. Heberer T, Dunnbier U, Reilich C, Stan HJ (1997) Detection of drugs and drug metabolites in ground water samples of a drinking water treatment plant. Fresenius Environ Bull 6:438–443
89. Heberer T, Mechlinski A, Fanck B, Knappe A, Massmann G, Pekdeger A, Fritz B (2004) Field studies on the fate and transport of pharmaceutical residues in bank filtration. Ground Water Monit Rem 24:70–77
90. Heberer T, Stan HJ (1997) Determination of clofibric acid and N-(Phenylsulfonyl)-Sarcosine in sewage, river and drinking water. Int J Environ Anal Chem 67:113–124

91. Stan HJ, Heberer T, Linkerhägner M (1994) Occurrence of clofibric acid in the aquatic system – is the use in human medical care the source of the contamination of surface ground and drinking water? Vom Wasser 83:57–68
92. Heberer T (2002) Occurrence, fate, and removal of pharmaceutical residues in the aquatic environment: a review of recent research data. Toxicol Lett 131:5–17
93. Zuccato E, Calamari D, Natangelo M, Fanelli R (2000) Presence of therapeutic drugs in the environment. Lancet 355:1789–1790
94. Tauber R (2003) Quantitative analysis of pharmaceuticals in drinking water from ten Canadian cities. Enviro-Test Laboratories, Xenos Division, Ontario, Canada
95. Togola A, Budzinski H (2008) Multi-residue analysis of pharmaceutical compounds in aqueous samples. J Chromatogr A 1177:150–158
96. Rabiet M, Togola A, Brissaud F, Seidel J-L, Budzinski H, Elbaz-Poulichet F (2006) Consequences of treated water recycling as regards pharmaceuticals and drugs in surface and ground waters of a medium-sized mediterranean catchment. Environ Sci Technol 40:5282–5288
97. Vieno NM, Tuhkanen T, Kronberg L (2005) Seasonal variation in the occurrence of pharmaceuticals in effluents from a sewage treatment plant and in the recipient water. Environ Sci Technol 39:8220
98. Loraine GA, Pettigrove ME (2005) Seasonal variations in concentrations of pharmaceuticals and personal care products in drinking water and reclaimed wastewater in Southern California. Environ Sci Technol 40:687–695
99. Reddersen K, Heberer T, Dünnbier U (2002) Identification and significance of phenazone drugs and their metabolites in ground- and drinking water. Chemosphere 49:539–544
100. Zühlke S, Dünnbier U, Heberer T (2004) Detection and identification of phenazone-type drugs and their microbial metabolites in ground and drinking water applying solid-phase extraction and gas chromatography with mass spectrometric detection. J Chromatogr A 1050:201–209
101. Waggot A (1981) Trace organic substances in the River Lee (Great Britain). In: Cooper WJ (ed) Chemistry in water reuse, 1st edn. Ann Arbour Science, pp 55–99
102. Vanderford BJ, Snyder SA (2006) Analysis of pharmaceuticals in water by isotope dilution liquid chromatography/tandem mass spectrometry†. Environ Sci Technol 40:7312–7320
103. Jones OA, Lester JN, Voulvoulis N (2005) Pharmaceuticals: a threat to drinking water? Trends Biotechnol 23:163–167
104. Clara M, Strenn B, Kreuzinger N (2004) Carbamazepine as a possible anthropogenic marker in the aquatic environment: investigations on the behaviour of carbamazepine in wastewater treatment and during groundwater infiltration. Water Res 38:947–954
105. Escher BI, Bramaz N, Richter M, Lienert J (2006) Comparative ecotoxicological hazard assessment of beta-blockers and their human metabolites using a mode-of-action-based test battery and a QSAR approach†. Environ Sci Technol 40:7402–7408
106. Ramil M, El Aref T, Fink G, Scheurer M, Ternes TA (2009) Fate of beta blockers in aquatic-sediment systems: sorption and biotransformation. Environ Sci Technol 44:962–970
107. Lee H-B, Sarafin K, Peart TE (2007) Determination of [beta]-blockers and [beta]2-agonists in sewage by solid-phase extraction and liquid chromatography-tandem mass spectrometry. J Chromatogr A 1148:158–167
108. Gros M, Petrovic M, Ginebreda A, Barceló D (2010) Removal of pharmaceuticals during wastewater treatment and environmental risk assessment using hazard indexes. Environ Int 36:15–26
109. Heberer T, Fuhrmann B, Schmidt-Baumier K, Tsipi D, Koutsouba V, Hiskia A (2001) Occurrence of pharmaceutical residues in sewage, river, ground, and drinking water in Greece and Berlin (Germany). In: Daughton C, Jones-Lepp T (eds) Pharmaceuticals and personal care products in the environment. American Chemical Society, Washington, DC, Scientific and Regulatory Issues

110. Kümmerer K (2003) Significance of antibiotics in the environment. J Antimicrob Chemother 52:5–7
111. Huang CH, Renew JE, Smeby KL, Pinkerston K, Sedlak DL (2001) Assessment of potential antibiotic contaminants in water and preliminary occurrence analysis. Water Resour Update 120:30–40
112. Finnegan DP, Simonson LA, Meyer MT (2010) Occurrence of antibiotic compounds in source water and finished drinking water from the Upper Scioto River Basin, Ohio, 2005–6 (Scientific Investigations Report 2010)
113. Ye Z, Weinberg HS, Meyer MT (2004) Occurrence of antibiotics in drinking water. p. 138–142. In Proceedings of the 4th International Conference on Pharmaceuticals and Endocrine Disrupting Chemicals in Water (CD-ROM). 13–15 October 2004, Minneapolis, MN.
114. Watkinson AJ, Murby EJ, Kolpin DW, Costanzo SD (2009) The occurrence of antibiotics in an urban watershed: from wastewater to drinking water. Sci Total Environ 407:2711–2723
115. Fent K, Weston AA, Caminada D (2006) Ecotoxicology of human pharmaceuticals. Aquat Toxicol 76:122–159
116. Kümmerer K (2001) Introduction: pharmaceuticals in the environment. In: Pharmaceuticals in the environment: sources, fate, effects and risks. Springer, Berlin, pp 1–8
117. Touraud E, Roig B, Sumpter JP, Coetsier C (2011) Drug residues and endocrine disruptors in drinking water: risk for humans? Int J Hyg Environ Health 214: 437–441

Sulfonamide Antibiotics in Natural and Treated Waters: Environmental and Human Health Risks

María Jesús García Galán, M. Silvia Díaz-Cruz, and Damià Barceló

Abstract Concern regarding the environmental presence of sulfonamides and other species of antibiotics has focused mainly on the potential spread of antimicrobial resistance. However, their biological activity and high resistance to biodegradation may lead to long residence times in both water and soil matrices. Treated waters represent one of the main entrance pathways of these antimicrobials into the environment, and their potential impact in the aquatic ecosystems should be fully understood and investigated. Long-term ecological risks and unpredicted effects can result from unintentional exposure of different organisms and even human health could be negatively affected. This chapter aims to review the current knowlegde regarding sulfonamides ecotoxicity and to highlight the need for further data on the fate and ecotoxicity fo this family of antibiotics.

Keywords Bacterial resistance, Environmental risk assessment, Soil, Sulfonamide, Surface waters, Toxicity

M.J. García Galán (✉) and M.S. Díaz-Cruz
Department of Environmental Chemistry, IDAEA-CSIC, Jordi Girona 18-26, 08034 Barcelona, Spain
e-mail: mggqam@cid.csic.es

D. Barceló
Department of Environmental Chemistry, IDAEA-CSIC, Jordi Girona 18-26, 08034 Barcelona, Spain

Catalan Institute for Water Research (ICRA), Parc Científic i Tecnològic de la Universitat de Girona. C/Emili Grahit, 101 Edifici H2O, 17003 Girona, Spain

King Saud University, P.O. Box 2455, Riyadh 11451, Saudi Arabia

D. Barceló (ed.), *Emerging Organic Contaminants and Human Health*,
Hdb Env Chem (2012) 20: 71–92, DOI 10.1007/698_2011_129,
© Springer-Verlag Berlin Heidelberg 2012, Published online: 14 January 2012

Contents

1   Sulfonamides in the Environment: An Introduction

Nowadays, tons of different classes of pharmaceuticals products enter the environment on a regular basis after their usage and excretion [1]. In a society exerting an increasing pressure on the natural resources, water ecosystems are considered one of the most vulnerable elements. As basically, nearly any human activity results in the generation and emission of substances of anthropogenic origin which eventually end up in natural waters, to preserve and maintain its quality is one of the greatest environmental challenges to be faced. The increasing demand of quality drinkable water makes a sustainable water resource management essential, requiring protection of water resources from those persistent or toxic anthropogenic compounds. In Europe (EU), the EU Water Framework Directive (WFD) has aimed to gather these concerns and objectives and specifies the need to monitor different organic pollutants in surface waters as an informative step to protect and improve the quality of the European water resources [2].

The quantity and quality of synthetic chemicals are continually changing. Amongst them, pharmaceuticals (PhPs) have become one of the most intensively studied categories in the last decades [1, 3–5]. Developed to perform a biological effect in the patient or animals at very low doses, some of their general physico–chemical properties, such as polarity, liposolubility (they can go through biological membranes), and persistence in order to stay active can make them very prone to bioaccumulate and capable to provoke undesired biological effects in the environment, even at small concentrations [6]. At present, approximately 3,000 pharmaceutical ingredients are used in the European Union (EU), including antibiotics, beta-blockers, lipid regulators, antidepressants, and many more [1]. Special attention has been devoted to the environmental risks associated to the widespread occurrence of antibiotics in the aquatic environment, due mainly to their high potency and also to their high consumption rates. Although information on their usage is not available to the general public either in the United States (US) or in the European Union (EU), estimations indicate sales over the 16,000 t in US in 2001, of which 9,300 t are used in animal feeding operations [2]. The annual consumption of antibiotics in the EU in 1999 was 13,288 t in total, with 29% for veterinary medicine, 6% as antibiotic feed additives, and 65% in human medicine [7].

Sulfonamides (SAs) constitute one of the most consumed antimicrobial families. They are synthetic antimicrobial agents, derivatives of sulfanilamide, which are used mainly in aquaculture [8–10] and intensive livestock farming [11, 12] but also in human therapies, to treat many kinds of infection caused by bacteria and certain other microorganisms (urinary tract infections, ear infections, bronchitis, bacterial meningitis, certain eye infections, *Pneumocystis carinii* pneumonia, traveler's diarrhea, and more) [13, 14]. However, more relevant quantities are now being used in veterinary medicine [9, 15]. In the European Union (EU), they are the second most widely used family of veterinary antibiotics, after tetracyclines, accounting for 11–23% of the sales in several other European countries [2]. Their mode of action consist on the inhibition of the biosynthetic pathway of folate (an essential molecule required by all living organisms) [13]. Many research publications have demonstrated that SAs are ubiquitous in the environment, in both aqueous and solid matrices [15–21]. After administration, residues of SAs that have not been completely metabolized, together with their transformation products and metabolites, can reach the environment by several pathways [6]. In the case of human intake, SAs may enter the environment indirectly through the discharge of effluents of wastewater treatment plants (WWTPs), disposal of unused or expired products (directly into the domestic sewage system), and burial in landfills and accidental spills during manufacture [22–27]. Another entrance pathway is the application of biosolids from WWTPs in agriculture fields as nutrient amendment, as they may contain certain amounts of sequestered SAs and other PhPs [28]. Regarding veterinary use, animal excreta is considered one of the major entrance pathways in the environment for SAs (Fig. 1). Residues of these antimicrobials have been detected in manure from medicated animals at concentrations ranging from 8.7 mg/kg for sulfamethazine (SMZ) to 12.4 mg/kg for sulfathiazole (STZ) [29]. Depending on the drug and the animal age, up to 50–90% of the administered dose can be excreted in the parental form (9–30%) or metabolized [4, 30]. Manure is regarded as a very valuable fertilizer, as it contains essential nutrients for plant growth such as nitrogen, phosphorous, organic carbon, or potassium, and its organic material can improve soil quality. Therefore, it is frequently applied as nutrient amendment in agriculture and this extensive use is among the major routes by which veterinary SAs, and veterinary antibiotics in general, enter the environment on a cyclic basis [11, 19, 31–35].

After being released in the environment, and due to their high solubility and generally low weak sorption to soil tendency (K_d), SAs become very mobile contaminants [36–38]. Once on the topsoil, the excreted residues of SAs may reach surface waters during run off episodes or percolate and contaminate the groundwater bodies underneath [19, 39]. This possibility has already been proved in several publications, showing the presence of SAs at different concentrations in groundwater from various sites close to animal farming facilities [40–47]. This diffuse pollution from agricultural fields is difficult to prevent and deal with due to the large areas of application.

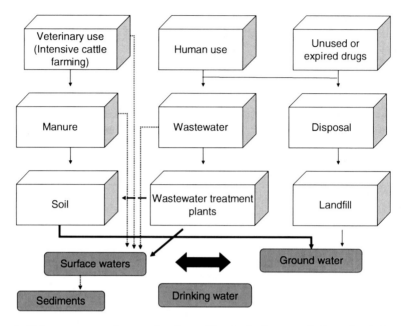

Fig. 1 Main entrance pathways of sulfonamides (and pharmaceutical products) into the environment

2 Environmental Occurrence of Sulfonamides

2.1 Presence of Sulfonamides in Wastewater Treatment Plants

In urban ecosystems, wastewater treatment plants (WWTPs) are considered the main entrance pathways for SAs (amongst other PhPs). In the last few years, there has been a noticeable increase in the number of scientific works devoted to evaluate not only their presence but also their fate in influent and effluent wastewaters [48–53]. The SAs loads in these effluents and indirectly the estimation of removal efficiencies (RE%) of the WWTPs provide very valuable information about the potential impact of these substances on the receiving streams and aquatic ecosystems. As mentioned in the introduction, incomplete removal of many of these compounds after secondary biologic treatment (conventional activated sludge (CAS)) can be directly linked to their presence in surface waters, sediments, and soils. Due to their high mobility, they could also eventually reach groundwater bodies. Furthermore, application of WWTP sewage sludge as organic amendment in croplands represents another environmental input for these compounds [54, 55].

Data on the removal of SAs during wastewater treatment is still scarce. Sulfamethoxazole (SMX), sulfapyridine (SPY), and sulfadiazine (SDZ), commonly used in human medicine, are the SAs most frequently detected in most of the WWTP monitoring studies, specially SMX and SPY, which usually account for most of the influent load and present the highest concentrations in both influent and effluent samples (maximum concentrations detected were up to 650 ng/L for SMX and 532 ng/L for SPY [49, 51, 56, 57]). Thus, the removal of these two SAs during the different wastewater treatments is critical to reduce the total SAs concentration in the effluents. RE% values found in the literature are hard to interpret, as they range from negative removals to 100% elimination. SDZ was in average the SA eliminated most efficiently, whereas SPY showed intermediate to high removal efficiency (RE%) values. SMX showed both RE% higher than 50% and also negative values in many WWTPs, meaning that higher concentrations are detected in the effluents than in the influent wastewaters. This fact is usually attributed to the presence of SA conjugates and metabolites, which are not comprised within the scope of the different studies; these conjugates can be transformed back during treatment into the original compound, as demonstrated recently [56] and could therefore explain higher concentrations of the different SAs in effluents than in influent waters [51, 58]. Only recently some acetylated metabolites of SAs (which are the majority metabolites) have been included within the scope of different monitoring studies. As expected, the acetylated moieties of SMX and SPY have been frequently detected and in high concentrations (up to 522 ng/L for N^4-acetylsulfaprydine (AcSPY) and 94.6 ng/L for N^4-acetylsulfamethoxazole (AcSMX)), together with other minor acetylated metabolites such as N^4-acetylsulfadiazine (AcSDZ) or N^4-acetylsulfamerazine (AcSMR) [49, 51, 56, 57, 59, 60]. The relevance of including these metabolites in future surveys is evident and its oversight would lead to an underestimation of the real total SAs concentration. Alternative secondary treatments, such as membrane bioreactors (MBRs) have proved not to be especially good in SAs removal, in particular for SMX and SPY, the two most relevant SAs in terms of frequencies of detection and concentration. Recent works demonstrated that although elimination rates for SMX were higher in the MBRs than in CAS, removal was only partial as nearly half of the SMX input could still be detected in their respective effluents [59, 61–63]. But again, MBRs showed to be more effective than CAS for other SAs, such as SDZ, which was completely removed after MBR treatment, whereas it was removed only 49% during the CAS treatment. Regarding acetylated metabolites, N^4-acetylsulfamethazine (AcSMZ) showed a similar pattern with a 54% removal in the CAS reactor and 100% in MBRs [59]. Tertiary treatments such as ozonation and nanofiltration have demonstrated high efficiencies in SAs removal [64–68], but still its application in WWTPs is scarce and the fate of the transformation products generated unknown [69].

2.2 Presence of Sulfonamides in Surface Waters

The intensification of wastewater inputs on the receiving streams is one of the direct consequences of an increasing population worldwide. As a result, an excess of nutrients together with a wide range of pollutants are constantly reaching natural

waters and threatening their natural ecological equilibrium. The low natural biodegradation of SAs and low tendency to adsorb to solid matrices (from the river
bed) [5, 70] have resulted in their frequent detection in river waters and, generally,
to concentration gradients from the source to the mouth of the water course, as
rivers usually receive SAs inputs, both agricultural and urban, all along the basin.
When interpreting the obtained data, seasonal changes should also be taken into
account. Generally, the highest concentrations of human SAs (SMX, SPY) are
expected during the dry seasons, as the dilution exerted by the receiving streams
is lower. During the rainy season, whereas these concentrations would be more
diluted, run off from irrigated rural areas may increase the concentrations in
freshwater of veterinary SAs (sulfadimethoxine (SDM), SMZ, or SDZ), denoting
its run off origin from crop lands after heavy rain periods [71–74]. In some
occasions, the release of untreated wastewaters due to strong rainfall events can
also lead to higher concentrations of human SAs than expected [75]. Cold
conditions can also contribute to higher concentrations due to reduced biodegradation of these contaminants in water [76]. SMX was detected in the main European
water courses: in the Ebro River Basin (1.9–35.6 ng/L,), in the Douro River in
Portugal (53.3 ng L $^{-1}$ maximum concentration), in the Seine River (37–140 ng/L),
and in the Elbe River (30–70 ng/L) [49, 74, 75, 77, 78]. The presence of SAs not
only in surface water samples but also in their sediments [16, 18, 72, 76, 79, 80],
despite their low K_d values, highlights the vulnerability of the river ecosystems
against these antimicrobials. Furthermore, as mentioned in section 2.2 the presence
of SAs metabolites such as their acetylated or glucuronidated moieties has been
already demonstrated and should be considered when estimating the total amount of
SAs present in this matrix. For instance, AcSMX has been detected in natural
streams at higher frequencies and concentrations than its parent molecule [81].
Likewise, the potential bioactivity and ecotoxicity of these metabolites against the
different ecosystem communities should also be taken into account. It has been
recently proved that AcSPY is more toxic than the parent compound to aquatic
bacteria [57], and the same could apply for other metabolites and transformation
products.

Some rivers, such as those in the Mediterranean region, usually present low
natural base flows due to the characteristic long draught periods. The Llobregat
River is an illustrative example of the hydrological pattern of Mediterranean rivers.
It is subjected to heavy anthropogenic pressure, receiving extensive industrial and
urban discharges from more than 50 WWTPs (137 Hm^3/year; 92% from wastewater-
treatment plants) [82]. These inputs are only partially diluted by its natural flow
(0.68–6.5 m^3/s basal flow) and play a major role in both the hydrology and the
presence of pollutants in the basin. Furthermore, 30% of the annual discharge of
the river (693 Hm^3) is used for drinking water supply, including the city of
Barcelona. SAs levels detected in this river were over two orders of magnitude
above the average values obtained in continental rivers (in the range of μg/L).
Highest concentrations were detected in the low course of the river and near its
mouth [60].

3   Environmental Concerns

Although environmental levels detected so far are generally low and regarded as harmless, SAs are being continuously introduced into the environment and concentrations previously considered innocuous may pose a risk to the different environmental compartments, even when these compounds have been evaluated as safe for human and veterinary use. So far, environmental research on antibacterial compounds has focused mainly on the bacterial resistance acquired against antibiotics in aquatic and soil bacterial communities, which has become a well documented fact. SMX, as the most ubiquitous SA, has been investigated in a higher number of studies. It turned out to be one of the less active antibiotics against *Aeromonas spp.*, typical waterborne bacteria, isolated from two rivers and also against different Enterobacteriaceae strains (representative of the human and animal flora) [83]. A recent study, on the contrary, reported higher frequencies of SMX resistant bacteria belonging to *Aeromonas spp.*, than macrolides-resistant bacteria (up to 94.44%) [84]. The *Acinetobacter* genera were also affected by the presence of SMX amongst other SAs, and a clear correlation was observed between SMX environmental concentration and occurrence of SMX-resistant bacteria. This correlation was also established by Luo et al. [85], who found SAs resistant genes in surface water and sediments samples. The concentration of these genes was up to 1,200 times higher in sediments, indicating that they can be considered as important antibiotic resistance genes (ARGs) reservoirs. SAs may have both qualitative and quantitative effects upon the resident microbial community found in sediment, which can in turn affect the degradation of organic matter. In sediments beneath fish farms the exposure is direct, as antibiotics are directly added to receiving waters and up to 70–80% of the dose administered enter the environment. WWTP effluents have been considered as ARG sources in different works too [86–88]. The use of reclaimed water or surface water for irrigation purposes could mean the transfer of these ARG to crop fields and the corresponding ecosystems [89]. Multi-resistance has also been reported for different SAs being simultaneously present in the same sampling site [72]. Regarding soils, veterinary SAs have been more thoroughly studied. They can accumulate in this matrix due to repeated application of manure (during the growth period of the crops) and concentrations detected, although in the subtherapeutic range, can reach the level of the minimum inhibitory concentrations for relevant soil bacterial species [35, 70, 90, 91]. Sorption in soils for SAs varies considerably and concentrations detected range from 0.5 to 6.5 µl/kg [70]. The antibiotic potency of SAs could be diminished or changed when degradation takes place. Temperature and moisture content of the soil, the timing of manure application, as well as prevailing weather conditions can determine the mobility and fate of SAs in the environment [92, 93]. Other factors that can influence the mobility of SAs and veterinary medicines in general are preferential flow via desiccation cracks and worm channels to the tile drains, as recently demonstrated in a field study in the United Kingdom [37, 94, 95]. As it is generally assumed that SAs enter the soil with manure and the pH of manure is usually alkaline, SAs would occur mainly as anionic species, resulting in a decrease in antibacterial activity as it takes a few days before soil adjusts and returns to the previous pH value [96]. Whether the antibacterial activity of a given SA reaches an ecologically significant level will therefore depend not only on the absolute concentration but also on speciation which can be altered, as can their tendency to sorb to soil, their bioavailability and their antimicrobial activity [97]. As a competitive process, sorption to soil increased with decreasing pH and resulted in decreased bioavailability [98].

3.1   Human Exposure to Sulfonamides

SAs act as structural analogs of *p*-aminobenzoic acid, inhibiting the biosynthetic pathway of folate in bacterial cells and therefore limiting their growth. In mammals, folate is acquired from diet and therefore SAs are not known to have any major toxic effect. The risk to humans from direct exposure to SAs residues in food products from medicated animals has been thoroughly assessed. EU Council Regulation EEC 2377/90 lays down a procedure through the European Medicines Evaluation Agency [99] to establish maximum residue limits (MRLs) of veterinary medicinal products in foodstuffs of animal origin. For SAs, a definitive MRL has been established for parent drugs – 100 μg/kg in target tissues (muscle, fat, liver and kidney) and also in milk from bovine, ovine, and caprine species. The sum of all SAs residues detected should not exceed 100 μg/kg [100]. However, the potential extent of any indirect human exposure to SAs present in the environment, with the potential consequences for human health, has not been legally established yet. The lack of ecotoxicological data may be one of the main reasons for the absence of European regulation regarding adverse effects of SAs or other antimicrobial substances. Parallel studies have been published recently aiming to propose risk hierarchies for pharmaceutical products. For instance, Capleton et al. created an interesting scheme to prioritize the risk posed by veterinary medicines on the basis of their potential for indirect human exposure via the environment and their toxicity profile [101]. SDZ was classified within the substances group requiring priority detailed risk assessment, whereas SMZ was considered a substance of very low priority. Similar works have followed, aiming to establish a ranking of veterinary medicines based on their environmental risk. Kools et al. [102] used the data on SAs consumption as an indicator of its potency to predict ecotoxicological effects. In this study, SDZ was the only SA showing with a very high risk index in soil ecosystems near intensive rearing facilities and pasture, although no direct effects on humans were highlighted.

Humans may be exposed to residues of veterinary medicines in the environment (i.e., soil, water, and sediment) by four main exposure routes:

(a) Crops that have accumulated substances from soils as a result of exposure to contaminated manure and slurry: the potential uptake of PPCPs into crops is increasingly being recognized, and calculations by Boxall et al. [104] indicate that foodborne exposure may be much more significant than drinking water. Because of their stability, SAs are supposed to maintain significant residual activity and their potential toxicity in animal manure for long periods of time. As mentioned in previous sections, after application of manure, SAs are very likely to reach surface waters and groundwaters. These contaminated waters may be used for irrigation and consequently taken up directly or indirectly by plants to meet their evapotranspiration and photosynthesis requirements [105]. Although it has been demonstrated that SAs showed a strong initial sorption to soil [96], these sorbed amounts will eventually be released to the soil solution and may be taken up by plants. Both toxicity and bioaccumulation of SDM were studied for different terrestrial plants by Migliore et al. [106–108]. SDM affected the post-germinative development of barley, millet, corn, and cosmopolitan weeds. Bioaccumulation in roots and stems was also demonstrated, being higher in roots. In all cases, concentrations tested were higher than those environmentally relevant (300 mg/L) and are unlikely to occur in soil. Forni et al. [109] observed that SDM altered the morphology of *Azolla spp.*, but was capable of removing up to a 88.5% of the total amount of the initial SDM concentration. Uptake investigation with carrots or lettuce leaves showed that these vegetables were not affected by the presence of SAs [104]. SMZ was taken up by crops, with concentrations in plant tissue ranging from 0.1 to 1.2 mg/kg dry weight. After 45 days of growth, this concentration represented less than 0.1% of the amount applied to soil in manure [110]. These results manifested that there was little evidence of an appreciable risk, although it would directly depend on the acceptable daily intake values of the crop. Prolonged periods of exposure or different simultaneous routes of exposure may eventually lead to negative effects on the plant. However, all these works suggest that for the study compounds exposure of consumers to veterinary medicines in soils via plants is likely to be considerably below the acceptable daily intake and that the risk to human health is probably low.

(b) Fish exposed to treatments used in aquaculture: data on SAs bioaccumulation is scarce. It was demonstrated that SDM tended to accumulate in brine shrimp *Artemia* with the potential implications for the rest of the food chain in the marine community [107]. Hou et al. [111] studied SMZ bioconcentration and elimination in sturgeon (*Acipenser schrenkii*), and results indicated the little environmental relevance. Bioaccumulation of SMZ was considered to be of little environmental concern and the drug was therefore not expected to bioconcentrate in tissues consumed by humans or to biomagnify in fish consumed by other fish predators. Similar results were found in mosquito fish or in fish residing in effluent dominated or influenced water bodies, in which SMX was not detected in either fish tissues or liver [112].

(c) Livestock that have been medicated: other possibility was that exposed by Hamscher et al. regarding exposure via inhalation of dust emitted from intensively reared livestock facilities [113].

(d) Abstracted groundwater and surface waters containing veterinary medicines: as mentioned in the previous sections, the occurrence of veterinary antibiotics in groundwater close to livestock feeding operation facilities or to soils fertilized with liquid manure have been frequently reported [20, 21, 96, 114, 115] with concentrations in the µg/L.

The only legally established restrictive measure up to date comes from the European Medicines Agency (EMEA), which requires that an environmental risk assessment (ERA) is presented together with an application for marketing authorization for a new medicinal product for human use [99]. Although this process was designed as a part of the process for registering new drugs, risk-assessment guidelines that include information on predicting levels of drugs in the environment can be used to prioritize the risk from drugs that are already in use and to assess the potential impact of drugs yet to be released [49, 52, 116–119]. The ERA protocol is a two-phase tiered process consisting of a first phase assessment of the environmental exposure to the drug substance, and a second phase in which information about the physical/chemical, pharmacological, and toxicological properties are obtained and assessed in relation to the extent of exposure of the environment. It begins with an approximate calculation of the predicted environmental concentration (PEC) of the new drug in surface water (see Fig. 2). If the PEC estimation is below 0.01 µg/L and no other environmental concerns are apparent, it is assumed that the medicinal product is unlikely to represent a risk for the environment. However, if the PEC surface water value is above 0.01 µg/L a Phase II environmental effect analysis should be performed, during which standard

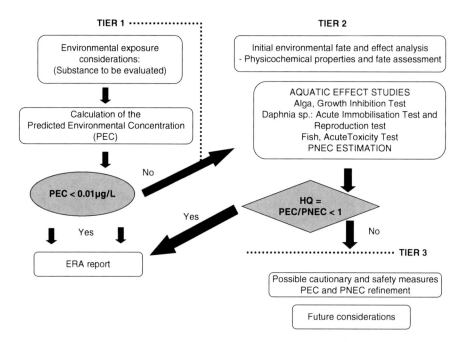

Fig. 2 Scheme of the tiered approach of the European Medicines Evaluation Agency (EMEA) for environmental risk assessment

acute toxicity tests will be carried out in order to estimate predicted no-effect concentration (PNEC) or non-observed effect concentration (NOEC) [99]. Both end points are the estimation of the concentration of the drug for which adverse environmental effects are not expected. Finally, the ratio of the PEC to PNEC, known as the hazard quotient (HQ), indicates whether a potential environmental impact is implicit or not. It is so when HQ > 1, meaning that further testing might be needed to refine both PEC and PNEC values. If HQ < 1, no further testing is required. It is also recommended that, when the total concentration of metabolites is greater than the 10% of the concentration of the corresponding parent drug, the metabolites are also to be further investigated (phase II tier B) in order to determine their ecotoxicological effects. PNEC is established using data from relevant environmental toxicity tests. When PNEC values are not available, an alternative PNEC can be derived by dividing the half maximal effective concentrations (EC_{50}) or median lethal concentration (LC_{50}) values (acute toxicity data) by an uncertainty factor of up to 1,000 [120], and so converting acute to chronic toxicity values, since data on chronic toxicity for SAs is lacking. Likewise, measured environmental concentrations (MECs) are used in the calculation instead of PECs. In order to set up a worst case scenario, MEC values used corresponded to the maximum values detected in this study whereas EC_{50}–LC_{50} values used were the lowest found in the literature. The taxonomic levels tested are intended to serve as indicators for the range of species present in the aquatic environment. Table 1 summarizes the HQ values reported to date in the literature. SMX, as one of the most consumed SAs in human medicine, has been reported most frequently and is usually considered ecologically harmful.

The EMEA Committee for Medicinal Products for Veterinary Use (CVMP) established similar guidelines to those for human pharmaceuticals, to assess the potential for veterinary medicines to affect nontarget species in the environment, including both aquatic and terrestrial species [123].

3.2 Ecosystems Vulnerability to Sulfonamides

Undoubtedly, the decline of vulture populations in Pakistan due to the uptake of the anti-inflammatory diclofenac has been the best documented example of toxicity of PhPs in the environment. This anti-inflammatory drug caused major kidney diseases in the birds scavenging on dead cattle medicated with this drug [124]. Although in the EU there is no registry of all veterinary medicinal active substances, the increase in the number of confined animal-feeding operations (CAFOs), which often lack proper waste management practices, can be considered the main reason for the increase in the use of SAs and other veterinary antibiotics and, therefore, the greater occurrence of these substances in the environment.

Table 1 Estimation of hazard quotients (HQ) for different sulfonamides following the EMEA guidelines

SA	MEC	REF$_{MEC}$	PEC	Taxa	PNEC$_{acute}$	REF$_{PNEC}$	HQ
SMX	–		0.95		0.146	[76]	*6.3*
	0.4	119			0.03	[119]	*13.4*
	–	–	0.31		0.027		*11.4*
	–	–	0.31		0.59[a]		*0.5*
	–	–	1.6	Blue green algae	0.027		*59.3*
	–	–	1.6		0.59[a]	[101]	*2.7*
	0.0356	49	–		0.027		**1.32**
	0.284	57	–		0.027		**10.52**
	4.3	61	–		0.027		**159.26**
	0.0356	49	–		78.1		<0.001
	0.284	57	–	V. fischerii	78.1	[36]	0.004
	4.3	61	–		78.1		0.005
	0.0356	49	–				0.001
	0.284	57	–	Daphnids	25.2	[121]	0.011
	4.3	61	–				0.171
	0.0356	49	–				<0.001
	0.284	57	–	Fish	562.5	[125]	<0.001
	4.3	61	–				0.008
AcSMX	0.094	57		Blue green algae	101	[122]	<0.001
SPY	0.042	49	–				0.002
	0.177	57	–	V. fischerii	27.4	[57]	0.006
	0.092	61	–				0.003
ACSPY	0.522	57		V. fischerii	8.2	[57]	0.064
SMZ	0.065	49	–				<0.001
	0.0364	57	–	V. fischerii	344.7	[36]	<0.001
	2.48	61	–				0.007
	0.065	49	–				<0.001
	0.0364	57	–	Daphnids	147.5	[108]	<0.001
	2.48	61	–				0.017
	0.065	49	–				<0.001
	0.0364	57	–	Fish	110.7	[76]	<0.001
	2.48	61	–				0.022
STZ	0.009	49	–				<0.001
	0.07	57	–	Blue green algae	16.32	[108]	0.004
	0.96	61	–				0.059
	0.009	49	–				<0.001
	0.07	57	–	V. fischerii	1,001	[36]	<0.001
	0.96	61	–				0.001
	0.009	49	–				<0.001
	0.07	57	–	Daphnids	78.9	[108]	<0.001
	0.96	61	–				0.012
	0.009	49	–				<0.001
	0.07	57	–	Fish	101	[107]	<0.001
	0.96	61	–				0.010
SDZ	0.286	57	–	Blue green algae	1.225	[108]	0.233

(continued)

Table 1 (continued)

SA	MEC	REF$_{MEC}$	PEC	Taxa	PNEC$_{acute}$	REF$_{PNEC}$	HQ
AcSDZ	0.067	57	–	Blue green algae	101	[122]	<0.001
SDM	0.023	49	–				0.010
	0.001	57	–	Blue green algae	2.3	[122]	<0.001
	0.136	61	–				0.059
	0.023	49	–				<0.001
	0.001	57	–	V. fischerii	501	[36]	<0.001
	0.136	61	–				<0.001
	0.023	49	–				<0.001
	0.001	57	–	Daphnids	204.5	[36]	<0.001
	0.136	61	–				<0.001
	0.023	49	–				<0.001
	0.001	57	–	Fish	101	[76]	<0.001
	0.136	61	–				0.001

MEC measured environmental concentrations (μg L^{-1}), *PNEC* predicted no effect concentration, *REF*$_{MEC}$ literature reference for the MEC value, *REF*$_{PNEC}$ literature reference for the PNEC value

3.2.1 Toxicity of Sulfonamides in Aquatic Ecosystems

Whether or not SAs are biodegraded in the aquatic environment would settle the very first step for a complete environmental risk assessment. As described in the previous section, SAs seem to resist biodegradation in WWTPs even in media with high microbial activity (activated sludge). Wastewater bacteria could be considered the first organisms to be potentially affected by antibiotics and SAs from human-medicine treatments. For instance, the presence of SMX resulted toxic for the CAS colony forming units [126], inhibited nitrite-oxidizing bacteria cultures [127], and also the methanogenes process to a certain extent [128]. Whereas SAs are probably not harmful to humans at the environmental concentrations detected so far (usually at the ng/L level), once discharged from WWTPs they become potential micropollutants to key living organisms in aquatic ecosystems (e.g., fish, aquatic invertebrates, and unicellular algae). All these diverse species belonging to different trophic levels may be exposed and negatively affected. Toxic effects in primary producers may imply loss of the whole food-chain structure, as they represent a significant portion of the total biomass of the ecosystem and are important as a source of carbon for the rest of the aquatic biosphere. Table 2 provides an overview of the estimated values for the most usual toxicity indicators found in the literature. Despite the lack of toxicity data available, it has been demonstrated that microalgae are more sensitive than crustaceans and fish to antibacterial agents. However, at concentrations reported in surface water (ng/L level), SAs are generally considered unlikely to be toxic to algal growth. On the other hand, belonging to the same family of compounds implies similar molecular structure and modes of action, so "concentration addition" or synergistic effects could be expected. It is necessary to take into account that the degradation products of SAs may also be involved in the

Table 2 Summary of the ecological risk estimated for different sulfonamides (based on [103])

Taxon	Species	Time	Toxicity endpoint	SPY	SDZ	SDM	SMX	SMZ	SCP	STZ	SCT	AcSMX	AcSDM	AcSDZ	AcSPY
Algae	Mycrocistis aeruginosa	7 d	EC_{50}	–	135	–	–	–	–	–	–	–	–	–	–
	Selenastrum capricornotum	72 h	EC_{50}	–	2.19	2.3	1.53	–	–	–	–	>100	>100	>100	–
		72 h	NOEC	–	<1.00	0.529	0.614	–	–	–	–	>100	80	>100	–
		7 d	EC_{50}	–	7.8	–	–	–	–	–	–	–	–	–	–
	Rhodomonas salina	7 d	EC_{50}	–	403[a]	–	–	–	–	–	–	–	–	–	–
	Chlorela vulgaris	72 h	EC_{50}	–	–	11.2	–	–	–	–	–	–	–	–	–
		72 h	NOEC	–	–	<20.3	–	–	–	–	–	–	–	–	–
		48 h	EC_{50}	–	–	–	–	–	>2.000	–	–	–	–	–	–
		24–48 h	EC_{50}	–	0.0049[b]	–	0.0062[b]	–	–	0.064[b]	0.062[a]	–	–	–	–
	Pseudokirchmeriella subcapitata	96 h	NOEC	–	–	–	0.09	–	–	–	–	–	–	–	–
		96 h	EC_{50}	–	–	–	0.146	–	–	–	–	–	–	–	–
		72 h	IC_{50}	–	–	–	0.0019–0.103	0.0087–0.103	–	–	–	–	–	–	–
		72 h	NOEC	–	–	–	<0.0005–0.103	0.001–0.103	–	–	–	–	–	–	–
		72 h	LOEC	–	–	–	0.0008–0.103	0.008–0.103	–	–	–	–	–	–	–
		30 m	EC_{50}	–	–	–	23.3	–	–	–	–	–	–	–	–
		3 d	EC_{50}	–	–	–	0.52	–	–	–	–	–	–	–	–
	Synechococcus leopolensis	96 h	EC_{50}	–	–	–	0.0268	–	–	–	–	–	–	–	–
Aquatic plant	Lemna gibba	96 h	NOEC	–	–	–	0.059	–	–	0.1	–	–	–	–	–
		3 d	EC_{50}	–	–	–	0.52	–	–	–	–	–	–	–	–
		7 d	LOEC	–	–	–	1	–	–	–	–	–	–	–	–
		7 d	$EC_{50\text{-chlorophyll b}}$	–	–	3.552	0.682	>1	–	3.552	–	–	–	–	–
		7 d	$EC_{50\text{-wet mass}}$	–	–	0.248	0.081	1.277	–	–	–	–	–	–	–

		Duration	Endpoint							
		7 d	EC$_{50}$-frond number	–	0.445	0.249	>1	–	–	–
		7 d	EC$_{50}$-frond number	–	–	0.132	–	–	–	–
		7 d	EC$_{50}$-fresh weight	–	–	0.0616	–	–	–	–
		7 d	EC$_{50}$-total ρ ABA	–	–	0.00336	–	–	–	–
		7 d	LOEC-frond number	–	–	0.108	–	–	–	–
		7 d	LOEC-fresh weight	–	–	0.028	–	–	–	–
		7 d	EC$_{50}$-total ρ ABA	–	–	0.009	–	–	–	–
Bacterium	Vibrio fischeri	5 m	EC$_{50}$	–	>500	74.2	303	53.7	>100	–
		15 m	EC$_{50}$	–	>500	78.1	344.7	26.4	>100	–
		15 m	EC$_{50}$	27.4	–	–	–	–	–	8.2
		30 m	EC$_{50}$	–	–	84	–	–	–	–
		30 m	LC$_{50}$	–	–	23.3	–	–	–	–
Crustacean	Daphnia magna	48 h	EC$_{50}$	–	248	–	174.4	375.3	149.3	–
		96 h	EC$_{50}$	–	204.5	–	158.8	233.5	85.4	–
		24 h	EC$_{50}$	–	639.8	>200	506.3	–	616.7	–
		48 h	EC$_{50}$	–	123.1	–	215.9	–	–	–
		21 h	LOEC	–	–	–	–	–	35	–
		21 h	NOEC	–	–	–	–	–	11	–
		48 h	EC$_{50}$	–	–	>100	–	–	–	–
		48 h	LC$_{50}$	–	–	>100	–	–	–	–
	Daphnids	48 h	EC$_{50}$	–	–	–	–	250	–	–
		24 h	LC$_{50}$	–	–	25.2	–	–	–	–
		24 h	EC$_{50}$	–	–	25.2	–	–	–	–
		48 h	EC$_{50}$	–	–	205.2	185.3	–	135.7	–
		96 h	EC$_{50}$	–	–	177.6	147.5	–	78.9	–
		48 h	EC$_{50}$-UVB	–	–	181	109.5	–	61.4	–
		96 h	EC$_{50}$-UVB	–	–	145.1	93.9	–	34.3	–
		48 h	EC$_{50}$-s unlight	–	–	96.7	31.4	–	8.2	–
	Ceriodaphnia dubia	48 h	EC$_{50}$	–	–	>100	–	–	–	–
		7 d	NOEC	–	–	250	–	–	–	–
		7 d	NOEC	–	–	0.25	–	–	–	–
		48 h	LC$_{50}$	–	–	15.51	–	–	–	–
		7 d	LC$_{50}$	–	–	0.21	–	–	–	–

(continued)

Table 2 (continued)

Taxon	Species	Time	Toxicity endpoint	SPY	SDZ	SDM	SMX	SMZ	SCP	STZ	SCT	AcSMX	AcSDM	AcSDZ	AcSPY
		48 h	EC_{50}	–	–	–	15.51	–	–	–	–	–	–	–	–
	Thamnocephalus platyurus	24 h	LC_{50}	–	–	–	35.36	–	–	–	–	–	–	–	–
Rotifer	Brachionus calyciflorus	48 h	NOEC	–	–	–	>25	–	–	–	–	–	–	–	–
		24 h	EC_{50}	–	–	–	26.27	–	–	–	–	–	–	–	–
		48 h	EC_{50}	–	–	–	9.63	–	–	–	–	–	–	–	–
Invertebrate	Artemia salina	24 h	LC_{50}	–	–	1.866	–	–	–	1.866	–	–	–	–	–
		48 h	LC_{50}	–	–	851	–	–	–	851	–	–	–	–	–
		72 h	LC_{50}	–	–	537	–	–	–	–	–	–	–	–	–
		96 h	LC_{50}	–	–	19.5	–	–	–	–	–	–	–	–	–
Fish	Moina macrocopa	24 h	EC_{50}	–	–	296.6	84.9	310.9	–	430.1	–	–	–	–	–
		48 h	EC_{50}-EROD activity	–	–	183.9	70.4	110.7	–	391.1	–	–	–	–	–
		8 d	LOEC	–	–	–	–	–	–	–	–	–	–	–	–
		8 d	NOEC	–	–	–	–	–	–	–	–	–	–	–	–
	Onchorynchus mykiss	24 h	EC_{50}	–	–	–	–	–	–	–	–	–	–	–	–
		24 h	EC_{50}-EROD activity	–	–	108	–	–	–	–	–	–	–	–	–
		48 h	LC_{50}	–	–	–	–	–	>1,000	>100	–	–	–	–	–
	Fish	96 h	LC_{50}	–	–	–	–	–	–	–	–	–	–	–	–
	Danio rerio	10 d	NOEC	–	–	–	>8	–	–	–	–	–	–	–	–
		96 h	LC_{50}	–	–	–	>1,000	–	–	–	–	–	–	–	–
	Lepomis punctatus	24 h	EC_{50}	–	–	–	–	–	–	>100	–	–	–	–	–

Lepomis macrochirus	48 h	EC$_{50}$	–	–	–	–	–	–	>100	–	–	–
Salmo trutta	8 d	LOEC	–	–	–	–	–	–	>100	–	–	–
Salvelinus fontinalis	8 d	NOEC	–	–	–	–	–	–	>100	–	–	–
Salvelinus namaycush	–	–	–	–	–	–	–	–	>100	–	–	–
Oryzias latipes	48 h	LC$_{50}$	–	>100	>750	>100	>500	589.3	>500	–	–	–
	96 h	LC$_{50}$	–	>100	562.5	>100	>500	535.7	>500	–	–	–
	48 h	LC$_{50}$	–	>500	–	>500	–	–	–	–	–	–
	96 h	LC$_{50}$	–	>500	–	>500	–	–	–	–	–	–

SPY sulfapyridine, SDZ sulfadiazine, SMX sulfamethoxazole, SMZ sulfamethazine, SCP sulfachloropyridazine, STZ sulfathiazole, SCT sulfacetamide, AcSMX N^4-acetylsulfamethoxazole, AcSDM N^4-acetylsulfadimethoxine, AcSDZ N^4-acetylsulfadiazine, EC$_{50}$ Median effective concentration (mg/L), LC$_{50}$ Median lethal concentration (mg/L), NOEC No-observed-effect concentration (mg/L), LOEC lowest-observed-effect concentration (mg/L), h hours, min minutes, d days

[a]Out of the measured range

[b]Millimol (mM)

final toxic effects on the algae, making the interpretation of the toxic data more complex. Eguchi et al. [122] considered the synergetic effects of combined SAs (SDM, SMX, and SDZ) with their acetylated metabolites and demonstrated that inhibition of the algae growth in the presence of the metabolites was enhanced. Antibiotics in general are known to have antichloroplastic properties due to the cyanobacterial nature of the plastids, being consequently susceptible as potential antibiotic targets. SDZ and SMX exposure led to the reduction of chlorophyll synthesis in blue-green algae [121], and similar effects were observed for aquatic higher plant *Lemna gibba* in the presence of SMX and SDM, of which SMX was especially toxic [129]. Besides algae, there are typical indicator species for toxicity at the different trophic levels. The most commonly used are marine luminescent bacterium *Vibrio fischerii*, freshwater zooplankton invertebrates *Daphnia magna* and *Moina macrocopa*, and fish such as Japanese medaka (*Oryzias latipes*).

Abiotic factors (e.g., UV irradiation) may also be influential in the potential ecotoxicity of SAs. The different transformation products of SAs should also be taken into account when evaluating their ecotoxicity. If these generated products are biodegradable, they can be removed during wastewater treatment using biological methods. If they are persistent or not readily biodegradable, risks of ecotoxicity should be considered. Under natural sunlight, the toxicity of SAs (STZ, SMX, and SMZ) against *D. magna* increased considerably [130], especially for STZ, which turned out to be 7.8-fold more toxic. It must be mentioned that STZ was the most vulnerable SA to photodegradation (39% degradation of the parent compound after 48-h exposure), suggesting that more toxic by-products were produced during photodegradation. On the contrary, whereas sulfacetamide, STZ, SMX, and SDZ were toxic for *Chlorella vulgaris*, their photodegradation products were less toxic against the algae, and showed both inhibitory and stimulatory effects on the algae cultures [131]. Ozonation by-products of SMX showed acute toxicity against *Daphnia magna* and the blue-green algae *Pseudokirchneriella subcapicata* [132]. For SMX ozonation by products, it was also found that they can be bioactive and change the morphology of mammalian cultured cells, also with no further negative effect [133].

4 Conclusions

The presence of SAs in the environment is a reality. Whether or not the concentrations detected pose a real health risk to humans is still unclear. Toxicology data and legal regulation so far focus exclusively in food products, neglecting the possibility of exposure via water or air. Regarding the potential adverse effects against the different ecosystems exposed, the ecotoxicological studies up to date show diverse results often for the same single SA and organism. However, the scarcity of available environmental exposure data and the lack of studies performed with realistic environmental concentrations (laboratory scale experiments) makes these results insufficient. The estimations of SAs ecological risks reported up to datc are based on limited information and future research studies should be focused in the improvement of data regarding fate and ecotoxicity of this substances.

Acknowledgements This work has been funded by the Spanish Ministry of Science and Innovation through the projects CEMAGUA (CGL2007-64551/HID) and SCARCE (Consolider Ingenio 2010 CSD2009-00065). MJ García acknowledges AGAUR (Generalitat de Catalunya, Spain) for economic support through an FI predoctoral grant.

References

1. Ternes T, Joss A (2006) Human Pharmaceuticals, Hormones and Fragrances: The Challenge of micropollutants in urban water management. IWA Publishing
2. Sarmah AK, Meyer MT, Boxall ABA (2006) Chemosphere 65:725
3. Besse JP, Garric J (2008) Toxicol Lett 176:104
4. Halling-Sörensen B, Nors Nielsen S, Lanzky PF, Ingerslev F, Holten Lützhöft HC, Jörgensen SE (1998) Chemosphere 36:357
5. Kolpin DW, Furlong ET, Meyer MT, Thurman EM, Zaugg SD, Barber LB, Buxton HT (2002) Environ Sci Technol 36:1202
6. Halling-Sorensen B, Nors Nielsen S, Lanzky PF, Ingerslev F, Holten Luftzhoft HC, Jorgensen SE (1998) Chemosphere 36:357
7. Kümmerer KE (2004) Pharmaceuticals in the environment – Sources, fate, effects, and risks. 2nd edn. Springer, Berlin.
8. Thuy H, Nga L, Loan T (2011) Environ Sci Pollut Res 18:835
9. Chafer-Pericas C, Maquieira T, Puchades R, Company B, Miralles J, Moreno A (2010) Aquacult Res 41:e217
10. Hamscher G, Priess B, Nau H (2006) Arch Lebensmittelhyg 57:97
11. Pan X, Qiang Z, Ben W, Chen M (2011) Chemosphere 84:695
12. Wei R, Ge F, Huang S, Chen M, Wang R (2011) Chemosphere 82:1408
13. Pérez-Trallero E, Iglesias L (2003) Tetraciclinas, sulfamidas y metronidazol 21:520
14. Miao XS, Bishay F, Chen M, Metcalfe CD (2004) Environ Sci Technol 38:3533
15. Hamscher G, Priess B, Nau H (2006) Untersuchung von Teichwässern und -sedimenten in niedersächsischen aquakulturen im sommer 2005 auf sulfonamide und tetracycline 57:97
16. Cai-Ming T, Qiu-Xin H, Yi-Yi Y, Xian-Zhi P (2009) Fenxi Huaxue/Chinese J Anal Chem 37:1119
17. Díaz-Cruz MS, García-Galán MJ, Barceló D (2008) J Chromatogr A 1193:50
18. Tamtam F, Le Bot B, Dinh T, Mompelat S, Eurin J, Chevreuil M, Bonté P, Mouchel JM, Ayrault S (2011) J Soils Sediments 1
19. Kwon JW (2011) Bull Environ Contam Toxicol 87:40
20. García-Galán MJ, Garrido T, Fraile J, Ginebreda A, Díaz-Cruz MS, Barceló D (2010) J Hydrol 383:93
21. Campagnolo ER, Johnson KR, Karpati A, Rubin CS, Kolpin DW, Meyer MT, Esteban JE, Currier RW, Smith K, Thu KM, McGeehin M (2002) Sci Total Environ 299:89
22. Jia A, Hu J, Wu X, Peng H, Wu S, Dong Z (2011) Environ Toxicol Chem 30:1252
23. Massey LB, Haggard BE, Galloway JM, Loftin KA, Meyer MT, Green WR (2010) Ecol Eng 36:930
24. Brown KD, Kulis J, Thomson B, Chapman TH, Mawhinney DB (2006) Sci Total Environ 366:772
25. Joss A, Zabczynski S, Göbel A, Hoffmann B, Löffler D, McArdell CS, Ternes TA, Thomsen A, Siegrist H (2006) Water Res 40:1686
26. Ahel M, Mikac N, Cosovic B, Prohic E, Soukup V (1998) Water Sci Technol 37:203

27. Schwarzbauer J, Heim S, Brinker S, Littke R (2002) Water Res 36:2275
28. Topp E, Monteiro SC, Beck A, Coelho BB, Boxall ABA, Duenk PW, Kleywegt S, Lapen DR, Payne M, Sabourin L, Li H, Metcalfe CD (2008) Sci Total Environ 396:52
29. Haller MY, Müller SR, McArdell CS, Alder AC, Suter MJF (2002) J Chromatogr A 952:111
30. Parfitt KE (1999) The complete drug reference. 32nd edn. Pharmaceutical Press, London, UK
31. Motoyama M, Nakagawa S, Tanoue R, Sato Y, Nomiyama K, Shinohara R (2011) Chemosphere 84:432
32. Hamscher G, Pawelzick HT, Hoper H, Nau H (2005) Environ Toxicol Chem 24:861
33. Boxall ABA, Kolpin DW, Halling-Sorensen B, Tolls J (2003) Environ Sci Technol 37:286A
34. Schauss K, Focks A, Heuer H, Kotzerke A, Schmitt H, Thiele-Bruhn S, Smalla K, Wilke BM, Matthies M, Amelung W, Klasmeier J, Schloter M (2009) TRAC-Trend Anal Chem 28:612
35. Kotzerke A, Sharma S, Schauss K, Heuer H, Thiele-Bruhn S, Smalla K, Wilke BM, Schloter M (2008) Environ Pollut 153:315
36. Boxall ABA, Blackwell P, Cavallo R, Kay P, Tolls J (2002) Toxicol Lett 131:19
37. Kay P, Blackwell PA, Boxall ABA (2004) Environ Toxicol Chem 23:1136
38. Kay P, Blackwell PA, Boxall ABA (2005) Environ Pollut 134:333
39. Diaz-Cruz MS, Garcia-Galan MJ, Barcelo D (2008) J Chromatogr A 1193:50
40. Batt AL, Snow DD, Aga DS (2006) Chemosphere 64:1963
41. Lindsey ME, Meyer M, Thurman EM (2001) Anal Chem 73:4640
42. Sacher F, Lang FT, Brauch HJ, Blankenhorn I (2001) J Chromatogr A 938:199
43. Blackwell PA, Lutzhoft HCH, Ma HP, Halling-Sorensen B, Boxall ABA, Kay P (2004) J Chromatogr A 1045:111
44. Diaz-Cruz MS, Barcelo D (2006) Anal Bioanal Chem 386:973
45. Karthikeyan KG, Meyer MT (2006) Sci Total Environ 361:196
46. Batt AL, Aga DS (2005) Anal Chem 77:2940
47. Watanabe N, Bergamaschi BA, Loftin KA, Meyer MT, Harter T (2010) Environ Sci Technol 44:6591
48. García Galán MJ, Díaz-Cruz MS, Barceló D (2010) Talanta 81:355
49. García-Galán MJ, Díaz-Cruz MS, Barceló D (2011) Environ Int 37:462
50. Gobel A, McArdell CS, Joss A, Siegrist H, Giger W (2007) Sci Total Environ 372:361
51. Gobel A, McArdell CS, Suter MJF, Giger W (2004) Anal Chem 76:4756
52. Gros M, Petrovic M, Ginebreda A, Barceló D (2010) Environ Int 36:15
53. Reemtsma T, Weiss S, Mueller J, Petrovic M, González S, Barcelo D, Ventura F, Knepper TP (2006) Environ Sci Technol 40:5451
54. Lapen DR, Topp E, Metcalfe CD, Li H, Edwards M, Gottschall N, Bolton P, Curnoe W, Payne M, Beck A (2008) Sci Total Environ 399:50
55. Sabourin L, Beck A, Duenk PW, Kleywegt S, Lapen DR, Li H, Metcalfe CD, Payne M, Topp E (2009) Sci Total Environ 407:4596
56. García Galán MJ, Frömel T, Müller J, Peschka M, Knepper T, Díaz Cruz S, Barceló D (2011) Anal Bioanal Chem (submitted)
57. García Galán MJ, Díaz-Cruz MS, Barceló D (2011) Water Res (submitted)
58. Gros M, Petrovic M, Ginebreda A, Barcelo D (2010) Environ Int 36:15
59. García Galán MJ, Díaz-Cruz MS, Barceló D (2011) J Hazard Mat (submitted).
60. Garcia-Galan MJ, Villagrasa M, Silvia Diaz-Cruz M, Barcelo D (2010) Anal Bioanal Chem 397:1325
61. Tambosi JL, de Sena RF, Favier M, Gebhardt W, Jose HJ, Schroder HF, Moreira R (2010) Desalination 261:148
62. Tambosi JL, de Sena RF, Gebhardt W, Moreira R, Jose HJ, Schroder HF (2009) Ozone-Sci Eng 31:428
63. Radjenovic J, Petrovic M, Barceló D (2007) Anal Bioanal Chem 387:1365
64. Dodd MC, Huang CH (2004) Environ Sci Technol 38:5607
65. Huber MM, Canonica S, Park GY, Von Gunten U (2003) Environ Sci Technol 37:1016
66. Nakada N, Shinohara H, Murata A, Kiri K, Managaki S, Sato N, Takada H (2007) Water Res 41:4373
67. Kosutic K, Dolar D, Asperger D, Kunst B (2007) Sep Purif Technol 53:244

68. Garoma T, Umamaheshwar SK, Mumper A (2010) Chemosphere 79:814
69. Zwiener C (2007) Anal Bioanal Chem 387:1159
70. Thiele-Bruhn S, Aust MO (2004) Arch Environ Contam Toxicol 47:31
71. Kolpin DW, Skopec M, Meyer MT, Furlong ET, Zaugg SD (2004) Sci Total Environ 328:119
72. Zheng S, Qiu X, Chen B, Yu X, Liu Z, Zhong G, Li H, Chen M, Sun G, Huang H, Yu W, Freestone D (2011) Chemosphere 84(11): 1677
73. Lin AYC, Tsai YT (2009) Sci Total Environ 407:3793
74. Madureira TV, Barreiro JC, Rocha MJ, Rocha E, Cass QB, Tiritan ME (2010) Sci Total Environ 408:5513
75. Tamtam F, Mercier F, Le Bot B, Eurin J, Tuc Dinh Q, Clement M, Chevreuil M (2008) Sci Total Environ 393:84
76. Kim SC, Carlson K (2007) Environ Sci Technol 41:50
77. Tamtam F, Mercier F, Eurin J, Chevreuil M, Le Bot B (2009) Anal Bioanal Chem 393:1709
78. Wiegel S, Aulinger A, Brockmeyer R, Harms H, Löffler J, Reincke H, Schmidt R, Stachel B, von Tümpling W, Wanke A (2004) Chemosphere 57:107
79. Yang JF, Ying GG, Zhao JL, Tao R, Su HC, Chen F (2010) Sci Total Environ 408:3424
80. Löffler D, Römbke J, Meller M, Ternes TA (2005) Environ Sci Technol 39:5209
81. Ashton D, Hilton M, Thomas KV (2004) Sci Total Environ 333:167
82. Muñoz I, Lopez-Doval JC, Ricart M, Villagrasa M, Brix R, Geiszinger A, Ginebreda A, Guasch H, Lopez De Alda MJ, Romani AM, Sabater S, Barcelo D (2009) Environ Toxicol Chem 28:2706
83. Goñi-Urriza M, Capdepuy M, Arpin C, Raymond N, Pierre Caumette CQ (2000) Appl Environ Microbiol 66:125
84. Hoa PTP, Managaki S, Nakada N, Takada H, Shimizu A, Anh DH, Viet PH, Suzuki S (2011) Sci Total Environ 409:2894
85. Luo Y, Mao D, Rysz M, Zhou Q, Zhang H, Xu L, Alvarez PJJ (2010) Environ Sci Technol 44:7220
86. Costanzo SD, Murby J, Bates J (2005) Mar Pollut Bull 51:218
87. Storteboom H, Arabi M, Davis JG, Crimi B, Pruden A (2010) Environ Sci Technol 44:7397
88. Storteboom H, Arabi M, Davis JG, Crimi B, Pruden A (2010) Environ Sci Technol 44:1947
89. Barker-Reid F, Fox EM, Faggian R (2010) J Water Health 8:521
90. Heuer H, Schmitt H, Smalla K (2011) Curr Opin Microbiol 14:236
91. Schmitt H, Haapakangas H, van Beelen P (2005) Soil Biol Biochem 37:1882
92. Halling-Sorensen B, Sengelov G, Ingerslev F, Jensen LB (2003) Arch Environ Contam Toxicol 44:7
93. Sengelov G, Agerso Y, Halling-Sorensen B, Baloda SB, Andersen JS, Jensen LB (2003) Environ Int 28:587
94. Kay P, Blackwell PA, Boxall ABA (2005) Chemosphere 60:497
95. Kay P, Blackwell PA, Boxall ABA (2005) Chemosphere 59:951
96. Stoob K, Singer HP, Mueller SR, Schwarzenbach RP, Stamm CH (2007) Environ Sci Technol 41:7349
97. Tappe W, Zarfl C, Kummer S, Burauel P, Vereecken H, Groeneweg J (2008) Chemosphere 72:836
98. Zarfl C, Matthies M, Klasmeier J (2008) Chemosphere 70:753
99. EMEA CHMP (2006) Guideline on the Environmental Risk Assessment of Medicinal Products for Human Use.
100. EU Comission Regulation EC 281/96
101. Capleton AC, Courage C, Rumsby P, Holmes P, Stutt E, Boxall ABA, Levy LS (2006) Toxicol Lett 163:213
102. Kools SAE, Boxall ABA, Moltmann JF, Bryning G, Koschorreck J, Knacker T (2008) Integr Environ Assess Manag 4:399
103. García Galán MJ, Díaz-Cruz MS, Barceló D (2008) Trac-Trends Anal Chem 28(6):804–819
104. Boxall ABA, Johnson P, Smith EJ, Sinclair CJ, Stutt E, Levy LS (2006) J Agric Food Chem 54:2288

105. Jjemba PK (2002) Agric Ecosyst Environ 93:267
106. Migliore L, Brambilla G, Cozzolino S, Gaudio L (1995) Agric Ecosyst Environ 52:103
107. Migliore L, Brambilla G, Grassitellis A, Dojmi di Delupis G (1993) Int J Salt Lake Res 2:141
108. Migliore L, Civitareale C, Brambilla G, Cozzolino S, Casoria P, Gaudio L (1997) Agric Ecosyst Environ 65:163
109. Forni C, Cascone A, Fiori M, Migliore L (2002) Water Res 36:3398
110. Dolliver H, Kumar K, Gupta S (2007) J Environ Qual 36:1224
111. Hou XL, Shen JZ, Zhang SX, Jiang HY, Coats JR (2003) J Agric Food Chem 51:7725
112. Ramirez AJ, Brain RA, Usenko S, Mottaleb MA, O'Donnell JG, Stahl LL, Wathen JB, Snyder BD, Pitt JL, Perez-Hurtado P, Dobbins LL, Brooks BW, Chambliss CK (2009) Environ Toxicol Chem 28:2587
113. Hamscher G, Pawelzick HT, Sczesny S, Nau H, Hartung J (2003) Environ Health Perspect 111:1590
114. Hirsch R, Ternes T, Haberer K, Kratz KL (1999) Sci Total Environ 225:109
115. García-Galán MJ, Garrido T, Fraile J, Ginebreda A, Díaz-Cruz MS, Barceló D (2011) Anal Bioanal Chem 399:795
116. Ginebreda A, Muñoz I, de Alda ML, Brix R, López-Doval J, Barceló D (2010) Environ Int 36:153
117. Ferrari B, Mons R, Vollat B, Fraysse B, Paxéus N, Lo Giudice R, Pollio A, Garric J (2004) Environ Toxicol Chem 23:1344
118. Grung M, Källqvist T, Sakshaug S, Skurtveit S, Thomas KV (2008) Ecotoxicol Environ Saf 71:328
119. Park S, Choi K (2008) Ecotoxicology 17:526
120. Sanderson H, Johnson DJ, Wilson CJ, Brain RA, Solomon KR (2003) Toxicol Lett 144:383
121. Lutzhoft HHC, Halling-Sorensen B, Jorgensen SE (1999) Arch Environ Contam Toxicol 36:1
122. Eguchi K, Nagase H, Ozawa M, Endoh YS, Goto K, Hirata K, Miyamoto K, Yoshimura H (2004) Chemosphere 57:1733
123. EMEA CVMP (2004). Guideline on environmental impact assessment for veterinary medicine products Phase II.
124. Oaks JL, Gilbert M, Virani MZ, Watson RT, Meteyer CU, Rideout BA, Shivaprasad HL, Ahmed S, Chaudhry MJI, Arshad M, Mahmood S, Ali A, Khan AA (2004) Nature 427:630
125. Choi KJ, Kim SG, Kim CW, Kim SH (2007) Chemosphere 66(6):977
126. Al-Ahmad A, Daschner FD, Kümmerer K (1999) Arch Environ Contam Toxicol 37:158
127. Dokianakis SN, Kornaros ME, Lyberatos G (2004) Water Sci Technol 50:341
128. Fountoulakis M, Drilla P, Stamatelatou K, Lyberatos G (2004) Water Sci Technol 50(5):335
129. Brain RA, Ramirez AJ, Fulton BA, Chambliss CK, Brooks BW (2008) Environ Sci Technol 42:8965
130. Jung J, Kim Y, Kim J, Jeong DH, Choi K (2008) Ecotoxicology 17:37
131. Baran W, Sochacka J, Wardas W (2006) Chemosphere 65:1295
132. Gómez-Ramos MDM, Mezcua M, Agüera A, Fernández-Alba AR, Gonzalo S, Rodríguez A, Rosal R (2011) J Hazard Mater 192:18
133. Yargeau V, Huot JC, Rodayan A, Rouleau L, Roy R, Leask RL (2008) Environ Toxicol 23:492

Drinking Water Disinfection By-products

Susan D. Richardson and Cristina Postigo

Abstract Drinking water disinfection by-products (DBPs) are an unintended consequence of using chemical disinfectants to kill harmful pathogens in water. DBPs are formed by the reaction of disinfectants with naturally occurring organic matter, bromide, and iodide, as well as from anthropogenic pollutants. Potential health risks of DBPs from drinking water include bladder cancer, early-term miscarriage, and birth defects. Risks from swimming pool DBP exposures include asthma and other respiratory effects. Several DBPs, such as trihalomethanes (THMs), haloacetic acids (HAAs), bromide, and chlorite, are regulated in the U.S. and in other countries, but other "emerging" DBPs, such as iodo-acids, halonitromethanes, haloamides, halofuranones, and nitrosamines, are not widely regulated. DBPs have been reported for the four major disinfectants: chlorine, chloramines, ozone, and chlorine dioxide (and their combinations), as well as for newer disinfectants, such as UV treatment with post-chlorination. Each disinfectant can produce its own suite of by-products. Several classes of emerging DBPs are increased in formation with the use of alternative disinfectants (e.g., chloramines), including nitrogen-containing DBPs ("N-DBPs"), which are generally more genotoxic and cytotoxic than those without nitrogen. Humans are exposed to DBPs not only through ingestion (the common route studied), but also through other routes, including bathing, showering, and swimming. Inhalation and dermal exposures are now being recognized as important contributors to the overall human health risk of DBPs. Analytical methods continue to be developed to measure known DBPs, and

S.D. Richardson (✉)
National Exposure Research Laboratory, U.S. Environmental Protection Agency, 960 College
Station Rd., Athens, GA, U.S.
e-mail: richardson.susan@epa.gov

C. Postigo
Institute for Environmental Assessment and Water Research - Spanish National Research Council
(IDAEA-CSIC), Barcelona, Spain
e-mail: cprqam@idaea.csic.es

D. Barceló (ed.), *Emerging Organic Contaminants and Human Health*,
Hdb Env Chem (2012) 20: 93–138, DOI 10.1007/698_2011_125,
© Springer-Verlag Berlin Heidelberg 2011, Published online: 14 December 2011

research continues to uncover new highly polar and high-molecular-weight DBPs that are part of the missing fraction of DBPs not yet accounted for. New studies are now combining toxicology and chemistry to better understand the health risks of DBPs and uncover which are responsible for the human health effects.

Keywords Chloramination, Chlorination, Chlorine dioxide, DBPs, Disinfection by-products, Drinking water, Occurrence, Ozonation, Swimming pools, Toxicity

Contents

1 Introduction

The disinfection of drinking water has been rightly hailed as a public health triumph of the twentieth century. Before its widespread use, millions of people died from waterborne diseases. Now, people in developed nations receive quality drinking water every day from their public water systems. However, chemical disinfection has also produced an unintended health hazard: the potential for cancer and reproductive and developmental effects (including early-term miscarriages and birth defects) that are associated with chemical disinfection by-products (DBPs) [1–6]. Research is being conducted worldwide to solve these important human health issues.

Chemical disinfectants, such as chlorine, ozone, chloramines, and chlorine dioxide, are used to kill harmful pathogens in drinking water, and produce safe, potable water. However, these disinfectants are also powerful oxidants, and can chemically react with the naturally occurring organic matter (NOM), mostly present from decaying leaves and other plant matter, and also with bromide and iodide salts naturally present in some source waters (rivers, lakes, and groundwaters). NOM, which is mostly comprised of humic and fulvic acids, serves as the primary precursor to DBP formation. Anthropogenic contaminants can also react with disinfectants to form contaminant DBPs. These contaminants mostly enter drinking water sources from treated wastewater, but can also enter from other sources, such as agricultural run-off. Contaminant DBPs have been reported from pharmaceuticals, antibacterial agents, estrogens, pesticides, textile dyes, bisphenol A, alkylphenol surfactants, UV filters (used in sunscreens), and diesel fuel [1]. Many times, the contaminants react directly with the disinfectants, but sometimes, it is an environmental degradation product of these initial contaminants that react to form DBPs.

Chlorine, ozone, chlorine dioxide, and chloramines are the most common chemical disinfectants in use today, and each produces its own suite of DBPs in drinking water. Two nonchemical means of disinfecting drinking water – UV light and reverse osmosis (RO) membranes – are also gaining in popularity for disinfecting water, and these technologies may hold promise in reducing levels of DBPs formed in drinking water. However, these other nonchemical disinfectants may not be without drawbacks. For example, there is some early evidence that medium-pressure UV can react with NOM to form higher levels of some DBPs after post-treatment with chlorine [7]. And, the use of RO has issues with disposal of the resulting brines left over from treatment. RO is increasingly being used at desalination plants that convert seawater into potable drinking water. Iodine point-of-use treatments were also recently investigated for formation of iodo-DBPs [8]. These point-of-use treatments are used by the military in remote locations (iodine tincture), by campers and hikers (iodine tablets), and for rural consumers in developing countries (e.g., the new Lifestraw, a reusable device that uses an iodinated anion exchange resin material with activated carbon post-treatment).

Over the last 30 years, significant research efforts have been directed toward increasing our understanding of DBP formation, occurrence, and health effects. More than 600 DBPs have now been reported in the scientific literature [9, 10]. Examples of DBP chemical classes are shown in Table 1. However, only less than 100 have been addressed either in quantitative occurrence or toxicity studies. The DBPs that have been quantified in drinking water range from parts-per-trillion (ng/L) to parts-per-billion (μg/L) levels. However, more than 50% of the halogenated DBP material (containing chlorine, bromine, or iodine) formed during the chlorination of drinking water (Fig. 1), and more than 50% of the DBPs formed during ozonation of drinking water are still not accounted for [11, 13], and nothing is known about the potential toxicity of many of the DBPs present in drinking water. Much of the previous health effects research has focused on cancer, genotoxicity, mutagenicity, or cytotoxicity. There are concerns that the types of cancer observed in animal studies (primarily liver cancer) for the regulated DBPs do not correlate with the types observed in human epidemiology studies (primarily bladder cancer). It is possible that emerging, unregulated DBPs may be responsible. It is also possible that ingestion (the primary route included in animal studies) is not the only important route of exposure.

This chapter will provide an overview of regulated and emerging, unregulated DBPs, including discussion of their occurrence and formation from different disinfectants and their toxicity. Discussions will include classical DBPs formed by reactions of disinfectants with NOM and contaminant DBPs formed by reaction of disinfectants with anthropogenic contaminants. Analytical methods used in the discovery of new DBPs and for the measurement of known DBPs will also be discussed, as well as new research investigating other routes of exposure beyond ingestion.

2   Regulated DBPs

Chloroform and other trihalomethanes (THMs) were the first DBPs identified in chlorinated drinking water in 1974 [14, 15]. Soon after their discovery, the THMs were found to cause cancer in laboratory animals [16]. As a result, they became regulated in the U.S. in 1979 [17], and later in several other countries. A few additional DBPs are now regulated in the U.S., including five haloacetic acids (HAAs), chlorite, and bromate (Fig. 2). The regulated THMs are sometimes referred to as THM4, regulated HAAs as HAA5, and the nine commonly occurring chloro-bromo-HAAs as HAA9. THMs and HAAs are formed primarily by chlorine and chloramines; chlorite is a DBP from chlorine dioxide, and bromate is mostly from ozonation. Table 2 lists the DBPs currently regulated in the U.S. and Europe, as well the current World Health Organization (WHO) guidelines.

Of the four major disinfectants used today, chlorine generally produces the highest levels of THMs and HAAs. Because drinking water-treatment plants can have difficulty in meeting the regulatory limits, many plants have changed their

Table 1 Examples of DBP chemical classes

Halogenated DBPs

Halomethanes (Dichloroiodomethane)	Haloacids (Iodoacetic acid)	Haloaldehydes (Bromochloroacetaldehyde)

Haloketones (1,1,1-Trichloropropanone) — Halonitriles (Dibromoacetonitrile) — Haloamides (Dichloroacetamide)

Halofuranones (3-Chloro-4-(dichloromethyl)-5-hydroxy-2(5H)-furanone (MX)) — Halopyrroles (2,3,5-Tribromopyrrole) — Haloquinones (2,6-Dichloro-1,4-benzoquinone)

Oxyhalides (Chlorate) — Halonitromethanes (Dibromonitromethane)

Non-Halogenated DBPs

Nitrosamines (NDMA) — Aldehydes (Formaldehyde) — Ketones (Dimethylglyoxal)

Carboxylic acids (ethanedioic acid)

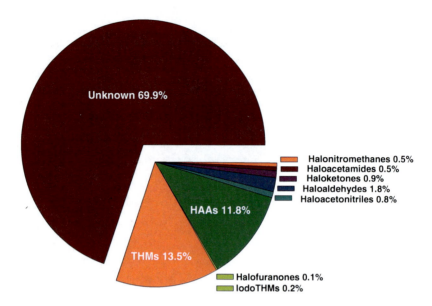

Fig. 1 Percentage of Total Organic Halogen (TOX) accounted for by quantified DBPs (data from [11, 12])

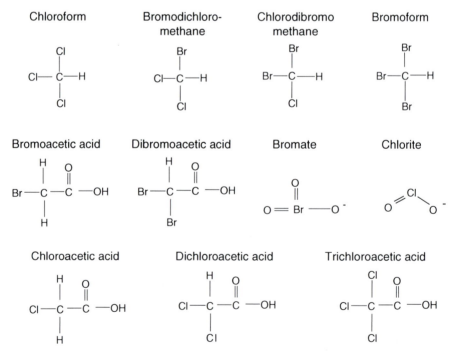

Fig. 2 Molecular structures of regulated DBPs (THMs, HAAs, bromate, and chlorite)

Table 2 DBP regulations and guidelines

U.S. EPA regulations	*MCL (mg/L)*
Total THMs (chloroform, bromodichloromethane, chlorodibromomethane, bromoform	0.080
5 Haloacetic acids (chloro-, bromo-, dichloro-, dibromo-, trichloroacetic acid)	0.060
Bromate	0.010
Chlorite	1.0
World Health Organization (WHO) guidelines	*Guideline value[a] (mg/L)*
Chloroform	0.3
Bromodichloromethane	0.06
Chlorodibromomethane	0.1
Bromoform	0.1
Carbon tetrachloride	0.004
Chloroacetic acid	0.02
Dichloroacetic acid	0.05[b]
Trichloroacetic acid	0.2
Bromate	0.01[b]
Chlorite	0.7[b]
Dichloroacetonitrile	0.02[b]
Dibromoacetonitrile	0.07
Cyanogen chloride	0.07
2,4,6-trichlorophenol	0.2
N-Nitrosodimethylamine (NMDA)	0.1
European Union Standards	*Standard value[a] (mg/L)*
Total THMs	0.1
Bromate	0.01[c]
Other regulations	*MCL (ng/L)*
NMDA	9[d], 10[e]

[a]World Health Organization (WHO) guidelines on THMs state that the sum of the ratio of the concentration of each THM to its respective guideline value should not exceed unity. WHO guidelines can be found at http://www.who.int/water_sanitation_health/dwq/gdwq3rev/en/. European Union drinking water standards can be found at www.nucfilm.com/eu_water_directive.pdf
[b]Provisional guideline value
[c]Where possible, without compromising disinfection, EU member states should strive for a lower value
[d]Ontario, Canada
[e]California, U.S.

disinfection practices. Often, the primary disinfectant is changed from chlorine to "alternative" disinfectants, including ozone, chlorine dioxide, chloramines, or UV. In some cases, chlorine is used as a secondary disinfectant following primary treatment with an alternative disinfectant, particularly for ozone, chlorine dioxide, and UV to maintain a disinfectant residual in the water distribution system. However, new issues and problems can result with changes in disinfection practices. For example, the use of ozone can significantly reduce or eliminate the formation of THMs and HAAs, but it can result in the formation of bromate, especially when elevated levels of bromide salts are present in the source waters. Bromide (and iodide) salts can be present in source waters (e.g., rivers) near coastal areas, due to salt water intrusion into the water supplies, and also in inland locations, due to "fossilized seawater," where salts from ancient seas impact surface water or groundwater. Bromate is a concern because it causes cancer in laboratory animals [18]. Several other DBPs, including nitrosamines, iodo-acids, iodo-THMs, and bromonitromethanes, can also be increased in formation with the use of alternative disinfectants. They will be discussed in detail in later sections on emerging DBPs. Differences in source water conditions, including concentrations of bromide or iodide salts, concentrations of NOM, and pH, can have a dramatic effect on the formation of various DBPs (chlorine-, bromine-, or iodine-containing) and the levels formed.

3   Emerging DBPs

3.1   Overview

Emerging DBPs beyond those that are currently regulated are becoming important. In general, brominated DBPs are now being recognized as toxicologically important because there is indication that brominated DBPs may be more carcinogenic than their chlorinated analogs, and new studies are indicating that iodinated compounds may be more toxic than their brominated analogs [19–21]. Brominated and iodinated DBPs form due to the reaction of the disinfectant (such as chlorine) with natural bromide or iodide present in source waters. Coastal cities, whose groundwaters and surface waters can be impacted by salt water intrusion, and some inland locations, whose surface waters can be impacted by natural salt deposits from ancient seas or oil-field brines, are examples of locations that can have high bromide and iodide levels. A significant proportion of the U.S. population and several other countries now live in coastal regions that are impacted by bromide and iodide; therefore, exposures to brominated and iodinated DBPs can be important. Early evidence in epidemiologic studies also gives indication that brominated DBPs may be associated with the new reproductive and developmental effects [3, 4], as well as cancer effects.

Specific DBPs that are of current interest include iodo-acids, bromonitromethanes, iodo-THMs, halooamides, halofuranones, halopyrroles, haloquinones, haloaldehydes, halonitriles, and nitrosamines. Many of these were predicted to be carcinogens [22] and were the subject of a nationwide occurrence study in the U.S., which reported the most extensive quantitative occurrence of priority, unregulated DBPs [11, 13]. In addition, many of these are nitrogen-containing DBPs (the so-called "N-DBPs"), which have recently been shown to be more genotoxic and cytotoxic than those without nitrogen [19]. N-DBPs can be increased in formation through the use of chloramination, and new research also indicates that algae and amino acids can serve as precursors in their formation [23–25].

3.2 Iodo-Acids and Iodo-THMs

Iodo-acids are a new and potentially toxicologically significant class of DBP identified as part of the U.S. Nationwide Occurrence Study [11, 12, 20] and quantified in a recent 23-city occurrence study in the U.S. [21]. Five iodo-acids have been identified in finished drinking water: iodoacetic acid, bromoiodoacetic acid, (Z)-3-bromo-3-iodopropenoic acid, (E)-3-bromo-3-iodopropenoic acid, and (E)-2-iodo-3-methylbutenedioic acid [13]. Iodo-acids, including diiodoacetic acid, were also recently found in waters treated with iodine [8]. They were initially discovered in chloraminated drinking water, and have been found up to 1.7 μg/L individually [21].

Iodoacetic acid is the most genotoxic DBP studied to-date in mammalian cells, approximately 2x more genotoxic than bromoacetic acid [20], which is regulated in drinking water, but rarely detected. The rank order for genotoxicity mono-haloacetic acids follows: iodo- > bromo- ≫ chloroacetic acid [20]. New research is revealing a potential mechanism for this rank order. Monohalogenated acids have been found to inhibit glyceraldehyde-3-phosphate dehydrogenase (GAPDH) activity in a concentration-dependent manner with the same rank order, and the rate of inhibition and toxic potency were highly correlated with their alkylating potential and their propensity of the halogen leaving group. Other iodo-acids have also been recently shown to be genotoxic, and iodo-acids are also highly cytotoxic in mammalian cells [21]. The rank order for genotoxicity is iodoacetic acid ≫ diiodoacetic acid > bromoiodoacetic acid > (E)-2-iodo-3-methylbutenedioic acid > (E)-3-bromo-3-iodopropenoic acid > (E)-3-bromo-2-iodopropenoic acid. Iodoacetic acid is also teratogenic, producing developmental effects (neural tube closures) in mouse embryos, at levels (nM) similar to levels that induce DNA damage in mammalian cells [26, 27].

Iodo-THMs have been known as DBPs since the mid-1970s [28] and dichloroiodomethane was even referred to as the "5th trihalomethane" (after the original four regulated THMs) [29]. They have since been measured in drinking waters treated with chlorination or chloramination [11, 21, 30, 31], with highest levels observed in chloraminated water (up to 15 μg/L individually). In chloraminated drinking water, iodo-THMs can be formed at levels comparable to the regulated THMs (THM4). In the U.S. Nationwide Occurrence Study, one location showed iodo-THMs at 81% of the THM4 levels in a chloraminated drinking water [11]. Iodo-THMs identified and measured include dichloroiodomethane, bromochloroiodomethane, dibromoiodomethane, chlorodiiodomethane, bromodiiodomethane, and iodoform. Point-of-use treatment with iodine was also recently shown to produce iodo-THMs, with highest levels observed in iodine tincture treatment [8].

Until recently, their major concern had to do with taste and odor problems in drinking water (due to a low threshold concentration of medicinal tastes and odors in drinking water – as low as 0.02–5 μg/L) [30]. It was not until 2008 that they were investigated for genotoxicity and cytotoxicity [21]. One iodo-THM (chlorodiiodomethane) is highly genotoxic in mammalian cells, and all six iodo-THMs are cytotoxic [21]. With the exception of iodoform, the iodo-THMs are less cytotoxic than the iodo-acids.

Iodo-DBPs are of concern not only for their potential health risks, but also because research indicates that they are formed at increased levels (along with iodo-THMs) in waters treated with chloramines. Chloramination has become a popular alternative to chlorination for water-treatment systems that have difficulty meeting the regulations with chlorine, and also for treatment plants with long distribution systems because chloramines can provide a more stable residual than chlorine. Chloramines are generated from the reaction of chlorine with ammonia, and it appears that the length of free chlorine contact time (before ammonia addition to form chloramines) is an important factor in the formation of iodo-DBPs [21].

Scheme 1 illustrates the reactions involved in the formation of iodo-DBPs from chloramination vs. chlorination and helps to explain their increased formation with chloramination. Analogous to the formation of brominated DBPs from naturally occurring bromide, iodo-DBPs can be formed by the reaction of disinfectants with naturally occurring iodide and NOM. With chlorine, reactions to form iodate are much faster than reactions to form other iodo-DBPs, but the corresponding reactions with monochloramine (to form iodite and iodate) are much slower, such that iodo-DBPs increase in formation [32, 33]. Because of chlorine's competing reaction to form iodate as a sink for the natural iodide, it is likely that treatment with significant free chlorine contact time before the addition of ammonia will not produce substantial levels of iodo-acids or iodo-THMs [21, 32, 33]. New research has also revealed that anthropogenic contaminants (i.e., compounds used for medical imaging) can also be a source of iodine in the

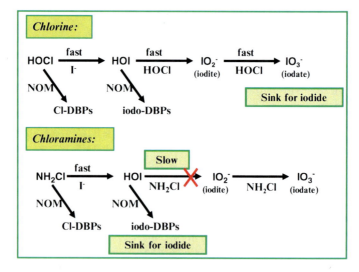

Scheme 1 Proposed mechanism for formation of iodo-DBPs with chlorine and chloramine disinfection (adapted from Bichsel and von Gunten [32, 33])

formation of iodo-DBPs [34]. This new work will be discussed in detail later in the section on Contaminant DBPs.

3.3   Halonitromethanes

Just as there are nine possible chloro-bromo haloacetic acids (HAA9) that can form in drinking water, nine halonitromethanes can be formed. Chloropicrin (trichloronitromethane) has been the most commonly measured example in this class, but has not been a concern for toxicity in drinking water. Bromonitro-methanes, however, have shown significant toxicity [35] and have been found in drinking water, particularly that treated with preozonation [11, 12, 35–37]. Bromonitromethanes are more cytotoxic and genotoxic than most DBPs currently regulated in drinking water [35]. Dibromonitromethane is more than an order of magnitude more genotoxic to mammalian cells than MX (3-chloro-4-(dichloromethyl)-5-hydroxy-2(5H)-furanone, a carcinogenic DBP), and is more genotoxic than all of the regulated DBPs, except for monobromoacetic acid. Other brominated forms are also potent in this assay. Halonitromethanes are also mutagenic in the *Salmonella* bacterial cell assay [38], with mutagenic potencies greater than that of the regulated THMs [39]. The halonitromethanes were also at least 10x more cytotoxic than the THMs, and the greater cytotoxic and mutagenic activities of the halonitromethanes was indicated to be likely due to the greater intrinsic reactivity conferred by the nitro group [39].

Bromonitromethanes are substantially increased in formation with the use of pre-ozonation before chlorine or chloramine treatment, and concentrations up to 3 µg/L individually have been reported [11, 12]. Laboratory-scale formation studies indicate that nitrite may play a role in the formation of the nitro group in these DBPs [40]. Tribromonitromethane (bromopicrin) and other trihalonitromethanes (which include bromodichloro- and chlorodibromonitromethane) require particular analytical conditions for their analysis. These compounds are thermally unstable and decompose under commonly used injection port temperatures during gas chromatography (GC) or GC/mass spectrometry (MS) analysis [41].

3.4   Nitrosamines

Nitrosamines have been of significant interest since they were discovered to be DBPs in 2002 [42, 43]. Their structures are shown in Fig. 3. *N*-Nitrosodimethylamine (NDMA) is a probable human carcinogen, and there are toxicological concerns regarding other nitrosamines. NDMA was initially discovered in chlorinated drinking waters from Ontario, Canada [44], and has since been found in other locations [42, 43, 45]. The detection of NDMA in drinking water is largely due to improved analytical techniques that have allowed its determination at low

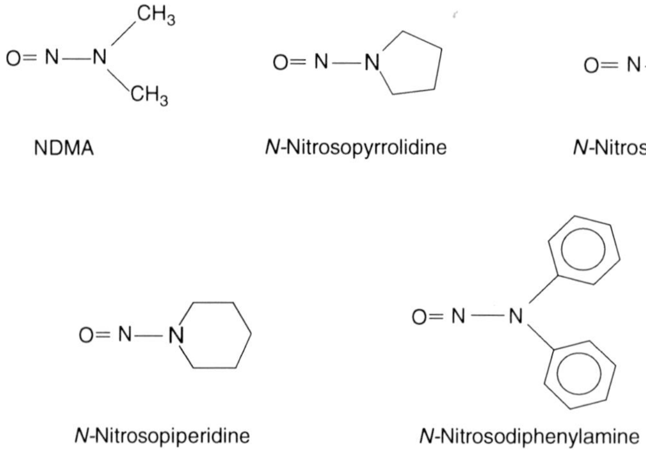

Fig. 3 Molecular structures of nitrosamine DBPs

ng/L concentrations. NDMA is generally found at highest levels in chloraminated drinking water, where the nitrogen in monochloramine (NH_2Cl) is incorporated into the structure of the NDMA by-product [42]. Chlorination can also form NDMA to some extent when nitrogen precursors are present (e.g., natural ammonia in the source water or nitrogen-containing coagulants or ion-exchange resins used in the water-treatment process) [46–48].

NDMA is regulated in California at 10 ng/L [49] and Ontario, Canada at 9 ng/L [50]. A Canadian national drinking water guideline is also under development [51], and the U.S. Environmental Protection Agency (EPA) has recently announced that they intend to regulate a group of nitrosamines in the U.S. NDMA was included in the U.S. EPA's second Unregulated Contaminant Monitoring Rule (UCMR-2), along with five other nitrosamines (N-nitrosodiethylamine, N-nitrosodibutylamine, N-nitrosopropylamine, N-nitrosomethylethylamine, and N-nitrosopyrrolidine), and national occurrence data are currently available [52]. This new national data reveals a maximum level of 530 ng/L for NDMA in chloraminated drinking water, which surpasses the previous highest level (180 ng/L) observed in chloraminated drinking water from Canada [53]. In addition, NDMA and four other nitrosamines are also on the U.S. EPA's final Contaminant Candidate List (CCL-3), a priority list of contaminants for potential future regulation in drinking water [54].

An EPA method was created for measuring NDMA and six additional nitrosamines in drinking water (EPA Method 521) [55]. This method uses GC/chemical ionization (CI)-MS/MS and enables the measurement of NDMA and six other nitrosamines (N-nitrosomethylethylamine, N-nitrosodiethylamine, N-nitroso-di-n-propylamine, N-nitroso-di-n-butylamine, N-nitrosopyrrolidine, and N-nitrosopiperidine) in drinking water at detection limits ranging from 1.2 to 2.1 ng/L. A liquid chromatography (LC)/MS/MS method [56] can also be used to measure nine nitrosamines, including N-nitrosodiphenylamine, which is thermally unstable and cannot be measured using the EPA Method.

NDMA (and other nitrosamines) can dramatically increase in concentration in distribution systems (relative to finished water at the drinking water-treatment plant). For example, an initial level of 67 ng/L in drinking water-treatment plant effluent was shown to increase to 180 ng/L in the distribution system [53]. As a result, measurements taken at water-treatment plants may substantially underestimate the public's exposure to this carcinogen.

While generally attributed to the use of chloramines or chlorine, NDMA was recently identified in ozonated drinking water from Germany [57]. An anthropogenic contaminant containing a dimethylamine group was discovered to be the precursor in its formation (discussed in more detail in the Contaminant DBP section).

3.5   Haloamides

Haloamides are formed primarily by chlorine or chloramine, and they were quantified for the first time in the Nationwide Occurrence Study. They have been measured in finished drinking waters from several U.S. states, up to 9.4 µg/L, individually [11, 13]. There is some indication that haloamides may be increased with chloramination. Because nitriles can hydrolyze to form amides [58, 59], it is possible that some of their formation is due to hydrolysis of the corresponding halonitriles, which are commonly found as DBPs. The first iodo-amide – bromoiodoacetamide – was recently identified in chloraminated drinking waters from several cities in the U.S. that had high bromide levels in their source waters [60]. This iodo-amide is highly cytotoxic and genotoxic in mammalian cells, as are other haloamides. As a class, haloamides are the most cytotoxic of all DBP classes measured to-date, and they are the second-most genotoxic DBP class, very close behind the halonitriles [19].

3.6   Halonitriles

Although they are not regulated in the U.S., haloacetonitriles (HANs) have been measured in several occurrence studies. Dichloro-, bromochloro-, dibromo-, and trichloroacetonitrile (HAN4) are the most commonly measured HAN species and have been included in a survey of 35 U.S. water utilities [61], a survey of 53 Canadian water utilities [62], and the US EPA's Information Collection Rule (ICR) effort [63]. In the ICR, HANs were found up to 41 µg/L and were generally present at 12% of the levels of the four regulated THMs. HANs are formed by treatment with chlorine, chloramine, chlorine dioxide, or ozone disinfection; plants using chloramines (with or without chlorine) had the highest levels in their finished drinking water. Several other HANs, including a number of brominated species, were also measured in the U.S. Nationwide Occurrence Study. Total HAN levels reached a maximum of 14 µg/L, and were approximately 10% of the levels of the four regulated THMs combined, although a maximum of 25% was observed. When higher bromide levels were present in the source waters, more brominated HAN species were formed. This shift in speciation was observed in another study of high-bromide waters in Israel, which also provided evidence that chlorine dioxide disinfection can form HANs (dibromoacetonitrile) [31]. Two other halonitriles, cyanogen chloride (CNCl) and cyanogen bromide (CNBr), can be formed by chlorine or chloramines, but are generally found with chloramination [61]. CNCl was measured at several chloramination plants as part of the ICR effort, with levels ranging from submicrogram per liter to 21 µg/L. CNBr can also be formed with ozonation when source waters contain natural bromide [64]. Other halonitriles, including three- and four-carbon halonitriles, have also been identified, but have not been quantified in drinking water.

As a class, halonitriles are the most genotoxic of the DBPs studied in mammalian cells [19], and they are third in cytotoxicity, similar to other N-DBPs, halo-amides, and halonitromethanes.

3.7 Halofuranones

Before it was discovered to be a drinking water DBP, MX (3-chloro-4-(dichloromethyl)-5-hydroxy-2($5H$)-furanone) was originally identified in pulp mill effluent; subsequently, it was found in chlorinated drinking water from a number of samples taken around the world. MX has both an open and closed form that is dependent on pH; the ring-opened, oxo-butenoic acid form is present at the pH of drinking water (ZMX, Fig. 4). Other analogs of MX were also later identified in chlorinated drinking water, including its geometric isomer (EMX) [65], oxidized and reduced forms of MX (ox-MX and red-MX), as well as brominated analogs (the so-called BMXs) [66]. Structures of several of these analogs are shown in Fig. 4.

Bacterial mutagenicity tests were the original cause of concern for MX, as MX was found to be a potent mutagen in the *Salmonella* Ames assay, and MX can account for as much as 20–50% of the total mutagenicity in chlorinated drinking water [67]. At the time it was identified, MX was the most mutagenic DBP ever identified in drinking water, and in 1997, it was found to be a carcinogen in rats [68]. However, the genotoxic effects in mammalian cells are relatively moderate, such that several other classes of DBPs (including iodo-acids, halonitromethanes, haloamides, and halonitriles) show greater genotoxicity [19]. The concentration of MX required to produce a genotoxic effect in vivo is usually very high, around 100 mg/kg mouse oral administration [69]. Mutagenicity studies with transgenic medaka fish showed that MX did not induce mutations in the liver (for 96 h exposures) [70].

In the few occurrence studies that had been previously carried out, measured concentrations of MX were generally 60 ng/L or lower. In 2002, Wright et al. reported levels as high as 80 ng/L of MX found in drinking waters from

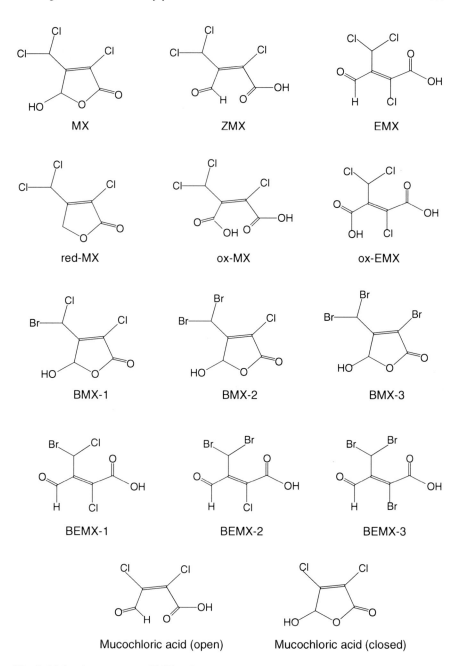

Fig. 4 Molecular structures of MX analogs

Massachusetts [71], and in the U.S. Nationwide Occurrence Study, which specifically focused on waters high in natural organic matter and/or bromide, much higher
levels were found (frequently >100 ng/L and as high as 850 ng/L) in finished
drinking waters across the U.S. [11, 12]. Halogenated furanones were highest at a
plant that disinfected with chlorine–chloramines (2.38 µg/L in plant effluent drinking water) and at a plant that disinfected with (1.02 µg/L in the distribution system).
In drinking water plant effluents, a maximum level of 0.31 µg/L was observed for
MX; maximum levels of brominated MX analogs included measurements of 0.72
and 0.81 µg/L for BEMX-1 and BEMX-2, respectively.

It is also interesting to note that the halogenated furanones were often stable in
the distribution system and in simulated distribution system tests. Previous controlled laboratory studies had suggested that halogenated furanones, particularly
MX, may not be stable in distribution systems. In at least five instances, MX levels
actually increased in concentration from the plant effluent to the distribution system
point sampled [11, 12]. Occasionally, MX levels decreased in the distribution
system, but in these instances, it was still generally present at detectable levels.

3.8   Haloaldehydes

Haloaldehydes are formed primarily with chlorine or chloramine disinfection, but they
are increased in formation with preozonation. In the Nationwide Occurrence Study,
haloaldehydes were the third largest DBP class by weight (behind THMs and HAAs)
of all the DBPs studied. Dichloroacetaldehyde was the most abundant of these
haloaldehydes, with a maximum concentration of 16 µg/L. Before this study, chloral
hydrate (trichloroacetaldehyde) was the only commonly measured haloaldehyde, and
it was included in the ICR. Chloral hydrate and monochloroacetaldehyde are
mutagenic in vitro [1], and tribromoacetaldehyde and chloral hydrate were recently
found to be genotoxic in human cells [72].

New work on the entire class of haloaldehydes indicates that many are highly
cytotoxic and genotoxic in mammalian cells [19].

3.9   Halopyrroles

In 2003, a new halogenated pyrrole – 2,3,5-tribromopyrrole (structure in Table 1) –
was identified in drinking water [31]. This represents the first time that a halogenated
pyrrole has been observed as a drinking water DBP for any disinfectant. This
halopyrrole was found in finished drinking water from a full-scale drinking water-
treatment plant in Israel that used pre-chlorination (at an initial reservoir) followed by
primary treatment with combined chlorine dioxide–chlorine or combined chlorine
dioxide–chloramine to treat a high bromide source water (approximately 2 ppm).
This identification resulted from the first study of chlorine dioxide DBPs formed

under high bromide/iodide conditions. Bromide levels in U.S. source waters generally range up to a maximum of approximately 0.5 ppm, and so to-date, this tribromopyrrole has not been identified in drinking waters from the U.S.

Mammalian cell toxicity testing revealed tribromopyrrole to be 8x more cytotoxic than dibromoacetic acid (a regulated DBP) and to have about the same genotoxic potency as MX. When the formation of tribromopyrrole was investigated using isolated humic and fulvic acid fractions collected from the source waters (as NOM precursors), tribromopyrrole was found to be formed primarily from humic acid, whereas the THMs, HAAs, and aldehydes were mostly formed from fulvic acid. It is interesting to note that a soil humic model proposed by Schulten and Schnitzer that was based on ^{13}C NMR, pyrolysis, and oxidative degradation data, includes a pyrrole group in its structure [73]. In addition, the elementary analysis (C, H, N, X) for these natural humic and fulvic acids showed a greater contribution from N in the humic acid as compared to that in the fulvic acid. In none of the samplings from this research was tribromopyrrole found in pre-chlorinated waters (with chlorine treatment only). Thus, the combination of chlorine dioxide and chlorine (or chloramines) may be necessary for its formation. It is also possible that chloramination alone may also be important for its formation.

3.10 Haloquinones

In 2010, the first haloquinone DBP was reported in drinking water – 2,6-dichloro-1,4-benzoquinone – using SPE and LC/MS/MS [74]. Quantitative structure-toxicity relationship (QSTR) analysis had predicted that haloquinones are highly toxic and may be formed during drinking water treatment. The chronic lowest observed adverse effect levels (LOAELs) of haloquinones are predicted to be in the low µg/kg body weight per day range, which is 1,000× lower than most regulated DBPs, except bromate. This new DBP was found in drinking water treated with chlorine and chloramines, as well as chloramines and UV irradiation, at levels ranging from 5.3 to 54.6 ng/L. It has a predicted LOAEL of 49 µg/kg body weight per day.

Following this initial discovery, three additional haloquinones were identified in drinking water using LC/ESI-MS/MS: 2,6-dichloro-3-methyl-1,4-benzoquinone, 2,3,6-trichloro-1,4-benzoquinone, and 2,6-dibromo-1,4-benzoquinone [75]. Following their discovery in chlorinated drinking water, they were quantified, along with 2,6-dichloro-1,4-benzoquinone. Levels ranged from 0.5 to 165 ng/L. An unusual feature about these compounds is that, using negative ion-ESI, they form $(M+H)^-$ ions through a reduction step, rather than the classic $(M–H)^-$ ions that are typically observed. The authors used tandem-MS and accurate mass measurements to confirm the identity of these unusual ions. The structures of the haloquinone DBPs are shown in Fig. 5.

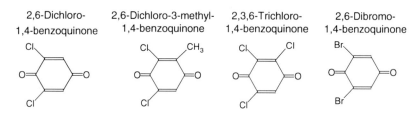

| 2,6-Dichloro-1,4-benzoquinone | 2,6-Dichloro-3-methyl-1,4-benzoquinone | 2,3,6-Trichloro-1,4-benzoquinone | 2,6-Dibromo-1,4-benzoquinone |

Fig. 5 Molecular structures of new haloquinone DBPs

4   Other DBPs

4.1   Other Haloacids

There are four bromochloro-HAAs that are not currently regulated in the U.S., bromochloroacetic acid, bromodichloroacetic acid, chlorodibromoacetic acid, and tribromoacetic acid. Many laboratories routinely measure them as part of the nine total bromochloro-HAAs (HAA9). A recent study by Singer and colleagues makes the case that measuring all nine bromochloro-HAAs is important because measuring only the five regulated ones can significantly underestimate the total exposure, especially for water systems that contain appreciable levels of bromide in their source waters. The additional four unregulated HAAs are bromine-containing species that can be found at increased levels in drinking waters that have high bromide in their source waters, and their concentrations can be similar to the five regulated HAAs. Also, because bromine-containing DBPs are generally more toxic than chlorine-containing DBPs, knowing their concentrations can be important.

Other haloacids with longer carbon chains can also be formed in drinking water, mostly with chlorine and chloramine. One of these, 3,3-dichloropropenoic acid, was included as a priority DBP measured in the U.S. Nationwide Occurrence Study [11, 12]. It was found at a maximum of 4.7 µg/L and was present in nearly all of the water-treatment plants studied. The corresponding brominated acid, 3,3-dibromopropenoic acid, has also been identified as a DBP in drinking water, as well as several other three-, four-, and five-carbon acids and diacids [10].

Two of the more unusual bromoacids include the bromo-oxoacids 3,3-dibromo-4-oxopentanoic acid and 3-bromo-3-chloro-4-oxopentanoic acid. So far, there are no quantitative data on these other brominated acids, but preliminary toxicity data indicate that they may be toxicologically important.

4.2   Haloketones

Haloketones can be formed in waters treated with chlorine, chloramines, chlorine dioxide, as well as ozone–chlorine and ozone–chloramine combinations.

Two haloketones, 1,1-dichloropropanone and 1,1,1-trichloropropanone, were measured in the ICR effort [63], where they ranged up to 10.0 and 17.0 μg/L, respectively. Other haloketones, including chloropropanone, 1,3-dichloropropanone, 1,1-dibromopropanone, 1,1,3-trichloropropanone, 1-bromo-1, 1-dichloropropanone, 1,1,1-tribromopropanone, 1,1,3-tribromopropanone, 1,1,3, 3-tetrachloropropanone, 1,1,1,3-tetrachloropropanone, and 1,1,3,3-tetrabromopropanone, were also measured in the U.S. Nationwide Occurrence Study [11, 12], and were found in drinking waters treated with a variety of disinfectants, though generally at sub-μg/L levels. To-date, they have not been investigated for toxicity, but some were predicted to cause cancer in the prioritization effort mentioned earlier [22].

4.3 Chlorate and Iodate

Chlorate is a DBP from chlorine dioxide and can also be present in chlorinated drinking water when hypochlorite bleach solutions are used for treatment (due to decomposition of the hypochlorite upon storage). In chlorine dioxide-treated drinking water, chlorate levels can approach 20% of the original chlorine dioxide dose. Chlorate concentrations in drinking water are typically much higher than other DBPs, including THMs. From the U.S. EPA ICR data, which represents the most extensive data for chlorate, the median level of chlorate was 120 μg/L at plants using chlorine dioxide for disinfection, but plants can sometimes exceed the health reference level of 210 μg/L. Recent measurements of chlorate included a study of full-scale treatment plants in Israel, in which chlorate was found up to 52 μg/L [31]; a full-scale treatment plant in Virginia, where chlorate was found at a median level of 14 μg/L [76]; and full-scale treatment plants in Quebec, where chlorate was present at a maximum of 190 μg/L [77].

Chlorate is mutagenic in Salmonella and induces chromosome aberrations and micronuclei in mammalian cells [78]. It has also been shown to induce thyroid tumors in laboratory animals [79]. The U.S. EPA has placed chlorate on the current CCL-3 [54], as well as its Unregulated Contaminant Monitoring Rule-3 (UCMR-3) [80] to collect further national data, and is currently considering it for regulation.

Iodate can be formed as a chlorination or ozonation DBP when elevated levels of natural iodide are present in the source waters [32, 33]. Unlike bromate, iodate is not a concern for toxicity because it is reduced to iodide in the body.

4.4 Aldehydes and Ketones

Several nonhalogenated aldehydes were measured in the U.S. ICR effort, including formaldehyde, acetaldehyde, glyoxal, and methyl glyoxal [63]. These aldehydes are DBPs produced primarily by ozone, although both chlorine and chlorine dioxide treatment can also form low parts per billion levels of formaldehyde. In the ICR, these aldehydes were detected at higher concentrations in water-treatment systems using ozone (up to 30.6 µg/L) than chlorine dioxide. Formaldehyde was detected at more than 50% of the treatment plants using chlorine dioxide at a mean of 5.3 µg/L and 90th percentile of 9.0 µg/L. Acetaldehyde, glyoxal, and methyl glyoxal were observed at maximum levels of 11, 16, and 6 µg/L, respectively, in ozonated drinking water, but were generally below the detection limit (5 µg/L) in chlorine dioxide-treated waters. Pentafluorobenzylhydroxylamine (PFBHA) derivatization is important for measuring these polar aldehydes and ketones by GC/MS because, without derivatization, they are almost impossible to extract from water. This derivatization process is discussed later in this chapter.

Additional aldehydes and ketones were also included in the U.S. Nationwide Occurrence Study: dimethylglyoxal (2,3-butanedione), cyanoformaldehyde, 2-butanone (methyl ethyl ketone), trans-2-hexanal, 5-keto-1-hexanal, and 6-hydroxy-2-hexanone [11, 13]. Dimethylglyoxal was the most consistently detected of these carbonyl compounds (up to 3.5 µg/L) and was found at higher levels in plants using ozone. Maximum levels of 0.3, 5.0, and 0.7 µg/L were observed for cyanoformaldehyde, 2-butanone, and trans-2-hexanal, respectively; 6-hydroxy-2-hexanone and 5-keto-1-hexanal were only detected in early stages of treatment, and not in finished waters.

4.5 Carboxylic Acids

Nonhalogenated carboxylic acids are also common DBPs from chlorine, chloramines, ozone, and chlorine dioxide [10]. In addition to halogenation reactions that can occur (primarily with chlorine and chloramine), oxidation reactions also occur, and can produce carboxylic acids. There is generally not a concern for toxicity for them, as many are naturally present in foods.

5 Discovery Research for New Highly Polar and High-Molecular-Weight DBPs

As mentioned earlier, more than 50% of the total organic halogen (TOX) formed in chlorinated drinking water remains unidentified, and much less is accounted for ozone, chloramine, and chlorine dioxide treated water. Because DBPs are typically present at nanogram per liter to microgram per liter levels, they are usually extracted into an organic solvent (with SPE or liquid–liquid extraction) and concentrated before measurement by GC or GC/MS. This means that most previous DBP research has focused on low molecular weight, volatile and semivolatile DBPs that are easy to extract from water. As a result, high-molecular-weight DBPs and highly polar DBPs are likely to be found in the "missing" DBP fraction.

Ultrafiltration (UF) studies indicate that >50% of the TOX in chlorinated drinking water is >500 Da in molecular weight [81], and new research is revealing that highly polar DBPs are also part of this "missing" fraction. For example, new LC/MS/MS research using precursor scans of bromine (*m/z* 79 and 81) and iodine (*m/z* 127) has allowed the identification of new polar compounds, such as 1,1,2-tribromo-1,2,2-tricarboxylethane, 1-bromoamino-1,2-dibromo-1,2,2-tricarboxylethane, chloroiodoacetic acid, (*E*)- and (*Z*)-iodobutenedioic acid, 4-iodobenzoic acid, 3-iodophthalic acid, 2,4-diiodobenzoic acid, 5,6-diiodosalicylic acid, and 5,6-diiodo-3-ethylsalicylic acid [82, 83]. The iodo-acids were found at higher levels in chloraminated drinking water, consistent with previous results for other iodo-acids [21].

High-molecular-weight DBPs, which are not possible to measure using GC/MS, are also being investigated using ESI-MS/MS. Most of this work is very preliminary, due to the complexity of the mass spectra obtained. ESI-MS/MS has been used to generate chlorine and bromine fragment ions that can be used to select halogenated DBPs from the complex mixture of high-molecular-weight DBPs. In addition, radiolabeled chlorine (^{36}Cl) has been used to further probe high-molecular-weight DBPs formed on chlorination of drinking water [84].

6   Contaminant DBPs

All of the DBPs previously discussed are "classical DBPs", formed by the reaction of disinfectants with natural organic matter in source waters. However, source waters are also impacted by municipal and industrial emissions [85], and recent investigations have shown that some of these water contaminants can also react with disinfectants used in drinking water treatment to form their own by-products, some of which are toxic or estrogenic. Contaminant DBPs have been reported for several classes of drugs, pesticides, personal care products, estrogens, bisphenol A, alkylphenol surfactants, and algal toxins. Most of these contaminant DBPs were found in controlled laboratory studies and not in actual drinking water, but the potential is there for their formation in drinking water treatment. It is not surprising that DBPs can form from these contaminants because many of them have activated aromatic rings that can readily react with oxidants like chlorine, chloramines, ozone, and chlorine dioxide.

6.1   Pharmaceutical DBPs

Several classes of antibiotics, e.g., tetracyclines [86], fluoroquinolones [87, 88], and β-lactams [89] were observed to react with chemical oxidants such as chlorine dioxide (ClO_2) and free chlorine. Oxidation with ClO_2 yields hydroxylated and oxygenated products in the case of tetracyclines, and leads to dealkylation, hydroxylation, and intramolecular ring closure at the piperazine moiety of the fluoroquinolones [86, 88].

Reaction of these antibiotics with chlorine mostly generated chlorinated and OH-substituted by-products [86, 87]. Unlike fluroquinolones, whose quinolone ring is left mostly intact, disinfection with ClO_2 may diminish the antibiotic capacity of tetracyclines because it leads to cleavage of the tetracyclines' ring system [86, 88]. On the other hand, oxidation of β-lactam antibiotics such as penicillin, amoxicillin, and cefadroxil with ClO_2 leads to the formation of hydroquinone and a wide range of substituted phenols [89].

Adachi and Oka investigated the formation of cyanide during the reaction of chlorine with 20 different pharmaceuticals containing nitrogen in their molecular structure [90]. High levels of cyanide were generated by chlorination of hexamine and losartan potassium aqueous solutions. Other precursors of cyanide included metronidazol, dacarbazine, and allopurinol.

The antibacterial agent sulfamethoxazole produced chlorinated and nonchlorinated DBPs when reacted with chlorine [91]. A ring-chlorinated product was formed via halogenation of the aniline moiety at sub-stoichiometric concentrations of chlorine. 3-Amino-5-methylisoxazole, sulfate, and N-chloro-p-benzoquinone imine were formed via rupture of the sulfamethoxazole sulfonamide moiety at stoichiometric excess of chlorine.

Carbadox, a veterinary antibacterial agent, also formed oxidation products when reacted with chlorine [92]. These products are believed to maintain the antibacterial activity because they kept the biologically active N-oxide group in their structure.

Chlorination of the antacid cimetidine leads to the formation of four major DBPs: cimetidine sulfoxide, 4-hydroxylmethyl-5-methyl-*1H*-imidazole, 4-chloro-5-methyl-*1H*-imidazole, and a product proposed to be either a β- or δ-sulfam. The formation of the last three products resulted from unexpected reactions and more substantial structural changes than those typically observed in chlorination [93].

The reaction of the lipid-regulator gemfibrozil with free chlorine yielded four chlorinated derivatives of this compound [94]. Chlorination of acetaminophen (paracetamol) generated 11 discernible DBPs, including the toxic compounds 1,4-benzoquinone and N-acetyl-p-benzoquinone imine and two ring chlorination products, chloro-4-acetamidophenol and dichloro-4-acetamidophenol [95].

Shen and Andrews investigated several pharmaceuticals containing dimethylamine or diethylamine in their structures as potential precursors of NDMA and N-nitrosodiethylamine during chloramination [96]. Eight out of 19 pharmaceuticals yielded molar conversions higher than 1%. Ranitidine, one of the most prescribed drugs in the world, showed the strongest potential to form NDMA, as previously reported [97]. NDMA-related compounds are also suggested to be generated when controlled drugs like amphetamine-type drugs react with chloramines [98]. Although the latter is still to be proven, two chlorinated ring products, 4-chloro-1,3-benzodioxole and 1-chloro-3,4-dihydroxybenzene were identified during the chlorination of amphetamine-type drugs [98].

Ozonation of the antiepileptic drug carbamazepine resulted in the formation of three main DBPs: 1-(2-benzaldehyde)-4-hydro-(*1H,3H*)-quinazoline-2-one, 1-(2-benzaldehyde)-(*1H,3H*)-quinazoline-2,4-dione, and 1-(2-benzoic acid)-(*1H,3H*)-quinazoline-2,4-dione [99]. Acridine, a compound with known carcinogenic properties, acridine-9-carbaldehyde, and 9-hydroxy-acridine were DBPs observed during the treatment of carbamazepine solutions with chlorine dioxide [100]. Acridine and 1-(2-benzaldehyde)-(*1H,3H*)-quinazoline-2,4-dione were also identified as major DBPs of ozonation and chlorination of oxcarbazepine, a keto analog of carbamazepine [101].

Major oxidation products of propanolol and metoprolol formed during ozonation in aqueous solution were investigated by Benner et al. [102, 103]. In the case of propanolol, the main ozonation product is a ring-opened compound with two aldehyde moieties, which results from ozone attack to the naphthalene ring [103]. Formation of aldehyde moieties was also one of the main oxidation routes during metoprolol ozonation, together with hydroxylation [102].

Recently, Duirk et al. [34] showed evidence that iodinated X-ray contrast media (ICM), such as iopamidol, constitute an iodine source to form iodo-THM DBPs, e.g., dichloroiodomethane, and iodo-acid DBPs, e.g., iodoacetic acid, in chlorinated and chloraminated drinking waters. However, the complete reaction pathway is not fully understood yet, and it is under further investigation. Chloraminated and chlorinated source waters with iopamidol were genotoxic and cytotoxic in mammalian cells. This is in agreement with the previously reported high genotoxicity and cytotoxicity of the iodo-acids and iodo-THMs [20, 21].

6.2   Estrogen DBPs

As reviewed by Pereira et al. [104], the reaction of estrogens, that is estrone, estradiol, and ethinylestradiol, with free chlorine occurs mainly via an electrophilic substitution at the *ortho* and *para* positions, which results eventually in cleavage of the aromatic structure. Several authors have reported that dichlorinated derivatives present less estrogenic activity than monochlorinated derivatives, and in most cases, estrogen DBPs are less potent in terms of estrogenicity than the parent compounds.

Molecular ozone not only reacts easily with double bonds, activated aromatic structures, or hetero-atoms, but it can also form highly reactive and nonselective free radicals, e.g., HO$^{\bullet}$. Therefore, and due to the latter reaction mechanism, some of the estrogens DBPs generated during the ozonation of estradiol water solutions are common to those formed during diverse photocatalytic processes (O_3/UV, TiO_2/UV, and photo-Fenton). In addition to forming hydroxylated derivatives from estrogens, ozone can also form dicarboxylic acids via the opening of an aromatic ring. This transformation route was also identified during the heterogeneous photocatalysis with TiO_2 of estradiol [104].

6.3   Pesticide DBPs

Oxidation of triazine herbicides with chlorine and chlorine dioxide has been widely studied [105–108]. In the case of sulfur-containing triazines, oxidation occurs mainly via cleavage of the weakened R–S–CH$_3$ bond rather than by addition of chlorine. Reactions of S-triazines with chlorine are faster than with chlorine dioxide, and form sulfoxide, sulfone, and a sulfone hydrolysis product. Chlorination with chlorine dioxide only produced sulfoxide [108]. Lopez et al. identified the formation of sulfonate esters during the chlorination of ametryn and terbutryn [106, 107]. Triazine DBPs identified by Brix et al. exhibited higher toxicities than the parent compounds [105]. Similar to triazines, clethodim, a cyclohexanedione herbicide, is oxidized by hypochlorite and chloramines to clethodim sulfoxide and then to sulfone [109].

Chlorpyrifos reacted with free chlorine to form chlorpyrifos oxon, which is more toxic than the parent compound. Both compounds further hydrolyze to a more stable product, 3,5,6-trichloro-2-pyridinol [110].

Chlorination products of glyphosate, one of the most widely used herbicide in the world, and glycine, one of the intermediates in glyphosate chlorination, were investigated by Mehrsheikh et al. [111]. Both compounds followed a similar degradation route, with the final glyphosate chlorination products identified as methanediol and other small molecules, such as phosphoric acid, nitrate, CO_2, and N_2.

Isoxaflutole is an isoxazole herbicide that, in the presence of hypochlorite, hydrolyzed to a stable and phytotoxic metabolite, diketonitrile. This intermediate further degraded to yield benzoic acid as the major end product, which is nonphytotoxic [112].

Chlorination of waters containing two phenylurea-type herbicides, isoproturon and diuron, results in the formation of THMs. The reaction of the phenylurea-type herbicide isoproturon with chlorine produced compounds that still contained the aromatic ring of the herbicide with the urea side-chain unmodified. The formation of chlorinated and brominated derivatives was related to the bromide concentration present in the water [113].

Zambodin et al. [114] studied the DBPs of the herbicide chlortoluron generated during chlorination. In this case, halogenation (chlorination) and hydroxylation reactions were the main transformation routes observed, taking place exclusively on the aromatic ring of the molecule. Xu et al. [115] reported the formation of six volatile DBPs, including chloroform, dichloroacetonitrile, 1,1-dichloropropanone, 1,1,1-trichloropropanone, dichloronitromethane, and trichloronitromethane.

Ozonation of organophosphorous pesticides led to the formation of oxon intermediates (diazooxon, methyl paraoxon, and paraoxon for diazinon, methyl parathion, and parathion, respectively) [116]. These compounds accumulated to a different extent as a function of the solution pH.

The fungicide tolylfluanide was recently shown to form a new microbial transformation product, N,N-dimethylsulfamide, which subsequently reacts with ozone to form the carcinogenic NDMA [57]. This was discovered after high ng/L levels of

NDMA were observed in ozonated drinking water from Germany and came as a surprise because ozone does not form NDMA by reaction with natural organic matter. The chlorination products of *N,N*-dimethylsulfamide have not been investigated yet.

6.4   Personal Care Product DBPs

Parabens are widely used as preservatives in the cosmetic and pharmaceutical industries and also as food additives, due mainly to their bactericidal and fungicidal properties. Terasaki and Makino identified 14 monochloro- and dichloro-parabens formed by chlorination of parabens [117]. Ozonation of parabens in aqueous solutions produced paraben DBPs mainly through hydroxylation of their aromatic ring and/or their ester chain [118].

Chlorination and chloramination of a widely used antibacterial additive, triclosan, which is used in many household personal care products, results in the formation of chloroform, 5,6-dichloro-2-(2,4-dichlorophenoxy)phenol, 4,5-dichloro-2-(2,4-dichlorophenoxy)phenol, 4,5,6-trichloro-2-(2,4-dichlorophenoxy)phenol, 2, 4-dichlorophenol, and 2,4,6-trichlorophenol [119]. The reaction of triclosan with monochloramine is slow, however, compared to chlorine [120]. The chlorophenoxyphenols are formed via bimolecular electrophilic substitution of triclosan.

Two UV filters (used to block UV-rays in sunscreens and other products), octyl-*p*-methoxycinnamate and octyl-dimethyl-*p*-aminobenzoate, reacted with chlorine, producing chlorine-substituted compounds as intermediates that finally cleaved to smaller ester products [121]. Some of the identified octyl-*p*-methoxycinnamate DBPs showed weak mutagenic properties. Chlorinated and brominated intermediates were formed during chlorination of 2-ethylhexyl-4-(dimethylamine)benzoate and 2-hydroxy-4-methoxybenzophenone, with trichloromethoxyphenol the most abundant DBP [122].

Chlorine DBPs of the polycyclic musks 6-acetyl-1,1,2,4,4,7-hexamethyltetraline (AHTN) and 1,3,4,6,7,8-hexahydro-4,6,6,7,8,8-hexamethylcyclopenta-γ-2-benzopyran (HHCB), which are widely used fragrances in cosmetics, daily care products, and cleaning products for household and industry, were investigated by Kuhlich et al. [123]. This study evidenced chlorination of HHCB as a potential new source of HHCB-lactone in the environment, other than biological transformation.

Terpenoid DBPs were investigated by Joll et al. [124] and Qi et al. [125]. The main ozonation product of 2-methylisoborneol was camphor, which was further oxidized to formaldehyde, acetaldehyde, propanal, buntanal, glyoxal, and methyl glyoxal [125]. Chlorination of β-carotene, retinol, β-ionone, and geranyl acetate resulted in the formation of THMs [124].

6.5 Alkylphenol Surfactant and Bisphenol A DBPs

Alkylphenol ethoxylate surfactants are widely used in laundry detergents. Chlorination of these compounds results in the formation of halogenated nonylphenolic DBPs, most of them brominated acidic compounds [126].

Bisphenol A, a compound highly used in the production of epoxy resins and polycarbonate plastics, forms monochloro-, dichloro, trichloro-, and tetrachloro derivatives when chlorinated [127]. Its reaction with ozone produces as major transformation products, catechol, orthoquinone, muconic acid derivatives of bisphenol A, benzoquinone, and 2-(4-hydroxyphenyl)-propan-2-ol [128].

6.6 Algal Toxin DBPs

Cyanobacterial toxins are toxins produced by certain species of blue-green algae that have become a major environmental and public health concern. The behavior of cyanotoxins during chlorination treatment has been recently reviewed by Merel et al. [129]. Chlorination DBPs have been reported only for the hepatotoxins microcystin-LR and cylindrospermopsin. Other cyanotoxins, such as nodularins, saxitoxins, and anatoxins, have yet to be investigated. Different isomers of six chlorination products of microcystin-LR have been characterized: dihydroxy-microcystin, monochloro-microcystin, monochloro-hydroxy-microcystin, monochloro-dihydroxy-microcystin, dichloro-dihydroxy-microcystin, and trichloro-hydroxy-microcystin. Only two chlorination DBPs have been reported so far for cylindrospermopsin: 5-chloro-cylindrospermopsin and cylindrospermopsic acid [129]. Chlorination of microcystin, cylindrospermopsin, and nodularins seems to reduce the mixture toxicity; however, this aspect has not been extensively studied [129].

7 Human Exposure

New research also indicates that exposures from other activities, including showering, bathing, and swimming in chlorinated swimming pools can increase exposures to certain DBPs [130–147]. DBPs are not only ingested by drinking the water, but some can also be inhaled or can penetrate the skin [130–132, 134, 135]. In particular, volatile DBPs that easily transfer from the water to the air (including THMs) can be inhaled during showering or visiting an indoor chlorinated swimming pool – either through active swimming or from sitting near the pool, breathing in the pool vapors. THMs, HAAs, and haloketones have been measured in human blood, urine, or exhaled breath after showering, bathing, or swimming [130, 132, 134, 135, 141, 142, 145, 148]. These exposure routes are now being recognized in human exposure and human epidemiologic studies.

Recent results indicate that these other exposure routes may increase the risk of bladder cancer [149]. There is also new evidence that genetic susceptibility may play a role in bladder cancer. A recent epidemiologic study conducted in Spain revealed that people who carry a particular glutathione S-transferase zeta-1 (*GSTZ1*) polymorphism and are missing one or both copies of glutathione S-transferase theta-1 (*GSTT1*) were particularly susceptible to bladder cancer when exposed to >49 μg/L THMs in drinking water [6]. Approximately 29% of the Spanish study population had this genetic susceptibility, and approximately 25% of the U.S. population would also have this genetic susceptibility.

Chlorinated swimming pool exposures have also been linked with respiratory effects, including asthma [140, 150–152]. Trichloramine, which is formed by the reaction of chlorine with urea (from sweat and urine), has been suspected in these cases of asthma. In addition to sweat and urine, pool waters also contain other DBP precursors, such as skin cells, hair, and lotions/sunscreens.

To-date, only two efforts to comprehensively identify DBPs in swimming pools have been reported. In the first, 19 DBPs were identified in outdoor swimming pools [136]. In the second, >100 DBPs were identified in indoor chlorinated and brominated pools, including many nitrogenous DBPs (haloamides, halonitriles, haloanilines, haloamines, haloanisoles, and halonitro-compounds), likely due to the nitrogen-containing precursors from swimmers (urine, sweat, etc.) [143]. Trichloramine and THMs were also measured in the pool air [143]. Nitrosamines have been measured in chlorinated pools and hot tubs, up to a maximum of 429 ng/L [137]. Levels observed were up to $500\times$ greater than the level (0.7 ng/L) associated with a one in a million lifetime cancer risk. Volatile DBPs, such as trichloramine, dichloromethylamine, and dichloroacetonitrile, have also been measured in pool waters using membrane introduction mass spectrometry (MIMS) [133]. Brominated DBPs from sunscreens have been reported [136], as have DBPs from the reaction of chlorine with parabens used in lotions, cosmetics, and sunscreens [117].

The mutagenicity, genotoxicity, and cytotoxicity of swimming pool waters have recently been reported [143, 144, 146]). One study showed that pool waters were significantly more toxic than their tap water sources [146]. Because THM concentrations are similar between tap waters and pool waters, using THMs to monitor exposure in epidemiological studies may not be the best metric. Pools treated with a combination of UV light and chlorine disinfection indoors, or outdoor sunlight exposure exhibited lower cytotoxicity than their indoor counterparts disinfected with chlorine [146].

8   Combining Chemistry with Toxicology

More studies are combining DBP identification/measurement efforts with toxicology to understand their potential health effects. For example, a large integrated multidisciplinary study (called the Four Lab Study) was recently published [13, 153, 154]. This effort involved the collaboration of chemists, toxicologists, engineers, and risk assessors from the four National Research Laboratories of the U.S. EPA, as well as collaborators from academia and the water industry. For this study, a new procedure using reverse osmosis was developed for producing chlorinated drinking water concentrates for animal toxicology experiments. DBPs were then comprehensively identified (resulting in the identification of >100 DBPs), and 75 priority and regulated DBPs were quantified to assess what DBPs the animals were exposed to. An extensive battery of in vivo and in vitro toxicity assays were used, with an emphasis on reproductive and developmental effects. When the NOM was concentrated first and disinfected with chlorine afterward, DBPs (including volatiles and semivolatiles) were formed and maintained in a water matrix suitable for animal studies. DBPs were relatively stable over the course of the animal studies (125 days) with multiple chlorination events, and a significant proportion of the TOX was accounted for through a comprehensive identification approach. Many DBPs were reported for the first time, including previously undetected and unreported haloacids and haloamides. The new concentration procedure not only produced a concentrated drinking water suitable for animal experiments but also provided a greater TOC concentration factor (136x), enhancing the detection of trace DBPs that are often below detection using conventional approaches.

9 Analytical Methods for Identifying and Quantifying DBPs

Experiments to identify disinfection by-products (DBPs) have been carried out using two different procedures. In the first, natural waters (e.g., river, lake) are reacted with the disinfectant, either in a pilot plant, an actual treatment plant, or in a controlled laboratory study. In the second type of procedure, aquatic humic material is isolated and reacted with the disinfectant in purified water in a controlled laboratory study. This latter type of study is relevant because humic material is an important precursor of THMs and other DBPs. Aquatic humic material is present in nearly all natural waters, and isolated humic material reacts with disinfectants to produce most of the same DBPs found from natural waters. Because DBPs are typically formed at low levels (ng/L-μg/L), samples are usually concentrated to allow for DBP detection. Concentration methods that are commonly used include solid phase extraction (SPE), solid phase microextraction (SPME), liquid–liquid extraction, and XAD resin extraction (for larger quantities of water) [9].

9.1 GC/MS

GC/MS was the primary tool for identifying the first DBPs, and it remains an important tool for measuring and identifying new DBPs. Large mass spectral libraries (NIST and Wiley databases, which contain >200,000 spectra) enable rapid identifications. When DBPs are not present in these databases, high-resolution

MS, chemical ionization-MS, and sometimes GC/infrared spectroscopy (IR) have been used with GC/MS to obtain structural information. Examples of the use of GC/ MS for identifying new DBPs include the recent identification of iodo-acids. The iodo-acids were discovered in drinking water treated with chloramination through the use of full-scan GC/MS on the methylated extracts. Empirical formula information for both the molecular ions and the fragment ions was obtained by high-resolution electron ionization (EI)-MS, and the spectra were interpreted to yield tentative identifications of five new iodo-acids (iodoacetic acid, bromoiodoacetic acid, (*E*)-3-bromo-3-iodopropenoic acid, (*Z*)-3-bromo-3-iodopropenoic acid, and (*E*)-2-iodo-3-methylbutenedioic acid). Structural assignments were then confirmed by the match of mass spectra and GC retention times to authentic chemical standards, several of which had to be synthesized.

GC/MS(/MS) is also popular for quantifying DBPs. Selected ion monitoring (SIM) or multiple reaction monitoring (MRM) mode are used with GC/MS and GC/ MS/MS, respectively, to maximize the sensitivity and provide low detection limits. Some EPA Methods utilize GC/MS, including EPA Method 524.2, which uses GC/ EI-MS for THM analysis [155], and EPA Method 521, which uses for GC/CI-MS/ MS for nitrosamine analysis [55]. In addition, many priority unregulated DBPs have been measured using GC/MS in a U.S. Nationwide Occurrence Study [11, 12].

9.2   GC/ECD

GC/electron capture detection (ECD) is also used to measure DBPs. In particular, EPA Method 552.2 and 552.3 are commonly used to measure haloacetic acids in drinking water [156, 157]. ECD is very sensitive toward halogenated compounds and allows low-level detection for HAAs (0.012–0.17 µg/L detection limits for EPA Method 552.3).

9.3   LC and UPLC/MS/MS

LC/MS/MS and ultraperformance liquid chromatography (UPLC)/MS/MS are increasingly being used to identify and quantify highly polar DBPs and probe high-molecular-weight DBPs [158]. For example, LC/MS/MS was used to discover the first haloquinone DBP found in drinking water: 2,6-dichloro-1,4-benzoquinone [74]. In addition, a new nitrosamine method was created using LC/MS/MS, and with this method, two new nitrosamine DBPs were found in drinking water – nitrosopiperidine and nitrosodiphenylamine [56]. LC/MS/MS was essential for detecting nitrosodiphenylamine, as it is thermally unstable and cannot be measured by GC/MS. Derivatizing agents, such as 2,4-dinitrophenylhydrazine (DNPH), have also been used with LC/MS to enable the detection of highly polar DBPs; these are discussed in the later section on derivatizing agents.

The presence of high-molecular-weight DBPs had been indicated in research using UF membranes and TOX analysis. This research revealed that >50% of the total halogenated material in chlorinated drinking water may be >1,000 Da in molecular size [81]. Subsequent LC/ESI-MS/MS studies have been used to probe its chemical composition [159]. ^{36}Cl-labeled HOCl (aqueous chlorine) has also been used to react with NOM to enhance the detection of chlorine-containing DBPs in the high-molecular-weight fractions [84, 159]. Results revealed a highly dispersed molecular weight distribution, and an average molecular mass of 2,000 Da.

Precursor ion scans of chlorine (*m/z* 35), bromine (*m/z* 79 and 81), and iodine (*m/z* 127) have also been used to target chlorinated, brominated, and iodinated DBPs, respectively [82, 83, 160]. For example, precursor scans of *m/z* 127 were used with UPLC/ESI-MS/MS to provide a more comprehensive picture of polar iodinated DBPs formed in drinking water [83]. This recently allowed the detection of 17 iodo-DBPs, including a few that had not been previously reported.

9.4 IC/ICP-MS and IC/ESI-MS

A few DBPs, such as bromate, chlorate, iodate, and chlorite, are present as anions in drinking water. As a result, they are not volatile and cannot be analyzed by GC/MS. They are also difficult to separate by LC, but will separate nicely using ion chromatography (IC). At neutral pH, HAAs are also anions and can be separated using IC. A number of methods have been created for these DBPs using both IC/ inductively coupled plasma (ICP)-MS and IC/ESI-MS. Pretreatment to remove interfering ions (e.g., sulfate and chloride), along with the use of a suppressor column prior to introduction into the MS interface, is beneficial for trace-level measurement.

Several EPA Methods have been created for measuring bromate, a carcinogenic DBP that is currently regulated in U.S. drinking waters at 10 μg/L. Two of these use mass spectrometry: EPA Method 321.8 and EPA Method 557. EPA Method 321.8 uses IC/ICP-MS and can achieve 0.3 μg/L detection limits [161], EPA Method 557 uses IC/ESI-MS/MS and can achieve 0.02 μg/L detection limits [162]. EPA Method 557 can also be used to measure the commonly occurring chloro-bromo-HAAs (HAA9) and dichloropropanoic acid, with detection limits ranging from 0.015 to 0.20 μg/L. Roehl et al. published a good review covering the use of IC/ESI-MS for analyzing HAAs and bromate [163], and Paull and Barron published a nice review of IC applications (including IC/ESI-MS and IC/ICP-MS) for measuring HAAs [164].

Shi and Adams recently created a rapid IC/ICP-MS method for simultaneously measuring iodoacetic acids, bromoacetic acids, iodate, and bromate in drinking water, groundwater, surface water, and swimming pool water [165]. Method detection limits were sub-μg/L for iodinated DBPs, and low-μg/L for brominated DBPs.

However, mono-, di-, and tri-chlorinated species could not be detected because the sensitivity of ICP-MS for chlorine is poor.

9.5 IC/Conductivity

Bromate has also been measured using IC with conductivity detection. For example, EPA Method 302.0 uses two-dimensional IC with suppressed conductivity detection to measure bromate at 0.12 μg/L detection limits [166]. Bromate, chlorite, and chlorate can also be measured by an earlier EPA Method (Method 300.1), which uses IC with conductivity detection [167]. Method detection limits ranging from 0.45 to 1.28 μg/L can be achieved.

9.6 MIMS

MIMS is a technique that uses a semipermeable membrane for directly introducing analytes into the mass spectrometer. This allows analytes to be measured in real-time with little or no sample preparation. MIMS has been previously used to measure the stability of CNCl in chlorinated and chloraminated drinking water [168], to quantify CNCl and CNBr in drinking water [169], to measure chloramines and chlorobenzenes in water samples [170], and investigate the mechanism and kinetics of chloroform formation in drinking water [171]. More recently, it has been used to measure volatile DBPs in indoor swimming pools [138, 172].

9.7 FAIMS-MS

High-field asymmetric waveform ion mobility spectrometry (FAIMS)-MS offers an additional degree of separation of analytes, based on the differences in the ratio of ion mobility at high electric field vs. low field [173]. When used with ESI-MS, FAIMS can significantly reduce chemical backgrounds and enhance the detection of DBPs. Low ng/L detection limits can often be achieved without preconcentration, extraction, or derivatization. ESI-FAIMS-MS has been used to measure HAAs in drinking water and human urine [174, 175]; bromate, chlorate, and iodate in drinking water [176]; and nitrosamines in drinking water [177, 178]; and in wastewater-treatment plant effluents [179].

9.8   Derivatization Techniques

For some classes of compounds, derivatizations are performed to enable their detection (Table 3). For example, methylations enable the detection and measurement of carboxylic acids (including haloacids) by GC/MS. Derivatization with *o*-(2,3,4,5,6-pentafluorobenzyl)hydroxylamine (PFBHA) is popular for the GC/MS analysis of polar aldehydes and ketones that are difficult or impossible to extract from water without derivatization. And, *N*-methyl-bis-trifluoroacetamide (MBTFA) derivatization with GC/ion trap-MS/MS was recently shown to offer improved detection limits for measuring MX (3-chloro-4-(dichloromethyl)-5-hydroxy-2(5*H*)-furanone) in drinking water [180]. Silylating agents, such as bis (trimethylsilyl)trifluoroacetamide (BSTFA) and *N*-methyltrimethylsilyltrifluoroacetamide (MSTFA), are sometimes used to derivatize alcohols/phenols to enable their detection by GC/MS. For example, BSTFA derivatization has been used recently with GC/MS to identify several DBPs formed by the chlorination of parabens (*para*-hydroxybenzoate esters), which are water contaminants used as preservatives in a wide variety of personal care products (e.g., sunscreens, bath gels, shampoos, and toothpaste) [117].

Several newer derivatization techniques have also been developed for enabling the identification of new, highly polar DBPs with GC/MS or LC/MS(/MS). For example, chloroformate derivatizing agents have been developed for extracting highly polar DBPs with multiple hydroxyl, carboxyl, and amino substituents for analysis with GC/negative chemical ionization-MS [181, 182]. DNPH [183], *O*-(carboxymethyl hydroxylamine) (CMHA) [184], and 4-dimethylamino-6-(4-methoxy-1-naphthyl)-1,3,5-triazine-2-hydrazine (DMNTH) [185] have been used with LC/MS(/MS) to extract and identify highly polar carbonyl DBPs in drinking water.

Table 3 Derivatizing agents used to identify and measure DBPs

Derivatizing agent	Target functional group	MS mode
Diazomethane	Carboxylic acids	GC/MS
BF₃/Methanol	Carboxylic acids	GC/MS
1-(Pentafluorophenyl) diazoethane	Carboxylic acids	GC/MS
Pentafluorobenzylhydroxylamine (PFBHA)	Carbonyls (aldehydes, ketones)	GC/MS
Bis(trimethylsilyl)trifluoroacetamide (BSTFA)	Alcohols, phenols	GC/MS
N-methyltrimethylsilyltrifluoroacetamide (MSTFA)	Alcohols, phenols	GC/MS
5-Chloro-2,2,3,3,4,4,5,5-octafluoropentyl chloroformate (ClOFPCF)	Alcohols, amines, carboxylic acids	GC/NCI-MS
2,4-Dinitrophenylhydrazine (DNPH)	Carbonyls (aldehydes, ketones)	LC/MS
4-Dimethylamino-6-(4-methoxy-1-naphthyl)-1,3,5-triazine-2-hydrazine (DMNTH)	Carbonyls (aldehydes, ketones)	LC/MS
O-(carboxymethyl hydroxylamine) (CMHA)	Carbonyls (aldehydes, ketones)	LC/MS

9.9   Near Real-Time Methods

Researchers continue to pursue the development of new instruments to enable real-time measurements of DBPs in drinking water, which would be a tremendous benefit to drinking water utilities and to epidemiologists, who could obtain more accurate exposure information for their studies. A recent development includes a new instrument that can selectively measure THMs and HAAs in near real-time directly from drinking water distribution systems [186]. The instrument uses a capillary membrane sampler-flow injection analyzer and is based on the fluorescence of the reaction of nicotinamide in basic solution with THMs and HAAs. The analyzer alternates sampling between two sample loops connected to a capillary membrane sampler, which discriminates between the volatile THMs and the non-volatile HAAs. Low μg/L detection limits are possible for the four regulated THMs and five regulated HAAs (chloro-, dichloro-, trichloro-, bromo-, and dibromoacetic acid), and results compare favorably to EPA Methods. This method provided automated online sampling with hourly sample analysis rates.

9.10   Total Organic Chlorine, Bromine, and Iodine

A few years ago, Minear's group at the University of Illinois pioneered the development of a method to speciate TOX, such that total organic chlorine (TOCl), total organic bromine (TOBr), and total organic iodine (TOI) could be differentiated [187]. This method involves the sorption of analytes onto activated carbon, followed by removal of inorganic analytes, combustion of the activated carbon, bubbling the combustion gas into ultrapure water, and injection of this water onto an ion chromatograph for measurement of chloride, bromide, and iodide. The original TOX measurement served a useful purpose in providing an idea of the total halogenated material formed in chlorinated and other disinfected waters, so that it could be determined how much of the halogenated DBPs were being accounted for through quantification of targeted DBPs. This measurement has been widely used and has revealed that more than 50% of the halogenated DBPs in drinking water are still not accounted for. The development of the TOCl/TOBr/TOI method allowed an even finer distinction of these DBPs, and has become an important measurement because of increased toxicity among the brominated and iodinated DBPs. This method was recently used to measure the contribution of chlorinated, brominated, and iodinated DBPs to the mixture of halogenated DBPs formed in iodine point-of-use treatments [8]. This method was also recently used to follow the formation of TOCl and TOBr over time in a kinetic study of DBPs from chlorination [188].

10 Conclusions

Through more than 30 years of research, many DBPs have been identified, and we have a greater understanding of how they are formed, as well as ways to reduce or eliminate many of them. However, despite much research, more than 50% of the halogenated DBPs in chlorinated drinking water remains unaccounted for, and much less is accounted for with ozone, chloramine, and chlorine dioxide treatment. It is especially important to investigate DBPs formed by alternative disinfectants because more water-treatment plants in the U.S. are changing from chlorine to alternative disinfectants to meet requirements of the new regulations. Beyond the three most popular alternative disinfectants (chloramines, ozone, and chlorine dioxide), there is also a trend toward nonchemical disinfection, such as UV irradiation and membrane technology. UV irradiation is sometimes presented as a DBP-free disinfectant, but it has the potential to form hydroxyl radicals in water (as ozone does), which can produce oxygen-containing DBPs and has been shown to activate NOM to make it more reactive toward chlorine. The use of membranes in desalination plants can cause shifts to brominated DBPs when the disinfectant is added (due to the considerable amount of bromide that can traverse the membrane). It will be important to continue to investigate these new treatments to determine their relative safety compared to existing treatment technologies.

In addition, it is important to continue research on contaminant DBPs. With increased drought and increased populations in many parts of the world, our rivers contain increasingly higher concentrations of anthropogenic contaminants, which can also form hazardous DBPs. It is important to study their formation and devise wastewater and drinking water-treatment methods that will remove them.

Finally, it is paramount to determine which DBPs are responsible for human health effects observed and eliminate or minimize them in drinking water. As mentioned earlier, it is still not known which DBPs are responsible for the bladder cancer observed in human epidemiologic studies or which DBPs are responsible for the reproductive/developmental effects observed. Investigating new, emerging DBPs that show a toxic response is an important element in solving this important human health issue, as is investigating human health effects from routes of exposure beyond ingestion. In this regard, it will be important to consider inhalation and dermal exposure in future toxicity studies, so it can be determined which exposure route(s) are responsible for the adverse human health effects and also which DBPs are responsible.

Acknowledgments This manuscript has been reviewed in accordance with the U.S. EPA's peer and administrative review policies and approved for publication. Mention of trade names or commercial products does not constitute endorsement or recommendation for use by the U.S. EPA. Cristina Postigo acknowledges support from the EU through the FP7 program: Marie Curie International Outgoing Fellowship.

References

1. Richardson SD, Plewa MJ, Wagner ED, Schoeny R, DeMarini DM (2007) Occurrence, genotoxicity, and carcinogenicity of regulated and emerging disinfection by-products in drinking water: a review and roadmap for research. Mutat Res 636:178–242
2. Villanueva CM, Cantor KP, Cordier S, Jaakkola JJK, King WD, Lynch CF, Porru S, Kogevinas M (2004) Disinfection byproducts and bladder cancer: a pooled analysis. Epidemiology 15(3):357–367
3. Waller K, Swan SH, DeLorenze G, Hopkins B (1998) Trihalomethanes in drinking water and spontaneous abortion. Epidemiology 9:134–140
4. Savitz DA, Singer PC, Hartmann KE, Herring AH, Weinberg HS, Makarushka C, Hoffman C, Chan R, Maclehose R (2005) Drinking water disinfection by-products and pregnancy outcome. AWWA Research Foundation, Denver, CO
5. Nieuwenhuijsen MJ, Toledano MB, Eaton NE, Fawell J, Elliott P (2000) Chlorination disinfection byproducts in water and their association with adverse reproductive outcomes: a review. Occup Environ Med 57(2):73–85
6. Cantor KP, Villanueva CM, Silverman DT, Figueroa JD, Real FX, Garcia-Closas M, Malats N, Chanock S, Yeager M, Tardon A, Garcia-Closas R, Serra C, Carrato A, Castaño-Vinyals G, Samanic C, Rothman N, Kogevinas M (2010) Polymorphisms in GSTT1, GSTZ1, AND CYP2E1, disinfection by-products, and risk of bladder cancer in Spain. Environ Health Perspect 118(11):1545–1550
7. Liu W, Cheung LM, Yang X, Shang C (2006) THM, HAA and CNCl formation from UV irradiation and chlor(amination) of selected organic waters. Water Res 40:2033–2043
8. Smith EM, Plewa MJ, Lindell CL, Richardson SD (2010) Comparison of byproduct formation in waters treated with chlorine and iodine: relevance to point-of-use treatment. Environ Sci Technol 44:8446–8452
9. Richardson SD (1998) Drinking water disinfection by-products. In: Meyers RA (ed) The encyclopedia of environmental analysis and remediation, vol 3. Wiley, New York, pp 1398–1421
10. Richardson SD (2011) Disinfection by-products: formation and occurrence in drinking water. In: Nriagu JO (ed) Encyclopedia of environmental health, vol 2. Elsevier, Burlington, M. A., pp 110–136
11. Krasner SW, Weinberg HS, Richardson SD, Pastor SJ, Chinn R, Sclimenti MJ, Onstad G, Thruston AD Jr (2006) The occurrence of a new generation of disinfection by-products. Environ Sci Technol 40:7175–7185
12. Weinberg HS, Krasner SW, Richardson SD, Thruston AD Jr (2002) The occurrence of disinfection by-products (DBPs) of health concern in drinking water: results of a nationwide DBP occurrence study. EPA/600/R02/068. Available at www.epa.gov/athens/publications/reports/EPA_600_R02_068.pdf
13. Richardson SD, Thruston AD Jr, Krasner SW, Weinberg HS, Miltner RJ, Schenck KM, Narotsky MG, McKague AB, Simmons JE (2008) Integrated disinfection by-products mixtures research: comprehensive characterization of water concentrates prepared from chlorinated and ozonated/postchlorinated drinking water. J Toxicol Environ Health A 71 (17):1165–1186
14. Rook JJ (1974) Formation of haloforms during chlorination of natural waters. Water Treat Examination 23:234–243
15. Bellar TA, Lichtenberg JJ, Kroner RC (1974) Occurrence of organohalides in chlorinated drinking waters. J Am Water Works Assoc 66(12):703–706
16. National Cancer Institute (1976) Report on the carcinogenesis bioassay of chloroform (CAS no. 67-66-3), MD: National Cancer Institute. TR-000 NTIS Rpt No PB264018, Bethesda, MD

17. U.S. Environmental Protection Agency (1979) National interim primary drinking water regulations: control of trihalomethanes in drinking water: Final rules. Fed Regist 44:68624–68705

18. Kurokawa Y, Aoki S, Matsushima Y (1986) Dose–response studies on the carcinogenicity of potassium bromate in F344 rats after long-term oral administration. J Nat Cancer Inst 77(4):977–982

19. Plewa MJ, Wagner ED, Muellner MG, Hsu KM, Richardson SD (2008) Comparative mammalian cell toxicity of N-DBPs and C-DBPs. In: Karanfil T, Krasner SW, Westerhoff P, Xie Y (eds) Occurrence, formation, health effects and control of disinfection by-products in drinking water, vol 995. American Chemical Society, Washington DC, pp 36–50

20. Plewa MJ, Wagner ED, Richardson S, Thruston ADJ, Woo YT, McKague AB (2004) Chemical and biological characterization of newly discovered iodoacid drinking water disinfection byproducts. Environ Sci Technol 38:4713–4722

21. Richardson SD, Fasano F, Ellington JJ, Crumley FG, Buettner KM, Evans JJ, Blount BC, Silva LK, Waite TJ, Luther GW, McKague AB, Miltner RJ, Wagner ED, Plewa MJ (2008) Occurrence and mammalian cell toxicity of iodinated disinfection byproducts in drinking water. Environ Sci Technol 42:8330–8338

22. Woo YT, Lai D, McLain JL, Manibusan MK, Dellarco V (2002) Use of mechanism-based structure-activity relationships analysis in carcinogenic potential ranking for drinking water disinfection by-products. Environ Health Perspect 110(suppl 1):75–87

23. Chu WH, Gao NY, Deng Y, Krasner SW (2010) Precursors of dichloroacetamide, an emerging nitrogenous DBP formed during chlorination or chloramination. Environ Sci Technol 44(10):3908–3912

24. Fang J, Yang X, Ma J, Shang C, Zhao Q (2010) Characterization of algal organic matter and formation of DBPs from chlor(am)ination. Water Res 44(20):5897–5906

25. Yang X, Fan C, Shang C, Zhao Q (2010) Nitrogenous disinfection byproducts formation and nitrogen origin exploration during chloramination of nitrogenous organic compounds. Water Res 44(9):2691–2702

26. Hunter ES III, Rogers EH, Schmid JE, Richard A (1996) Comparative effects of haloacetic acids in whole embryo culture. Teratology 54:57–64

27. Hunter ES III, Tugman JA (1995) Inhibitors of glycolytic metabolism affect neurulation-stage mouse conceptuses in vitro. Teratology 52:317–323

28. Glaze WH, Henderson JE, Smith G (1975) Analysis of new chlorinated organic compounds formed by chlorination of municipal wastewater. In: Jolley RJ (ed) Water chlorination: environmental impact and health effects. Ann Arbor Science, Ann Arbor, MI, pp 139–159

29. Thomas RF, Weisner MJ, Brass HJ (1980) The fifth trihalomethane: dichloroiodomethane. Its stability and occurrence in chlorinated drinking water. In: Jolley RL, Brungs WA, Cumming RB, Jacobs VA (eds) Water chlorination: environmental impact and health effects, vol 3. Ann Arbor Science, Ann Arbor, MI, pp 161–168

30. Cancho B, Ventura F, Galceran M, Diaz A, Ricart S (2000) Determination, synthesis and survey of iodinated trihalomethanes in water treatment processes. Water Res 34:3380–3390

31. Richardson SD, Thruston AD Jr, Rav-Acha C, Groisman L, Popilevsky I, Juraev O, Glezer V, McKague AB, Plewa MJ, Wagner ED (2003) Tribromopyrrole, brominated acids, and other disinfection byproducts produced by disinfection of drinking water rich in bromide. Environ Sci Technol 37(17):3782–3793

32. Bichsel Y, Von Gunten U (1999) Oxidation of iodide and hypoiodous acid in the disinfection of natural waters. Environ Sci Technol 33(22):4040–4045

33. Bichsel Y, Von Gunten U (2000) Formation of iodo-trihalomethanes during disinfection and oxidation of iodide-containing waters. Environ Sci Technol 34:2784–2791

34. Duirk SE, Lindell C, Cornelison CC, Kormos J, Ternes TA, Attene-Ramos M, Osiol J, Wagner ED, Plewa MJ, Richardson SD (2011) Formation of toxic iodinated disinfection by-products from compounds used in medical imaging. Environ Sci Technol 45:6845–6854

35. Plewa MJ, Wagner ED, Jazwierska P, Richardson SD, Chen PH, McKague AB (2004) Halonitromethane drinking water disinfection byproducts: chemical characterization and mammalian cell cytotoxicity and genotoxicity. Environ Sci Technol 38(1):62–68
36. Krasner SW, Chinn R, Hwang CJ, Barrett S (1991) Analytical methods for brominated organic disinfection by-products. In: Proceedings of the 1990 American Water Works Association water quality technology conference, American Water Works Association, Denver, CO
37. Richardson SD, Thruston AD Jr, Caughran TV, Chen PH, Collette TW, Floyd TL, Schenck KM, Lykins BW Jr, Sun G, Majetich G (1999) Identification of new drinking water disinfection byproducts formed in the presence of bromide. Environ Sci Technol 33:3378–3383
38. Kundu B, Richardson SD, Swartz PD, Matthews PP, Richard AM, DeMarini DM (2004) Mutagenicity in Salmonella of halonitromethanes: a recently recognized class of disinfection by-products in drinking water. Mutat Res 562(1–2):39–65
39. Kundu B, Richardson SD, Granville CA, Shaughnessy DT, Hanley NM, Swartz PD, Richard AM, DeMarini DM (2004) Comparative mutagenicity of halomethanes and halonitro-methanes in Salmonella TA100: structure-activity analysis and mutation spectra. Mutat Res 554(1–2):335–350
40. Choi J, Richardson SD (2005) Formation of halonitromethanes in drinking water. In: Proceedings of the international workshop on optimizing the design and interpretation of epidemiologic studies to consider alternative disinfectants of drinking water, U.S. EPA, Raleigh, NC
41. Chen PH, Richardson SD, Krasner SW, Majetich G, Glish GL (2002) Hydrogen abstraction and decomposition of bromopicrin and other trihalogenated disinfection byproducts by GC/MS. Environ Sci Technol 36(15):3362–3371
42. Choi J, Valentine RL (2002) Formation of N-nitrosodimethylamine (NDMA) from reaction of monochloramine: a new disinfection by-product. Water Res 36(4):817–824
43. Mitch WA, Sedlak DL (2002) Formation of N-nitrosodimethylamine (NDMA) from dimethylamine during chlorination. Environ Sci Technol 36:588–595
44. Jobb DB, Hunsinger R, Meresz O, Taguchi VY (1992) A study of the occurrence and inhibition of formation of N-nitrosodimethylamine (NDMA) in the Ohsweken water supply. In: Proceedings of the fifth national conference on drinking water, Winnipeg, Manitoba, Canada
45. Boyd JM, Hrudey SE, Li XF, Richardson SD (2011) Solid-phase extraction and high-performance liquid chromatography mass spectrometry analysis of nitrosamines in treated drinking water and wastewater. Trends Anal Chem 30(9):1410–1421
46. Mitch WA, Sharp JO, Trussell RR, Valentine RL, Alvarez-Cohen L, Sedlak DL (2003) N-nitrosodimethylamine (NDMA) as a drinking water contaminant: a review. Environ Eng Sci 20:389–404
47. Wilczak A, Assadi-Rad A, Lai HH, Hoover LL, Smith JF, Berger R, Rodigari F, Beland JW, Lazzelle LJ, Kincannon EG, Baker H, Heaney CT (2003) Formation of NDMA in chloraminated water coagulated with DADMAC cationic polymer. J Am Water Works Assoc 95(9):94–106
48. Kemper JM, Wale SS, Mitch WA (2010) Quaternary amines as nitrosamine precursors: a role for consumer products? Environ Sci Technol 44:1224–1231
49. California Department of Publich Health. Drinking water notification levels and response levels: an overview. Available at: http://www.cdph.ca.gov/certlic/drinkingwater/Documents/Notificationlevels/notificationlevels.pdf
50. Ontario Regulation 169/03. Ontario drinking water quality standards. Safe drinking water act, 2002. Available at: www.e-laws.gov.on.ca/html/regs/english/elaws_regs_030169_e.htm
51. Health Canada. N-nitrosodimethylamine (NDMA) in drinking water. Document for public comment. Prepared by the Federal-Provincial-Territorial Committee on drinking water.

Available at: http://www.hc-sc.gc.ca/ewh-semt/alt_formats/hecs-sesc/pdf/consult/_2010/ndma/draft-ebauche-eng.pdf

52. U.S. Environmental Protection Agency. Occurrence data accesing unregulated contaminant monitoring data – UCMR 2. Available at: http://water.epa.gov/lawsregs/rulesregs/sdwa/ucmr/data.cfm#ucmr2010

53. Charrois JW, Arend MW, Froese KL, Hrudey SE (2004) Detecting N-nitrosamines in drinking water at nanogram per liter levels using ammonia positive chemical ionization. Environ Sci Technol 38:4835–4841

54. U.S. Environmental Protection Agency. Contaminant Candidate List 3 – CCL. Available at http://water.epa.gov/scitech/drinkingwater/dws/ccl/ccl3.cfm

55. Munch JW, Bassett MV (2004) EPA Method 521. Determination of nitrosamines in drinking water by solid-phase extraction and capillary column gas chromatography with large volume injection and chemical ionization tandem mass spectrometry (MS/MS), U.S. EPA, Cincinnati, OH. Available at: www.epa.gov/nerlcwww/m_521.pdf

56. Zhao YY, Boyd J, Hrudey SE, Li XF (2006) Characterization of new nitrosamines in drinking water using liquid chromatography tandem mass spectrometry. Environ Sci Technol 40 (24):7636–7641

57. Schmidt CK, Brauch HJ (2008) N, N-dimethosulfamide as precursor for N-nitrosodimethylamine (NDMA) formation upon ozonation and its fate during drinking water treatment. Environ Sci Technol 42:6340–6346

58. Glezer V, Harris B, Tal N, Iosefzon B, Lev O (1999) Hydrolysis of haloacetonitriles: linear free energy relationship kinetics and products. Water Res 33:1939–1948

59. Reckhow DA, MacNeill AL, Platt TL, MacNeill AL, McClellan JN (2001) Formation and degradation of dichloroacetonitrile in drinking waters. J Water Supp Res Technol Aqua 50 (1):1–13

60. Plewa MJ, Muellner MG, Richardson SD, Fasano F, Buettner KM, Woo YT, McKague AB, Wagner ED (2008) Occurrence, synthesis, and mammalian cell cytotoxicity and genotoxicity of haloacetamides: an emerging class of nitrogenous drinking water disinfection byproducts. Environ Sci Technol 42:955–961

61. Krasner SW, McGuire MJ, Jacangelo JG, Patania NL, Reagan KM, Aieta EM (1989) The occurrence of disinfection by-products in United States drinking water. J Am Water Works Assoc 81:41–53

62. Williams DT, LeBel GL, Benoit FM (1997) Disinfection by-products in Canadian drinking water. Chemosphere 34:299–316

63. McGuire MJ, McLain JL, Obolensky A (2002) Information collection rule data analysis. American Water Works Association Research Foundation, Denver, CO

64. Najm IN, Krasner SW (1995) Effects of bromide and NOM on by-product formation. J Am Water Works Assoc 87(1):106–115

65. Kronberg L, Vartiainen T (1988) Ames mutagenicity and concentration of the strong mutagen 3-chloro-4-(dichloromethyl)-5-hydroxy-2(5H)-furanone and of its geometric isomer E-2-chloro-3-(dichloromethyl)-4-oxo-butenoic acid in chlorine-treated tap waters. Mutat Res 206:177–182

66. Suzuki N, Nakanishi J (1995) Brominated analogues of MX (3-chloro-4-(dichloromethyl)-5-hydroxy-2 (5H)-furanone) in chlorinated drinking water. Chemosphere 30:1557–1564

67. Kronberg L, Holmbom B, Reunanen M, Tikkanen L (1988) Identification and quantification of the Ames mutagenic compound 3-chloro-4-(dichloromethyl)-5-hydroxy-2(5H)-furanone and of its geometric isomer (E)-2-chloro-3-(dichloromethyl)-4-oxobutenoic acid in chlorine-treated humic water and drinking water extracts. Environ Sci Technol 22:1097–1103

68. Komulainen H, Kosma VM, Vaittinen SL, Vartiainen T, Kaliste-Korhonen E, Lotjonen S, Tuominen RK, Tuomisto J (1997) Carcinogenicity of the drinking water mutagen 3-chloro-4-(dichloromethyl)-5-hydroxy-2(5H)-furanone in the rat. J Nat Cancer Inst 89:848–856

69. Sasaki YF, Nishidate E, Izumiyama F, Watanabe-Akanuma M, Kinae N, Matsusaka N, Tsuda S (1997) Detection of in vivo genotoxicity of 3-chloro-4-(dichloromethyl)-5-hydroxy-2[5H]-

furanone (MX) by the alkaline single cell gel electrophoresis (Comet) assay in multiple mouse organs. Mutat Res 393:47–53

70. Geter DR, Winn RN, Fournie JW, Norris MB, DeAngelo AB, Hawkins WE (2004) MX [3-chloro-4-(dichloromethyl)-5-hydroxy-2[5H]-furanone], a drinking-water carcinogen, does not induce mutations in the liver of cII transgenic medaka (Oryzias latipes). J Toxicol Environ Health A 67(5):373–383

71. Wright JM, Schwartz J, Vartiainen T, Mäki-Paakkanen J, Altshul L, Harrington JJ, Dockery DW (2002) 3-Chloro-4-(dichloromethyl)-5-hydroxy-2(5H)-furanone (MX) and mutagenic activity in Massachusetts drinking water. Environ Health Perspect 110(2):157–164

72. Liviac D, Creus A, Marcos R (2010) DNA damage induction by two halogenated acetaldehydes, byproducts of water disinfection. Water Res 44:2638–2646

73. Schulten HR, Schnitzer MA (1993) State of the art structural concept for humic substances. Naturwissenschaften 80:29–30

74. Qin F, Zhao YY, Zhao Y, Boyd JM, Zhou W, Li XF (2010) A toxic disinfection by-product, 2,6-dichloro-1,4-benzoquinone, identified in drinking water. Angew Chem Int Ed 49(4):790–792

75. Zhao Y, Qin F, Boyd JM, Anichina J, Li XF (2010) Characterization and determination of chloro- and bromo-benzoquinones as new chlorination disinfection byproducts in drinking water. Anal Chem 82:4599–4605

76. Hoehn RC, Ellenberger CS, Gallagher DL, Wiseman ETV Jr, Benninger RW, Rosenblatt A (2003) ClO2 and by-product persistence in a drinking water system. J Am Water Works Assoc 95(4):141–150

77. Baribeau H, Prevost M, Desjardins R, Lafrance P, Gates DJ (2002) Chlorite and chlorate ion variability in distribution systems. J Am Water Works Assoc 94(7):96–105

78. Kurokawa Y, Takayama S, Konishi Y (1986) Long-term in vivo carcinogenicity tests of potassium bromate, sodium hypochlorite, and sodium chlorite conducted in Japan. Environ Health Perspect 69:221–235

79. National Toxicology Program (2005) Toxicology and carcinogenesis studies of sodium chlorate (CAS no. 7775-09-9) in F344/N rats and B6C3F1 mice (drinking water studies). Natl Toxicol Program Tech Rep, Ser 517, pp 1–255

80. U.S. Environmental Protection Agency. Unregulated Contaminant Monitoring Rule 3 (UCMR 3). http://water.epa.gov/lawsregs/rulesregs/sdwa/ucmr/ucmr3/index.cfm

81. Khiari D, Krasner SW, Hwang CJ, Chinn R, Barrett S (1997) In: Proceedings of the 1996 American Water Works Association water quality technology conference, American Water Works Association, Denver, CO

82. Zhang X, Talley JW, Boggess B, Ding G, Birdsell D (2008) Fast selective detection of polar brominated disinfection byproducts in drinking water using precursor ion scans. Environ Sci Technol 42(17):6598–6603

83. Ding G, Zhang X (2009) A picture of polar iodinated disinfection byproducts in drinking water by (UPLC/)ESI-tqMS. Environ Sci Technol 43(24):9287–9293

84. Zhang X, Minear RA (2006) Formation, adsorption and separation of high molecular weight disinfection byproducts resulting from chlorination of aquatic humic substances. Water Res 40(2):221–230

85. Kolpin DW, Furlong ET, Meyer MT, Thurman EM, Zaugg SD, Barber LB, Buxton HT (2002) Pharmaceuticals, hormones, and other organic wastewater contaminants in U.S. streams, 1999–2000: A national reconnaissance. Environ Sci Technol 36:1202–1211

86. Wang P, He Y-L, Huang C-H (2011) Reactions of tetracycline antibiotics with chlorine dioxide and free chlorine. Water Res 45:1838–1846

87. Dodd MC, Shah AD, Von Gunten U, Huang C-H (2005) Interactions of fluoroquinolone antibacterial agents with aqueous chlorine: reaction kinetics, mechanisms, and transformation pathways. Environ Sci Technol 39:7065–7076

88. Wang P, He Y-L, Huang C-H (2010) Oxidation of fluoroquinolone antibiotics and structurally related amines by chlorine dioxide: reaction kinetics, product and pathway evaluation. Water Res 44:5989–5998

89. Navalon S, Alvaro M, Garcia H (2008) Reaction of chlorine dioxide with emergent water pollutants: Product study of the reaction of three β-lactam antibiotics with ClO$_2$. Water Res 42(8–9):1935–1942
90. Adachi A, Okano A (2008) Generation of cyanide ion by the reaction of hexamine and losartan potassium with sodium hypochlorite. J Health Sci 54(5):581–583
91. Dodd MC, Huang C-H (2004) Transformation of the antibacterial agent sulfamethoxazole in reactions with chlorine: kinetics, mechanisms, and pathways. Environ Sci Technol 2004:5607–5615
92. Shah AD, Kim JH, Huang C-H (2006) Reaction kinetics and transformation of carbadox and structurally related compounds with aqueous chlorine. Environ Sci Technol 40:7228–7235
93. Buth JM, Arnold WA, McNeill K (2007) Unexpected products and reaction mechanisms of the aqueous chlorination of cimetidine. Environ Sci Technol 41(17):6228–6233
94. Krkosek WH, Koziar SA, White RL, Gagnon GA (2011) Identification of reaction products from reactions of free chlorine with the lipid-regulator gemfibrozil. Water Res 45:1414–1422
95. Bedner M, MacCrehan WA (2006) Transformation of acetaminophen by chlorination produces the toxicants 1,4-benzoquinone and N-acetyl-p-benzoquinone imine. Environ Sci Technol 40:516–522
96. Shen R, Andrews SA (2011) Demonstration of 20 pharmaceuticals and personal care products (PPCPs) as nitrosamine precursors during chloramine disinfection. Water Res 45:944–952
97. Schmidt CK, Sacher F, Brauch HJ (2006) Strategies for minimizing formation of NDMA and other nitrosamines during disinfection of drinking water. In: Proceedings of the AWWA Water Quality Technology Conference, Denver, CO
98. Huerta-Fontela M, Galcerán MT, Ventura F (2011) Presence and removal of illicit drugs in conventional drinking water treatment plants. In: Castiglioni S, Zuccato E, Fanelli R (eds) Illicit drugs in the environment: occurrence, analysis, and fate using mass spectrometry. Wiley, Hoboken, New Jersey, pp 205–222
99. McDowell DC, Huber MM, Wagner M, Von Gunten U, Ternes TA (2005) Ozonation of carbamazepine in drinking water: Identification and kinetic study of major oxidation products. Environ Sci Technol 39(20):8014–8022
100. Kosjek T, Andersen HR, Kompare B, Ledin A, Heath E (2009) Fate of carbamazepine during water treatment. Environ Sci Technol 43(16):6256–6261
101. Li Z, Fenet H, Gomez E, Chiron S (2011) Transformation of the antiepileptic drug oxcarbazepine upon different water disinfection processes. Water Res 45:1587–1596
102. Benner J, Ternes TA (2009) Ozonation of metoprolol: Elucidation of oxidation pathways and major oxidation products. Environ Sci Technol 43(14):5472–5480
103. Benner J, Ternes TA (2009) Ozonation of propanolol: formation of oxidation products. Environ Sci Technol 43(13):5086–5093
104. Pereira RO, Postigo C, López de Alda M, Daniel LA, Barceló D (2011) Removal of estrogens through water disinfection processes and formation of by-products. Chemosphere 82:789–799
105. Brix R, Bahi N, López de Alda MJ, Farre M, Fernandez JM, Barceló (2009) Identification of disinfection by-products of selected triazines in drinking water by LC-Q-ToF-MS/MS and evaluation of their toxicity. J Mass Spectrom 44(3):330–337
106. Lopez A, Mascolo G, Tiravanti G, Passino R (1997) Degradation of herbicides (ametryn and isoproturon) during water disinfection by means of two oxidants (hypochlorite and chlorine dioxide). Water Sci Technol 35:129–136
107. Lopez A, Mascolo G, Tiravanti G, Passino R (1998) Formation of herbicide degradation by-products during groundwater disinfection: an LC-MS investigation. J Anal Chem 53:856–860
108. Mascolo G, Lopez A, Passino R, Ricco G, Tiravanti G (1994) Degradation of sulphur containing S-triazines during water chlorination. Water Res 28:2499–2506
109. Sandin-España P, Magrans JO, Garcia-Baudin JM (2005) Study of clethodim degradation and by-product formation in chlorinated water by HPLC. Chromatographia 62(3–4):133–137

110. Duirk SE, Collette TW (2006) Degradation of chlorpyrifos in aqueous chlorine solutions: pathways, kinetics and modeling. Environ Sci Technol 40:546–551

111. Mehrsheikh A, Bleeke M, Brosillon S, Laplanche A, Roche P (2006) Investigation of the mechanism of chlorination of glyphosate and glycine in water. Water Res 40(16): 3003–3014

112. Lin CH, Lerch RN, Garrett HE, George MF (2003) Degradation of isoxaflutole (Balance) herbicide by hypochlorite in tap water. J Agric Food Chem 51:8011–8014

113. Mascolo G, Lopez A, James H, Fielding M (2001) By-products formation during degradation of isoproturon in aqueous solution. II: chlorination. Water Res 35(7):1705–1713

114. Zambodin CG, Losito II, Palmisano F (2000) Liquid chromatography/electrospray ionisation sequential mass spectrometric identification of the main chlortoluron by-products during water disinfection using chlorine. Rapid Commun Mass Spectrom 14:824–828

115. Xu B, Tian F-X, C-y Hu, Lin Y-L, S-j X, Rong R, Li D-P (2011) Chlorination of chlortoluron: kinetics, pathways and chloroform formation. Chemosphere 83:909–916

116. Wu J, Chongyu L, Sing Chan GY (2009) Organophosphorous pesticide ozonation and formation of oxon intermediates. Chemosphere 76:1308–1314

117. Terasaki M, Makino M (2008) Determination of chlorinated by-products of parabens in swimming pool water. Int J Environ Anal Chem 88(13):911–922

118. Tay KS, Rahman NA, Bin Abas MR (2010) Ozonation of parabens in aqueous solutions: kinetics and mechanisms of degradation. Chemosphere 81:1446–1453

119. Rule KL, Ebbett VR, Vikesland PJ (2005) Formation of chloroform and chlorinated organics by free-chlorine-mediated oxidation of triclosan. Environ Sci Technol 39:3176–3185

120. Greyshock AE, Vikesland PJ (2006) Triclosan reactivity in chloraminated waters. Environ Sci Technol 40:2615–2622

121. Nakajima M, Kawakami T, Niino T, Takahashi Y, Onodera S (2009) Aquatic fate of sunscreen agents octyl-4-methoxycinnamate and octyl-4-dimethylaminobenzoate in model swimming pools and the mutagenic assays of their chlorination byproducts. J Health Sci 55(3):363–372

122. Negreira N, Canosa P, Rodriguez I, Ramil M, Rubi E, Cela RJ (2008) Study of some UV filters stability in chlorinated water and identification of halogenated by-products by gas chromatography–mass spectrometry. J Chromatogr A 1178(1–2):206–214

123. Kuhlich P, Göstl R, Teichert P, Piechotta C, Nehls I (2011) Transformation of polycyclic musks AHTN and HHCB upon disinfection with hypochlorite: two new chlorinated disinfection by-products (CDBP) of AHTN and a possible source for HHCB-lactone. Anal Bioanal Chem 399:3579–3588

124. Joll CA, Alessandrino MJ, Heitz A (2010) Disinfection by-products from halogenation of aqueous solutions of terpenoids. Water Res 44(1):232–242

125. Qi F, Xu BB, Chen ZL, Ma J, Sun DZ, Zhang LQ (2009) Efficiency and products investigations on the ozonation of 2-methylisoborneol in drinking water. Water Environ Res 81:2411–2419

126. Petrovic A, Diaz A, Ventura F, Barceló D (2003) Occurrence and removal of estrogenic short-chain ethoxy nonylphenolic compounds and their halogenated derivatives during drinking water production. Environ Sci Technol 37:4442–4448

127. Hu JY, Aizawa T, Ookubo S (2002) Products of aqueous chlorination of bisphenol A and their estrogenic activity. Environ Sci Technol 36:1980–1987

128. Deborde M, Rabouan S, Mazellier P, Duguet J-P, Legube B (2008) Oxidation of bisphenol A by ozone in aqueous solutions. Water Res 42:4299–4308

129. Merel S, Clement M, Thomas O (2010) State of the art on cyanotoxins in water and their behaviour towards chlorine. Toxicon 55:677–691

130. Ashley DL, Blount BC, Singer PC, Depaz E, Wilkes C, Gordon S, Lyu C, Masters J (2005) Changes in blood trihalomethane concentrations resulting from differences in water quality and water use activities. Arch Environ Occup Health 60(1):7–15

131. Haddad S, Tardif GC, Tardif R (2006) Development of physiologically based toxicokinetic models for improving the human indoor exposure assessment to water contaminants: trichloroethylene and trihalomethanes. J Toxicol Environ Health A 69(23):2095–2136

132. Leavens TL, Blount BC, DeMarini DM, Madden MC, Valentine JL, Case MW, Silva LK, Warren SH, Hanley NM, Pegram RA (2007) Disposition of bromodichloromethane in humans following oral and dermal exposure. Toxicol Sci 99(2):432–445

133. Li J, Blatchley ER III (2007) Volatile disinfection byproduct formation resulting from chlorination of organic - Nitrogen precursors in swimming pools. Environ Sci Technol 41 (19):6732–6739

134. Xu X, Weisel CP (2005) Dermal uptake of chloroform and haloketones during bathing. J Exp Anal Environ Epidemiol 15(4):289–296

135. Xu X, Weisel CP (2005) Human respiratory uptake of chloroform and haloketones during showering. J Exp Anal Environ Epidemiol 15(1):6–16

136. Zwiener C, Richardson SD, DeMarini DM, Grummt T, Glauner T, Frimmel FH (2007) Drowning in disinfection byproducts? Assessing swimming pool water. Environ Sci Technol 41:363–372

137. Walse SS, Mitch WA (2008) Nitrosamine carcinogens also swim in chlorinated pools. Environ Sci Technol 42:1032–1037

138. Weaver WA, Li J, Wen YL, Johnston J, Blatchley MR, Blatchley ER III (2009) Volatile disinfection by-product analysis from chlorinated indoor swimming pools. Water Res 43:3308–3318

139. Weisel CP, Richardson SD, Nemery B, Aggazzotti G, Baraldi E, Blatchley ER III, Blount BC, Carlsen KH, Eggleston PA, Frimmel FH, Goodman M, Gordon G, Grinshpun SA, Heederik D, Kogevinas M, LaKind JS, Nieuwenhuijsen MJ, Piper FC, Sattar SA (2009) Childhood asthma and environmental exposures at swimming pools: state of the science and research recommendations. Environ Health Perspect 117:500–507

140. LaKind JS, Richardson SD, Blount BC (2010) The good, the bad, and the volatile: can we have both healthy pools and healthy people? Environ Sci Technol 44(9):3205–3210

141. Font-Ribera L, Kogevinas M, Zock J, Gómez FP, Barreiro E, Nieuwenhuijsen MJ, Fernandez P, Lourencetti C, Perez-Olabarria M, Bustamante M, Marcos R, Grimalt JO, Villanueva CM (2010) Short-term changes in respiratory biomarkers after swimming in a chlorinated pool. Environ Health Perspect 118(11):1538–1544

142. Kogevinas M, Villanueva CM, Font-Ribera L, Liviac D, Bustamante M, Espinoza F, Nieuwenhuijsen MJ, Espinosa A, Fernandez P, DeMarini DM, Grimalt JO, Grummt T, Marcos R (2010) Genotoxic effects in swimmers exposed to disinfection by-products in indoor swimming pools. Environ Health Perspect 118(11):1531–1537

143. Richardson SD, DeMarini DM, Kogevinas M, Fernandez P, Marco E, Lourencetti C, Balleste C, Heederik D, Meliefste K, McKague AB, Marcos R, Font-Ribera L, Grimalt JO, Villanueva CM (2010) What's in the pool? A comprehensive identification of disinfection by-products and assessment of mutagenicity of chlorinated and brominated swimming pool water. Environ Health Perspect 118(11):1523–1530

144. Liviac D, Wagner ED, Mitch WA, Altonji MJ, Plewa MJ (2010) Genotoxicity of water concentrates from recreational pools after various disinfection methods. Environ Sci Technol 44(9):3527–3532

145. Cardador MJ, Gallego M (2011) Haloacetic acids in swimming pools: swimmer and worker exposure. Environ Health Perspect 118:1545–1550

146. Plewa MJ, Wagner ED, Mitch WA (2011) Comparative mammalian cell cytotoxicity of water concentrates from disinfected recreational pools. Environ Sci Technol 45:4159–4165

147. Amer K, Karanfil T (2011) Formation of disinfection by-products in indoor swimming pool water. The contribution from filling water natural organic matter and swimmer body fluids. Water Res 45:926–932

148. Erdinger L, Kuhn KP, Kirsch F, Feldhues R, Frobel T, Nohynek B, Gabrio T (2004) Pathways of trihalomethane uptake in swimming pools. Int J Hygiene Environ Health 207(6):571–575

149. Villanueva CM, Cantor KP, Grimalt JO, Malats N, Silverman D, Tardon A, Garcia-Closas R, Serra C, Carrato A, Castaño-Vinyals G, Marcos R, Rothman N, Real FX, Dosemeci M, Kogevinas M (2007) Bladder cancer and exposure to water disinfection by-products through ingestion, bathing, showering, and swimming in pools. Am J Epidemiol 165(2):148–156

150. Bernard A, Nickmilder M, Voisin C, Sardella A (2009) Impact of chlorinated swimming pool attendance on the respiratory health of adolescents. Pediatrics 124(4):1110–1118

151. Goodman M, Hays S (2008) Asthma and swimming: a meta-analysis. J Asthma 45 (8):639–647

152. Voisin C, Sardella A, Marcucci F, Bernard A (2010) Infant swimming in chlorinated pools and the risks of bronchiolitis, asthma and allergy. Eur Respiratory J 36(1):41–47

153. Simmons JE, Richardson SD, Teuschler LK, Miltner RJ, Speth TF, Schenck KM, Hunter ES III, Rice G (2008) Research issues underlying the four-lab study: integrated disinfection by-products mixtures research. J Toxicol Environ Health A 71:1125–1132

154. Pressman JG, Richardson SD, Speth TF, Miltner RJ, Narotsky MG, Hunter ES III, Rice GE, Teuschler LK, McDonald A, Parvez S, Krasner SW, Weinberg HS, McKague AB, Parrett CJ, Bodin N, Chinn R, Lee CFT, Simmons JE (2010) Concentration, chlorination, and chemical analysis of drinking water for disinfection byproduct mixtures health effects research: U.S. EPA's four lab study. Environ Sci Technol 44:7184–7192

155. Munch JW (1995) EPA Method 524.2. Measurement of purgeable organic compounds in water by capillary column gas chromatography/mass spectrometry, rev 4.1, U.S. EPA, Cincinnati, OH. Available at: www.caslab.com/EPA-Methods/PDF/524_2.pdf

156. Munch DJ, Munch JW, Pawlecki AM (1995) EPA Method 552.2. Determination of haloacetic acids and dalapon in drinking water by liquid-liquid extraction, derivatization and gas chromatography with electron capture detection. U.S. EPA, Cincinnati, OH, Available at: http://www.caslab.com/EPA-Methods/PDF/552_2.pdf

157. Domino MM, Pepich BV, Munch DJ, Fair PS, Xie Y (2003) EPA Method 552.3. Determination of haloacetic acids and dalapon in drinking water by liquid-liquid microextraction, derivatization, and gas chromatography with electron capture detection. EPA 815-B-03-002. U.S. EPA, Cincinnati, OH. Available at: http://www.epa.gov/ogwdw/methods/pdfs/methods/met552_3.pdf

158. Zwiener C, Richardson SD (2005) Drinking water disinfection by-product analysis by LC/MS and LC/MS/MS. Trends Anal Chem 24:613–621

159. Zhang X, Minear RA (2002) Characterization of high molecular weight disinfection byproducts resulting from chlorination of aquatic humic substances. Environ Sci Technol 36(19):4033–4038

160. Zhang X, Minear RA, Barrett SE (2005) Characterization of high molecular weight disinfection byproducts from chlorination of humic substances with/without coagulation pretreatment using UF-SEC-ESI-MS/MS. Environ Sci Technol 39(4):963–972

161. Creed JT, Brockhoff CA, Martin TD (1997) EPA Method 321.8. Determination of bromate in drinking waters by ion chromatography inductively coupled plasma-mass spectrometry. U.S. EPA, Cincinnati, OH, Available at: http://www.epa.gov/microbes/m_321_8.pdf

162. Zaffiro AD, Zimmerman M, Pepich BV, Slingsby RW, Jack RF, Pohl CA, Munch DJ (2009) EPA Method 557. Determination of haloacetic acids, bromate, and dalapon in drinking water by ion chromatography electrospray ionization tandem mass spectrometry (IC-ESI-MS/MS). U.S. EPA, Cincinnati, OH, Available at: http://water.epa.gov/scitech/drinkingwater/labcert/upload/met557.pdf

163. Roehl R, Slingsby R, Avdalovic N, Jackson PE (2002) Applications of ion chromatography with electrospray mass spectrometric detection to the determination of environmental contaminants in water. J Chromatogr A 956(1–2):245–254

164. Paull B, Barron L (2004) Using ion chromatography to monitor haloacetic acids in drinking water: a review of current technologies. J Chromatogr A 1046(1–2):1–9
165. Shi HL, Adams C (2009) Rapid IC-ICP/MS method for simultaneous analysis of iodoacetic acids, bromoacetic acids, bromate, and other related halogenated compounds in water. Talanta 79:523–527
166. Wagner P, Pepich BV, Pohl C, Srinivasan K, De Borba B, Lin R, Munch DJ (2009) EPA Method 302.0. Determination of bromate in drinking water using two-dimensional ion chromatography with suppressed conductivity detection. U.S. EPA, Cincinnati, OH, Available at: http://water.epa.gov/scitech/drinkingwater/labcert/upload/met302_0.pdf
167. Pfaff JD, Hautman DP, Munch DJ (1997) EPA Method 300.1. Determination of inorganic anions in drinking water by ion chromatography. U.S. EPA, Cincinnati, OH, Available at: http://water.epa.gov/scitech/drinkingwater/labcert/upload/met300.pdf
168. Na C, Olson TM (2004) Stability of cyanogen chloride in the presence of free chlorine and monochloramine. Environ Sci Technol 38(22):6037–6043
169. Yang X, Shang C (2005) Quantification of aqueous cyanogen chloride and cyanogen bromide in environmental samples by MIMS. Water Res 39(9):1709–1718
170. Riter LS, Charles L, Turowski M, Cooks RG (2001) External interface for trap-and-release membrane introduction mass spectrometry applied to the detection of inorganic chloramines and chlorobenzenes in water. Rapid Commun Mass Spectrom 15:2290–2295
171. Rios RVRA, Da Rocha LL, Vieira TG, Lago RM, Augusti R (2000) On-line monitoring by membrane introduction mass spectrometry of chlorination of organics in water. Mechanistic and kinetic aspects of chloroform formation. J Mass Spectrom 35(5):618–624
172. Kristensen GH, Klausen MM, Hansen VA, Lauritsen FR (2010) On-line monitoring of the dynamics of trihalomethane concentrations in a warm public swimming pool using an unsupervised membrane inlet mass spectrometry system with off-site real-time surveillance. Rapid Commun Mass Spectrom 24(1):30–34
173. Guevremont R, Ding L, Ells B, Barnett DA, Purves RW (2001) Atmospheric pressure ion trapping in a tandem FAIMS-FAIMS coupled to a TOFMS: studies with electrospray generated gramicidin S ions. J Am Soc Mass Spectrom 12(12):1320–1330
174. Ells B, Barnett DA, Purves RW, Guevremont R (2000) Detection of nine chlorinated and brominated haloacetic acids at part-per-trillion levels using ESI-FAIMS-MS. Anal Chem 72(19):4555–4559
175. Gabryelski W, Wu F, Froese KL (2003) Comparison of high-field asymmetric waveform ion mobility spectrometry with GC methods in analysis of haloacetic acids in drinking water. Anal Chem 75(10):2478–2486
176. Barnett DA, Guevremont R, Purves RW (1999) Determination of parts-per-trillion levels of chlorate, bromate, and iodate by electrospray ionization/high-field asymmetric waveform ion mobility spectrometry/mass spectrometry. Appl Spectrosc 53(11):1367–1374
177. Planas C, Palacios O, Ventura F, Rivera J, Caixach J (2008) Analysis of nitrosamines in water by automated SPE and isotope dilution GC/HRMS - Occurrence in the different steps of a drinking water treatment plant, and in chlorinated samples from a reservoir and a sewage treatment plant effluent. Talanta 76:906–913
178. Yuan YZ, Liu X, Boyd JM, Qin F, Li J, Li XF (2009) Identification of N-nitrosamines in treated drinking water using nanoelectrospray ionization high-field asymmetric waveform ion mobility spectrometry with quadrupole time-of-flight mass spectrometry. J Chromatogr Sci 47(1):92–96
179. Zhao YY, Liu X, Boyd JM, Qin F, Li J, Li XF (2009) Identification of N-nitrosamines in treated drinking water using nanoelectrospray ionization high-field asymmetric waveform ion mobility spectrometry with quadrupole time-of-flight mass spectrometry. J Chromatogr Sci 47:92–96
180. Kubwabo C, Stewart B, Gauthier SA, Gauthier BR (2009) Improved derivatization technique for gas chromatography–mass spectrometry determination of 3-chloro-4-(dichloromethyl)-5-hydroxy-2(5H)-furanone in drinking water. Anal Chim Acta 649(2):222–229

181. Vincenti M, Biazzi S, Ghiglione N, Valsania MC, Richardson SD (2005) Comparison of highly-fluorinated chloroformates as direct aqueous sample derivatizing agents for hydrophilic analytes and drinking-water disinfection by-products. J Am Soc Mass Spectrom 16(6):803–813
182. Vincenti M, Fasano F, Valsania MC, Guarda P, Richardson SD (2010) Application of the novel 5-chloro-2,2,3,3,4,4,5,5-octafluoro-1-pentyl chloroformate derivatizing agent for the direct determination of highly polar water disinfection byproducts. Anal Bioanal Chem 397(1):43–54
183. Richardson SD, Caughran TV, Poiger T, Guo Y, Gene Crumley F (2000) Application of DNPH derivatization with LC/MS to the identification of polar carbonyl disinfection by-products in drinking water. Ozone Sci Engin 22(6):653–675
184. Zwiener C, Glauner T, Frimmel FH (2003) Liquid chromatography/electrospray ionization tandem mass spectrometry and derivatization for the identification of polar carbonyl disinfection by-products. In: Ferrer I, Thurman EM (eds) Liquid chromatography/mass spectrometry, MS/MS and time-of-flight MS, ACS symposium series, vol 850. American Chemical Society, Washington DC, pp 356–375
185. Richardson SD, Karst U (2001) A new tailor-made derivatizing agent for identifying polar carbonyl DBPs in drinking water. In: Proceedings of the American Chemical Society conference. American Chemical Society, Washington DC
186. Emmert GL, Geme G, Brown MA, Simone PS Jr (2009) A single automated instrument for monitoring total trihalomethane and total haloacetic acid concentrations in near real-time. Anal Chim Acta 656(1–2):1–7
187. Echigo S, Zhang X, Minear RA, Plewa MJ (2000) Differentiation of total organic brominated and chlorinated compounds in total organic halide measurement: a new approach with an ion chromatographic technique. In: Barrett SE, Krasner SW, Amy GL (eds) Natural organic matter and disinfection by-products: characterization and control in drinking water. American Chemical Society, Washington DC, pp 330–342
188. Zhai H, Zhang X (2011) Formation and decomposition of new and unknown polar brominated disinfection byproducts during chlorination. Environ Sci Technol 45(6):2194–2201

Micro-pollutants in Hospital Effluent: Their Fate, Risk and Treatment Options

Paola Verlicchi, Alessio Galletti, Mira Petrovic, and Damià Barceló

Abstract This chapter deals with the physicochemical characteristics of hospital effluents, focusing on both macropollutants (conventional contaminants) and micro-pollutants (pharmaceutical compounds, heavy metals and diagnostic agents). It compares their typical concentration ranges measured in hospital effluents and urban wastewaters and then discusses the behaviour of micro-pollutants, mainly pharmaceutical compounds during common treatment processes and upon their release into the environment, attempting to find correlations with their physical properties. It then goes on to provide an overview of the most commonly adopted strategies for managing and treating hospital effluents worldwide, focusing on the ability of the various treatments to remove the wide spectrum of pharmaceuticals contained therein. Some considerations are also made on the tools available for assessing the environmental risks posed by the discharge of hospital effluents, and, finally, guidelines for the best treatment options are proposed.

Keywords Environmental risk assessment, Hospital effluents, Management and treatment, Pharmaceuticals

P. Verlicchi (✉) and A. Galletti
Department of Engineering, University of Ferrara, Via Saragat 1, 44122 Ferrara, Italy
e-mail: paola.verlicchi@unife.it

M. Petrovic
Catalan Institute for Water Research (ICRA), c/Emili Grahit 101, 17003 Girona, Spain

Catalan Institution for Research and Advanced Studies (ICREA), Passeig Lluis Companys 23, 80010 Barcelona, Spain

D. Barceló
Catalan Institute for Water Research (ICRA), c/Emili Grahit 101, 17003 Girona, Spain

Department of Environmental Chemistry, IDAEA-CSIC, c/Jordi Girona 18-26, 08034 Barcelona, Spain

King Saud University, Box 2454, Riyadh, 11451, Saudi Arabia

D. Barceló (ed.), *Emerging Organic Contaminants and Human Health*,
Hdb Env Chem (2012) 20: 139–172, DOI 10.1007/698_2011_134,
© Springer-Verlag Berlin Heidelberg 2012, Published online: 17 January 2012

Contents

1   Introduction

Environmental contamination by pharmaceutical compounds (PhCs) has become an issue of great concern to many countries in recent years. Hence, the European Community has recently funded several projects to quantify their presence in water bodies and wastewaters, to evaluate the removal rates of compounds from major therapeutic classes during conventional and advanced treatments and to assess the risks they pose to the environment. Notable examples of these studies are:

- ERAVMIS (*Environmental Risk Assessment of Veterinary Medicines in Slurry*, 1999–2003), whose aim was to evaluate the impact of some veterinary antibiotics (mainly macrolides and sulphonamides)
- REMPHARMWATER (*Evaluation of the presence of pharmaceutical compounds*, 2000–2003), which studied the removal kinetics of some antibiotics in the major processes employed by wastewater treatment plants (WWTPs)
- POSEIDON (*Assessment of technologies for the removal of pharmaceuticals and personal care products in sewage and drinking water facilities to improve the indirect potable water reuse*, 2001–2004), which focussed on the removal capacity of WWTPs and waterworks for the main micro-pollutants
- ERAPharm (*Environmental Risk Assessment of Pharmaceuticals*, 2004–2007), which focussed on the fate of a wide range of pharmaceuticals in the environment
- PILLS (*Pharmaceuticals Input and Elimination from Local Sources*, 2007–2012), which investigated solutions for pharmaceutically burdened wastewater directly at source.

Although these studies aimed to *monitor* the occurrence of PhCs at the various stages in the water management cycle, i.e. generic urban wastewaters (UWWs) [1], conventional WWTP effluents (particularly conventional activated sludge (CAS) effluents) [2, 3], surface wastewater [4] and/or drinking water from different water-works, and to evaluate the removal capacity of the most common treatment steps [5, 6], studies generally confined themselves to the investigation of a limited number of PhCs, in particular antibiotics (mainly ciprofloxacin, norfloxacin, ofloxacin, sulfamethoxazole) and anti-inflammatories/analgesics (paracetamol, ibuprofen and diclofenac). Furthermore, few studies have thus far been devoted to monitoring these compounds in *specific* wastewaters (i.e. effluent from hospitals, livestock farms and pharmaceutical factories) (among them [7–10]), despite many authors (including [11, 12]) indicating hospitals in particular as hot spot sources of environmental contaminants. In fact, a great variety of micro-contaminants, including active principles of drugs and their metabolites, as well as other chemicals, heavy metals, disinfectants and sterilizing agents, specific detergents for endoscopes and other instruments, radioactive markers and iodinated contrast media (ICM), are produced through diagnostic, laboratory and research activities and/or excreted by medicated patients.

From a treatment perspective, despite their specific contents, hospital waste-waters (HWWs) are quite often considered to be of comparable pollutant nature to UWWs and, as such, are discharged directly into public sewers and co-treated with UWWs at the nearest WWTP. Unsurprisingly, this management strategy is cur-rently the subject of lively debate in the scientific community [12–14], which has recently begun to evaluate the hospital contribution to UWWs in terms of micro-pollutant load at a local level.

The issue of improving HWW management and treatment is by no means a simple one to resolve; in fact, in these early stages, not only do PhC concentrations in hospital effluents require accurate evaluation, but their annual loads also need to be determined and the efficacy of existing treatments assessed for each PhC present in HWW. In addition, the characteristics of each particular receiving water body (average, minimum and maximum flow rate, use) will also have an effect on manage-ment strategy outcomes. Moreover, any study into this issue must involve careful consideration, on a case-by-case basis, of the size of the hospital and its catchment area, its bed density (the number of beds per 1,000 inhabitants resident in the surrounding area), the characteristics of the local WWTP and the risks posed by the presence of micro-contaminant residues in the treated effluent discharged into the environment.

Below, an in-depth analysis of the state-of-the-art of physicochemical characteri-zation of HWWs is presented, comparing them with UWWs. The fate of PhCs during treatment and in the environment is then discussed, making an attempt to identify the properties that can best predict their behaviour. A brief review of the strategies commonly adopted for treating HWWs is then provided, followed by several considerations on the assessment of the environmental risks posed by these kinds of effluent. To complete the study, the previous considerations are extrapolated to suggest some guidelines for the management and treatment of HWWs.

2 Characteristics of Hospital Effluents

2.1 *Macropollutants*

Based on an in-depth review of the international literature on conventional pollutants (mainly BOD_5, COD, SS, TKN, NH_4, NO_3, total P, *E. coli*, surfactants and faecal and total coliforms) found in wastewaters from hospitals of different sizes (60–900 beds), ward types and locations, [9] identified their typical concentration ranges in hospital effluents. In particular, that study provided estimated average values and cumulative frequency bands (Figs. 1–3) for BOD_5, COD and SS in HWWs. This was performed by grouping the literature minimum and maximum values of each of these conventional parameters reported in hospital effluents, and subsequent elaboration yielded the minimum and maximum cumulative frequency curves for each contaminant. This approach revealed that the specific daily contribution of each hospital patient is 150–170 g BOD_5 patient^{-1} d^{-1}, 260–300 g COD patient^{-1} d^{-1} and 120–150 g SS patient^{-1} d^{-1}. Comparison of these values with those generally adopted for a person equivalent (p.e.) in UWW, respectively 50–60 g BOD_5 p.e.$^{-1}$ d^{-1}, 100–120 g COD p.e.$^{-1}$ d^{-1}, 90–120 g SS p. e.$^{-1}$ d^{-1}, shows that a *hospital patient* is responsible for a mean two- to three fold greater contribution than a generic *inhabitant*.

Other authors [8, 13, 15–23] have reported the ranges of variability and corresponding average values of a wider range of conventional contaminants in

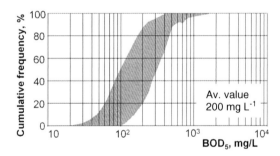

Fig. 1 Cumulative frequency curves for BOD_5 in HWWs

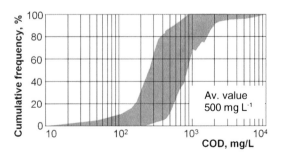

Fig. 2 Cumulative frequency curves for COD in HWWs

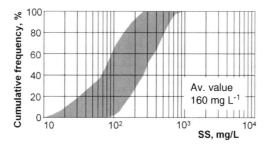

Fig. 3 Cumulative frequency curves for SS in HWWs

Table 1 Range of variability in concentrations of the main physical, chemical and microbiological parameters in HWWs and UWWs and their corresponding average values in brackets

Parameters	HWWs	UWWs
pH	6.3–9.2 (8)	7.5–8.5
Conductivity, $\mu S\ cm^{-1}$	297–1,000 (850)	420–1,340
SS, mg L^{-1}	120–400 (160)	120–350
VSS/SS	0.45–0.75 (0.58)	0.65–0.8
COD, mg L^{-1}	450–2,300 (650)	500–600
BOD_5, mg L^{-1}	150–603 (200)	100–400
TKN, mg L^{-1}	30–100 (45)	20–70
NH_4 mg L^{-1}	10–68 (30)	12–45
Total P, mg L^{-1}	0.2–8 (5)	4–10
Chlorides, mg L^{-1}	80–400 (220)	30–100
Fats and oils, mg L^{-1}	13–60 (25)	50–150
Total detergents[a], mg L^{-1}	3–7.2 (4.5)	4–8
Disinfectants, mg L^{-1}	2–200	–
COD/BOD_5	1.4–6.6 (2.5)	1.7–2.4
Total coliforms, MPN/100 mL	10^6–10^9 (10^6)	10^7–10^8
Faecal coliforms, MPN/100 mL	10^3–10^7 (10^5)	10^6–10^7
E. coli, MPN/100 mL	10^3–10^6 (10^4)	10^6–10^7
Streptococci, MPN/100 mL	10^3–10^5 (10^4)	10^3–10^5

[a]Detergents include ionic and non-ionic compounds

hospital and urban wastewaters. These data are summarized in Table 1, and further details of quali-quantitative characterization of hospital and urban effluents and their variability over the course of a day and a year can be found in [24].

The most common *disinfectants* in use nowadays are sodium hypochlorite, peracetic acid, ammonium quaternary salts, aldehydes, alcohols and phenol compounds. However, according to [25], all disinfectants require special attention. Indeed, glutaraldehyde, used widely in the past but now generally discarded in favour of compounds with a smaller impact on WWTP biological processes, has been found in concentrations ranging from 0.5 to 3.72 mg L^{-1} [15]. Moreover, triclosan, a common detergent and antimicrobial agent found in many personal care

products (detergents, disinfectants and pharmaceuticals), was detected in the range
0.12–210 µg L^{-1} [2, 26].

As regards *microbiological parameters*, no consistent differences in variability
ranges between hospital and urban effluents were revealed, as reported in Table 1.
Nevertheless, over the last decade, [27–29] have detected the presence of several
bacterial strains carrying different resistance genes in hospital effluents. Chitnis et al.
[30] found that these patterns of resistance included concomitant resistance to
ampicillin, cephalosporin-aminoglycosides, quinolones, co-trimoxazole, tetracycline
and chloramphenicol, i.e. to the majority of existing antibiotics. Furthermore, they
found that residential sewage samples harboured far smaller multi-drug-resistant
(MDR) populations than the hospital effluents investigated. In fact, samples from
the municipal sewage stream just upstream of the hospital effluent outlet failed to
reveal MDR bacteria. Resistant populations were however detected after the hospital
effluent had been mixed with the UWW, and 0.5% persisted 100 m downstream and
0.06% at a distance of 2 km from the hospital effluent discharge point.

Similarly, Prado et al. [31] reported that HWWs may contain, on average,
approximately 2-log higher levels of enteric viruses, mainly rotavirus A (RV-A)
and human adenovirus (HAdV), than raw UWWs.

2.2 Micro-pollutants

Hospital effluent also contains a large variety of micro-contaminants, which are
generally present in concentrations ranging from ng L^{-1} to µg L^{-1}. Examples of
these chemicals include PhCs, AOX (organic halogens adsorbable onto active
carbon), volatile halogenated organic compounds and other organic compounds
including acetaldehyde, ketones, alcohols acetates and phenols [23].

Sources of AOX include PhCs, their metabolites, chlorine-forming disinfectants,
used in cleaning activities, and halogen-containing solvents, employed in
laboratories, as well as other chemical substances like ethidium bromide [17].
AOX levels have been found to range from 150 to 7,760 µg L^{-1} in HWWs, in
stark contrast to the 0.04–0.2 µg L^{-1} range measured in UWWs.

Heavy metals have also been discovered in HWWs [23, 32], particularly,
platinum, excreted by oncology patients treated with cis-platinum, carboplatinum
or other cytostatic agents [33, 34]; mercury, usually found in diagnostic agents and
active ingredients of disinfectants and diuretics [17]; gadolinium, present in
iodinated contrast media (ICM) and, due to its high magnetic moment, used in
magnetic resonance imaging (MRI) [35]. Silver, nickel, zinc, copper, lead and
arsenic, whose organic complexes are very quickly excreted as their parent
compounds following administration, have also been detected in hospital wastewater samples, at concentration ranges reported in Table 2 [13, 23, 35, 36]. These
metals are non-degradable and highly toxic in some oxidative states.

These micro-pollutants are also considered *emerging contaminants*: unregulated
pollutants which may be candidates for future regulation depending on their
potential health effects and the results of monitoring of their occurrence.

Table 2 Typical concentration ranges of the main heavy metals in hospital effluent

Heavy metal	Concentration range, $\mu g\ L^{-1}$
Hg	0.04–5.03
Gd	1–100
Pt	0.01–41.3
Ag	150–437
As	0.8–11
Cu	50–230
Ni	7–71
Pb	3–19
Zn	70–670

Their main characteristic is that they do not need to persist in the environment to cause negative effects, as high transformation/removal rates are cancelled out by their continuous introduction into the environment [37].

Medicaments are composed of one or more active pharmaceutical ingredients (APIs) also called parent compounds, designed to provoke the desired effect, together with an adjuvant component, which may include pigments and dyes. Much of the focus has thus far been on API concentrations in the aquatic environment; these molecules are generally complex, possessing different functional groups and physicochemical and biological properties; they are quite often polar and have a molecular weight ranging from 200 to 500/1,000 Da. APIs are generally grouped according to their therapeutic effect, i.e. analgesics/anti-inflammatories, antibiotics, anti-diabetics, anti-hypertensives, barbiturates, beta-agonists, beta-blockers, diuretics, lipid regulators, psychiatric drugs, receptor antagonists anti-neoplastics and X-ray ICMs, rather than their chemical structure, as even minute changes in the latter may cause significant difference in polarity, solubility and/or other important properties that influence and define their environmental fate.

After administration, the active substances of medicines are metabolized, but only to a certain extent. The unmetabolized active substances (varying between 10% and 95%) are excreted, largely through the renal system (urine) and partially through the biliary system (faeces), depending on the nature of the compound and the individual in question [38]. As a consequence, API residues join wastewater and enter the water cycle through the sewage system as unchanged substances (parent compounds), a mixture of metabolites or conjugated with an inactivating compound attached to the molecule [39, 40], whose fate in the environment will be governed by their characteristics (mainly solubility, volatility, adsorbability, absorbability, biodegradability, hydrophilicity, lipophilicity, polarity and stability) as well as environmental conditions (temperature, pH, aerobic/anaerobic/anoxic conditions).

Verlicchi et al. [36] found that antibiotics, analgesics/anti-inflammatories and steroid compounds are the therapeutic classes most abundantly administered in Italian hospitals, but consistent consumption of cardiovascular compounds, tranquilizers, anti-neoplastics and anaesthetics was also detected. According to other authors in other localities [32, 41], PhCs detected in the largest amounts in hospital effluents are antibiotics, anti-epileptics, beta-blockers, lipid regulators and anti-neoplastics. Thus, PhC consumption varies from country to country, especially given worldwide variation in restrictions on the use of some of these compounds (e.g. vancomycin is widely used as a first-line antibiotic in the United States, whereas its use in European countries is highly restricted), and from year to year, due to continual progress in drug development. Furthermore, PhC concentrations in effluents may change over the course of a single year. Indeed, analysis of the distribution of PhC consumption throughout the year reveals the existence of *critical* months, featuring greater administered quantities, especially of antibiotics [36].

The range of variability of the concentrations of most prevalent compounds detected in hospital effluent, grouped according to their therapeutic class, is reported in Table 3 [8, 23, 42–47]; ICMs are included in the list, even though they are more diagnostic than pharmaceutical agents, as they are denoted micro-pollutants and emerging contaminants.

Table 3 Concentration ranges of the most abundant PhCs measured in hospital effluent, grouped according to their therapeutic class

Therapeutic class	Investigated compounds	Range in HWW µg L^{-1}
Analgesics/anti-inflammatories	Codeine	0.02–50
	Diclofenac	0.24–15
	Ibuprofen	0.07–77
	Indomethacin	0.3–6.1
	Ketoprofen	0.8–10
	Ketorolac	0.2–60
	Naproxen	9.8–18
	Paracetamol	1.4–1,368
	Salicylic acid	2–70.1
Antibiotics	Azithromycin	0.11–1.1
	Cephalexin	0.1–2.5
	Cephazolin	0.2–6.2
	Ciprofloxacin	0.03–125
	Clarithromycin	0.2–3
	Coprofloxacin	0.85–2
	Doxycycline	0.1–6.7
	Erythromycin	27–83
	Lincomycin	0.3–2
	Metronidazole	0.1–90
	Norfloxacin	0.029–44
	Ofloxacin	0.353–37
	Oxytetracycline	0.01–3.75
	Penicillin G	0.85–5.2
	Sulfamethoxazole	0.04–83
	Tetracycline	0.01–4.2
	Trimethoprim	0.01–15
Psychiatric drugs	Carbamazepine	0.54–2.1
	Fluoxetine	0.018–0.19
	Lorazepam	0.17–0.80
Anti-hypertensives	Diltiazem	0.71–1.6
	Enalapril	0.07–0.40

(continued)

Table 3 (continued)

Therapeutic class	Investigated compounds	Range in HWW $\mu g \, L^{-1}$
	Hydrochlorothiazide	0.54–5.5
	Lisinopril	0.08–0.61
Anti-neoplastics	5-fluorouracil	5–124[a]
	Methotrexate	1
	Cyclophosphamide	0.019–4.5
	Ifosfamide	0.01–1.9
	Tamoxifen	0.001–0.04
Beta-blockers	Atenolol	2.2–6.6
	Metoprolol	0.42–25
	Pindolol	0.032–0.26
	Propranolol	0.030–6.2
	Sotalol	0.61–6.7
Hormones	17b-estradiol, E2	0.028–0.043
	Estriol, E3	0.353–0.502
	Estrone, E1	0.017–0.034
	Ethinylestradiol, EE2	0.015–0.018
Contrast media	Iobitridol	0.1–3213
	Iopromide	0.2–2,500
	Iomeprol	0.01–1392
Anti-diabetics	Glibenclamide	0.048–0.113
Barbiturates	Butalbital	0.014–0.48
	Pentobarbital	0.011–0.15
	Phenobarbital	0.013–0.36
Beta-agonists	Clenbuterol	0.86–1.19
	Salbutamol	0.04–0.14
Diuretics	Furosemide	5.3–18
Lipid regulators	Atorvastatin	0.062–0.31
	Bezafibrate	0.042–2.9
	Clofibric acid	0.01–0.043
	Gemfibrozil	0.014–0.064
	Mevastatin	0.068–2.0
	Pravastatin	0.064–1.1
Receptor antagonists	Cimetidine	0.019–0.26
	Famotidine	0.048–0.087
	Loratadine	0.015–0.026
	Ranitidine	0.24–4.1

[a]In the sewage of an oncology ward

Reported concentrations generally refer to those measured in 24-h composite samples, giving their average values over the course of the day, thereby "balancing" the fluctuations in their concentrations during the 24-h monitoring period. PhC concentration trends have been investigated by several authors; to give a general idea of their findings, data reported for paracetamol [4] and ciprofloxacin [7] are shown in Fig. 4. Profiles of other PhCs are documented in [9, 24].

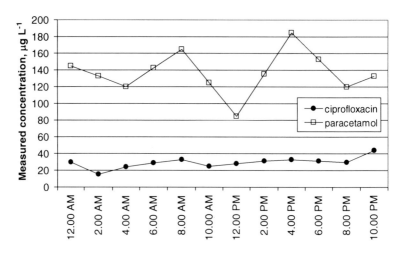

Fig. 4 Variation in hospital effluent concentrations of two common pharmaceutical compounds over a 24-h period

Unfortunately, not only is chemical detection of these micro-pollutants costly, time-consuming and complex, it also requires very expensive instrumentation and specialized personnel; furthermore, authorization for sampling hospital water is not always forthcoming. Hence, some authors have suggested the use of *predicted* environmental concentrations (PECs) as an alternative method of characterizing hospital effluents [34, 48, 49]. PECs may be calculated from annual consumption values [50, 51] using formulae such as the simplified model given in (1); this estimates the PEC on the basis of the medicament consumption M (g) in a given reference period (generally a year), excretion values $f_{excreted}$ (%) and water volume V consumed inside the structure during the same reference period (m³).

$$PEC = \frac{M\, f_{excreted}}{V} \tag{1}$$

Consumption of medicaments is usually expressed in terms of number of daily defined doses (DDD) administered, n_{DDD}, as set by the World Health Organization [52]. The correlation between M and n_{DDD} is expressed by (2):

$$M = n_{DDD}\, DDD_{API} \tag{2}$$

where DDD_{API} is the daily defined dose of the pharmaceutical compound in question. As an example, consider that for medicament P, the corresponding DDD_{API} is equal to 3 mg of its active ingredient (API) and that a packet contains 60 mg of this active ingredient. This means that each packet contains $60/3 = 20$ DDD_{API}. If 2,000 packets are consumed in the reference period, the corresponding consumption expressed as n_{DDD} is equal to $2,000 \times 20$, i.e. 40 000, while the mass of API consumed in the same period is $40,000 \times 3$ or 120,000 mg.

However, predicted concentrations do not always accurately reflect those measured; this is due to several reasons, including variation in consumption patterns throughout the year, the use of an inaccurate excretion rate $f_{excreted}$, UV exposure setting off unavoidable photodegradation processes during sampling and transportation, inappropriate sample conservation methods and adsorption onto particulates present in wastewater [34, 49].

3 Fate of Pharmaceutical Compounds

As mentioned earlier on, after administration, PhCs are eliminated as a mixture of unchanged parent compounds and their metabolites. Although it would seem obvious that a highly metabolized compound (i.e. with a low excretion rate) is more easily degraded in the environment, a negative correlation has in fact been found between the proportion of pharmaceuticals excreted and their concentrations in the environment, thereby suggesting that poorly excreted pharmaceuticals may have an inherently low environmental degradability [38]. In fact, once in the sewage network, pharmaceuticals may embark on different pathways, perhaps exhibiting great stability and persistence in the environment or perhaps being subjected to volatilization, biological or chemical degradation and/or sorption onto solids or other particles.

The complex behaviour of pharmaceuticals in the sewage network and during subsequent wastewater treatment is correlated to the nature of their molecular structure, which may contain concomitant acidic and basic functional groups, as in the case of both ciprofloxacin and ceftazidime. This implies that these molecules may be considered as neutral, cationic, anionic or zwitterionic [53], according to the particular environmental conditions, which will consequently affect their behaviour.

Hence, a working knowledge of the physicochemical properties of PhCs may help provide a (rough) prediction of the processes occurring during their passage through WWTPs. These processes may involve sorption onto solids, biodegradation or chemical transformation, and, after their discharge into a surface water body, residual PhCs may be subjected to photolysis and photodegradation, which may result in a reduction of their potential environmental impact [3, 53].

PhC properties most investigated by scientists to date are their water solubility (s, mg/mL), volatility (correlated to the Henry constant H) ($\mu g\ m^{-3}$ air/$\mu g\ m^{-3}$ wastewater), biodegradability (correlated to pseudo-first-order degradation constant k_{biol} L $gSS^{-1}\ d^{-1}$), acid dissociation constant (K_a), distribution and sorption (through the sludge-water distribution coefficient K_d, expressed in L gSS^{-1} or the octanol-water partition coefficient K_{ow}). The main focus has been to find any correlations between these parameters and to determine PhC removal rates during the different treatment steps. Thus, different properties have been quantified for many compounds, and software, such as EPI Suite 4.00 [54], consenting their estimation, is available.

One property, volatilization, does not seem to be a removal pathway of any significance for these micro-pollutants. This is due to the fact that PhCs in general are characterized by extremely low H values, in the range 10^{-6}–10^{-27} µg m^{-3} air/µg m^{-3} wastewater [55]; the amount of compound being stripped from the water into the gas phase is consistent if its Henry constant H is > 0.1.

Biodegradability, on the other hand, has even been proposed as a means of classifying PhCs by Ternes and Joss and Joss et al. [56, 57]. In this system, compounds with $k_{biol} > 10$ L g SS^{-1} d^{-1} are classified as highly biodegraded, those with k_{biol} ranging between 0.1 and 10 L g SS^{-1} d^{-1} are moderately biodegraded, and finally, those with $k_{biol} < 0.1$ L gSS^{-1} d^{-1} are scarcely removed by biological transformation. The parameter k_{biol} is influenced by several factors: the reactor operating conditions, in particular temperature; the biochemical versatility of the sludge and hence the sludge retention time (SRT) or sludge age; the availability of a co-substrate; and the fraction of inert matter contained within the sludge. Several studies have reported that the k_{biol} of membrane biological reactors (MBRs) is quite often higher than that of CAS for many PhCs [56, 58]; this is presumably primarily due to the higher SRT achievable in an MBR. In fact, longer SRTs consent the development of slower growing bacteria (i.e. nitrifying bacteria), which in turn provide a more diverse community of micro-organisms with broader physiological capabilities, enhancing metabolic and co-metabolic processes and thereby helping to degrade recalcitrant compounds and promote complete mineralization [59, 60].

Sorption of pharmaceuticals onto the surface of particulate matter or their distribution between two phases (water and either sludge, sediment or soil) depends on many factors, the most important being liquid phase pH and redox potential, the stereochemical structure and chemical nature of both the pharmaceutical compound and the sorbent, the lipophilicity of the sorbed molecules (excellent sorption at log $K_{ow} > 4$, low sorption at log $K_{ow} < 2.4$), the sludge-water distribution coefficient K_d ($K_d > 2$ L g SS^{-1} good sorption, $K_d \leq 0.3$ L g SS^{-1} low sorption), the extent of neutral and ionic species present in the wastewater and the characteristics of the suspended particles. Moreover, the presence of humic and fulvic substances may alter the surface properties of the sludge, as well as the number of sites available for sorption and reactions, thereby enhancing or suppressing sorption of PhCs [38, 55, 61].

The sorption behaviour of antibiotics, in particular, can be very complex and therefore difficult to assess. As an example, ciprofloxacin has a log K_{ow} equal to 0.28, conferring it with a small tendency to leave the aquatic phase; however, it does sorb well onto active sludge or sediments [62, 63] in WWTP. Tetracyclines (log K_{ow} equal to 1.4), on the other hand, form complexes with double cations (calcium and magnesium) present in the water [64] and also tend to adsorb onto the surface of complexes between humic acids and hydrous Al oxide [65].

To enhance the removal of substances with high sorption properties, for example diclofenac (log $K_d = 2.7$ and log $K_{ow} = 4.5$–4.8), ferric and aluminium salts may be added, increasing their removal rate to as much as 50–70% [66]. This strategy can also improve the removal of acidic compounds, for example naproxen, by ionic or chelating interactions [55].

Parent compounds or modified forms of drugs in wastewaters tend to be either hydrolysed or conjugated. Hydrolysis of derivatives may also result in the production of the parent compounds at a later stage, during sewage treatment or once discharged into a receiving body, thereby representing an additional indirect (endogenous) source of environmental release of the drug [67]. Carbamazepine, for example, is excreted as glucuronides, which may act as a reservoir from which a later yield of the parent substance can occur [36, 68].

To further complicate matters, concentrations of certain PhCs have been found between the limit of quantification and the limit of detection of the investigation technique and unfailingly below the predicted environmental concentration. This occurred, for instance, with the anti-neoplastic tamoxifen [45], and in this case could be ascribed to one or more of several factors. One hypothesis is that tamoxifen could be degraded prior to analysis; indeed, it is known to be sensitive to UV light, with up to 90% being degraded in 5 days [69]. In this scenario, even if analysis was performed as soon as possible and the samples were protected from light in the interim, the effect of photodegradation could not entirely be eliminated. Another possible culprit is tamoxifen's high lipophilicity (estimated value of $\log K_{ow} = 6.3$); this makes it easy to adsorb onto particulate matter, which would settle to the bottom in sewer systems and therefore escape analysis. Another reason for the discrepancy between measured and predicted values of tamoxifen concentrations could be overestimation of its PEC due to the adoption of an erroneously exaggerated excretion rate for the unchanged compound.

Once treated effluent leaves the WWTP, further degradation of the residual PhC APIs may still occur in surface water bodies. In fact, as mentioned above, if a substance is light sensitive, photodecomposition may contribute to its further removal once in the environment. Phototransformation easily takes place in clear surface water, and the effectiveness of this process is strictly correlated to the intensity and frequency of available light [53, 70]. Nonetheless, this process may be affected by other parameters, specifically pH, water hardness, location, season and latitude [71, 72].

Nevertheless, although antibiotics such as quinolones, tetracyclines and sulphonamides are light sensitive, not all compounds are photodegradable [71, 73, 74]; indeed, the significance and extent of direct and indirect photolysis of antibiotics in the aquatic environment are unique for each compound. It is therefore evident that data regarding indirect photolysis of PhCs and their interaction with dissolved organic matter (humic and fulvic acids) would improve knowledge regarding the fate of antibiotics and other emerging contaminants in surface waters; however, thus far, very few studies have been devised to investigate this process [75].

It is not only the properties of the compound in question that influence its behaviour but also the environmental and operational conditions it is subjected to (temperature, pressure, pH, redox conditions), as well as the particular configurations of the (biological) reactors (in particular, their partitioning into compartments featuring different conditions: mainly aerobic, anoxic and anaerobic), SRT and, to variable extents, HRT. Moreover, in chemical processes, reagent dosage and contact time are fundamental parameters in determining PhC behaviour [24, 56, 60], as will be discussed in the following paragraphs.

4 Hospital Effluent Management Strategies and PhC Removal Efficiencies of Various Treatments

According to [12], four possible scenarios for the management and treatment of HWWs can be envisaged, namely: (a) direct discharge into a surface water body, (b) co-treatment in a municipal WWTP, (c) on-site treatment and subsequent discharge into a surface water body and, finally, (d) on-site treatment prior to co-treatment at a municipal WWTP. The choice of the strategy to adopt should be based not only on economic constraints but also on technical data, including assessment of the ecotoxicological risk posed by particular hospital wastewaters. Furthermore, different operational set-ups should be tested, in order to provide meaningful information about the financial aspects and overall risks associated with proposed future HWW treatment strategies.

At the present time, most developing countries deal with hospital effluents by direct discharge into the environment without any treatment (Option *a* from the list above); this unsurprisingly results in widespread contamination of the water cycle by both micro- and macropollutants [76]. In contrast, in most European countries, hospital effluents are discharged into the public sewage system and conveyed to municipal WWTPs (Option *b*); these were originally built and have more recently been upgraded, with the aim of removing nutrients (carbon, nitrogen and phosphorus compounds) and micro-organisms and pollutants, which commonly arrive at the plant in concentrations of the order of several milligram per litre, or at least 10^5 CFU/100 mL, and are therefore not sensitive enough to remove micropollutants, typically present in quantities of $\mu g\ L^{-1}$, effectively [9].

In contrast, Option *d* promises the highest risk reductions but implies the highest costs; furthermore, in the case of a co-treatment at a municipal WWTP, HWW flow rates quite often amount to only a small percentage of the total influent flow rate. Consequently, although dilution of HWWs with UWWs usually results in a decrease in the PhC content in the final effluent (from $\mu g\ L^{-1}$ to $ng\ L^{-1}$), the total load, that is to say the quantity released daily into the receiving water body, is not affected. It is evident, therefore, that scenario *c* could be the best option, as an expensive, highly effective small-scale technology could eventually prove to be more eco- and cost-effective than a relatively cheap large-scale solution with less effects on the diluted hospital effluent.

All that being said, cost is not the only issue, and decision-makers also need to bear in mind several scientific considerations. Unfortunately, few studies have thus far looked into treatment and treatability of hospital effluents, and therefore the majority of PhC removal rate data reported below are based on studies of UWWs passing through conventional and (biological and chemical) advanced treatment plants. Attention has recently been focussed on the risks posed by residual PhCs in the treated effluents, but ecotoxicological data are not always available for the single compounds under study, and even less is known about their behaviour as part of a mixture, as will be discussed in detail later on. Nonetheless, lessons have been learned from several in-depth reviews, among them [9, 55] and experimental investigations (cited case by case), and are reported below, with reference to the different treatment steps.

4.1 Coagulation/Flocculation and Flotation

- Coagulation/flocculation/precipitation of HWWs by means of $FeCl_3$ or $Al_2(SO_4)_3$ seems to be a suitable option for removing lipophilic compounds, such as diclofenac, although it is unable to eliminate many other common hydrophilic PhCs, including carbamazepine, iopromide, diazepam and antibiotics (i.e. roxythromycin, erythromycin, trimethoprim), from the liquid phase [66].
- Suspended solids, which were found in up to three times higher concentrations in hospital effluent than municipal sewage, can be very efficiently removed by coagulation/flocculation [8, 16], thereby averting their accumulation on primary and secondary sludge.
- Persistent compounds such as carbamazepine seem to be eliminated to a moderate extent (about 20%) by flotation [66].

4.2 Biological Treatments

- Secondary biological treatments are considered to be an effective barrier to most PhCs due to the metabolic and co-metabolic processes at work in these systems [58, 77, 78].
- Removal efficiencies are enhanced at longer SRTs (>25–30 d), although some compounds feature thresholds beyond which removal rates do not improve [77].
- With respect to CAS, MBRs exhibit similar removal rates for simple PhC molecules, such as ibuprofen [79].
- MBR outperforms CAS for many compounds, furnishing in some cases a 30–50% higher removal rate. Moreover, MBR has consistently shown 40–65% improvement in the elimination of some compounds recalcitrant to CAS treatment (mefenamic acid, indomethacin, diclofenac, gemfibrozil) [58, 80].
- No relationship has been found between PhC chemical structure and removal rate. However, the range of variation in removal rate seen in MBR is small for most compounds, while in CAS greater fluctuations are observed [58, 80].
- The superiority of MBR processes has also been suggested in terms of the removal of pathogenic micro-organisms, including several viruses [81, 82].

- Separation of a biological reactor into reactor cascades can improve performance [5, 6, 56].
- Nitrifying bacteria play a key role in the biodegradation of pharmaceuticals in WWTPs operating at higher SRTs [60].
- Biological systems featuring nitrogen treatment seem to achieve higher removal rates for PhCs with respect to other treatments, such as submerged biofilters or fixed biomass reactors [60, 68, 83].

4.3   Ozonation and Advanced Oxidation Processes

- These techniques are promising candidates for efficient degradation of PhCs in wastewaters. They commonly produce an increase in the BOD_5/COD ratio and improve the biodegradability of persistent substances such as antibiotics, cytostatic agents, hormones, X-ray contrast media, carbamazepine and some acidic drugs like clofibric acid. Ozone-based advanced oxidation processes (AOPs) (O_3/H_2O_2, O_3/UV), Fenton-type processes and photochemical AOPs are generally more effective than ozonation alone, due to enhanced generation of hydroxyl radicals and photon-initiated cleavage of carbon-halogen bonds, and are therefore recommended for the treatment of these recalcitrant substances [84]. The dose of ozone commonly applied ranges between 5 and 15 mg L^{-1}, depending on the COD in the wastewater in question, at a contact time of about 15–30 min [56].
- The extent of PPCP removal is, in general, >90%. The presence of particulate matter at concentrations regularly measured in the secondary effluent does not influence the removal efficiency of soluble compounds showing high reaction rates with ozone.

4.4   Adsorption onto Activated Carbon

- Adsorption onto both powdered and granular forms of activated carbon, PAC and GAC, respectively, shows great potential for the removal of trace emerging contaminants, in particular, non-polar compounds with a log K_{ow} >2. PAC dose or GAC regeneration and replacement are critical for excellent removal rates [85].
- According to [86], only 9 of the 66 PPCPs examined had a removal efficiency of less than 50% at a dose of 5 mg/L PAC at 5-h contact time.
- Nevertheless, the carbon regeneration/disposal issue cannot be ignored; PAC must be disposed of in landfills or other solid handling methods, while spent GAC may be regenerated. However, as thermal regeneration of GAC requires a significant quantity of energy, which may lead indirectly to greater environmental risks than the presence of trace micro-pollutants, a cost/benefits assessment should be performed to take these factors into account.

4.5   Reverse Osmosis and Nanofiltration

- Reverse osmosis (RO) removes greater than 90% of many compounds, and in numerous cases, RO is able to reduce their concentration below their detection limit. Nonetheless, lower removal rates have been noted in the case of diclofenac (55.2–60%) and ketoprofen (64.3%) [79].
- Nanofiltration (NF) also seems to be a promising alternative for eliminating pharmaceuticals, as it is able to achieve removal rates greater than 90% [87].
- The removal of PhCs by NF membranes occurs via a combination of three mechanisms: adsorption, sieving and electrostatic repulsion. Removal efficiency can vary widely from compound to compound, as it is strictly correlated to (a) the physicochemical properties of the micro-pollutant in question, (b) the properties of the membrane itself (permeability, pore size, hydrophobicity and surface charge) and (c) the operating conditions, such as flux, transmembrane pressure, rejections/recovery and water feed quality.
- However, once again, the issue of RO/NF brine disposal needs to be taken on board; in general, brine is far more toxic than the influent water, so it cannot be disposed of in natural water bodies.

4.6   Disinfection

- Ozone amounts required for PhC oxidation lead to partial disinfection. However, it is expected that, as for sorbed compounds, micro-organisms incorporated into particulate matter would be significantly shielded from ozone or OH radicals. Nonetheless, concentrations of 5–10 mg O_3 L^{-1} and contact times of 15–20 min are sufficient to obtain a reduction of 2–3 log units (99–99.9%). Good rates of bacteria and virus removal in the effluent from an infectious disease ward were achieved by the addition of 10 mg L^{-1} of ClO_2 at a contact time of 30 min [88]. That study also investigated the toxicological effects of disinfection using a chlorine derivative, NaClO, revealing that doses of 1–8 mg L^{-1} can greatly reduce bacterial pollution but tend to contribute to the formation of AOX in HWWs, thereby giving rise to toxic effects on aquatic organisms.

4.7   Natural Polishing by Constructed Wetlands

- Constructed wetlands (CWs) can promote removal of PhCs through a number of different mechanisms, including photolysis, plant uptake, microbial degradation and sorption to the soil. The main benefits of horizontal and vertical subsurface flow systems are the existence of aerobic, anaerobic and anoxic redox conditions in proximity to plant rhizomes; this provides an ideal environment for reducing concentrations of a variety of drug compounds, as some PhCs are best reduced under aerobic conditions (ibuprofen), whereas removal of others is favoured by anaerobic conditions (clofibric acid, diclofenac), and halogenated pollutants are eliminated at a higher rate in anoxic conditions [89].

- Experimental studies conducted by Galletti et al. and Matamoros et al. [89, 90] have shown that aerobic conditions are generally more efficient than anaerobic pathways in removal of the majority of emerging contaminants.
- In addition, certain PhCs (like ketoprofen and diclofenac) can also be eliminated from aquatic environments by the photodegradation processes that take place in surface flow systems [70, 91]. The photo-and biodegradation reactions involved in contaminant removal are promoted by high hydraulic retention times.

In order to complete this snapshot of the processes currently available for treating HWW, several case studies from the literature are reported below; they all detail recent dedicated WWTPs constructed specifically for HWWs in different countries.

4.8   Waldbrol Hospital (350 Beds) (Cologne, Germany)

The hospital is of medium size, with 342 beds. It is situated near Cologne, Germany, in a catchment area of roughly 10,000 inhabitants and therefore features a bed density of 33.5 beds per 1,000 inhabitants, quite high with respect to the German national average of 6.2 beds per 1,000 inhabitants. Effluent from the hospital ($130 \ m^3 \ d^{-1}$) is treated in a dedicated plant, which consists of a screening (mesh of 1 mm); a primary sedimentation stage ($V = 21 \ m^3$ and a minimum retention time of 1 h); an MBR, in stable operation since 2007; and an anoxic-aerobic tank ($V = 56 \ m^3$, SRT = 100 d and mixed liquor concentration $10–12 \ g \ L^{-1}$) [92]. It is equipped with 5 Kubota EK 400 flat sheet *micro-filtration* membrane modules.

Its performance was monitored via nine target compounds, chosen according to their prevalence, persistence in the environment, frequency of administration and medical relevance of their active ingredients. The removal rates achievable using the MBR was then monitored for these nine compounds, revealing poor performance (<20%) for diclofenac, good for ciprofloxacin, moxifloxacin and tramadol (50–80%) and very good (>80%) for bisoprolol, clarithromycin, ibuprofen and metronidazole. Carbamazepine was detected at a higher concentration in the permeate, presumably due to its re-formation in the wastewater from its metabolites. The permeate concentrations of the different monitored compounds exceeded quite often the value of 100 ng L^{-1}. This value is the target one set by the International Association of Waterworks in the Rhine catchment area and also the recommendation of the German Federal Ministry for the Environment. It should be considered that this value applies to water systems that are part of drinking water supply networks, and for this reason, Beier et al. [93] recommended the addition of an advanced technology such as activated carbon adsorption, ozone treatment or a further membrane step (nanofiltration or reverse osmosis) to remove residual PhCs. In this regard, experimental investigations on RO and NF pilot plants have shown that RO is sufficient to remove all traces of PhCs present in the MBR permeate. The recommended yield of a two-stage RO is 70%, which results in a retentate side-stream of 9%.

4.9   Cona Hospital (900 Beds) (Near Ferrara, Italy)

Another example of dedicated HWW treatment was reported by Verlicchi et al. [13] and describes a new 900-bed hospital situated in a small catchment area (about 2,000 p.e.) in the town outskirts, with a bed density of 450 beds per 1,000 inhabitants. The treatment employed consists of a full-scale MBR (SRT 30 d and biomass concentration 8–10 kg m^{-3}) equipped with *ultrafiltration* membranes (pore size 0.01 μm and surface flux 15–25 L m^{-2} h^{-1}), followed by an advanced oxidation step using O_3 (7.5–10 mg O_3 L^{-1} and 16 min contact time, with reference to the average flow rate) and UV (100 mJ cm^{-2} and exposure time 6 s, with reference to the average flow rate). This last phase was designed to guarantee good removal of various kinds of PhCs, and a multi-barrier treatment was chosen because the treated effluent is discharged into an effluent-dominant river used for irrigation in the summer season. To address this issue, the following technologies able to greatly reduce sludge production and disposal frequency were adopted: aerobic digestion; ozonation (50 g O_3/kg TSS treated) of the digested sludge to reduce the excess sludge and favour oxidation of the PhCs adsorbed onto the solid phase, increasing the content of readily biodegradable COD; and mechanical thickening.

4.10   Other Cases

Direct chlorination, or primary treatment followed by chlorination, is still the most widely used basic method of treating HWWs in some developing countries. This has the aim of preventing the spread of pathogenic micro-organisms, the causal agents of nosocomial infections [13, 94]. The widespread medical use of chlorine as a disinfectant is due to its (potential) capacity to remove bacteria, virus and fungi and to its ease of application. However, high quantities of this type of disinfectant are required to treat sewage, and only a modest elimination rate can be achieved due to the organic load of the wastewater being treated. Moreover, due to the formation of chloramines and other undesirable disinfection by-products (DBPs), residual toxicity in the treated effluent is unavoidable, as reported by Emmanuel et al. [88]. Bearing this in mind, in the last decade, China, due to its application of wastewater re-use strategies and accordingly stringent regulations, has seen the introduction of

MBR plants with capacities ranging from 20 to 2,000 m^3 d^{-1}, mostly exploiting submerged technology, for the treatment of HWWs [94].

5 Environmental Risk Assessment of Hospital Effluents

The wide variety in PhC chemical structure combined with the specific biological activities they are generally designed to trigger, their surface-active nature and their persistence, as well as the persistence of some of their known metabolites, make this a group of environmental pollutants whose behaviour and environmental risk need to be promptly addressed [95, 96], not only with respect to the impact on the WWTP charged with treating the effluent but also as regards the receiving water body. To this end, the environmental fate and effects of the most administered and most persistent pharmaceuticals compounds need to be assessed. In fact, this necessity was clearly stated in European Council Directive 93/39/EEC [97]: *if applicable, reasons for any precautionary and safety measures (have) to be taken for the storage of the medicinal product, its administration to patients and for disposal of waste products, together with an indication of any potential risks presented by medicinal product for the environment.*

Further to this directive, the European Medicines Agency (EMEA) began, in 2001 [98], to outline a feasible risk assessment scheme, which was later published in a draft form in 2003 [50]; in 2006 [51], the following five-step procedure was adopted, with a view to guaranteeing that possible long-term and low-level effects of the new substances are not overlooked; this procedure is also used in the licensing of new compounds (including pharmaceuticals).

1. *Hazard identification*, consisting of the assessment of the potential environmental risk of a substance based on its physicochemical properties and expected environmental pathways.
2. *Exposure assessment*, quantification or prediction of environmental concentrations of the substance based on on-site measurements (MEC) or its projected fate and behaviour (PEC), respectively.
3. *Effects assessment*, by, as in the case of risk assessment for chemicals and pesticides, determining a set of marker organisms (including algae, zebrafish, insect larvae, benthic worm, water flea, etc.) that represent ecosystem components and food networks and are used to indicate acute and chronic effects. This step is also used to define the predicted no-effect concentrations (PNECs).
4. *Risk characterization* based on comparing either predicted or measured environmental concentrations with effects data for the most sensitive organisms (PEC/PNEC or MEC/PNEC, respectively). An environmental risk is considered unacceptable if the ratio equals or exceeds 1. In general, this phase considers the worst-case scenario.
5. *Risk management*, i.e. the identification of the most appropriate actions necessary to reduce or mitigate the identified risks.

Despite this clearly outlined procedure, scientific progress in the field has been slow, as very few authors have as yet attempted environmental risk assessment of PhCs in water, focussing their attention, furthermore, mainly on *parent compounds* rather than their *metabolites*, on the effects of *individual substances* rather than *mixtures* on *target organisms* and on *acute* rather than *chronic toxicity*. In particular, metabolite analysis tended to be disregarded as their exposure is very difficult to assess due to a lack of consensus in the literature regarding excreted metabolite fractions; moreover, analysis has shown that their relative contribution to the overall risk is typically low [99].

Target organisms are in general simultaneously exposed to great numbers of different substances, which may exert additional antagonist or synergic effects. Hence, the potential effects of the *mixtures* of compounds on the organisms need in-depth analysis [1], particularly as this aspect is still unresolved in currently implemented risk assessment schemes. What is more, few attempts have even been made to propose a strategy capable of evaluating this factor. However, if one assumes that mixtures of components exerting the same mode of toxic action act according to the concentration addition model, the combined effect of the components is expected to be equal to the sum of the concentrations of each substance, expressed as a fraction of its own individual toxicity [100]. If all components act according to a strictly different mode of action, however, they must be assumed to act accord to the model of independent action, rather than concentration addition [101, 102].

As regards toxicity, once again, few studies dealing with this issue and very little data are available at the present time. It should also come as no surprise that more short- than long-term data have been reported, making chronic toxicity more difficult to evaluate. Moreover, it is worth noting that it is generally the risk of chemicals resulting from anthropogenic activities affecting organisms in the environment that is assessed, rather than the danger to humans posed by chemicals in the environment (e.g. via consumption of polluted drinking water).

The environmental risk assessment approach most commonly adopted consists of estimation of the *risk quotient* (RQ) (as suggested by Hernando et al. [103]), which is defined as the ratio between the environmental concentration (measured or predicted, respectively MEC and PEC) and the predicted no-effect concentration (PNEC), and can be used to collocate compounds in one of three risk bands: $RQ < 0.1$, minimal risk to aquatic organisms; $0.1 \leq RQ < 1$, median risk; and $RQ \geq 1$, high risk [103–105]. In their risk assessment calculations, [106], further to [107], estimated PNEC values at 1,000 times lower than the most sensitive species assayed, so as to take into account the effect on other, potentially more sensitive, aquatic species to those used in toxicity studies.

PNECs themselves are preferentially extrapolated from chronic toxicity data or, if no long-term data are available, acute toxicity data [51]; they generally refer to algae, daphnia or fish toxicity. Sanderson et al., Stuer-Lauridsen et al., Boillot and Ferrari et al. [108–111] have reported PNECs for a large list of PhCs, and [49, 112] recently proposed a model for the predicting PNEC of a substance, taking into consideration its acidic or basic behaviour in the environment. This procedure provides relationships that *rescale* previous toxicity values through a correction parameter (hydrophobicity descriptor), which corrects for speciation at pH 7 in organic acids and bases.

Escher et al. [49] also proposed that PNEC calculations for mixtures could be based on a common species, assuming concentration addition for risk assessment. They suggested assessing the risk to algae, daphnia and fish individually for each of the n monitored compounds present in the water and then selecting the species with the highest resulting RQ_{mix} defined by (3). In this way, the risk quotients of the individual pharmaceuticals can be added up to yield a sum risk quotient (RQ_{mix}).

$$RQ_{mix} = \sum_{i=1}^{n} RQ_i = \sum_{i=1}^{n} \frac{PEC_i}{PNEC_i} \qquad (3)$$

Recent technical literature [109, 113, 114] has shown that the concentrations of pharmaceutical and personal care products (PPCPs) in aquatic environments may exceed their PNECs. The majority of these studies were performed on WWTP effluent discharged into the surface water body, and very few studies have been conducted on treated hospital effluent.

One study that has tackled this issue, however, was recently published by de Souza et al. [104], who set up an environmental risk assessment of the 21 intravenous antibiotics most used in an intensive care unit of a hospital in Curitiba (Brazil). They evaluated the RQ_i based on PEC_i both in the raw effluent and after a dedicated conventional biological treatment. They found that, in the raw effluent from the ward, the environmental risk was high for 15 compounds, medium for 4 and low for 2; similarly, the treated effluent was labelled as high risk in terms of 14 compounds, medium for 5 and low for 2.

Likewise, Verlicchi et al. [24] monitored the effluent of a large Italian hospital (900 beds) and the influent and effluent of the nearby municipal biological wastewater treatment plant (average influent flow rate of 28,000 m^3/day) where the hospital effluent is treated (its flow rate contributes roughly 2% on the influent hydraulic load) in terms of 73 pharmaceuticals belonging to 12 different therapeutic classes. They performed an environmental risk analysis using an RQ value calculated using the maximum MEC and the PNEC as a risk marker. These analyses revealed that, in the hospital effluent, 9 of the 73 substances tested (the four analgesics/anti-inflammatories: acetaminophen, ibuprofen, naproxen and salicylic acid; the four antibiotics: clarithromycin, erythromycin, ofloxacin and sulfamethoxazole; and the psychiatric drug fluoxetine) posed a potentially high ecotoxicological risk, while five (codeine, indomethacin, clenbuterol, atenolol, metoprolol and propranolol) posed a medium risk. Only five compounds were found to pose a high level of risk in the WWTP influent and effluent (the same antibiotics and the psychiatric drug), whereas RQ classification showed that a moderate risk was posed by the concentrations of acetaminophen, ibuprofen, naproxen, salicylic acid, clenbuterol, metoprolol, propranolol and gemfibrozil in the WWTP influent and,

more importantly, salicylic acid, clenbuterol, propranolol, fenofibrate and gemfibrozil in its effluent.

In a study by Escher et al. [49], environmental risk assessment of 31 of the 100 most
commonly administered pharmaceutical compounds was based on the ratio between
PEC and PNEC. PECs were estimated in the effluent of a medium-sized hospital (338
beds) in Switzerland and in the effluent discharged from a subsequent dedicated (ideal)
conventional biological treatment, whose quality was predicted based on removal rates
reported in the literature for the top 100 pharmaceuticals administered in the hospital;
PNECs were defined on the basis of the procedure set out in [49]. This study revealed
that in the untreated hospital effluent, risk was high for ten compounds (amiodarone,
clotrimazole, ritonavir, progesterone, meclozine, atorvastatin isoflurane, tribenoside,
ibuprofen and clopidogrel), medium for 8 (amoxicillin, diclofenac, 4-methylaminoan-
tipyrine, floxacillin, salicylic, paracetamol, azithromycin and thiopental) and low for 13
(oxazepam, valsartan, clarithromycin, rifampicin, tramadol, carbamazepine, tetra-
caine, sevelamer, metoclopramide, dipyridamole, pravastatin, prednisolone and erythromycin). After the biological treatment, the risk remained high for 8 of the previous
10 substances, namely amiodarone, clotrimazole, ritonavir, meclozine, atorvastatin,
isoflurane, tribenoside and clopidogrel; the risk was medium for 2 out of the 8, i.e.
diclofenac and thiopental, and low for the remaining 21 substances under evaluation.

This simulation and the two case studies reveal that conventional biological
treatment is not able to consistently reduce PhC content in the treated effluent to
low risk levels. This is significant, as CAS is the most treatment step most common
adopted in municipal WWTPs, in which HWWs are often co-treated with UWWs.

Furthermore, it is not only PhCs that pose a significant ecotoxicological risk.
Indeed, Boillot et al. [23] investigated the presence of specific pollutants other than
drugs in hospital effluent, namely AOX, glutaraldehydes, free chlorine, detergents,
Freon 113, alcohols, acetone, formaldehyde, acetaldehyde, ammonium, phenols
and several metals (copper, zinc, lead and arsenic), finding that many of them
(in particular, free chlorine, 2-propanol, ammonium, ethanol, Cu, Pb and Zn) make
a considerable contribution to its ecotoxicity. The study also showed that ecotoxicological characteristics of HWW presented hour-by-hour fluctuations and that the
highest values occurred during the day.

Emmanuel et al. [19] proposed a framework for ecotoxicological risk assessment
of raw hospital wastewater directly discharging into an urban sewer network, based
on the effects of the presence of conventional parameters (BOD_5, COD, TSS, NH_4,
faecal bacteria), chlorides, heavy metals and AOX on both the receiving WWTP
and the natural aquatic ecosystem. They then applied the proposal methodology to
the effluent of an infectious and tropical diseases department of a hospital of a large
city in France, finding that the most critical of these parameters are ammonia,
BOD_5, COD and AOX.

Expanding the assessment parameters somewhat, however, the individual substance risk quotients reported in the literature show that the most critical compounds
in hospital effluents are certain antibiotics, anti-neoplastics and disinfectants. Among
these, antibiotics merit special attention due to their biological activity, which leads to
their potential to generate multi-resistant bacteria, even in the presence of weak

antibiotic concentrations [21, 115, 116]. Furthermore, cytostatins and fluoroquinolone antibiotics found in hospital effluents can possess mutagenic and bacteriotoxic properties [117, 118], and, due to the wide spectrum of antibiotic classes in use, they can adversely affect many species [119]. In addition, both antibiotics and disinfectants can disturb and inhibit biological treatment processes [120].

In conclusion, the authors of the cited studies all agree that further research into environmental risk assessment of hospital effluents, incorporating different types of substances used in care and diagnostic activities, as well as cleaning operations (pharmaceuticals, detergents, disinfectants, heavy metals, macropollutants), is vital. Moreover, further studies need to be focussed on evaluating the risk posed by pollutant mixtures, and work is needed to validate the predictive models proposed thus far [19, 49], to evaluate chronic toxicity due to PhCs and their mixtures and to provide experimental data pertaining to specific case studies.

6   Best Treatment Options for Hospital Effluents and Conclusions

Some guidelines for decision-makers charged with finding optimal solutions for dealing with HWW are suggested below; they are based on the expected/observed fate, behaviour and environmental risk of PhCs, and the ability of proposed treatments to guarantee the highest removal rates and to preserve aquatic environment from persistent compounds.

First and foremost, proper management of hospital effluent should take into consideration the characteristics of both the hospital structure and its catchment area. In case of a *new* healthcare structure, the main aspects to consider when devising its wastewater management strategy are:

• Evaluation of the hospital in question, paying particular attention to:

1. Number of beds
2. Type of wards
3. Any diagnostic and research activities
4. Any services (kitchen, laundry, etc.)
5. Average and maximum flow rate
6. Type of sewage (combined or separate)

• Assessment of the catchment area, in particular:

1. Size
2. Number of residents and non-residents (p.e.)
3. Average and maximum urban flow rates
4. Type of sewage (combined or separate)

5. Any industrial activities present in the area (type, WW flow rate, adopted pre-treatments within the factory battery limits, final disposal of the effluent, co-treatment with other kind of WWs, etc.)
6. Characteristics of existing WWTPs (nominal capacity, residual capacity, treatment sequence, authorized limits for the final discharge)
7. Features of the receiving water body (hydraulic regime, auto-depurative capacity, irrigation, recreational and industrial uses, etc.)
8. Legal and regulatory constraints for the discharge into public sewage network and in the environment.

It would be erroneous to consider hospital effluents as having the same pollutant nature as urban wastewaters; as a rule of thumb, a hospital bed is equivalent to 100 p.e. in terms of PhCs excreted annually (2.3 kg PhCs bed^{-1} year^{-1} against 23 g p.e. $^{-1}$ year^{-1}) and 2–3 p.e. in terms macropollutant yield [9, 49]. Furthermore, Hartemann et al. [21] postulated that if a hospital of 1,000 beds has an internal laundry, it is as polluting as a town with a population of 10,000 people.

Local assessment of the mass balance of the most critical PhCs (the most frequently administered antibiotics, analgesics/anti-inflammatories and the psychiatric drug carbamazepine, considered an anthropogenic marker in wastewaters [121]) also provides useful information about the PhC contribution of the hospital effluent with respect to that of the catchment area. The extent of this contribution will differ between compound, but Beier et al. [92], in particular, reported that it can reach as high as 94% for some antibiotics (ciprofloxacin), although Kummerer [53], on the other hand, stated that only up to 25% of the antibiotics administered in Germany are used in hospitals.

When considering the type of treatment strategy to adopt, it is generally advisable for a multi-barrier system to be installed; this will enable different kinds of conventional and persistent compounds (PhCs) to be removed or eliminated by different steps. In fact, different chemicals require different operational conditions (aerobic, anoxic, anaerobic ones) to be efficiently removed from the water phase.

A *biological* step is always necessary to remove the carbonaceous fraction from the influent wastewater; suspended biomass treatments are the most common. These entail long SRTs (>25–30 d), and compartmentalization of the biological reactor is necessary for the removal of recalcitrant compounds. Furthermore, as many micro-pollutants tend to adsorb/absorb to the biomass flocks, efficient solid/liquid separation can greatly improve their removal from wastewater and, at the same time, guarantee consistently good effluent quality. MBRs have been suggested for this purpose by many authors [9, 58, 80, 93], some of whom found that ultrafiltration (UF) membranes are more efficient than MF membranes [9, 93].

Powerful *oxidant* treatments are also necessary to remove multi-drug-resistant bacteria [122] and residues of medicaments with ecotoxicological effects from hospital effluent, in order to reduce their impact on the final receiving water bodies. Advanced oxidation processes, including the Fenton reaction, are likely candidates [123].

A dedicated treatment for HWWs is always desirable, especially for large hospitals in rural areas, where treated effluent may be indirectly re-used for irrigation after its discharge into a surface water body. In fact, although co-treatment with UWWs at a municipal WWTP is a common practice, it has several fundamental drawbacks. In particular, dilution of HWWs with UWWs tends to defeat the object, as some substances in the hospital effluents may cause inhibition of the treatment plant biomass and thereby reduce its removal efficiency.

However, should building a new treatment plant not be feasible, there are two strategies available for removing micro-pollutants at conventional WWTPs: either to optimize existing technologies or to upgrade the plant with new technologies [49]. The most time-effective and, as a consequence, the most often adopted of these approaches are end-of-pipe treatments, including ozonation of the secondary effluent [124] or powdered activated carbon [86]. Indeed, although WWTPs must be reliable and "robust", their treatment sequences should be designed bearing in mind the future necessity of upgrading it in response to regulatory changes and scientific advice. Particular attention should also be paid to the treatment and disposal of chemical and biological sludge produced by the treatment processes. Source separation of urine and faeces is still under discussion. Pharmaceuticals are excreted to a great extent in urine, but according to Escher et al. [49], high-risk pharmaceuticals are excreted mainly in the faeces; source separation is therefore not a viable option for reducing the potential risk of hospital effluent.

Based on our experience and the results of the most recently published studies [76], we suggest that the following the measures are adopted in the following scenarios:

1. *Large hospital in a small catchment area*: a dedicated treatment should be adopted, featuring advanced biological and oxidation processes (AOPs), such as ultrafiltration membrane biological reactors (MBRs), and then ozone/UV, which also guarantees efficient disinfection.
2. *Large hospital in a densely populated catchment area*: verify the type of the sewer system for both hospital and urban settlement and evaluate the treatment capacity of the existing municipal WWTP to determine whether a dedicated line could be added for the HWW. Alternatively, in the case of co-treatment, upgrade the existing treatment sequence to guarantee removal of the most persistent compounds: AOP is the best technology available to date.
3. *Small hospital in a densely populated catchment area*: evaluate the specific contribution of HWWs to the total WWTP influent, in particular, the most critical PhCs administered in the health structure. Evaluate the technical and economic feasibility of adopting dedicated specific treatments for HWWs. Evaluate advantages and drawbacks of co-treatment.
4. *Small hospital in a small urban catchment area*: a local mass balance analysis of micro- and macropollutant loads can provide useful information about the contribution of the different users. Environmental risk assessment of the expected final effluent and analysis of the characteristics of the local receiving water body will guide selection of the advanced treatment sequence (MBR, ozone, UV).

5. *Basic treatment*: in developing countries where UWWs are also barely treated and only few WWTPs are in function, HWWs should be dealt with on site through a basic treatment sequence (pre-treatment, biological treatment and disinfection) in order to safeguard the receiving water body from the spread of micro-pollutants and micro-organisms. Municipal WWTPs should be conceived and constructed, taking into consideration the necessity of controlling pollution from hospital effluents. In this context, administrators and technicians will need to evaluate the possibility of adopting dedicated lines for specific WWs (including HWWs) in the same placement area as the municipal WWTP, especially in the case of large hospitals.

References

1. Kolpin DW, Furlong ET, Meyer MT, Thurman EM, Zaugg SD, Barber LB, Buxton HT (2002) Pharmaceuticals, hormones, and other organic wastewater contaminants in U.S. streams, 1999-2000: a national reconnaissance. Environ Sci Technol 36:1202–1211
2. Foster AL (2007) Occurrence and fate of endocrine disruptors through the San Marco Wastewater treatment plant. Thesis of Master of Science, Texas State University, USA
3. Fatta-Kassinos D, Meric S, Nikolaou A (2011) Pharmaceutical residues in environmental waters and wastewater: current status of knowledge and future research. Anal Bioanal Chem 399:251–275
4. Khan S, Ongerth J (2005) Occurrence and removal of pharmaceuticals at an Australian sewage treatment plant. Water 32:35–39
5. Joss A, Andersen H, Ternes T, Richle PR, Siegrist H (2004) Removal of estrogen in municipal wastewater treatment under aerobic and anaerobic conditions: consequences for plant optimization. Environ Sci Technol 38:3047–3055
6. Joss A, Keller E, Alder AC, Gobel A, McArdell CS, Ternes T, Siegrist H (2005) Removal of pharmaceuticals and fragrances in biological wastewater treatment. Water Res 39:3139–3152
7. Duong H, Pham N, Nguyen H, Hoang T, Pham H, Ca Pham V, Berg M, Giger W, Alder A (2008) Occurrence, fate and antibiotic resistance of fluoroquinolone antibacterials in hospital wastewaters in Hanoi, Vietnam. Chemosphere 72:968–973
8. Suarez S, Lema JM, Omil F (2009) Pre-treatment of hospital wastewater by coagulation-flocculation and flotation. Bioresour Technol 100:2138–2146
9. Verlicchi P, Galletti A, Petrovic M, Barceló D (2010) Hospital effluents as a source of emerging pollutants: an overview of micropollutants and sustainable treatment options. J Hydrol 389:416–428
10. Sim WJ, Lee JW, Lee ES, Shin SK, Hwang SR, Oh JE (2011) Occurrence and distribution of pharmaceuticals in wastewater from households, livestock farms, hospitals and pharmaceutical manufactures. Chemosphere 82:179–186
11. Hawkshead JJ (2008) Hospital wastewater containing pharmaceuticals active compounds and drug-resistant organisms: a source of environmental toxicity and increased antibiotic resistance. J Residual Sci Tech 5:51–60
12. Pauwels B, Verstraete W (2006) The treatment of hospital wastewater: an appraisal. J Water Health 4:405–416
13. Verlicchi P, Galletti A, Masotti L (2010) Management of hospital wastewaters: the case of the effluent of a large hospital situated in a small town. Water Sci Technol 61:2507–2519
14. Schuster A, Hadrick C, Kummerer K (2008) Flows of active pharmaceutical ingredients originating from health care practises on a local, regional, and national level in Germany – is

hospital effluent treatment an effective approach for risk reduction? Water Air Soil Pollut Focus 8:457–471

15. Leprat P (1999) Caractéristiques et impacts des rejets liquides hospitaliers. Tech Hosp 634:56–57

16. Altin A, Altin S, Degirmenci M (2003) Characteristics and treatability of hospital (medical) wastewaters. Fresenius Environ Bull 12:1098–1108

17. Kummerer K, Erbe T, Gartiser S, Brinker L (1998) AOX-emissions from hospital into municipal wastewater. Chemosphere 36:2437–2445

18. Wen X, Ding H, Huang X, Liu R (2004) Treatment of hospital wastewater using a submerged membrane bioreactor. Process Biochem 39:1427–1431

19. Emmanuel E, Perrodin Y, Keck G, Blanchard JM, Vermande P (2005) Ecotoxicological risk assessment of hospital wastewater: a proposed framework for raw effluents discharging into urban sewer network. J Hazard Mater A117:1–11

20. Emmanuel E, Pierre MG, Perrodin Y (2009) Groundwater contamination by microbiological and chemical substances released from hospital wastewater: health risk assessment for drinking water consumers. Environ Int 35:718–726

21. Hartemann P, Hautemaniere A, Joyeux M (2005) La problématique des effluents hospitaliers. Hygiène 13:369–374

22. Tsakoma M, Anagnostopoulou E, Gidarakos E (2007) Hospital waste management and toxicity evaluation: a case study. Waste Manag 27:912–920

23. Boillot C, Bazin C, Tissot-Guerraz F, Droguet J, Perraud M, Cetre JC, Trepo D, Perrodin Y (2008) Daily physicochemical, microbiological and ecotoxicological fluctuations of a hospital effluent according to technical and care activities. Sci Total Environ 403:113–129

24. Verlicchi P, Galletti A, Al Aukidy M (2011) Hospital wastewaters: quali-quantitative characterization and strategies for their treatment and disposal. In: Sharma SK, Sanghi R (eds) Water treatment and pollution prevention: advances in research. Springer (in press)

25. Lopez N, Deblonde T, Hartemann Ph (2010) Les effluents liquid hospitaliers. Hygiène 18:405–410

26. Gomez MJ, Aguera A, Mezcua M, Hurtado J, Mocholi F, Fernandez-Alba AR (2007) Simultaneous analysis of neutral and acidic pharmaceuticals as well as related compounds by gas chromatography-tandem mass spectrometry in wastewater. Talanta 73:314–320

27. Kummerer K, Henninger A (2003) Promoting resistance by the emission of antibiotics from hospital and households into effluent. Clin Microbiol Infect 9:1203–1214

28. Schwartz T, Kohnen W, Jansen B, Obst U (2003) Detection of antibiotic-resistant bacteria and their resistance genes in wastewater, surface water, and water biofilms. FEMS Microbiol Ecol 43:325–335

29. Adelowo O, Fagade OE, Oke AJ (2008) Prevalence of co-resistance to disinfectants and clinically relevant antibiotics in bacterial isolates from three hospital laboratory wastewaters in South-western Nigeria. World J Microbiol Biotechnol 24:1993–1997

30. Chitnis V, Chitnis S, Vaidya K, Ravikant S, Patil S, Chitnis DS (2004) Bacterial population changes in hospital effluent treatment plant in central India. Water Res 38:441–447

31. Prado T, Silva DM, Guilayn WC, Rose TL, Gaspar AMC, Miagostovich MP (2011) Quantification and molecular characterization of enteric viruses detected in effluents from two hospital wastewater treatment plants. Water Res 45:1287–1297

32. Kummerer K (2001) Drugs in the environment: emission of drugs, diagnostic aids and disinfectants into wastewater by hospital in relation to other sources - a review. Chemosphere 45:957–969

33. Kummerer K, Helmers E, Hubner P, Mascart G, Milandri M, Reinthaler F, Zwakenberg M (1999) European hospitals as a source for platinum in the environment in comparison with other sources. Sci Total Environ 225:155–165

34. Lenz K, Mahnik SN, Weissenbacher N, Mader RM, Krenn P, Hann S, Koellensperger G, Uhl M, Knasmuller S, Ferk F, Bursch W, Fuerhacker M (2007) Monitoring, removal and risk assessment of cytostatic drugs in hospital wastewater. Water Sci Technol 56:141–149

35. Kummerer K, Helmers E (2000) Hospital effluents as a source of gadolinium in the aquatic environment. Environ Sci Technol 34:573–577
36. Verlicchi P, Galletti A, Masotti L (2008) Caratterizzazione e trattabilità di reflui ospedalieri: in-dagine sperimentale (con sistemi MBR) presso un ospedale dell'area ferrarese, Proc. Sidisa Conference Florence, Italy
37. Barceló D (2003) Emerging pollutants in water analysis. Trac-Trend Anal Chem 22:xiv–xvi
38. Jjemba PK (2006) Excretion and ecotoxicity of pharmaceuticals and personal care products in the environment. Ecotox Environ Safe 63:113–130
39. Halling-Sorensen B, Nors Nielsen S, Lanzky PF, Ingerslev F, Holten Lutzhoft HC, Jorgensen SE (1998) Occurrence, fate and effects of pharmaceutical substances in the environment- a review. Chemosphere 36:357–393
40. Lienert J, Burki T, Escher BI (2007) Reducing micropollutants with source control: substance flow analysis of 212 pharmaceuticals in faeces and urine. Water Sci Technol 55:87–96
41. Heberer T (2002) Occurrence, fate, and removal of pharmaceutical residues in the aquatic environment: a review of recent research data. Toxicol Lett 131:5–17
42. Verlicchi P, Al Aukidy M, Galletti A, Petrovic M, Barcelò D (2011) Hospital effluent: investigation of the concentrations and distribution of pharmaceuticals and environmental risk assessment (in press)
43. Lin AYC, Yu TH, Lin CF (2008) Pharmaceutical contamination in residential, industrial, and agricultural waste streams: risk to aqueous environments in Taiwan. Chemosphere 74:131–141
44. Galletti A, Verlicchi P, Al Aukidy M, Petrovic M, Barcelò D (2011) Hospital as a source of emerging contaminants (pharmaceuticals): results of an investigation on its final effluent chemical characteristics. Proc. CEST 2011, Rhodes A578–A587
45. Tauxe-Wuersch A, De Alencastro LP, Grandjean D, Tarradellas J (2006) Trace determination of tamoxifen and 5-fluorouracil in hospital and urban wastewaters. Int J Environ Anal Chem 86:473–485
46. Gomez MJ, Petrovic M, Fernandez-Alba AR, Barcelò D (2006) Determination of pharmaceuticals of various therapeutic classes by solid-phase extraction and liquid chromatography-tandem mass spectrometry analysis in hospital effluent wastewaters. J Chromatogr A 1114:224–233
47. Nagarnaik P, Batt A, Boulanger B (2011) Source characterization of nervous active pharmaceutical ingredients in healthcare facility wastewaters. J Environ Manage 92:872–877
48. Heberer T, Feldman D (2005) Contribution of effluents from hospital and private households to the total loads of diclofenac and carbamazepine in municipal sewage effluents – modeling versus measurements. J Hazard Mater 122:211–218
49. Escher BI, Baumgartner R, Koller M, Treyer K, Lienert J, McArdell CS (2011) Environmental toxicology and risk assessment of pharmaceuticals from hospital wastewater. Water Res 45:75–92
50. EMEA (2003) Note for guidance on environmental risk assessment of medicinal products for human use. DRAFT. CPMP/SWP/4447/00 draft corr. The European Agency for the Evaluation of Medical Products
51. EMEA (2006) Guideline on the environmental risk assessment of medicinal products for human use CHMP/SWP/4447/00. The European Agency for the Evaluation of Medicinal Products, London
52. DDD data web site URL: http://www.whocc.no/atc_ddd_index/. Accessed 2 Nov 2011
53. Kummerer K (2009) Antibiotics in the aquatic environment – a review – Part I. Chemosphere 75:417–439
54. Epi: suite program available at the website (accessed 28 Oct 2011) http://www.epa.gov/oppt/exposure/pubs/episuite.htm
55. Suarez S, Carballa M, Omil F, Lema JM (2008) How are pharmaceutical and personal care products (PPCPs) removed from urban wastewaters? Rev Environ Sci Biotechnol 7:125–138

56. Ternes TA, Joss A (2006) Human Pharmaceuticals, Hormones and Fragrances. The challenge of micropollutants in urban water management. IWA Publishing, London

57. Joss A, Zabczynsk S, Gobel A, Hoffmann B, Loffler D, McArdell CS, Ternes TA, Thomsen A, Siegrist H (2006) Biological degradation of pharmaceuticals in municipal wastewater treatment: proposing a classification scheme. Water Res 40:1686–1696

58. Radjenovic J, Petrovic M, Barceló D (2009) Fate and distribution of pharmaceuticals in wastewater and sewage sludge of the conventional activated sludge (CAS) and advanced membrane bioreactor (MBR) treatment. Water Res 43:831–841

59. Clara M, Kreuzinger N, Strenn B, Gans O, Kroiss H (2005) The solids retention time—a suitable design parameter to evaluate the capacity of wastewater treatment plants to remove micropollutants. Water Res 39:97–106

60. USEPA (2009) Nutrient Control Design Manual, State of Technology Review report, EPA/600/R-09/012, U.S. Environmental Protection Agency, Washington, D.C.

61. Kummerer K (2009) The presence of pharmaceuticals in the environment due to human use – present knowledge and future challenges. J Environ Manage 90:2354–2366

62. Golet EM, Alder AC, Giger W (2002) Environmental exposure and risk assessment of fluoroquinolone antibacterial agents in wastewater and river water of the Glatt Valley Watershed, Switzerland. Environ Sci Technol 36:3645–3651

63. Gu C, Karthikeyan KG (2005) Sorption of the antimicrobial ciprofloxacin to aluminum and iron hydrous oxides. Environ Sci Technol 39:9166–9173

64. Christian T, Schneider RJ, Färber HA, Skutlarek D, Meyer MT, Goldbach HE (2003) Determination of antibiotic residues in manure, soil, and surface waters. Acta Hydrochim Hydrobiol 31:36–44

65. Gu C, Karthikeyan KG (2008) Sorption of the antibiotic tetracycline to humic-mineral complexes. J Environ Qual 37:704–711

66. Carballa M, Omil F, Lema JM (2005) Removal of cosmetic ingredients and pharmaceuticals in sewage primary treatment. Water Res 39:4790–4796

67. Bendz D, Paxéus NA, Ginn TR, Loge FJ (2005) Occurrence and fate of pharmaceutically active in the environment, a case study: Hoje River in Sweden. J Hazard Mater 122:195–204

68. Vieno N, Tuhkanen T, Kronberg L (2007) Elimination of pharmaceuticals in sewage treatment plants in Finland. Water Res 41:1001–1012

69. Jalonen HG (1988) Simultaneous determination of tamoxifen citrate and its E-isomer impurity in bulk drug and tablets by high-performance liquid-chromatography. J Pharm Sci 77:810–813

70. Andreozzi R, Raffaele M, Nicklas P (2003) Pharmaceuticals in STP effluents and their solar photodegradation in aquatic environment. Chemosphere 50:1319–1330

71. Werner JJ, Arnold WA, McNeill K (2006) Water hardness as a photochemical parameter: tetracycline photolysis as a function of calcium concentration, magnesium concentration, and pH. Environ Sci Technol 40:7236–7241

72. Kallenborn R, Fick J, Lindberg R, Moe M, Nielsen KM, Tysklind M, Vasskog T (2008) Pharmaceutical residues in Northern European environments: consequences and perspectives. In: Kümmerer K (ed) Pharmaceuticals in the environment. Sources, fate, effects and risk, 3rd edn. Springer, Berlin, pp 61–74

73. Turiel E, Bordin G, Rodríguez AR (2005) Study of the evolution and degradation products of ciprofloxacin and oxolinic acid in river water samples by HPLC–UV/MS/MS–MS. J Environ Monit 7:189–195

74. Hu L, Flanders PM, Miller PL, Strathmann TJ (2008) Oxidation of sulfamethoxazole and related antimicrobial agents by TiO2 photocatalysis. Water Res 41:2612–2626

75. Sukul P, Lamshöft M, Zühlke S, Spiteller M (2008) Photolysis of (14)C-sulfadiazine in water and manure. Chemosphere 71:717–725

76. Verlicchi P, Galletti A, Al Aukidy M (2011) Best practices in the management of hospital effluents. Proc. VI EWRA International Symposium Water Engineering and Management in a Changing Environment, Catania, Italy

77. Kreuzinger N, Clara M, Droiss H (2004) Relevance of the Sludge Retention Time (SRT) as de-sign criteria for wastewater treatment plants for the removal of endocrine disruptors and pharmaceuticals from wastewater. Water Sci Technol 50:149–156
78. Kimura K, Hara H, Watanabe Y (2007) Elimination of selected acidic pharmaceuticals from municipal wastewaters by activated sludge systems and membrane bioreactors. Environ Sci Technol 41:3708–3714
79. Oppenheimer J, Wert EC, Yoon Y (2007) Role of membranes and activated carbon in the removal of endocrine disruptors and pharmaceuticals. Desalination 202:156–181
80. Radjenovic J, Petrovic M, Barceló D (2007) Analysis of pharmaceuticals in wastewater and removal using a membrane bioreactor. Anal Bioanal Chem 387:1365–1377
81. Ottoson J, Hansen A, Bjorlenius B, Norder H, Stenstrom TA (2006) Removal of viruses, parasitic protozoa and microbial indicators in conventional and membrane processes in a wastewater pilot plant. Water Res 40:1449–1457
82. Zhang K, Farahbakhsh K (2007) Removal of native coliphages and coliform bacteria from municipal wastewater by various wastewater treatment processes: implications to water reuse. Water Res 41:2816–2824
83. Batt AL, Kim S, Aga DS (2007) Comparison of the occurrence of antibiotics in four-full scale wastewater treatment plants with varying designs and operations. Chemosphere 68:428–435
84. Ikehata K, Naghashkar NJ, El-Din MG (2006) Degradation of aqueous pharmaceuticals by ozonation and advanced oxidation processes: a review. Ozone-Sci Eng 28:353–414
85. Bolong N, Ismail AF, Salim MR, Matsuura T (2009) A review of the effects of emerging contaminants in wastewaters and options for their removal. Desalination 239:229–246
86. Snyder SA, Adham S, Redding AM, Cannon FS, DeCarolis J, Oppenheimer J, Wert EC, Yoon Y (2006) Role of membranes and activated carbon in the removal of endocrine disruptors and pharmaceuticals. Desalination 202:156–181
87. Watkinson AJ, Murby EJ, Costanzo SD (2007) Removal of antibiotics in conventional and advanced wastewater treatment: implications for environmental discharge and wastewater recycling. Water Res 41:4164–4174
88. Emmanuel E, Keck G, Blanchard J, Vermande P, Perrodin Y (2004) Toxicological effects of disinfections using sodium hypochlorite on aquatic organisms and its contribution to AOX formation in hospital wastewater. Environ Int 30:891–900
89. Galletti A, Al Aukidy M, Verlicchi P, Petrovic M, Barcelò D (2010) Pharmaceuticals removal in a H-SSF treating a secondary domestic wastewater – an experimental investigation. Proc. IWA Conference Wetland Systems for Water Pollution Control, Venice, 382–384
90. Matamoros V, Garcia J, Bajona JM (2008) Organic micropollutant removal in a full-scale surface flow constructed wetland fed with secondary effluent. Water Res 42:653–660
91. Bartels P, von Tumpling W (2008) The environmental fate of the antiviral drug oseltamivir carboxylate in different waters. Sci Total Environ 405:215–225
92. Beier S, Cramer C, Koster S, Mauer C, Palmowski L, Schroeder HF, Pinnekamp J (2011) Full scale membrane bioreactor treatment of hospital wastewater as forerunner for hot-spot wastewater treatment solutions in high density urban areas. Water Sci Technol 63:66–71
93. Beier S, Koster S, Veltmann K, Schroder HFr, Pinnekamp J (2010) Treatment of hospital wastewater effluent by nanofiltration and reverse osmosis. Water Sci Technol 61:1691–1698
94. Liu Q, Zhou Y, Chen L, Zheng X (2010) Application of MBR for hospital wastewater treatment in China. Desalination 250:605–608
95. Bound JP, Voulvoulis N (2006) Predicted and measured concentrations for selected pharmaceuticals in UK rivers: implications for risk assessment. Water Res 40:2585–2592
96. Furhacker M (2008) The Water Framework Directive – can we reach the target? Water Sci Technol 57:9–17
97. Council Directive 93/39/EEC 14 June 1993 amending Directives 65/65/EEC, 75/318/EEC and 75/319/EEC in respect of medicinal products

98. EMEA (2001) Discussion paper on environmental risk assessment on non-genetically modified organisms (non-GMO) containing medicinal products for human use. CPMP/ SWP/4447/00 draft corr. The European Agency for the Evaluation of Medical Products
99. Lienert J, Gudel K, Escher BI (2007) Screening method for ecotoxicological hazard assessment of 42 pharmaceuticals considering human metabolism and excretory routes. Environ Sci Technol 41:4471–4478
100. Brown VM (1968) The calculation of the acute toxicity of mixture of poisons to rainbow trout. Water Res 2:723–733
101. Altenburger R, Nendza M, Schuurmann G (2003) Mixture toxicity and its modeling by quantitative structure-activity relationships. Environ Toxicol Chem 22:1900–1915
102. Altenburger R, Walter H, Grote M (2004) What contributes to the combined effect of a complex mixture? Environ Sci Technol 38:6353–6362
103. Hernando MD, Mezcua M, Fernandez-Alba AR, Barceló D (2006) Environmental risk assessment of pharmaceutical residues in wastewater effluents, surface waters and sediments. Talanta 69:334–342
104. de Souza SML, de Vasconcelos EC, Dziedzic M (2009) Environmental risk assessment of antibiotics: an intensive care unit analysis. Chemosphere 77:962–967
105. Zhao JL, Ying GG, Liu YS, Chen F, Yang JF, Wang L, Yang XB, Stauber JL, Warne MSt J (2010) Occurrence and screening level risk assessment of human pharmaceutical in the Pearl River system, South China. Environ Toxicol Chem 29:1377–1384
106. Tauxe-Wuersch A, De Alencastro LF, Grandjean D, Tarradellas J (2005) Occurrence of several acidic drugs in sewage treatment plants in Switzerland and risk assessment. Water Res 39:1761–1772
107. EC (2003) Technical guidance on risk assessment, TGD Part II. Technical Report, Institute for health and consumer protection, European Chemicals Bureau, European Commission (EC)
108. Sanderson H, Johnson DJ, Wilson CJ, Brain RA, Solomon KR (2003) Probabilistic hazard assessment of environmentally occurring pharmaceuticals toxicity to fish, Daphnia and algae by ECOSAR screening. Toxicol Lett 144:383–395
109. Stuer-Lauridsen F, Birkved M, Hansen LP, Holten Lutzhoft H-C, Halling-Sùrensen B (2000) Environmental risk assessment of human pharmaceuticals in Denmark after normal therapeutic use. Chemosphere 40:783–793
110. Boillot C (2008) Évaluation des risques écotoxicologiques liés aux rejets d'effluents hospitaliers dans les milieux aquatiques. Contribution à l'amélioration de la phase « caractérisation des effets ». PhD Thesis. Institut National des Sciences Appliquées de Lyon, France, N° d'ordre ISAL 0021
111. Ferrari B, Paxeus N, Lo Giudice R, Pollio A, Garric J (2003) Ecotoxicological impact of pharmaceuticals found in treated wastewaters: study of carbamazepine, clofibric acid, and diclofenac. Ecotox Environ Safe 55:359–370
112. Escher BI, Baumgartner R, Lienert J, Fenner K (2009) In: Boxall ABA (ed) Reaction and processes, Part P – transformation products of synthetic chemicals in the environment, vol 2. Springer, Berlin, pp 205–244. doi:210.1007/1978-1003-1540-88273-88272
113. Santos JL, Aparicio I, Alonso E (2007) Occurrence and risk assessment of pharmaceutically active compounds in wastewater treatment plants. A case study: Seville city (Spain). Environ Int 33:596–601
114. Santos LHMLM, Araujo AN, Fachini A, Pena A, Delerue-Matos C, Montenegro MCBSM (2010) Ecotoxicological aspects related to the presence of pharmaceuticals in the aquatic environment. J Hazard Mater 175:45–95
115. Chitnis V, Chitnis D, Patil S, Kant R (2000) Hospital effluent as a source of multiple drug-resistant bacteria. Curr Sci India 79:989–991
116. Kummerer K (2004) Resistance in the environment. J Antimicrob Chemother 54:311–320

117. Jolibois B, Guerbet M (2005) Evaluation of industrial, hospital and domestic wastewater genotoxicity with the Salmonella fluctuation test and the SOS chromotest. Mutat Res 565 (2):151–162
118. Ferk F, Misik M, Grummt T, Majer B, Fuerhacker M, Buchmann C, Vital M, Uhl M, Lenz K, Grillitsch B, Parzefall W, Nersesyan A, Knasmuller S (2009) Genotoxic effects of wastewater from an oncological ward. Mutat Res 672:69–75
119. Hirsch R, Ternes TA, Haberer K, Kratz KL (1999) Occurrence of antibiotics in the aquatic environment. Sci Total Environ 225:109–118
120. Al-Almad A, Daschner FD, Kummerer K (1999) Biodegradability of cefotiam, ciprofloxacin, meropenem, penicillin G and sulfamethoxazole and inhibition of wastewater bacteria. Arch Environ Contam Toxicol 37:158–163
121. Clara M, Strenn B, Kreuzinger N (2004) Carbamazepine as a possible anthropogenic marker in the aquatic environment: investigations on the behaviour of carbamazepine in wastewater treatment and during groundwater infiltration. Water Res 38:947–954
122. Blaise C, Gagné F, Eulaffroy P, Férard J-F (2007) Ecotoxicology of selected pharmaceuticals of urban origin discharged to the Saint-Lauwence river (Québec, Canada): a review. Braz J Aquat Sci Technol 10:29–51
123. Berto J, Rochenbach GC, Barreiros MAA, Correa AXR, Silva SP, Radetski CM (2009) Physico-chemical microbiological and ecotoxicological evaluation of a septic tank/Fenton reaction combination for the treatment of hospital wastewaters. Ecotox Environ Safe 72: 1076–1081
124. Hollender J, Zimmermann SG, Koepke S, Krauss M, McArdell CS, Ort C, Singer H, von Gunten U, Siegrist H (2009) Elimination of organic micropollutants in a municipal wastewater treatment plant Upgraded with a full-scale Post ozonation followed by sand filtration. Environ Sci Technol 43:7862–7869

Antibiotic Resistance in Waste Water and Surface Water and Human Health Implications

Célia M. Manaia, Ivone Vaz-Moreira, and Olga C. Nunes

Abstract The utilization of antibiotics to control infectious diseases is one of the biggest advances in human and veterinary health care. However, the generalized use of antibiotics has been accompanied by a worrisome increase in the prevalence of antibiotic-resistant bacteria. This evidence motivated numerous studies on the diversity and distribution of antibiotic-resistant bacteria and resistance genetic determinants not only in clinic but also in different environmental compartments. Given the particular importance that the anthropic water cycle (waste water/surface water/drinking water) may have in the development and dissemination of antibiotic-resistant organisms, this chapter aims at summarizing the recent advances in this area. Sections 1 and 2 are an Introduction to antibiotic resistance, summarizing some mechanisms and modes of resistance acquisition. In Sect. 3, the contribution of the environmental pollution and other anthropic pressures for antibiotic resistance evolution is discussed. The use of different methodologies and the limitations to achieve general conclusions on the characterization and quantification of antibiotic resistance in aquatic environments are examined in Sects. 4 and 7. Sections 5–7 summarize recent evidences on the widespread distribution of antibiotic resistance in different compartments of the anthropic water cycle. The scarcity of studies giving evidences on the direct effect of anthropic pressures on antibiotic resistance acquisition and maintenance in treated waste/drinking waters is highlighted. The contribution of bacterial community rearrangement, imposed by water treatment processes, on the increase of antibiotic resistance is discussed.

C.M. Manaia (✉) and I. Vaz-Moreira
CBQF/Escola Superior de Biotecnologia, Universidade Católica Portuguesa, R. Dr. António Bernardino de Almeida, 4200-072 Porto, Portugal
e-mail: cmmanaia@esb.ucp.pt

O.C. Nunes
LEPAE – Departamento de Engenharia Química, Faculdade de Engenharia, Universidade do Porto, 4200-465 Porto, Portugal

D. Barceló (ed.), *Emerging Organic Contaminants and Human Health*,
Hdb Env Chem (2012) 20: 173–212, DOI 10.1007/698_2011_118,
© Springer-Verlag Berlin Heidelberg 2011, Published online: 20 October 2011

Keywords Resistance genes, Resistome, Selective pressure, Water

Contents

Abbreviations

16S rRNA	16S small subunit ribosomal RNA
30S	Small subunit of the 70S ribosome of prokaryotes
50S	Large subunit of the 70S ribosome of prokaryotes
A	Aminoglycosides
aac(3)-I	Genes encoding 3-*N*-aminoglycoside acetyltransferases; confer resistance to aminoglycosides
aac(6′)-Ib-cr	Gene encoding an aminoglycoside acetyltransferase; confers reduced susceptibility to ciprofloxacin and norfloxacin
*aad*A	Genes encoding aminoglycoside-3″-adenylyltransferases (AAD); confer resistance to streptomycin and spectinomycin
AIA	Antibiotic impregnated agar
*amp*C	Gene encoding chromosomal β-lactamase; confers resistance to β-lactams
Ap	Amphenicol
AP	Ampicillin
*aph*A	Gene encoding acid phosphatase/phosphotransferase; confers resistance to aminoglycosides

APUA	Alliance for the Prudent Use of Antibiotics
AVn	Average values of n plants
AX	Amoxicillin
bla(TEM,CTX-M/GES/OXA/PER/SHV/TLA/VEB)	Genes encoding extended spectrum β-lactamases; confer resistance to β-lactams
blaNDM-1	Gene encoding for the New Delhi metallo-β-lactamase-1; confers resistance to almost all β-lactams
BSAC	British Society for Antimicrobial Chemotherapy
C	Ciprofloxacin
CA-SFM	Antibiogram Committee of the French Society for Microbiology
cat	Genes encoding chloramphenicol acetyltransferases; confer resistance to chloramphenicol
CDC	Centres for Disease Control and Prevention
CFU	Colony forming units
CLSI	Clinical and Laboratory Standards Institute
cmr	Gene encoding a putative efflux pump; confers resistance to chloramphenicol
COST	European Cooperation in Science and Technology
DANMAP	The Danish Integrated Antimicrobial Resistance Monitoring and Research programme
DARE	Detecting evolutionary hot spots of antibiotic resistance in Europe
DDM	Disc diffusion method
dfr	Genes encoding dihydrofolate reductases; confers resistance to trimethoprim
DGGE	Denaturating gradient gel electrophoresis
DNA	Deoxyribonucleic acid
EARS-Net	European Antimicrobial Resistance Surveillance Network
ECDC	European Centre for Disease Prevention and Control
ECOFF	Epidemiological cut-off values

erm(A/E)	Genes encoding rRNA methylase; confers resistance to erythromycin
erm(B/C/F)	Genes encoding rRNA methylases; confer cross-resistance to macrolides, lincosamides and streptogramin B
ESAC	European Surveillance of Antimicrobial Consumption
EUCAST	European Committee on Antimicrobial Susceptibility Testing
FE	Final effluent
*flo*R	Gene encoding an exporter protein that specifically exports amphenicol antibiotics
G	Glycopeptides
*gyr*A	Gene encoding DNA gyrase subunit A; gene mutation confers resistance to ciprofloxacin
L	β-lactam
M	Macrolide
MD	Microdilution method
*mec*A	Gene encoding penicillin binding protein 2; confers resistance to penicillins
MMPN	Modified most probable number
*msr*A	Gene encoding methionine sulfoxide reductase A; confers resistance to erythromycin
n.a.	Not available
NA	Nalidixic acid
NARMS/USA	National Antimicrobial Resistance Monitoring System (United States of America)
*par*C	Gene encoding DNA topoisomerase IV subunit A; gene mutation confers resistance to quinolone
PCR	Polymerase chain reaction
Q	Quinolone
qac	Gene encoding multidrug transporters
qepA	Gene encoding an efflux pump; confers resistance to fluoroquinolone
qnr	Genes encoding Qnr proteins, capable of protecting DNA gyrase; confer resistance to quinolone
qPCR	Quantitative real time polymerase chain reaction

RNA	Ribonucleic acid
RV*n*	Range values for *n* plants
S	Sulfonamide
sat(1–2)	Genes encoding a nourseothricin *N*-acetyltransferase; confer resistance to aminoglycosides
str	Genes encoding phosphotransferases; confer resistance to streptomycin
sul(I–II)	Genes encoding a drug-resistant dihydropteroate synthase enzyme required for folate biosynthesis; confer resistance to sulfonamide
T	Tetracycline
tet(A–D/K/L/Y)	Genes encoding efflux pumps; confer resistance to tetracyclines
tet(M/O/Q/W)	Genes encoding proteins protecting the ribosome from the inhibiting effects of tetracycline
*tet*R	Gene encoding a repressor protein, which regulates the tetracycline efflux system genes
Ts	Trimethoprim
TT	Tertiary treatment implemented
van	Genes encoding D-alanine:D-alanine ligases with a broad substrate specificity; confer inducible resistance to the glycopeptides antibiotics, as vancomycin
WHO	World Health Organization
WT	Wild-type

1 Antibiotic Resistance Is a Natural Property of Bacteria

Most of the antibiotics commercially available nowadays are derivatives of natural compounds produced by bacteria or fungi. It is widely accepted that in nature these secondary metabolites can act as weapons for microbial cell defence, inhibiting the growth of competitors. However, it seems that antibiotics have, in nature, more sophisticated and complex functions [1–3]. Many environmental bacteria can not only cope with natural antimicrobial substances but also benefit from their presence. For instance, the use of antibiotics by bacteria as biochemical signals, modulators of metabolic activity or even carbon sources has been demonstrated [1, 2, 4]. In other cases, antibiotics can be tolerated because they have structures similar to the natural substrates of bacterial housekeeping enzymes and thus are inactivated, leading to a natural form of resistance [2]. These are just some examples that illustrate that antibiotic resistance is a natural property of bacteria, eventually as old as bacteria themselves [5–7]. Indeed, the genetic potential of bacteria to live in the presence of antibiotics has been consistently demonstrated [3, 8]. Before the massive introduction of antibiotics in human activities, the concentrations of these compounds were low and confined to the site of their production. However, this scenario of natural equilibrium changed dramatically in the middle of the twentieth century, when the chemotherapy of many bacterial infections became possible. Over the last 70 years, the widespread use of antibiotics and other substances with antimicrobial activity had inverted the natural equilibrium between fully susceptible and resistant bacteria [9, 10]. Gradually, antibiotic-resistant bacteria and their specific genetic determinants have reached new habitats, with evident increases on the prevalence of resistance and the extension of the spectrum of antimicrobial substances tolerated [10]. Dramatic situations of both high prevalence and multidrug resistance are reported, mainly in the clinical environment. But, unfortunately, these resistant bacteria are not confined to hospitals and health care facilities. Antibiotic resistance levels above the expectable are reported in wild animals, surface waters or agriculture soils, allegedly due to antibiotics use and anthropic pressure [11–14]. Nowadays, antibiotic resistance is considered a serious global public health problem, which control demands the efforts of clinicians, scientists and policy makers. Such a concern is shared by several international health agencies, which provide substantial attention to the topic antibiotic resistance, for instance the World Health Organization (WHO) [15], the European Centre for Disease Prevention and Control (ECDC) [16], the Centres for Disease Control and Prevention (CDC) [17] or the Alliance for the Prudent Use of Antibiotics (APUA) [18]. With an intervention more focussed on the interface environment-clinical settings, the action COST Detecting evolutionary hot spots of Antibiotic Resistance in Europe (DARE) [19] launched in 2009 counts with the participation of scientists of about 20 countries, committed to identify and characterize environmental hot spots for antimicrobial resistance emergence and spreading, aiming the development of control measures.

2   Action, Resistance and Dissemination

The success of antibiotics as therapeutic agents is due to the capacity of these molecules to interfere with structures and/or functions of the bacterial cell (prokaryotic), which are absent in the host cells (eukaryotic). In turn, resistance mechanisms are related with the ability that bacteria have or may develop to avoid such interference (Table 1). Not surprisingly, given the plasticity that characterize bacteria, antibiotic resistance mechanisms are much more diverse than the modes by which a drug can interfere with a cell. Some bacteria, owning given genes and physiological functions, are intrinsically resistant to one or more classes of antibiotics. This is an ancestral property within a group and thus is common to most or all representatives of a genus or species [10, 20]. In contrast, acquired resistance

Table 1 Examples of action and resistance mechanisms found in bacteria for different classes of antibacterial agents

Action mechanism		Classes of antibacterial agents	Example	Resistance mechanisms
Interference with cell wall synthesis		Lactams	• Penicillins • Cephalosporins • Carbapenems • Monobactams	• Hydrolysis • Efflux • Altered target
		Glycopeptides	• Vancomycin • Teicoplanin	• Reprogramming peptidoglycan biosynthesis
Protein synthesis inhibition – bind to ribosomal subunit	50S	Macrolides	• Chloramphenicol • Quinupristin–dalfopristin • Linezolid • Erythromycin • Azithromycin	• Hydrolysis • Glycosylation • Phosphorylation • Efflux • Altered target
	30S	Tetracyclines	• Minocycline • Tigecycline	• Monooxygenation • Efflux • Altered target
		Aminoglycosides	• Gentamicin • Streptomycin • Spectinomycin	• Phosphorylation • Acetylation • Nucleotidylation • Efflux • Altered target
Nucleic acid synthesis inhibition	DNA	Quinolones	• Nalidixic acid • Ciprofloxacin • Enrofloxacin	• Acetylation • Efflux • Altered target
	RNA	Rifamycins	• Rifampin	• ADP-ribosylation • Efflux • Altered target
Inhibition of metabolic pathway		Sulfonamides	• Sulfamethoxazole	• Efflux • Altered target
Disruption of bacterial membrane structure		Polymyxins	• Colistin	• Altered target • Efflux

Adapted from [10, 146, 169]

is observed only in some representatives of a species, in which most of the representatives are susceptible to that antimicrobial agent.

Acquired antibiotic resistance may result from gene mutation or genetic recombination (Fig. 1a). Gene mutations occur randomly in the genome, often potentiated by mutagens. Examples of resistance phenotypes emerging by mutation include altered targets for an antimicrobial agent (e.g. quinolones, rifampin, linezolid, clarithromycin, amoxicillin, and streptomycin) [21], limited access of the antimicrobial agent to the intracellular target (e.g. penicillin, cephalosporins, glycopeptides, and tetracyclines) or transformation and further broadening of the range of antimicrobial agents that can be inactivated (e.g. extended spectrum β-lactamases). Under favourable conditions, the clones harbouring the gene mutation may have advantage, achieving higher rates of cell division than the

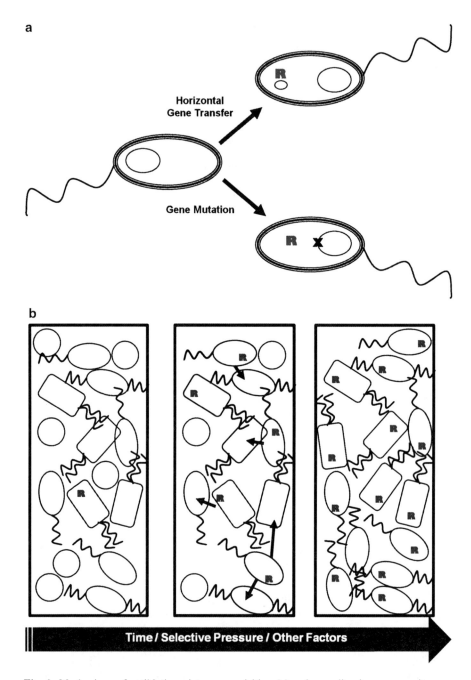

Fig. 1 Mechanisms of antibiotic resistance acquisition (**a**) and spreading in a community over time (**b**). *Arrows* indicate horizontal gene transfer processes and R acquired antibiotic resistance gene

non-mutated cells (higher fitness, i.e., the capacity of an individual to survive and reproduce) and, thus, become dominant. In such a situation the resistance genetic determinant is disseminated by vertical transmission.

In bacteria, genetic recombination is frequently referred to as horizontal gene transfer (Fig. 1b). This process, also named "bacterial sex", is very common among bacteria and represents a major driving force for bacterial evolution [22]. This form of genetic recombination involves the transfer of genetic material from a donor to a recipient and requires that both share the same space, but not necessarily the same species. Horizontal gene transfer can occur by (1) transformation, consisting on the uptake of naked DNA (on plasmids or as linear DNA), released by dead cells; (2) transduction, mediated by bacteriophages; (3) conjugation, involving cell-to-cell contact through a *pilus*. In general, horizontal gene transfer processes are potentiated by genetic elements which facilitate the mobilization and integration of exogenous DNA either between cells or between chromosomal DNA and extra-chromosomal genetic elements and vice versa. Examples of these genetic elements are plasmids, transposons and integrons, in which many of the known antibiotic resistance genes are inserted. In other words, some studies suggest that a considerable part of the genetic antibiotic resistome is associated to these genetic elements and thus has a high mobility potential [23–25]. Indeed, Fondi and Fani [26] concluded that apparent geographical or taxonomic barriers are not a limitation for the occurrence of horizontal gene transfer, as they observed that bacteria phylogenetically unrelated and/or inhabiting distinct environments had similar antibiotic resistance determinants.

3 A Big Environmental Bioreactor for Antibiotic Resistance Evolution

Essentially, three major lines of evidence contribute to demonstrate the gradual increase of antibiotic resistance – (1) annual reports on antibiotic resistance prevalence values of clinically relevant antibiotics for human pathogens [27, 28]; (2) the comparison of antibiotic resistance prevalence values in samples or bacterial cultures of our days with others archived from the pre-antibiotic era (e.g. [29, 30]); and (3) the establishment of significant correlations between antibiotics consumption and resistance increase [28, 31].

The generalized evidences on the increase of antibiotic resistance, allied with the development of analytical methods and genome exploring tools, motivated numerous studies on the environmental pollution produced by antimicrobials and other anthropogenic substances or on the diversity and distribution of antibiotic resistance genes (e.g. [3, 10, 26, 32]). Overall, these studies showed the complexity of antibiotic resistance dissemination in the environment. For instance, it was revealed that (1) not only bacterial pathogens but, very often environmental bacteria are important reservoirs of antibiotic resistance; (2) antibiotic resistance may have a reduced cost for its host and thus become stable once acquired; (3) the so-called "selective pressures", which supposedly contribute to enrich the antibiotic resistome, are diverse and act by different mechanisms, most of them still unclear.

Although it is still difficult to establish clear cause effect relationships, it is widely accepted that chemical pollution contributes for antibiotic resistance dissemination [10, 33, 34]. There are evidences that antibiotic resistance increase is related with environmental pollution and anthropic pressures. In this respect, antibiotics seem to be a major, although not the unique, form of pollution, mainly because it is estimated that about 75% of the antibiotics consumed by humans and animals are eliminated as active substances [35, 36]. In the environment, antibiotics can suffer adsorption, photolysis or biodegradation, reaching very low concentrations [37]. Nevertheless, at sub-inhibitory levels, as they are found in the environment, antibiotics can promote several alterations on housekeeping functions of the cells. Apparently, some of these alterations are not associated with antibiotic resistance. Even though, they contribute for the perturbation of the microbial community, leading, eventually, to an overall resistance increase [1, 34, 38].

Other substances, such as heavy metals, disinfectants and some pharmaceutical products, are also important pollutants on the promotion of resistance dissemination. Sometimes this effect can be mediated by co-selection due to genetic linkage between antibiotics and metals/biocides [39–42]. In other cases, it is associated with cellular transcriptomic alterations, sometimes induced by pharmaceutical products other than antibiotics [34, 38, 43]. Mutagenic agents (environmental pollutants and stress conditions) represent another class of presumable resistance promoters. Although bacteria have DNA repair systems (to minimize mutations), it has been argued that hypermutators (bacteria in which DNA repair systems are not fully operational) may increase under stress conditions [2]. In nature, bacteria face continuously several types of stress conditions – starvation (nutrients deprivation), DNA damage, temperature shift, oxidative stress or exposure to toxic compounds such as heavy metals, disinfectants or antibiotics (e.g. quinolones or β-lactams) [44]. Recently, Miyahara et al. [41] demonstrated the influence of environmental mutagens on antibiotic resistance increase. In this study, they showed that ethylmethanesulfonate and N-nitroso-N-methylurea induced resistance to ciprofloxacin and rifampicin in *Pseudomonas aeruginosa*.

Some potential environmental reservoirs of the observed generalized distribution of antibiotic-resistant bacteria and/or the respective genes have been proposed. In general, such reservoirs correspond to habitats rich in nutrients (organic matter), fed with anthropogenic chemical residues, such as antimicrobials, heavy metals and other substances, and with bacteria of human/animal origin. In these habitats, the abundance of nutrients favours bacterial growth, leading to high cell densities. Under these conditions, the anthropogenic residues may trigger different responses on bacteria, some of which are believed to favour horizontal gene transfer, while others may induce bacterial community rearrangements [2, 38, 45, 46]. Curiously, in spite of the vast number of publications that emphasize the importance of horizontal gene transfer on antibiotic resistance dissemination, the influence of external factors on these processes is still not much characterized. If, as many investigators suspect, horizontal gene transfer is indeed a major driving force on antibiotic resistance evolution, this is a serious gap in the current state of the knowledge.

Examples of sites with the characteristics mentioned above and, thus, potential reservoirs of antibiotic resistance are common in areas subjected to human activities (Fig. 2). Particularly, sites in which the use or disposal of antimicrobials is frequent, such as animal farms or municipal waste water treatment plants, are likely habitats to favour the spreading of antibiotic resistance. In Europe, the use of antibiotics as growing factors in animal husbandry is prohibited since 2006 [47]. However, animal farming still represents a critical point for antibiotic resistance development and dissemination. Among the major main routes of dispersal are manure, which sometimes is used directly or after composting in agriculture as fertilizers [48, 49], and animal farms waste water [50–53].

Municipal waste water treatment plants receive daily considerable amounts of non-metabolized antibiotics and metabolites thereof, as well as other environmental pollutants. Moreover, domestic waste water is a complex mixture of human commensal and environmental bacteria, which get in contact with considerable amounts of antimicrobial agents and other substances with unknown effects on bacterial metabolism and microbial communities [38, 54–57]. The free movement of bacteria between different environmental niches and the easiness with which bacteria become adapted to new conditions contribute to a rapid dissemination of antibiotic

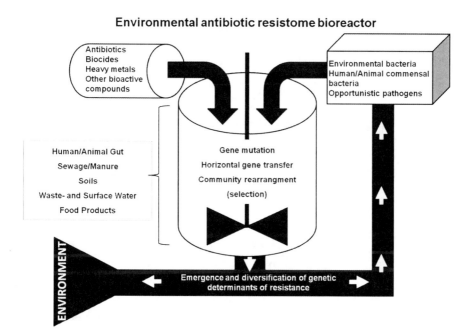

Fig. 2 Schematic representation of the environmental bioreactor for the antibiotic resistance enrichment

resistance to sites in which selective pressures are attenuated or absent (e.g. surface water and food products). For instance, the propagation of antibiotic-resistant bacteria from fish farms or waste water treatment plants to surface water is well documented [34, 52, 58–60]. Bacteria move between different environmental niches and, like stickers, drive antibiotic resistance determinants from heavily contaminated sites to other places in which selective pressures (no matter which they are) may be inexistent or negligible.

In principle, in the absence of selective pressures, acquired antibiotic resistance genes can be a dead weight for its host. In such a case, after some generations, these genes would be lost or contribute to the elimination of the host bacterial lineages. Nevertheless, in opposition to these expectations, some forms of acquired antibiotic resistance seem to be stable, with very slow reversibility rates [24, 61]. Depending on the genetic context and/or environmental conditions, resistance mutations can contribute to increase both resistance and fitness. In other words, if an acquired resistance determinant contributes for increased fitness, even in the absence of selective pressure, it will contribute for the selection of the hosting lineage, and therefore, to the increase of resistance rates. This was demonstrated, for instance, with fluoroquinolones [24].

Although in literature the experts need to organize bacteria into different compartments or virtual categories (waste water, hospital, humans, clinical, and environmental), there are no real boundaries for bacteria or their genes. Indeed, bacteria or the respective genetic determinants can move freely between different environments and contribute for the progressive enrichment of the resistance determinants at a global scale. This is clearly illustrated in the literature. For instance, the same antibiotic resistance genes are detected in hospital and animal husbandry waste waters, sewage, waste water treatment plants, surface water, ground water and drinking water [35]. A very recent example was given by the rapid spreading of the gene encoding for the New Delhi metallo-β-lactamase-1 (bla_{NDM-1}), presumably from India to the rest of the world. Plasmids containing the gene bla_{NDM-1} have also up to 14, other antibiotic resistance genes, transforming its host on a super-resistant bacterium and thus a serious threat for public health. Since its first report on 2008, the same gene was detected in several countries across Europe, America, Australia and Asia [62].

4 Methods to Assess and Characterize Antibiotic Resistance in the Environment: The Example of Waters

Antibiotic resistance testing was developed by and for clinical microbiologists aiming the therapy of bacterial infections. These methods, highly standardized worldwide, enable laboratories to assist the clinicians in the selection of the appropriate agent and the adequate doses to administrate in each particular situation [63–65]. Additionally, the use of standardized methods supports different

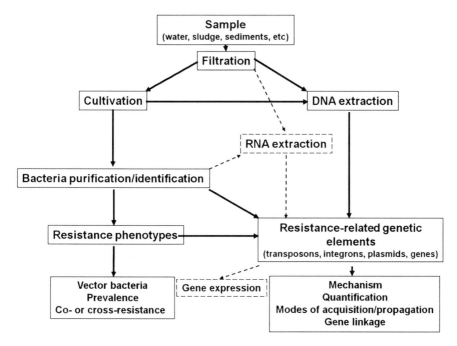

Fig. 3 Methods of characterization and quantification of antibiotic resistance in environmental samples

surveillance programs (e.g. [20, 27, 28]). The current knowledge on the general trends on the increase of antibiotic resistance and, consequently, on the emerging threats to human health is now stronger because of those surveillance programs.

These methods have been adapted to survey antibiotic resistance in the environment. Nevertheless, given the nature of the samples, the diversity of bacteria and even the different degrees of tolerance to antimicrobial agents, it is often difficult to implement the methods described for clinical applications. Because of the shortage on standardized methods to characterize antibiotic resistance in the environment, reliable comparisons aiming the recognition of temporal and geographical trends hardly can be made. Given the importance of water as a vehicle for the dissemination of organisms in the environment, the paragraphs below summarize some of the methods reported in the literature used to characterize and quantify antibiotic resistance in this type of environment (Fig. 3).

4.1 Culture-Dependent Methods

Culture-dependent methods to characterize antibiotic resistance in the environment are essentially based on the guidelines developed for clinical and veterinary microbiology (e.g. [20, 66–69]). Nevertheless, several adaptations have been introduced, including those that contribute to reduce time and costs involved. The membrane filtration method [70] is frequently the first step, allowing both bacteria enumeration and isolation. At this stage, the use of selective and differential culture media leads to the recovery of specific bacterial groups. Among the most frequently analysed bacteria are the indicators of microbiological water quality [70–72], coliforms and enterococci, for which a considerable number of selective culture-media are commercially available. The bacterial isolates can, then, be purified, identified, eventually typed to avoid repetitions and characterized for the antibiotic resistance phenotypes and genotypes. The selection of antibiotics to test, the procedures to use and the interpretative criteria can be handled as recommended by the standard guidelines developed for clinical microbiology. Commonly, the disc diffusion or micro-dilution methods are used to determine the antibiotic resistance/susceptibility phenotypes. If the objective is, for instance, to assess spatial or temporal variations of resistance patterns and percentages, it is necessary to compare a representative number of isolates, capable of supporting a reliable statistical analysis. This is a highly laborious and time consuming procedure, which requires some expertise in bacteriology – clearly, major weaknesses. In turn, through this procedure, it is possible to compare the role of the different species or of specific phylogenetic lineages in resistance dissemination or to track the presence of the same bacterium in the environment and clinical settings. The potential to make medium long term comparisons of antibiotic-resistant bacterial populations or to track relevant clones in the environment is an unquestionable strength of this approach.

The workload involved in the procedure described above may be impracticable for large numbers of samples, for instance, when a routine monitoring plan is to be implemented. To overcome such a limitation, some adaptations, which involve the enumeration and isolation of bacteria on antibiotic impregnated culture-media or modified most probable number estimates have been implemented [73, 74]. In these adaptations, selective culture media are supplemented with antibiotics at concentrations similar to or above those reported as inhibitory for the target bacteria. The ratio between the number of bacteria growing in the presence and in the absence of antibiotic gives an estimate of the percentage of resistance. Using this methodology, the definition of resistance differs from that established for clinical microbiology. Although a known concentration of antibiotic is added to the culture medium, the concentration that is effectively bioavailable is unknown and will depend on the culture medium used. Additionally, the fact that a bacterium can grow on culture medium supplemented with antibiotic does not imply necessarily that it is actually resistant to that substance. For instance, these bacteria can be persisters, or hold an unstable resistance phenotype, both with questionable relevance in terms of environmental resistance dissemination [61, 75]. Despite these weaknesses, the potential to analyse and compare large numbers of samples and to be implemented in non-specialized laboratories are important strengths of this kind of procedure.

4.2   Culture-Independent Methods

Every day novel microbial genomic information is generated and made available in public databases. This information brings a new dimension to the scale of microbial diversity and reveals that the microbial gene pool is considerably larger than it was though some years ago. It was this kind of information that contributed to demonstrate that most of the genes associated with antibiotic resistance has, actually, origin in soil and environmental bacteria [3, 8, 26, 76–78]. A vast majority of bacteria and of their genes, despite the importance that may have in the microbial community, are not detected using cultivation-dependent methods. Most of the studies on culture-dependent antibiotic resistance in waste- and surface water have focussed on human/animal commensal and ubiquitous bacteria for which cultivation methods are strongly implemented, such as *Enterococcus* spp. (*Firmicutes*), *Escherichia coli*, *Acinetobacter* spp. or *Aeromonas* spp. (*Proteobacteria*) [58, 60, 79–84]. However, these studies may be out of the step with reality, as culture-independent methods suggest that other bacterial groups may have an important role on antibiotic resistance dissemination. Different studies using culture-independent approaches showed that members of the phylum *Proteobacteria,* which includes numerous human/animal commensals and ubiquitous bacteria, are among the predominant groups in drinking, surface and waste waters, whereas *Firmicutes* and *Actinobacteria* are often minor representatives of these communities [57, 85–90]. Nevertheless, this cannot be taken as a rule as culture-independent methods are not exempt of bias. Actually, procedures such as the DNA extraction and DNA fragments analysis [polymerase chain reaction-denaturating gradient gel electrophoresis (PCR-DGGE), pyrosequencing, small subunit ribosomal RNA (16S rRNA) gene sequence library, etc.,] can impose serious bias. Another limitation of these methods is the level of identification, which frequently is above the taxonomic rank of family. In spite of those restraints, culture-independent methods, sometimes associated with cultivation stages, are a powerful approach to an in depth analysis of antibiotic resistance in the environment [91–95].

Surveys on the diversity of antibiotic resistance genes using culture-independent methods involve the extraction of the DNA from the total community, directly or after its enrichment in a culture medium. DNA extraction methods may be designed according to the fraction of the microbiome that is to be analysed (total DNA, plasmids) and must be optimized to render extracts with the desired quantity and quality. These are important aspects to assure the representativeness of the sample taken and also for a successful PCR amplification. Although metagenomic analyses may offer an overview of the diversity of genetic elements related with antibiotic resistance (genes, integrons, and transposons) [3, 93, 95], the use of specific or degenerated primers to target predefined antibiotic resistance genes is the most commonly used approach. A major limitation of this gene-targeted approach is the requirement of previous knowledge on the genetic elements that are searched for, at least to permit the primers design. Very often, such information has been obtained on the basis of culture-dependent methods and thus, although a direct DNA screening is made, it relies on a previous cultivation stage. Nevertheless, these gene-targeted methods have offered exhaustive and highly informative culture-independent surveys of antibiotic resistance genes and related genetic elements (integrons, transposons) in the plasmid or total DNA of waste waters (e.g. [55, 92, 93, 96–98]). Although much more informative than the culture-dependent methods, the mere survey of the resistance genes can be of limited value, for instance when the effect of external factors is to be assessed. In these cases, it is often necessary to use quantitative methods capable of supporting correlation analyses. For instance, if it is intended to assess the influence of a waste water treatment process or of the occurrence of antibiotic residues on the proliferation of resistance genes, it is necessary to use quantitative methods. In these cases, real-time quantitative PCR methods represent the most suitable approach [50, 98–101]. Quantitative methods represent a strong asset for the cause-effect assessment of antibiotic resistance dissemination.

In comparison with culture-dependent procedures, culture-independent methods are more sensitive and have an increased potential to survey the diversity of antibiotic resistance genes in the environment. A weakness of these methods is the impossibility to elucidate about the bacteria in the community that host specific resistance determinants. On the other hand, the possibility to explore the genetic environment (mobile element, associated genes, promoter, etc.) in which the resistance determinant is integrated offers relevant clues about the gene transfer potential and gene acquisition history.

The vast majority of the studies on antibiotic resistance in the environment have focussed on the survey of resistance genes. However, the mere detection of the antibiotic resistance genes may be insufficient to get a clear perspective of their function in the environment. If the role of antibiotic resistance genes in the environment is to be assessed, it is also important to determine the factors capable of triggering gene expression and to measure the expression levels [24]. Such an approach requires transcriptomic analyses supported, for instance, by reverse-transcription PCR or microarrays.

5   Antibiotic Resistance Phenotypes in Waste Waters

Over the last years, a renewed interest on the antibiotic resistance phenotypes in municipal waste water treatment plants became apparent in the scientific literature. The underlying hypothesis of these studies is that urban sewage treatment plants are potential reservoirs of antibiotic resistance, and, in general, it is aimed at contributing to assess the risks of dissemination, posed by the treated effluents discharged into natural water courses. As a general trend, these studies focus on human/animal commensal and environmental bacteria, frequently disseminated via faecal contamination, and which can survive in waters. The relevance of these bacteria, which may exhibit clinically relevant resistance phenotypes, as possible nosocomial agents seems also to be a motivation behind these studies.

The indicators of faecal contamination, coliforms and enterococci, are among the most studied groups (e.g. [46, 83, 102–105]). Other groups such as *Acinetobacter* spp., *Vibrio* spp. or staphylococci have been also examined (e.g. [106–108]). Table 2 presents some examples of studies on antibiotic resistance patterns in human commensal and environmental bacteria in urban waste water treatment plants. Despite the relevance these bacteria may have on the propagation of antibiotic resistance in the environment, the current state of art on waste water bacterial diversity suggests the predominance of a myriad of other bacteria, frequently referred to as unculturable, which may have an important role in antibiotic resistance dissemination. Unculturability is a broad sense condition that comprises three categories, not necessarily related with specific taxonomic groups, and which include bacteria: (1) for which the specific growth requirements (nutritional, temperature, aeration, etc.) are not available; (2) with very low growth rates, outcompeted in the presence of fast-growing microorganisms; and (3) unable to cope with the stressful conditions imposed by cultivation, due to physiological or genetic injuries. Frequently, unculturability is not a permanent condition and can be reverted. Nevertheless, most of the times, fast-growing and aerobic or facultative bacteria, without complex growth requirements, are the major targets in culture-dependent studies. In contrast, some bacterial physiological groups, important and prevalent in waste water, as for example, strict anaerobes, chemolithotrophs and phototrophs, have been almost ignored as potential vectors of antibiotic resistance. Culture-independent waste water diversity studies reveal that an important fraction of the bacterial community is represented by phototrophic and anaerobic *Proteobacteria*, *Planctomycetes*, *Verrucomicrobia*, *Bacteroidetes/Sphingobacteria*, among others [57, 87, 109]. Although these bacterial groups, in contrast to others referred to above, are not supposed to get in close contact with humans and animals, they may have a pivotal role as vectors of antibiotic resistance in waste water systems (sludge, biofilm, etc.). This topic is currently almost a black box.

In spite of the previously mentioned bias and limitations, the use of varied cultivation-dependent approaches to assess antibiotic resistance in the environment allows the recognition of general patterns, which can be vaguely compared with the situation in the clinical settings. Additionally, this approach is still a major tool to drive the detection of antibiotic resistance genes in the environment (see Sect. 7). Table 2 gives some examples of antibiotic resistance prevalence values in urban waste water treatment plants. For the sake of comparability, some bacterial groups and antibiotics common to more than one study were selected. These few examples are enough to show how the use of different cultivation conditions, antibiotics and concentrations can hamper reliable comparisons. Indeed, this seems to be a major limitation on the understanding of the factors ruling the dissemination of antibiotic resistance in environments subjected to human action. The definition of guidelines, as those available for clinical microbiology, adapted to environmental samples (water, soil and sludge), could be a valuable tool, for instance, to assess the efficiency of antibiotic resistance removal by waste water treatment plants, the influence of climate conditions (e.g. temperature or precipitation) or the risks posed by the use of recycled water as watering system (mainly of

Table 2 Examples of studies on antibiotic resistance patterns in human commensal and environmental bacteria in urban waste water treatment plants. Unless otherwise stated, values refer to average values of raw and treated waste water

	Percentage of resistance (concentration of antibiotic, μg or μg mL⁻¹)											
	Total heterotrophs	Acinetobacter		Coliforms	E. coli				Enterococci			
Aminopenicillins Ampicillin (AP) Amoxicillin (AX)	22.4–26.1 (32)^{AX}	0.0–0.6 (10)^{AP}	19.5 (30)^{AX,1}	31.4–50.2 (32)^{AX}	30.9 (25)^{AX}	15 (10)^{AP}	36.0–41.0 (32)^{AP}	16.5–23.5 (32)^{AP}	34 (2–16)^{AP}	7.0 (0.5–32.0)^{AP}	0.4–0.9 (32)^{AX}	2.0 (25)^{AX}
Tetracycline	1.1–1.8 (16)	5.5–10.4 (30)	n.a.	2.7–3.4 (16)	34.1 (30)	10.0 (30)	19–23 (16)	6.8 (4)	23.5 (1–8)	20.0 (0.5–8.0)	14.5–22.8 (16)	32.0 (30)
Sulfamethoxazole (+trimethoprim, Ts)	n.a.	0.6–2.4 (25)	11.2 (250)^2	n.a.	22.3 (25)^T	17.0 (300)^3	34–37 (350)	17.5 (256)	11.1 (10–40)^T	n.a.	n.a.	(25)^T
Quinolones Nalidixic acid (NA) Ciprofloxacin (C)	1.7–2.8 (4)^C	0.0–5.5 (5)^C 0.6–5.6 (30)^{NA}	7.5 (5)^C	1.0–2.8 (4)^C	5.5 (5)^C	5.0 (30)^{NA}	5.0–7.0 (4)^C	0.7–0.9 (4)^C	10.5 (0.5–2)^C	28.6 (0.5–4.0)^C	2.4–5.0 (4)^C	18.0 (5)^C
Gentamicin	n.a.	0.0–0.6 (10)	0.8 (10)	n.a.	4.4 m (10)	0.0 (10)	n.a.	n.a.	2.0 (1–8)	3.2 (synergy) (500)	n.a.	50.0 (10)
Erythromycin	n.a.	n.a.	n.a	n.a.	n.a.	n.a.	n.a.	n.a.	n.a.	44.3 (0.125–4.0)	n.a.	27.0 (15)
Method (antibiotic concentration)	AIA (μg mL)	DDM (μg)	DDM (μg)	AIA (μg mL⁻¹)	DDM (μg)	DDM (μg)	AIA (μg mL⁻¹)	MMPN (μg mL⁻¹)	MD (μg mL⁻¹)	MD (μg mL⁻¹)	AIA (μg mL⁻¹)	DDM (μg)
Number of isolates	–	290	241	–	274	100	–	–	153	185	–	133
Region (Observations)	Portugal FE, RV3	Denmark RV2, TT	USA TT	Portugal FE, RV3	Portugal	Australia AV5	Australia FE, RV2	Ireland FE, RV2	Poland	Poland	Portugal FE, RV3	Portugal FE, RV3
Reference	[111]	[112]	[82]	[111]	[81]	[60]	[73]	[74]	[110]	[110]	[111]	[80]

n.a. not/available, – not applicable, 1 with clavulanic acid, 2 Sulfisoxazole, 3 Sulfafurazole, AIA Antibiotic impregnated agar, DDM Disc diffusion method, MMPN Modified most probable number, MD Microdilution method, FE Final effluent, RVn Range values of n plants, AVn Average values of n plants, TT tertiary treatment implemented food products). The definition of antibiotic resistance itself may be inappropriate to describe acquired resistance in the environment. The most currently used definition of resistance refers to clinical breakpoints, which are defined for therapeutic uses. According to this, bacteria are defined as susceptible or resistant by a level of antimicrobial activity associated with a high likelihood of therapeutic success or failure, respectively. Clinical breakpoints differ between committees and have varied over time. Facing the global problem of antibiotic resistance, an alternative definition, without temporal, geographical or origin variation, was developed. Instead of using clinical breakpoints, the definition of epidemiological cut-off (ECOFF) values, proposed by EUCAST [20], assumes the distinction between wild-type (WT) bacteria of a given species and those which acquired a resistance mechanism by mutation or horizontal gene transfer. According to this, the ECOFF value corresponds to the limits in a WT population distribution (WT $\leq$ X mg L^{-1} or WT $\geq$ X mm) [20]. These cut-off values have been published for numerous bacterial groups and antimicrobials and may coincide or not with the clinical breakpoints. Although for some species and antibiotics the epidemiological cut-off value is identical to the clinical breakpoint, in other cases it can be lower or higher. Despite the importance that the clinical breakpoints may have, for environmental samples the use of ECOFF values will contribute to improve data analysis and interpretation.

The studies cited in Table 2, as well as others with similar experimental design, discuss the effects of waste water treatment and possible risks for human health posed by the discharge of antibiotic-resistant bacteria. Although a detailed comparative analysis is limited by methodological bias, it is possible to recognize some general trends. For instance, for *E. coli*, aminopenicillins, sulfonamides and tetracycline are among the antibiotics to which high rates of resistance are observed (10–40%). This contrasts with what is registered for quinolones or gentamycin, apparently, still belonging to a group of low resistance prevalence in *E. coli* (<10%). For enterococci, tetracycline and erythromycin are among the antibiotics with high resistance rates (20–40%), whereas aminopenicillins and sulfonamides are responsible for comparatively lower rates of resistance (1–7%). The increasing rates of enterococci resistance to quinolones are also evident in these and other studies [83], probably due to the environmental fitness of the species *Enterococcus faecalis* and mainly *Enterococcus faecium* in which quinolone resistance is highly prevalent [80, 110]. When compared with the commensal organisms, *Acinetobacter* show, in general, lower resistance prevalence values (Table 2).

The use of antibiotics incorporated into the culture medium (methods AIA, MMPN in Table 2) may indicate lower resistance percentages, for example, for tetracycline or ciprofloxacin, than the other methods. This was particularly evident when comparing the ciprofloxacin and tetracycline resistance rates of coliforms and *E. coli* in Novo and Manaia [111] and Ferreira da Silva et al. [81], examining the same waste water treatment plant (Table 2). The same difference can be inferred from the study of Galvin et al. [74], who, as Novo and Manaia [111], used a direct enumeration method. The same bias was observed for enterococci with the method of antibiotic impregnated agar (AIA) yielding much lower ciprofloxacin resistance rates than with the disc diffusion method (DDM) (Table 2, [83]). These differences may be due to the fact that the antibiotic concentrations used were too high to select all the resistant populations. Nevertheless, it is noteworthy that in some cases, bacteria isolated from culture-media supplemented with antibiotics may yield an intermediary resistance phenotype when tested by the disc diffusion method (our data, unpublished).

One of the ultimate aims of studying antibiotic resistance in the environment is the search for possible associations with clinical microbiology. In this respect, the comparison of resistance prevalence in clinical settings and in the environment can give some clues. In spite of the methodological bias, a first comparison of the data suggests that antibiotic resistance is more prevalent among clinical bacteria [27]. This is not surprising if one considers that clinical isolates have frequently been obtained from patients submitted to antibiotherapy and thus resistant strains were already selected. In the species *E. faecalis,* aminopenicillin resistance in waste water isolates was slightly below the values observed for clinical isolates in both Portugal and Poland for the corresponding years (0.0% vs. 2.6% and 2.3% vs. 4.4%, respectively). In contrast, in the species *E. faecium,* aminopenicillin resistance was clearly less prevalent in waste water than in the clinical settings (2.0% vs. 92.3% in Portugal and 10.7% vs. 97.5% in Poland, respectively). This suggests that different populations of *E. faecium* prevail in the human microbiota and in the environment. Clinical isolates of *E. coli* presented considerably higher values of aminopenicillin and fluoroquinolone resistance than waste water isolates (58.7% and 29.9% respectively, for Portugal; 64.7% and 23.0% respectively, for Poland; and 66.5% and 21.3%, respectively, for Ireland). Additionally, in comparison to what is observed for clinical isolates, in general, resistance to recent antibiotics is rare in waste water treatment facilities. Although this situation may change after some years of use, carbapenem-resistant *E. coli* was not detected by Łuczkiewicz et al. [110], in Poland, or by Figueira et al. [46], in Portugal, suggesting that if it exists, such strains are at very low numbers. Similarly, *E. coli* resistant to third generation cephalosporins are still rare in waste water; Łuczkiewicz et al. [110] reported a percentage of 1.3% and Figueira et al. [46] did not detect any resistant strain (in a total of 460 isolates of waste and surface water). More surprising is the data for vancomycin-resistant enterococci in Poland, with very low values in clinical isolates (0% in *E. faecalis* and 1.2% in *E. faecium*) whereas Łuczkiewicz et al. [110] reported percentages of 2.7% and 6.8%, respectively, in waste water. Indeed, the data of Łuczkiewicz et al. [110] confirm previous studies which show that vancomycin-resistant enterococci, mainly the species *E. faecium,* are inhabitants of different urban waste water treatment plants [83].

Waste water treatment is accompanied by bacterial community rearrangements which may lead to changes (increases or decreases) of the percentages of antimicrobial-resistant bacteria in the final effluent when compared with the raw inflow. Additionally, sewage treatment offers privileged conditions to favour antibiotic resistance acquisition and/or selection (see Sect. 3). Although some authors reported significant decreases of resistance to aminopenicillins (presumptive *Acinetobacter* and *Enterobacteriaceae*), cephaloporins (*Acinetobacter*) and tetracycline (total heterotrophs, enterobacteria and enterococci) [111, 112], the general trend seems to be the opposite. Significant increases of antibiotic resistance in the final (treated) effluent when compared with the raw inflow have been reported for different bacteria and antibiotics, often associated with the raising of multi-resistance levels (defined as resistance to different antimicrobial agents). Zhang et al. [82] showed that *Acinetobacter* spp. in the final effluent were significantly more multi-resistant and had higher percentages of resistance to rifampin, chloramphenicol and amoxicillin/clavulanic acid than in the raw inflow. Also working with *Acinetobacter*, Guardabassi et al. [112] concluded that in one of the plants examined, waste water treatment was accompanied by an increase on the prevalence of nalidixic acid resistance. Significant resistance increases to another quinolone (ciprofloxacin) were also reported for *E. coli* and enterococci in Portuguese waste water treatment plants [46, 80, 81, 111]. Similarly, in Poland, Łuczkiewicz et al. [110] observed significant increases of enterococci resistant to different quinolones. Based on the current state of art it is not possible to identify the factors that are responsible for significant variations of antibiotic resistance prevalence during waste water treatment. Nevertheless, for quinolones, the gathering of some different lines of evidence may contribute to understand the observed increases of resistance. Quinolones are not metabolized before excretion by mammals. In waste water, where sunlight cannot induce photodegradation, quinolones adsorb to sediments and may accumulate in the sludge, biosolids and soil particles [54, 113]. The accumulation of quinolones may contribute to the success of quinolone-resistant bacteria, which are in advantage in such conditions. On the other hand, in some bacteria, quinolone resistance is not associated with relevant fitness costs and thus can be well succeeded in the environment. For instance, in *E. coli*, it was demonstrated that strains harbouring resistance to fluoroquinolones, due to chromosome mutations in the DNA gyrase subunit A gene, *gyr*A, or also in the gene that encodes DNA topoisomerase IV subunit A, *par*C, had no substantial loss of fitness [24]. This explains the stability of these genotypes of resistance in the environment and supports the importance of the vertical antibiotic resistance transmission. Figueira et al. [46] observing a significant increase of ciprofloxacin-resistant *E. coli* due to waste water treatment examined the contribution of different population subsets. The authors concluded that the dynamics of the different population subsets during waste water treatment supported the significant increase of resistance observed. During waste water treatment, the ciprofloxacin susceptible population subset was eliminated more extensively than the resistant *E. coli* strains, leading to an apparent increase of quinolone resistance in the final effluent. Quinolone resistance in these isolates was due to mutations in the chromosomal genes *gyr*A and *par*C, whereas no plasmid encoded quinolone resistance was found. These observations were consistent with the fact that ciprofloxacin accumulation may favour the selection of antibiotic tolerant bacteria, and that those chromosome mutations have no costs for the cell maintenance even in the absence of selective pressures. The dynamics of bacterial populations was also a possible explanation for the increase of ciprofloxacin-resistant enterococci due to waste water treatment [80]. In this case, whereas members of the species *Enterococcus hirae*, with very low levels of antibiotic resistance, were extensively eliminated during waste water treatment, members of the species *E. faecium* and *E. faecalis*, which had higher levels of ciprofloxacin resistance, became prevalent in the final effluent [80]. These evidences show the importance of the resistance fitness costs on the vertical transmission of antibiotic resistance.

Waste water treatment is essential to prevent undesirable chemical and biological contamination of surface and ground waters. The undesirable biological contaminants include pathogenic microorganisms and/or their genes related with any form of virulence or hazardous potential (e.g. antibiotic resistance). Waste water treatment contributes to avoid the transmission of such biological agents, although it is insufficient to completely eliminate the hazardous potential of the effluents. Municipal sewage treatment, including tertiary (disinfection) treatment when available, implies bacterial removal rates ranging 1.2–2 log units [102, 111, 112, 114, 115]. These values correspond to the discharge of about 10^9 to 10^{12} Colony Forming Units (CFUs) per day per inhabitant equivalent in the final effluent, in which at least 10^7 to 10^{10} will have any kind of acquired antibiotic resistance [111]. These values, which may be underestimated considering the fraction of unculturable bacteria, suggest that waste water treatment plants have an important contribution for the accumulation and dispersal of antibiotic-resistant bacteria in the environment.

6 Closing the Cycle: Drinking Water

In spite of the consequences that pollution with antibiotics, antibiotic-resistant bacteria or antibiotic resistance genes may have in the environment, for instance in biodiversity, major concerns are, for multiple reasons, related with human health. In this respect, every form of transmission environment–human are critical points to control the spread of antibiotic resistance. Urban wild animals, food products and drinking and recreational waters represent important vectors in that transmission [11, 12, 106, 116–119]. Nevertheless, water, more precisely the different parts of the urban water cycle (waste, surface or drinking water) is a unique link among the different reservoirs and transmission vectors. Waste water treatment plant effluents can promote the contamination of soils and of irrigation waters, and thus, contribute to disseminate antibiotic resistance to food products. The consequent contamination of surface waters will contribute to spread the resistance to some urban wild animals and recreational water. Antibiotic-resistance in surface and ground waters poses also the serious risk of contaminating drinking water. Although disinfection processes contribute to minimize such risks, the persistence or re-colonization of antibiotic resistant bacteria in drinking waters is a reality, worsened by the high potential of many bacteria to produce biofilm in pipelines, reservoirs and taps.

Good practices recommend that drinking water must be produced from surface or ground water with suitable chemical and microbiological quality [63]. Although no requirements are normally made with respect to antibiotic-resistant bacteria, a low density of microorganisms and the absence of faecal contamination (indicated by *E. coli* and *Enterococcus* spp.) could be regarded as a precautionary principle in this regard. Thus, drinking water coping with legal recommendations would be expected to be free of antibiotic-resistant bacteria, except some intrinsic resistance phenotypes. Several studies have demonstrated that this is not true and that, indeed, treated drinking water can be an important mode of transmission of antibiotic-resistant bacteria to humans [106, 116, 120, 121].

In a study conducted in a drinking water treatment plant serving about 1.5 million of habitants, through a distribution network of 730 km and a total storage capacity of 130,000 m^3, it was observed that antibiotic-resistant bacteria could be found from the source to the taps. This water treatment plant is fed by surface (river) water, which goes through a disinfection process that includes filtration, ozonation, flocculation, flotation/filtration and chlorination. Water disinfection led to a drastic reduction of cultivable heterotrophic bacteria (from ~10^1–10^3 to ~10^{-1} CFU mL^{-1}), which, nevertheless, was not accompanied by a similar reduction of the total cell counts (including unculturable/viable bacteria). Indeed, total cells were reduced by only one order of magnitude (from ~10^5–10^6 to ~10^4 cells mL^{-1}). Not surprisingly, total heterotrophs counts were observed to increase after the disinfection stage and, in general, the numbers of cultivable heterotrophic bacteria found in tap water were similar to those of raw water. This trend seems to be common in these habitats [120, 122, 123]. Besides the increase in the cell density along the drinking water network, also changes in the microbial community composition are known to occur, mainly during the period of stagnation of water in taps [123]. *Proteobacteria* (mainly of classes *Alpha-*, *Beta-*, and *Gammaproteobacteria*), *Cyanobacteria* and *Bacteroidetes* are some of the predominant phyla observed in drinking water systems [124]. Genera such as *Pseudomonas, Flavobacterium, Aeromonas, Sphingomonas, Acinetobacter, Citrobacter, Enterobacter, Klebsiella, Serratia, Moraxella, Xanthomonas, Legionella* and *Mycobacterium* are frequently detected [46, 116, 125–135]. Many of these bacteria can survive in the environment and colonize humans and other animals causing serious diseases [136]. Such abilities make these water microbiota members potential vectors of antibiotic resistance dissemination [2, 29, 45].

Ubiquitous bacteria (e.g. enterobacteria, pseudomonads and aeromonads), able to survive in potential antibiotic resistance reservoirs as waste water treatment plants, can move freely between different aquatic environments, eventually reaching drinking water systems. Also the selective pressures imposed by the presence of the disinfectants used for drinking water treatment may favour the survival of organisms carrying resistance determinants. Indeed, many studies have reported the presence of antibiotic-resistant bacteria and antibiotic resistance determinants, as genes and mobile genetic elements, in drinking water [120, 121, 137–140]. For example, clinically relevant antibiotic resistance phenotypes have been observed in members of the genera *Enterobacter, Citrobacter, Klebsiella, Kluyvera, Aeromonas, Sphingomonas* and *Staphylococcus,* which comprise recognized opportunistic pathogenic species [35, 84, 116, 120, 121, 138, 141, 142].

The recovery of antibiotic-resistant *Salmonella* spp. and *Listeria* spp. from drinking water is also reported [35, 143, 144].

Water disinfection aims at reducing the charge of potential pathogens in drinking water. Nevertheless, disinfected water is not sterile and the possibility of re-colonization is also recognized. Moreover, disinfection processes impose stressful conditions which lead to rearrangements of bacterial populations. Such rearrangements may result on an increase in the percentage of antibiotic resistance [120]. Population alterations are apparently genera-, species- or even strain-dependent and, thus, difficult to predict. In a study aiming at assessing the dissemination of aeromonads in the urban water cycle, it was observed that members of this genus, commonly reported in waters worldwide, were found in waste and river water, but not in tap water. Water disinfection, which included a stage of ozonation, was observed to impose a bottleneck, originating a drop on prevalence of the species *Aeromonas veronii*, predominant in river water, and the raise of *Aeromonas hydrophila* subsp. *hydrophila*, which became predominant after ozonation. Moreover, after ozonation the strain diversity was drastically reduced. An additional stage of disinfection, by water chlorination, reduced significantly the remaining aeromonads, not detected in tap water [84]. In contrast, other genera, such as *Ralstonia*, *Pseudomonas* or *Acinetobacter*, isolated from raw river water could be detected, due to re-growth or re-colonization, in tap water.

Members of the family *Sphingomonadaceae* are highly ubiquitous and sporadically associated with opportunistic infectious episodes, but often disregarded as potential hazardous organisms in drinking water. Nevertheless, *Sphingomonadaceae* are reported worldwide among the predominant bacterial groups in drinking water systems. Recently, in a study with *Sphingomonadaceae* isolated from taps and cup fillers of dental chairs, it was demonstrated that these bacteria yielded high percentages of resistance to β-lactams, ciprofloxacin and co-trimoxazole, besides intrinsic resistance to colistin [116]. Such antibiotic resistance phenotypes allied to the fact that these bacteria were found in densities ranging from 10^1 to 10^3 CFU mL^{-1} in the taps and cup fillers suggest their hazardous potential in drinking water. Another group of ubiquitous bacteria are the coagulase negative staphylococci, also occurring in treated drinking water and harbouring resistance to erythromycin, β-lactams, tetracycline and clindamycin [106]. Ubiquitous *Enterobacteriaceae* represent another group of bacteria with expectable presence in drinking water, although often found in waters with faecal contamination. In a study comprising members of the genera *Citrobacter*, *Enterobacter*, *Klebsiella*, *Kluyvera* and *Raoutella* isolated from treated drinking water, it was observed that resistance to amoxicillin, tetracycline, ciprofloxacin and sulfamethoxazole/trimethoprim was in percentages that, most of the times, were not significantly different from those detected in waste or surface water isolates [121]. Moreover, cephalothin resistance was significantly more prevalent in drinking water than in waste- or surface water. Such difference was attributed to the predominance of *Enterobacter* spp. and *Citrobacter* spp. in drinking water, considered to be intrinsically resistant to cephalothin [64]. These results suggest that cephalosporin resistance favours the survival to the water disinfection processes.

The same results were previously observed by Murray et al. [145], who after sewage chlorination, in a pilot setup, found an increase in the proportion of bacteria resistant to cephalothin, mainly after bacterial re-growth. Besides the resistance phenotypes, class 1 or class 2 integrons were detected in more than 1% of the drinking water enterobacteria isolates [121]. These isolates were members of the genera *Citrobacter*, *Klebsiella* and *Kluyvera* and the integrons contained gene cassettes similar to those detected in surface and waste water enterobacteria. These findings suggest the potential of these bacteria to acquire additional resistance genes, mainly in the presence of selective pressures [121].

Most of the bacteria that colonize drinking water, for instance pseudomonads (non-*P. aeruginosa*), *Acinetobacter* spp. (non-*Acinetobacter baumannii*) or sphingomonads, have not been intensively studied in what respects antibiotic resistance genes and acquisition mechanisms. Information supporting further research, for example, complete genome annotation, is sometimes scarce and would be an important tool for further studies. Additionally, most of the studies on antibiotic resistance in drinking water have been performed with resource to cultivation techniques. As a consequence, little is known about the potential of the unculturable fraction of the drinking water microbiota as reservoir of resistance to antibiotics. For instance, functional metagenomic studies can bring important insights into the antibiotic resistome in drinking water.

7   The Footprint of Antibiotic Resistance Genes in Waters

Antibiotic resistance genes are the centrepiece of every discussion on the origins, evolution, physiology and modes of dissemination of resistance. In contrast to the ecological niches frequently defined for live beings, it is assumed that antibiotic resistance genetic elements have a widespread and unconfined distribution in microbial populations across the globe, designated as "antibiotic resistome" [8, 146]. The concept of antibiotic resistome is also closely related to the spreading of resistance genes via horizontal gene transfer [147]. Thus, the characterization of antibiotic resistance genes gives objective insights into the exchanges occurring between the environment and the clinic [148]. Water systems, particularly waste water treatment plants and the natural water courses to which the effluents are discharged, are regarded as important connectors between humans and the environment. This rationale has motivated numerous studies on antibiotic resistance genes in surface and waste waters over the last years. For the period between 2008 and mid-2011 more than 50 papers describing antibiotic resistance genes in water-related environments (excluding fish farms and animal production waste water treatment plants) were available in public databases of research literature [32, 35] (Table 3).

Recently, Zhang et al. [35] made a comprehensive literature review of genes conferring resistance to different classes of antibiotics, detected in different types of water (raw and treated waste water, surface water, drinking water or sediments).

Table 3 Examples of antibiotic resistance genes of clinical relevance distributed worldwide in aquatic environments and illustration of some methodological approaches commonly used to detect resistance determinants in the environment

Gene	Class	Type of water	Biological source	Detection method	World region	Reference
mecA	L	Recreational beach and seepage	Proteus vulgaris, Morganella morganni, Enterococcus faecallis	Enrichment in oxacillin and polymyxin B supplemented Mueller–Hinton broth/PCR	USA	[117]
tetK, tetM	T	Recreational beach	Methicillin resistant staphylococci	Polymyxin B supplemented Staphylococcus agar/PCR	USA	[119]
ermA, ermB, ermC	M					
bla_{SHV-12}, bla_{TLA-2}, bla_{PER-1}, bla_{PER-6}, bla_{GES-7}, bla_{VEB-1a}	L	River	Aeromonas spp.	Isolation on ceftazidime suplemented MacConkey agar/PCR	France	[170]
bla_{NDM-1}	L	Tap water and seepage	Escherichia coli, Klebsiella pneumoniae, Shigella boydii, Aeromonas caviae, Stenotrophomonas maltophilia, Vibrio cholerae, Citrobacter freundii, Pseudomonas aeruginosa	Vancomycin and cefotaxime or meropenem supplemented Mueller–Hinton agar/PCR/Probes hybridization	India	[166]
tetR, tetY	T	River	Plasmid pAB5S9 of Aeromonas bestiarum clone 5S9	Plasmid sequence analysis	France	[171]
sulII	S					
floR	Ap					
strA, strB	A					
mecA	L	River and waste water	DNA bacteriophages	qPCR	Spain	[94]
bla_{TEM}, bla_{CTX-M9}	L					
tetO	T	Artificial ground water	Total DNA	qPCR	Belgium, Spain, Italy	[163]
ermB	M					
ampC, bla_{SHV-5}, mecA	L				Spain	
sulI	S	River	Total DNA	PCR	Australia	[172]
aac(3)-I	A					
mecA	L					

Genes	Class	Source	Method		Country	Ref.
vanA, vanB	G					
aac(6')-Ib-cr, qepA, qnrA, qnrB, qnrS	Q	Wetlands	PCR	Total DNA	USA Mexico	[173]
tetL, tetM, tetO, tetQ, tetW	T	River	qPCR	Total DNA	Cuba	[34]
ermB, ermC, ermE, ermF	M					
bla$_{TEM-1}$, bla$_{CTX-M}$, bla$_{OXA-1}$, bla$_{SHV-1}$	L					
sulII	S	River sediments	Functional metagenomics	Total DNA	India	[95]
qnrD, qnrS, qnrVC	Q					
strA, strB	A					

Classes of antibiotics: *L* β-lactam, *T* tetracycline, *M* macrolide, *S* sulphonamide, *Ap* amphenicol, *A* aminoglycosides, *G* glycopeptides, *Q* quinolone

Their work clearly shows the diversity of resistance mechanisms as well as its widespread distribution by different bacterial groups and, above all, the overlap of these genetic determinants with those reported in clinical isolates. The overlap of resistance genes in waste water and in clinical settings was also strongly evidenced in the study of Szczepanowski et al. [32]. Using a set of 192 PCR primers specific for antibiotic resistance genes, these authors screened total plasmid DNA from activated sludge bacteria or from the final effluents of waste water treatment plants. The 140 amplicons obtained, with sizes around 200 base-pairs, indicated the presence of genes encoding resistance to aminoglycosides, β-lactams, chloramphenicol, fluoroquinolone, macrolides, rifampicin, tetracycline, trimethoprim and sulfonamide as well as efflux pumps. Additionally, some of the resistance genes detected had a recent description in clinical isolates.

Although the abundance of information on antibiotic resistance genes in water-related environments cannot be summarized easily, Table 3 furthers the data presented by Szczepanowski et al. [32] and Zhang et al. [35], giving some examples of potential environmental and biological reservoirs, and illustrating different methodological approaches. The multiple studies that have explored the antibiotic resistome in aquatic environments over the last years provide evidence that resident or transient bacteria in different habitats harbour resistance determinants of all classes, located either on the chromosome or associated with mobile genetic elements [32, 94, 95, 149]. Although the simple detection of a gene is not indicative of its expression in its host and in that environment, it evidences the stability and potential of that gene to spread to other environments or hosts. Of the three mechanisms of horizontal gene transfer in bacteria (Sect. 2), the conjugative transfer of plasmids have been considered of surmount importance in antibiotic resistance acquisition [26, 32, 149, 150]. Indeed, plasmid DNA extracted from waste waters reveal the presence of a myriad of antibiotic resistance genes of clinical relevance [32, 93]. Nevertheless, the importance of virus as vectors of housekeeping microbial functions in different biomes has been demonstrated [151]. These findings support the potential of phages to transfer antibiotic resistance genes, which, although less studied, deserve a special attention [94]. Among the genetic elements associated with the mobilization of the resistome, integrons are amongst the most studied. Integrons are gene capture elements, with a critical role in bacterial evolution and adaptation [152, 153]. These genetic elements encode an enzyme integrase which facilitates the capture of genes, including antibiotic resistance, in structured gene cassettes [42, 152]. The importance of these genetic elements in the antibiotic resistance spreading has been referred to consistently, with particular relevance in clinical environments [42, 152]. These evidences motivated the search and characterization of integrons in different water environments. Surprisingly, and in contrast to the evidences available for clinical isolates, many studies conducted worldwide in waste, surface and drinking waters suggest the occurrence of highly conserved gene cassette composition inserted in the variable region of the integrons. Typically, these gene cassettes contain aminoglycoside- (*aad*A, *sat* genes) and trimethoprim-resistance (*dfr*) genes [46, 96, 97, 121, 154–158]. Other genes, encoding oxacillinases, carbapenemases or quinolone resistance determinants (*qnr*) among others are also observed in integron gene cassettes [96, 97, 159]. Nevertheless, most of the times, the integron gene cassette content cannot explain the observed resistance phenotypes. In spite of these observations, the data available suggest that integrons are indeed related with resistance acquisition and multi-resistance evolution in the environment. Several studies report higher prevalence of integrons in habitats subjected to selective pressures or associated with multi-resistance phenotypes [46, 95, 155, 158, 160]. A possible explanation was proposed by Li et al. [155] who speculated that co-resistance to antibiotics of different classes can be due to the activity of multidrug transporters such as those encoded by *qac* genes found in the conserved region of class 1 integrons.

Numerous studies have demonstrated that waste water treatment processes, operating according to legal recommendations, cannot reduce efficiently the levels of antibiotic resistance (Table 2; [32]). Sewage composition and the treatment process itself may impose selective pressures capable of modulating either the bacterial populations or the antibiotic resistance pool. Using quantitative PCR, Lachmayr et al. [161] suggested that bacteria harbouring bla_{TEM} genes are more likely to survive waste water treatment, being introduced into the environment at densities significantly higher than they occur in nature. In this respect, waste water treatment plants of antibiotic-production industries and the rivers receiving these effluents are very interesting case studies [95, 155]. The discharge of these effluents strongly influences the composition of the bacterial populations in the receiving medium, promotes a generalized increase of antibiotic resistance, mainly, but not only, to the antibiotics being produced, and leads to an increase on the prevalence of integrons [95, 155, 160]. Furthermore, these studies demonstrate that environmental or ubiquitous bacteria are important targets of such selective pressures. For instance, in penicillin and tetracycline production waste water treatment plants, among the leading biological reservoirs of resistance genes appear environmental and ubiquitous bacteria, such as *Pseudomonas putida*, *Pseudomonas fragi*, *Pseudomonas libanensis*, *Stenotrophomonas maltophilia* or *Yersinia kristensenii* [155, 160], which presence in drinking water or in fresh food products would not be regarded as hazardous. A metagenomic study of river sediments collected downstream from an Indian waste water treatment plant, processing effluents from more than 90 bulk pharmaceuticals producers (including antibiotics), demonstrated also the dramatic effects of selective pressures [95]. In these areas, rivers can contain extremely high concentrations of antibiotics (for example, up to 6.5 mg L^{-1} of ciprofloxacin, 0.52 mg L^{-1} of norfloxacin and 0.16 mg L^{-1} of enoxacin) [162] and, thus, the effect of selective pressures is dramatic. These authors reported high levels of different resistance genes, including a surprisingly high diversity of mobile fluoroquinolone resistance genes, identified two abundant non-conjugative resistance plasmids and detected a significant increase of class 1 integrase [95] (Table 3). The study of Kristiansson et al. [95] was conducted in a river with exceptionally high concentrations of antibiotics, but this is not the rule. In general, antibiotics are found in waters at sub-inhibitory concentrations and, presumably, are not the unique factors responsible for the selection of antibiotic resistance genes. Using quantitative PCR, Graham et al. [34] demonstrated a strong correlation between the levels of copper, cobalt and lead (but not of cadmium and chromium) and the abundance of some of the tetracycline, macrolide or β-lactam resistance genes analysed. The generalized pollution of waste and surface waters seems to have unavoidable consequences for the spreading of antibiotic resistance genes throughout the urban water cycle. In some world regions, artificial ground water recharge systems are used to prevent the depletion of ground water levels, the intrusion of salt water or to store surface water. The risks of aquifers contamination with pathogenic bacteria or antibiotic resistance genes are high. Although such risks can be attenuated using expensive treatment systems (e.g. ultrafiltration or reverse osmosis) some resistance genes may persist and enter the food chain via drinking water [163]. In fact, several studies have demonstrated the occurrence of antibiotic resistance genes in drinking waters. Examples are the β-lactam (*ampC*, *bla*$_{TEM}$ and *bla*$_{SHV}$), chloramphenicol (*cat* and *cmr*), sulfonamide (*sul*I and *sul*II), tetracyline (*tet*A, *tet*B and *tet*D), aminoglycosides (*aph*A, *aad*A1, *aad*A2 and *sat*2), trimethoprim (*dfr*A12 and *dfr*A17), erythromycin (*msr*A, *erm*A and *erm*C) and vancomycin (*van*A) resistance genes [35, 92, 106, 120, 121, 164].

In spite of the scarcity of data, some world regions, in which infection control policies and environmental and clinical surveillance systems are inexistent or inadequate, may be important sources for resistance emergence and spreading [62, 95]. An emblematic example is given by the metallo-β-lactamase gene *bla*$_{NDM-1}$, firstly reported in India [62, 165]. This new metallo-β-lactamase is mainly hosted by bacteria spread via faecal–oral transmission, but its occurrence in ubiquitous bacteria of the species *Aeromonas punctata*, *S. maltophilia* or *Citrobacter freundii* is also reported [166]. Although the carriage of *bla*$_{NDM-1}$ by non-fermenters is frequently unstable and not associated with typical resistance [166], the role of those hosts as potential vectors should not be ignored. The search for antibiotic resistance genes in environmental samples evidences not only their occurrence in potential sources of human contamination, such as drinking and recreational waters, but also in unexpected hosts. For instance, Soge et al. [119] demonstrated that methicillin-resistant *Staphylococcus aureus* and coagulase-negative staphylococci, presumably of clinical origin, can be transmitted to visitors in public marine beaches. Moreover, staphylococci of the beach environment carried tetracycline and macrolide resistance genes on mobile elements, transferrable to other bacterial hosts. Kassem et al. [117] reported the occurrence of the gene *mec*A in recreational waters, in bacteria of the species *Proteus vulgaris*, *Morganella morganii* and *E. faecalis*. The high sequence similarity of the detected gene with the *mec*A determinant of the chromosomal cassette widely found in *S. aureus* and other staphylococci suggests that this gene, of highest concern in clinical settings, is, in the end, widely distributed and supports the hypothesis that it can be transferred among different bacteria phyla.

Despite the considerable amount of information published so far, it is important to remember that the current overview on antibiotic resistance genes in waters (and in the environment in general) is biased by some scientific and technical contingencies. That is to say, that, probably, only a very little fraction of the resistance genes occurring in waters is being characterized. As referred to in Sect. 2, acquired antibiotic resistance may result from mechanisms such as transformation of the antibiotic, the maintenance of a reduced access to the intracellular target or antibiotic target modification. However, most of the resistance genes surveyed are associated with the two first types of mechanism, i.e., the gain of a specific function by the resistant organism. Examples of these mechanisms are the acquired capacity to modify (e.g. hydrolysis or acetylation) or to extrude an antibiotic (e.g. efflux pumps). Although both PCR-based approaches and metagenomic analysis have been used, most of the antibiotic resistance genes surveyed so far in the environment are detected by PCR, using specific primers, followed by sequence similarity analysis. The utilization of specific primers seriously limits the range of antibiotic resistance genes that can be surveyed; for instance, resistance genes harboured by bacteria which genome is not deeply characterized, or genes still not sequenced might not be detected. Even though, the use of PCR-based approaches can be more effective than metagenomic analysis. At least this was the conclusion reached by Szczepanowski et al. [32] who, using a PCR-based approach, detected 59 additional resistance genes in activated-sludge bacteria that could not be identified in the plasmid metagenome dataset. Another bias is related with the fact that most of the genes screened were originally described in clinical pathogens, mainly culturable aerobic bacteria with fast and non-fastidious growth. For these reasons, the current perspective of the antibiotic resistome is still mainly culture-dependent. Nevertheless, functional metagenomic studies of the human microbiome showed that only a small fraction of the antibiotic resistance genes is currently identified [167, 168]. Moreover, those authors hypothesize on the existence of some barriers to horizontal gene transfer between unculturable bacteria and the readily cultured human pathogens [168]. There are no reasons to consider that such scenario is restricted to the human microbiome. Indeed, most probably the same immensity of antibiotic resistance genes is still to be discovered in the environment.

8   Final Remarks

The dramatic increase of severe or lethal infections caused by antibiotic-resistant bacteria triggered numerous studies on antibiotic resistance, not only from clinical but also from environmental sources. Nowadays it is clear that environment, and water in particular, plays a central role on antibiotic resistance dispersion to and from clinical settings. However, the current state of the art clearly suggests that *only a small fraction of the environmental resistome is known*. The *modes and mechanisms of emergence, evolution and transmission of resistance determinants are still not very well understood*. Although *environmental pollution* is recognized to *play an important role on antibiotic resistance evolution* and spreading, it is still *very difficult to draw cause–effect relationships*, which sometimes seems to be strain/species dependent.

References

1. Davies J, Spiegelman GB, Yim G (2006) The world of subinhibitory antibiotic concentrations. Curr Opin Microbiol 9(5):445–453
2. Martinez JL (2009) Environmental pollution by antibiotics and by antibiotic resistance determinants. Environ Pollut 157(11):2893–2902
3. Torres-Cortes G, Millan V, Ramirez-Saad HC et al (2011) Characterization of novel antibiotic resistance genes identified by functional metagenomics on soil samples. Environ Microbiol 13(4):1101–1114
4. Dantas G, Sommer MO, Oluwasegun RD et al (2008) Bacteria subsisting on antibiotics. Science 320(5872):100–103
5. Datta N, Hughes VM (1983) Plasmids of the same Inc groups in Enterobacteria before and after the medical use of antibiotics. Nature 306(5943):616–617
6. Hughes VM, Datta N (1983) Conjugative plasmids in bacteria of the 'pre-antibiotic' era. Nature 302(5910):725–726
7. Aminov RI (2010) A brief history of the antibiotic era: lessons learned and challenges for the future. Frontiers in, Microbiology, 1
8. D'Costa VM, McGrann KM, Hughes DW et al (2006) Sampling the antibiotic resistome. Science 311(5759):374–377
9. Larson E (2007) Community factors in the development of antibiotic resistance. Annu Rev Public Health 28:435–447
10. Davies J, Davies D (2010) Origins and evolution of antibiotic resistance. Microbiol Mol Biol Rev 74(3):417–433
11. Literak I, Dolejska M, Radimersky T et al (2010) Antimicrobial-resistant faecal *Escherichia coli* in wild mammals in central Europe: multiresistant *Escherichia coli* producing extended-spectrum beta-lactamases in wild boars. J Appl Microbiol 108(5):1702–1711
12. Simões LC, Simões M, Vieira MJ (2010) Influence of the diversity of bacterial isolates from drinking water on resistance of biofilms to disinfection. Appl Environ Microbiol 76(19):6673–6679
13. Storteboom H, Arabi M, Davis JG et al (2010) Identification of antibiotic-resistance-gene molecular signatures suitable as tracers of pristine river, urban, and agricultural sources. Environ Sci Technol 44(6):1947–1953
14. Thaller MC, Migliore L, Marquez C et al (2010) Tracking acquired antibiotic resistance in commensal bacteria of Galapagos land iguanas: no man, no resistance. PLoS One 5(2):e8989
15. World Health Organization (WHO) (2011) Available from: www.who.int/mediacentre/factsheets/fs194/en/ 30 Aug 2011
16. European Centre for Disease Prevention and Control (ECDC) (2011) Available from: http://ecdc.europa.eu/en/healthtopics/antimicrobial_resistance/Pages/index.aspx30 Aug 2011
17. Centers for Disease Control and Prevention (CDC) (2011) Available from: www.cdc.gov/drugresistance/index.html30 Aug 2011
18. Alliance for the Prudent Use of Antibiotics (APUA) (2011) Available from: www.tufts.edu/med/apua/30 Aug 2011
19. Detecting Evolutionary hot spots of Antibiotic Resistance in Europe (COST-DARE) (2011) Available from: http://www.cost-dare.eu/30 Aug 2011
20. European Committee on Antimicrobial Susceptibility Testing (EUCAST) (2011) Available from: http://www.eucast.org/30 Aug 2011
21. Woodford N, Ellington MJ (2007) The emergence of antibiotic resistance by mutation. Clin Microbiol Infect 13(1):5–18
22. Ochman H, Lawrence JG, Groisman EA (2000) Lateral gene transfer and the nature of bacterial innovation. Nature 405(6784):299–304
23. Partridge SR, Tsafnat G, Coiera E et al (2009) Gene cassettes and cassette arrays in mobile resistance integrons. FEMS Microbiol Rev 33(4):757–784

24. Andersson DI, Hughes D (2010) Antibiotic resistance and its cost: is it possible to reverse resistance? Nat Rev Microbiol 8(4):260–271
25. Parsley LC, Consuegra EJ, Kakirde KS et al (2010) Identification of diverse antimicrobial resistance determinants carried on bacterial, plasmid, or viral metagenomes from an activated sludge microbial assemblage. Appl Environ Microbiol 76(11):3753–3757
26. Fondi M, Fani R (2010) The horizontal flow of the plasmid resistome: clues from inter-generic similarity networks. Environ Microbiol 12(12):3228–3242
27. European Antimicrobial Resistance Surveillance Network (EARS-Net) Available from: http://www.ecdc.europa.eu/en/activities/surveillance/EARS-Net/Pages/index.aspx 30 Aug 2011
28. National Antimicrobial Resistance Monitoring System (NARMS) (2011) Available from: http://www.cdc.gov/narms/30 Aug 2011
29. Houndt T, Ochman H (2000) Long-term shifts in patterns of antibiotic resistance in enteric bacteria. Appl Environ Microbiol 66(12):5406–5409
30. Knapp CW, Dolfing J, Ehlert PA et al (2010) Evidence of increasing antibiotic resistance gene abundances in archived soils since 1940. Environ Sci Technol 44(2):580–587
31. European Surveillance of Antimicrobial Consumption – ESAC (2011) Available from: http://app.esac.ua.ac.be/public/index.php/en_gb/home30 Aug 2011
32. Szczepanowski R, Linke B, Krahn I et al (2009) Detection of 140 clinically relevant antibiotic-resistance genes in the plasmid metagenome of waste water treatment plant bacteria showing reduced susceptibility to selected antibiotics. Microbiology 155 (Pt 7):2306–2319
33. McArthur JV, Tuckfield RC (2000) Spatial patterns in antibiotic resistance among stream bacteria: effects of industrial pollution. Appl Environ Microbiol 66(9):3722–3726
34. Graham DW, Olivares-Rieumont S, Knapp CW et al (2011) Antibiotic resistance gene abundances associated with waste discharges to the Almendares River near Havana, Cuba. Environ Sci Technol 45(2):418–424
35. Zhang XX, Zhang T, Fang HH (2009) Antibiotic resistance genes in water environment. Appl Microbiol Biotechnol 82(3):397–414
36. Kummerer K, Henninger A (2003) Promoting resistance by the emission of antibiotics from hospitals and households into effluent. Clin Microbiol Infect 9(12):1203–1214
37. Halling-Sorensen B, Nors Nielsen S, Lanzky PF et al (1998) Occurrence, fate and effects of pharmaceutical substances in the environment – a review. Chemosphere 36(2):357–393
38. Yergeau E, Lawrence JR, Waiser MJ et al (2010) Metatranscriptomic analysis of the response of river biofilms to pharmaceutical products, using anonymous DNA microarrays. Appl Environ Microbiol 76(16):5432–5439
39. Baker-Austin C, Wright MS, Stepanauskas R et al (2006) Co-selection of antibiotic and metal resistance. Trends Microbiol 14(4):176–182
40. Hernandez A, Mellado RP, Martinez JL (1998) Metal accumulation and vanadium-induced multidrug resistance by environmental isolates of *Escherichia hermannii* and *Enterobacter cloacae*. Appl Environ Microbiol 64(11):4317–4320
41. Miyahara E, Nishie M, Takumi S et al (2011) Environmental mutagens may be implicated in the emergence of drug-resistant microorganisms. FEMS Microbiol Lett 317(2):109–116
42. Fluit AC, Schmitz FJ (2004) Resistance integrons and super-integrons. Clin Microbiol Infect 10(4):272–288
43. Riordan JT, Dupre JM, Cantore-Matyi SA et al (2011) Alterations in the transcriptome and antibiotic susceptibility of *Staphylococcus aureus* grown in the presence of diclofenac. Ann Clin Microbiol Antimicrob 10(1):30
44. Foster PL (2007) Stress-induced mutagenesis in bacteria. Crit Rev Biochem Mol Biol 42(5):373–397
45. Baquero F, Martinez JL, Canton R (2008) Antibiotics and antibiotic resistance in water environments. Curr Opin Biotechnol 19(3):260–265

46. Figueira V, Serra E, Manaia CM (2011) Differential patterns of antimicrobial resistance in population subsets of *Escherichia coli* isolated from waste- and surface waters. Sci Total Environ 409(6):1017–1023

47. 1831/2003 EN (2003) Regulation (EC) No 1831/2003 of the European Parliament and of the Council of 22 Sep 2003 on additives for use in animal nutrition. Official Journal of the European Union, L 268/29

48. Martin H (2005) Manure composting as a pathogen reduction strategy. In: Ontario Ministry of Agriculture, Food and Rural Affairs (OMAFRA) Factsheet (ISSN 1198-712X), Ontario Ministry of Agriculture, Food and Rural Affairs (OMAFRA), Toronto, ON, Canada

49. Klein M, Brown L, Ashbolt NJ et al (2011) Inactivation of indicators and pathogens in cattle feedlot manures and compost as determined by molecular and culture assays. FEMS Microbiol Ecol 77(1):200–210

50. Peak N, Knapp CW, Yang RK et al (2007) Abundance of six tetracycline resistance genes in waste water lagoons at cattle feedlots with different antibiotic use strategies. Environ Microbiol 9(1):143–151

51. Kazimierczak KA, Scott KP, Kelly D et al (2009) Tetracycline resistome of the organic pig gut. Appl Environ Microbiol 75(6):1717–1722

52. Ishida Y, Ahmed AM, Mahfouz NB et al (2010) Molecular analysis of antimicrobial resistance in gram-negative bacteria isolated from fish farms in Egypt. J Vet Med Sci 72(6):727–734

53. Sharma R, Munns K, Alexander T et al (2008) Diversity and distribution of commensal fecal *Escherichia coli* bacteria in beef cattle administered selected subtherapeutic antimicrobials in a feedlot setting. Appl Environ Microbiol 74(20):6178–6186

54. Kim S, Aga DS (2007) Potential ecological and human health impacts of antibiotics and antibiotic-resistant bacteria from waste water treatment plants. J Toxicol Environ Health B Crit Rev 10(8):559–573

55. Schluter A, Krause L, Szczepanowski R et al (2008) Genetic diversity and composition of a plasmid metagenome from a waste water treatment plant. J Biotechnol 136(1–2):65–76

56. Petrovic M, de Alda MJ, Diaz-Cruz S et al (2009) Fate and removal of pharmaceuticals and illicit drugs in conventional and membrane bioreactor waste water treatment plants and by riverbank filtration. Philos Transact A Math Phys Eng Sci 367(1904):3979–4003

57. Sanapareddy N, Hamp TJ, Gonzalez LC et al (2009) Molecular diversity of a North Carolina waste water treatment plant as revealed by pyrosequencing. Appl Environ Microbiol 75(6):1688–1696

58. Goni-Urriza M, Capdepuy M, Arpin C et al (2000) Impact of an urban effluent on antibiotic resistance of riverine *Enterobacteriaceae* and *Aeromonas* spp. Appl Environ Microbiol 66(1):125–132

59. Cabello FC (2006) Heavy use of prophylactic antibiotics in aquaculture: a growing problem for human and animal health and for the environment. Environ Microbiol 8(7):1137–1144

60. Watkinson AJ, Micalizzi GB, Graham GM et al (2007) Antibiotic-resistant *Escherichia coli* in waste waters, surface waters, and oysters from an urban riverine system. Appl Environ Microbiol 73(17):5667–5670

61. Johnsen PJ, Townsend JP, Bohn T et al (2009) Factors affecting the reversal of antimicrobial-drug resistance. Lancet Infect Dis 9(6):357–364

62. Walsh TR, Toleman MA (2011) The new medical challenge: why NDM-1? Why Indian? Expert Rev Anti Infect Ther 9(2):137–141

63. WHO (2008) Guidelines for drinking water 3rd edn. World Health Organization, Geneva

64. Clinical CLSI (2007) Performance standards for antimicrobial susceptibility testing; M100-S17. Wayne, PA, 27

65. Wiegand I, Hilpert K, Hancock RE (2008) Agar and broth dilution methods to determine the minimal inhibitory concentration (MIC) of antimicrobial substances. Nat Protoc 3 (2):163–175

66. Clinical and Laboratory Standards Institute (CLSI) (2011) Available from: http://www.clsi.org/ 30 August 2011

67. CA-SFM (SFdM) (2011) Available from: http://www.sfm-microbiologie.org/pages/?all=accueil 30 Aug 2011
68. Andrews JM (2009) BSAC standardized disc susceptibility testing method (version 8). J Antimicrob Chemother 64(3):454–489
69. Danish Integrated Antimicrobial resistance Monitoring and Research Programme (DANMAP-Denmark) Available from: http://www.danmap.org/ (30 August 2011)
70. APHA (2005) Standard methods for the examination of water and waste water, 21st edn. American Public Health Association, Washington, D.C
71. 98/83/EC CD (1998) of 3 Nov 1998 on the quality of water intended for human consumption as amended by regulation 1882/2003/EC
72. Ashbolt N, Grabow WOK, Snozzi M (2001) Indicators of microbial water quality. In: Fewtrell L, Bartram J (eds) Water quality guidelines standards and health assessment of risk and risk management for water-related infectious disease. IWA Publishing, London, pp 289–316
73. Watkinson AJ, Micalizzi GR, Bates JR et al (2007) Novel method for rapid assessment of antibiotic resistance in *Escherichia coli* isolates from environmental waters by use of a modified chromogenic agar. Appl Environ Microbiol 73(7):2224–2229
74. Galvin S, Boyle F, Hickey P et al (2010) Enumeration and characterization of antimicrobial-resistant *Escherichia coli* bacteria in effluent from municipal, hospital, and secondary treatment facility sources. Appl Environ Microbiol 76(14):4772–4779
75. Lewis K (2010) Persister cells. Annu Rev Microbiol 64:357–372
76. Riesenfeld CS, Goodman RM, Handelsman J (2004) Uncultured soil bacteria are a reservoir of new antibiotic resistance genes. Environ Microbiol 6(9):981–989
77. Monier JM, Demaneche S, Delmont TO et al (2011) Metagenomic exploration of antibiotic resistance in soil. Curr Opin Microbiol 14(3):229–235
78. Canton R (2009) Antibiotic resistance genes from the environment: a perspective through newly identified antibiotic resistance mechanisms in the clinical setting. Clin Microbiol Infect 15(Suppl 1):20–25
79. Goni-Urriza M, Arpin C, Capdepuy M et al (2002) Type II topoisomerase quinolone resistance-determining regions of *Aeromonas caviae*, *A. hydrophila*, and *A. sobria* complexes and mutations associated with quinolone resistance. Antimicrob Agents Chemother 46(2):350–359
80. Ferreira da Silva M, Tiago I, Verissimo A et al (2006) Antibiotic resistance of enterococci and related bacteria in an urban waste water treatment plant. FEMS Microbiol Ecol 55(2):322–329
81. Ferreira da Silva M, Vaz-Moreira I, Gonzalez-Pajuelo M et al (2007) Antimicrobial resistance patterns in *Enterobacteriaceae* isolated from an urban waste water treatment plant. FEMS Microbiol Ecol 60(1):166–176
82. Zhang Y, Marrs CF, Simon C et al (2009) Waste water treatment contributes to selective increase of antibiotic resistance among *Acinetobacter* spp. Sci Total Environ 407(12): 3702–3706
83. Martins da Costa P, Vaz-Pires P, Bernardo F (2006) Antimicrobial resistance in *Enterococcus* spp. isolated in inflow, effluent and sludge from municipal sewage water treatment plants. Water Res 40(8):1735–1740
84. Figueira V, Vaz-Moreira I, Silva M, Manaia CM (2011) Diversity and antibiotic resistance of *Aeromonas* spp. in drinking and waste water treatment plants. Water Research doi:10.1016/j.watres.2011.08.021
85. Vaz-Moreira I, Egas C, Nunes OC et al (2011) Culture-dependent and culture-independent diversity surveys target different bacteria: a case study in a freshwater sample. Antonie Van Leeuwenhoek 100(2):245–257
86. Kapley A, De Baere T, Purohit HJ (2007) Eubacterial diversity of activated biomass from a common effluent treatment plant. Res Microbiol 158(6):494–500

87. Du C, Wu Z, Xiao E et al (2008) Bacterial diversity in activated sludge from a consecutively aerated submerged membrane bioreactor treating domestic waste water. J Environ Sci (China) 20(10):1210–1217
88. Moura A, Tacão M, Henriques I et al (2009) Characterization of bacterial diversity in two aerated lagoons of a waste water treatment plant using PCR-DGGE analysis. Microbiol Res 164(5):560–569
89. Poitelon JB, Joyeux M, Welte B et al (2009) Assessment of phylogenetic diversity of bacterial microflora in drinking water using serial analysis of ribosomal sequence tags. Water Res 43(17):4197–4206
90. Xin J, Mingchao MA, Jun LI et al (2008) Bacterial diversity of active sludge in waste water treatment plant. Earth Sci Front 15(6):163–168
91. Stokes HW, Holmes AJ, Nield BS et al (2001) Gene cassette PCR: sequence-independent recovery of entire genes from environmental DNA. Appl Environ Microbiol 67(11):5240–5246
92. Schwartz T, Kohnen W, Jansen B et al (2003) Detection of antibiotic-resistant bacteria and their resistance genes in waste water, surface water, and drinking water biofilms. FEMS Microbiol Ecol 43(3):325–335
93. Szczepanowski R, Bekel T, Goesmann A et al (2008) Insight into the plasmid metagenome of waste water treatment plant bacteria showing reduced susceptibility to antimicrobial drugs analysed by the 454-pyrosequencing technology. J Biotechnol 136(1–2):54–64
94. Colomer-Lluch M, Jofre J, Muniesa M (2011) Antibiotic resistance genes in the bacteriophage DNA fraction of environmental samples. PLoS One 6(3):e17549
95. Kristiansson E, Fick J, Janzon A et al (2011) Pyrosequencing of antibiotic-contaminated river sediments reveals high levels of resistance and gene transfer elements. PLoS One 6(2): e17038
96. Tennstedt T, Szczepanowski R, Braun S et al (2003) Occurrence of integron-associated resistance gene cassettes located on antibiotic resistance plasmids isolated from a waste water treatment plant. FEMS Microbiol Ecol 45(3):239–252
97. Henriques I, Moura A, Alves A et al (2006) Analysing diversity among beta-lactamase encoding genes in aquatic environments. FEMS Microbiol Ecol 56(3):418–429
98. Castiglioni S, Pomati F, Miller K et al (2008) Novel homologs of the multiple resistance regulator marA in antibiotic-contaminated environments. Water Res 42(16):4271–4280
99. Volkmann H, Schwartz T, Bischoff P et al (2004) Detection of clinically relevant antibiotic-resistance genes in municipal waste water using real-time PCR (TaqMan). J Microbiol Methods 56(2):277–286
100. Auerbach EA, Seyfried EE, McMahon KD (2007) Tetracycline resistance genes in activated sludge waste water treatment plants. Water Res 41(5):1143–1151
101. Pei R, Cha J, Carlson KH et al (2007) Response of antibiotic resistance genes (ARG) to biological treatment in dairy lagoon water. Environ Sci Technol 41(14):5108–5113
102. Reinthaler FF, Posch J, Feierl G et al (2003) Antibiotic resistance of E. coli in sewage and sludge. Water Res 37(8):1685–1690
103. Boczek LA, Rice EW, Johnston B et al (2007) Occurrence of antibiotic-resistant uropathogenic Escherichia coli clonal group A in waste water effluents. Appl Environ Microbiol 73(13):4180–4184
104. Sabate M, Prats G, Moreno E et al (2008) Virulence and antimicrobial resistance profiles among Escherichia coli strains isolated from human and animal waste water. Res Microbiol 159(4):288–293
105. Araújo C, Torres C, Silva N et al (2010) Vancomycin-resistant enterococci from Portuguese waste water treatment plants. J Basic Microbiol 50(6):605–609
106. Faria C, Vaz-Moreira I, Serapicos E et al (2009) Antibiotic resistance in coagulase negative staphylococci isolated from waste water and drinking water. Sci Total Environ 407(12):3876–3882

107. Borjesson S, Melin S, Matussek A et al (2009) A seasonal study of the mecA gene and *Staphylococcus aureus* including methicillin-resistant *S. aureus* in a municipal waste water treatment plant. Water Res 43(4):925–932
108. Okoh AI, Igbinosa EO (2010) Antibiotic susceptibility profiles of some *Vibrio* strains isolated from waste water final effluents in a rural community of the Eastern Cape Province of South Africa. BMC Microbiol 10:143
109. Wagner M, Loy A, Nogueira R et al (2002) Microbial community composition and function in waste water treatment plants. Antonie Van Leeuwenhoek 81(1–4):665–680
110. Luczkiewicz A, Jankowska K, Kurlenda J et al (2010) Identification and antimicrobial resistance of *Enterococcus* spp. isolated from surface water. Water Sci Technol 62(2):466–473
111. Novo A, Manaia CM (2010) Factors influencing antibiotic resistance burden in municipal waste water treatment plants. Appl Microbiol Biotechnol 87(3):1157–1166
112. Guardabassi L, Lo Fo Wong DM, Dalsgaard A (2002) The effects of tertiary waste water treatment on the prevalence of antimicrobial resistant bacteria. Water Res 36(8):1955–1964
113. Strahilevitz J, Jacoby GA, Hooper DC et al (2009) Plasmid-mediated quinolone resistance: a multifaceted threat. Clin Microbiol Rev 22(4):664–689
114. Vilanova X, Manero A, Cerda-Cuellar M et al (2002) The effect of a sewage treatment plant effluent on the faecal coliforms and enterococci populations of the reception river waters. J Appl Microbiol 92(2):210–214
115. Blanch AR, Caplin JL, Iversen A et al (2003) Comparison of enterococcal populations related to urban and hospital waste water in various climatic and geographic European regions. J Appl Microbiol 94(6):994–1002
116. Vaz-Moreira I, Nunes OC, Manaia CM (2011) Diversity and antibiotic resistance patterns of *Sphingomonadaceae* isolated from drinking water. Appl Environ Microbiol 77(16):5697–5706
117. Kassem II, Esseili MA, Sigler V (2008) Occurrence of mecA in nonstaphylococcal pathogens in surface waters. J Clin Microbiol 46(11):3868–3869
118. Poeta P, Radhouani H, Igrejas G et al (2008) Seagulls of the Berlengas natural reserve of Portugal as carriers of fecal *Escherichia coli* harboring CTX-M and TEM extended-spectrum beta-lactamases. Appl Environ Microbiol 74(23):7439–7441
119. Soge OO, Meschke JS, No DB et al (2009) Characterization of methicillin-resistant *Staphylococcus aureus* and methicillin-resistant coagulase-negative *Staphylococcus* spp. isolated from US West Coast public marine beaches. J Antimicrob Chemother 64(6):1148–1155
120. Xi C, Zhang Y, Marrs CF et al (2009) Prevalence of antibiotic resistance in drinking water treatment and distribution systems. Appl Environ Microbiol 75(17):5714–5718
121. Figueira V, Serra EA, Vaz-Moreira I, Brandão TRS, Manaia CM/Comparison of ubiquitous antibiotic-resistant *Enterobacteriaceae* populations isolated from waste waters, surface waters and drinking waters. J Water Health (in press)
122. Niquette P, Servais P, Savoir R (2001) Bacterial dynamics in the drinking water distribution system of Brussels. Water Res 35(3):675–682
123. Lautenschlager K, Boon N, Wang Y et al (2010) Overnight stagnation of drinking water in household taps induces microbial growth and changes in community composition. Water Res 44(17):4868–4877
124. Kahlisch L, Henne K, Groebe L et al (2010) Molecular analysis of the bacterial drinking water community with respect to live/dead status. Water Sci Technol 61(1):9–14
125. Brenner DJ (1992) Introdution to the family *Enterobacteriaceae*. In: The prokaryotes, a handbook on the biology, ecophysiology, isolation, identification, applications, vol 3, 2nd edn. Springer-Verlag, Newyork, pp 2673–2695, Chapter 141
126. Kuhn I, Huys G, Coopman R et al (1997) A 4-year study of the diversity and persistence of coliforms and *Aeromonas* in the water of a Swedish drinking water well. Can J Microbiol 43(1):9–16

127. Koskinen R, Ali-Vehmas T, Kampfer P et al (2000) Characterization of *Sphingomonas* isolates from Finnish and Swedish drinking water distribution systems. J Appl Microbiol 89(4):687–696

128. Norton CD, LeChevallier MW (2000) A pilot study of bacteriological population changes through potable water treatment and distribution. Appl Environ Microbiol 66(1):268–276

129. Leclerc H, Mossel DA, Edberg SC et al (2001) Advances in the bacteriology of the coliform group: their suitability as markers of microbial water safety. Annu Rev Microbiol 55:201–234

130. Biscardi D, Castaldo A, Gualillo O et al (2002) The occurrence of cytotoxic *Aeromonas hydrophila* strains in Italian mineral and thermal waters. Sci Total Environ 292(3):255–263

131. Hoefel D, Monis PT, Grooby WL et al (2005) Profiling bacterial survival through a water treatment process and subsequent distribution system. J Appl Microbiol 99(1):175–186

132. Furuhata K, Kato Y, Goto K et al (2007) Identification of yellow-pigmented bacteria isolated from hospital tap water in Japan and their chlorine resistance. Biocontrol Sci 12(2):39–46

133. Blanch AR, Galofre B, Lucena F et al (2007) Characterization of bacterial coliform occurrences in different zones of a drinking water distribution system. J Appl Microbiol 102(3):711–721

134. Kampfer P, Nienhuser A, Packroff G et al (2008) Molecular identification of coliform bacteria isolated from drinking water reservoirs with traditional methods and the Colilert-18 system. Int J Hyg Environ Health 211(3–4):374–384

135. Pablos M, Rodriguez-Calleja JM, Santos JA et al (2009) Occurrence of motile *Aeromonas* in municipal drinking water and distribution of genes encoding virulence factors. Int J Food Microbiol 135(2):158–164

136. Rusin PA, Rose JB, Haas CN et al (1997) Risk assessment of opportunistic bacterial pathogens in drinking water. Rev Environ Contam Toxicol 152:57–83

137. Armstrong JL, Calomiris JJ, Seidler RJ (1982) Selection of antibiotic-resistant standard plate count bacteria during water treatment. Appl Environ Microbiol 44(2):308–316

138. Armstrong JL, Shigeno DS, Calomiris JJ et al (1981) Antibiotic-resistant bacteria in drinking water. Appl Environ Microbiol 42(2):277–283

139. Pathak SP, Gopal K (2008) Prevalence of bacterial contamination with antibiotic-resistant and enterotoxigenic fecal coliforms in treated drinking water. J Toxicol Environ Health A 71(7):427–433

140. Shehabi AA, Odeh JF, Fayyad M (2006) Characterization of antimicrobial resistance and class 1 integrons found in *Escherichia coli* isolates from human stools and drinking water sources in Jordan. J Chemother 18(5):468–472

141. Cordoba MA, Roccia IL, De Luca MM et al (2001) Resistance to antibiotics in injured coliforms isolated from drinking water. Microbiol Immunol 45(5):383–386

142. Messi P, Guerrieri E, Bondi M (2005) Antibiotic resistance and antibacterial activity in heterotrophic bacteria of mineral water origin. Sci Total Environ 346(1–3):213–219

143. Srinivasan V, Nam HM, Nguyen LT et al (2005) Prevalence of antimicrobial resistance genes in *Listeria monocytogenes* isolated from dairy farms. Foodborne Pathog Dis 2(3):201–211

144. Poppe C, Martin L, Muckle A et al (2006) Characterization of antimicrobial resistance of *Salmonella newport* isolated from animals, the environment, and animal food products in Canada. Can J Vet Res 70(2):105–114

145. Murray GE, Tobin RS, Junkins B et al (1984) Effect of chlorination on antibiotic resistance profiles of sewage-related bacteria. Appl Environ Microbiol 48(1):73–77

146. Morar M, Wright GD (2010) The genomic enzymology of antibiotic resistance. Annu Rev Genet 44:25–51

147. D'Costa VM, Griffiths E, Wright GD (2007) Expanding the soil antibiotic resistome: exploring environmental diversity. Curr Opin Microbiol 10(5):481–489

148. Wright GD (2010) Antibiotic resistance in the environment: a link to the clinic? Curr Opin Microbiol 13(5):589–594

149. Schluter A, Szczepanowski R, Puhler A et al (2007) Genomics of IncP-1 antibiotic resistance plasmids isolated from waste water treatment plants provides evidence for a widely accessible drug resistance gene pool. FEMS Microbiol Rev 31(4):449–477

150. Baquero F (2004) From pieces to patterns: evolutionary engineering in bacterial pathogens. Nat Rev Microbiol 2(6):510–518

151. Dinsdale EA, Edwards RA, Hall D et al (2008) Functional metagenomic profiling of nine biomes. Nature 452(7187):629–632

152. Rowe-Magnus DA, Mazel D (2002) The role of integrons in antibiotic resistance gene capture. Int J Med Microbiol 292(2):115–125

153. Huang L, Cagnon C, Caumette P et al (2009) First gene cassettes of integrons as targets in finding adaptive genes in metagenomes. Appl Environ Microbiol 75(11):3823–3825

154. Laroche E, Pawlak B, Berthe T et al (2009) Occurrence of antibiotic resistance and class 1, 2 and 3 integrons in *Escherichia coli* isolated from a densely populated estuary (Seine, France). FEMS Microbiol Ecol 68(1):118–130

155. Li D, Yang M, Hu J et al (2009) Antibiotic-resistance profile in environmental bacteria isolated from penicillin production waste water treatment plant and the receiving river. Environ Microbiol 11(6):1506–1517

156. Ozgumus OB, Sandalli C, Sevim A et al (2009) Class 1 and class 2 integrons and plasmid-mediated antibiotic resistance in coliforms isolated from ten rivers in northern Turkey. J Microbiol 47(1):19–27

157. Chen H, Shu W, Chang X et al (2010) The profile of antibiotics resistance and integrons of extended-spectrum beta-lactamase producing thermotolerant coliforms isolated from the Yangtze River basin in Chongqing. Environ Pollut 158(7):2459–2464

158. Chen B, Zheng W, Yu Y et al (2011) Class 1 integrons, selected virulence genes, and antibiotic resistance in *Escherichia coli* isolates from the Minjiang River, Fujian Province, China. Appl Environ Microbiol 77(1):148–155

159. Xia R, Guo X, Zhang Y et al (2010) qnrVC-like gene located in a novel complex class 1 integron harboring the ISCR1 element in an *Aeromonas punctata* strain from an aquatic environment in Shandong Province, China. Antimicrob Agents Chemother 54(8):3471–3474

160. Li D, Yu T, Zhang Y et al (2010) Antibiotic resistance characteristics of environmental bacteria from an oxytetracycline production waste water treatment plant and the receiving river. Appl Environ Microbiol 76(11):3444–3451

161. Lachmayr KL, Kerkhof LJ, Dirienzo AG et al (2009) Quantifying nonspecific TEM beta-lactamase (blaTEM) genes in a waste water stream. Appl Environ Microbiol 75(1):203–211

162. Fick J, Soderstrom H, Lindberg RH et al (2009) Contamination of surface, ground, and drinking water from pharmaceutical production. Environ Toxicol Chem 28(12):2522–2527

163. Bockelmann U, Dorries HH, Ayuso-Gabella MN et al (2009) Quantitative PCR monitoring of antibiotic resistance genes and bacterial pathogens in three European artificial groundwater recharge systems. Appl Environ Microbiol 75(1):154–163

164. Cernat RBC, Ivanescu D, Nedelcu D, Lazar V, Bucur M, Valeanu DTR, Mitache M, Dragoescu M (2007) Mechanisms of resistance in multiple-antibiotic-resistant *Escherichia coli* strains isolated from drinking and recreational, salmaster waters. Int J Antimicrob Agents 29:S274

165. Yong D, Toleman MA, Giske CG et al (2009) Characterization of a new metallo-beta-lactamase gene, bla(NDM-1), and a novel erythromycin esterase gene carried on a unique genetic structure in *Klebsiella pneumoniae* sequence type 14 from India. Antimicrob Agents Chemother 53(12):5046–5054

166. Walsh TR, Weeks J, Livermore DM et al (2011) Dissemination of NDM-1 positive bacteria in the New Delhi environment and its implications for human health: an environmental point prevalence study. Lancet Infect Dis 11(5):355–362

167. Sommer MO, Dantas G, Church GM (2009) Functional characterization of the antibiotic resistance reservoir in the human microflora. Science 325(5944):1128–1131

168. Sommer MO, Church GM, Dantas G (2010) The human microbiome harbors a diverse reservoir of antibiotic resistance genes. Virulence 1(4):299–303
169. Tenover FC (2006) Mechanisms of antimicrobial resistance in bacteria. Am J Med 119 (6 Suppl 1):S3–S10, discussion S62–70
170. Girlich D, Poirel L, Nordmann P (2011) Diversity of clavulanic acid-inhibited extended-spectrum beta-lactamases in *Aeromonas* spp. from the Seine River, Paris, France. Antimicrob Agents Chemother 55(3):1256–1261
171. Gordon L, Cloeckaert A, Doublet B et al (2008) Complete sequence of the floR-carrying multiresistance plasmid pAB5S9 from freshwater *Aeromonas bestiarum*. J Antimicrob Chemother 62(1):65–71
172. Barker-Reid F, Fox EM, Faggian R (2010) Occurrence of antibiotic resistance genes in reclaimed water and river water in the Werribee Basin, Australia. J Water Health 8(3):521–531
173. Cummings DE, Archer KF, Arriola DJ et al (2011) Broad dissemination of plasmid-mediated quinolone resistance genes in sediments of two urban coastal wetlands. Environ Sci Technol 45(2):447–454

Ecopharmacovigilance

L.J.G. Silva, C.M. Lino, L. Meisel, D. Barceló, and A. Pena

Abstract Active pharmaceutical ingredients (APIs) represent a group of emerging environmental contaminants. Albeit in trace amounts, they are of great concern since given their continuous introduction into the environment, their impact on ecosystems and human health is of great importance. As a result, the environmental risk assessment (ERA) of medicinal products has to be evaluated and appropriate legislation has been issued in the European Union (EU).

To accomplish these requirements, the concept of ecopharmacovigilance has been recently established. This chapter seeks to summarize the many aspects and nuances surrounding this issue. A comprehensive discussion of the different contamination sources, fate, occurrence, toxicological effects on non-target organisms and associated risks is presented. Perceived needs and sustainable strategies for minimizing APIs environmental impact will be identified. Such measures are imperative to improve awareness and encourage precautionary actions.

Keywords Active pharmaceutical ingredients (APIs), Ecopharmacovigilance, Environmental risk assessment (ERA)

L.J.G. Silva, C.M. Lino, and A. Pena (✉)
Group of Health Surveillance, Center of Pharmaceutical Studies, Faculty of Pharmacy, University of Coimbra, Polo III, Azinhaga de Santa Comba, 3000-548 Coimbra, Portugal
e-mail: apena@ff.uc.pt

L. Meisel
INFARMED, I.P. - National Authority of Medicines and Health Products, 1749-004 Lisboa, Portugal

Faculty of Pharmacy, University of Lisboa Av. Prof. Gama Pinto, 1649-003 Lisboa, Portugal

D. Barceló
Department of Environmental Chemistry, IIQAB-CSIC, c/ Jordi Girona 18-26, 08034 Barcelona, Spain

Institut Català de Recerca de l'Aigua (ICRA), c/ Pic de Peguera 15, 17003 Girona, Spain

D. Barceló (ed.), *Emerging Organic Contaminants and Human Health*,
Hdb Env Chem (2012) 20: 213–242, DOI 10.1007/698_2011_128,
© Springer-Verlag Berlin Heidelberg 2011, Published online: 20 December 2011

Contents

1   Introduction

Active pharmaceutical ingredients (APIs) represent a group of ubiquitous contaminants in the environment. Albeit in trace amounts, they are of concern since they are designed to be inherently biologically active and potentially stable under metabolic processes. Nowadays, it is recognized by the scientific community that the continual exposure of low doses of pharmaceuticals may produce subtle and long-term effects on the environment leading to possible irreversible damage of the ecosystem and human health. Given their continuous introduction into the environment, their environmental impact, both as stressors and as agents of change, is of great importance [1, 2].

According to the current knowledge, residues of pharmaceuticals and their metabolites are widespread in the aquatic systems. Surface water monitoring programmes in Europe and North America [3–8] have shown, as a result of improved analytical capabilities [9, 10], the presence of many different classes of pharmaceuticals, some of which are known to be environmentally persistent [10].

Since traditionally pharmaceuticals have not been considered as environmental contaminants, the study of their presence in the environment is in some ways a new area of research which has taken off recently. Nonetheless, so far, there is little knowledge on their occurrence, fate and environmental assessment, an essential tool for the implementation of minimizing measures and risk management. Potential risks associated with releases of pharmaceuticals into the environment have become an increasingly important issue for pharmaceutical industry and environmental regulators [7, 11].

Consequently, the environmental risk assessment (ERA) of medicinal products has to be evaluated and appropriate legislation and regulatory guidance has been issued in the European Union (EU). Recently, Directive 2001/83/EC [12], as amended by Directive 2004/27/EC [13], requires an evaluation of the potential environmental risks to be performed for every application for each active ingredient/excipients from every medicinal product to be authorized. However, the potential risk of various classes of pharmaceuticals is difficult to evaluate due to the lack or inadequate data.

To fulfil this requirement, the concept of ecopharmacovigilance has been recently established [14], and due to the many facets of this complex issue, the communication between medical/healthcare communities, pharmaceutical industry, medicine and environmental regulators, academia and environmental science researchers is primordial.

This chapter describes the main trends in ecopharmacovigilance, including an overview of the different contamination sources, fate, occurrence, toxicological effects on non-target organisms and associated risks. Moreover, perceived needs and gaps will be identified and the outcome of the discussion on this major issue will be provided.

2   Ecopharmacovigilance

Ecopharmacovigilance, a concept introduced by Daughton and Ruhoy [14], can be defined as *"the science and activities concerning detection, assessment, understanding and prevention of adverse effects or other problems related to the presence of pharmaceuticals in the environment, which affect both human and the other animal species (...)"* [15].

This is close to the World Health Organization (WHO) definition of pharmacovigilance [16], the science aiming to capture any adverse effects of pharmaceuticals in humans after use. Pharmacovigilance system involves collection, monitoring, researching, assessing and evaluating information received from health care workers such as doctors, dentists, pharmacists, nurses and other health professionals to be aware of the adverse drug reaction.

Could pharmacovigilance studies contribute for an improved ecopharmacovigilance? Several authors consider that the use of existing pharmacological, toxicological and pharmacokinetic data is likely to be helpful in assessing the environmental risk of pharmaceuticals, as they could provide a better knowledge of their fate and effect. Pharmacological data alone are not sufficient to assess the risk for the aquatic environment; however, such data can give information on the mode of action (MoA) and the toxicity of pharmaceuticals [17]. The International Society of Pharmacovigilance (ISOP) had a very active role on this issue and, today, ecopharmacovigilance is regarded as an integral part of pharmacovigilance [15].

In September 2010, the European Parliament adopted amendments to the pharmacovigilance legislation (Directive 2001/83/EC and Regulation EC No 726/2004 [12, 18]) in order to broaden its concept to environmental concerns. The concept paper was under public consultation, until 7 November 2011, to consider measures to monitor and evaluate the risk of environmental effects of medicinal products [19].

An ecopharmacovigilance vision should be oriented to allow the prediction of potential environmental problems and improve ERA. Some pharmaceutical industries, such as Astrazeneca, are committed to better understand this new concept, and have started to prepare an information pack for each of their drugs [20].

According to the EU project entitled "Knowledge and Need Assessment on Pharmaceutical Products in Environmental Waters" (KNAPPE), ecopharmacovigilance deals with APIs monitoring of sources, distribution, fate and biological impact on ecosystems and, ultimately, on human health [5], which environmental levels depend on the amounts sold/consumed, pharmacokinetic behaviour, degradation and wastewater treatment plants (WWTPs) removal efficiency.

2.1   World Market for Human Pharmaceuticals

For ecopharmacovigilance, it is important to know the worldwide data for drug use. The global pharmaceutical market reached, in 2010, 875 billion US dollars, while in 2000 only accounted for 365 billion US dollars [21], representing an increase of 240% in ten years.

In Portugal, the most consumed/sold pharmaceutical groups belong to the central nervous system, cardiovascular, anti-infective classes and cytostatics, being these groups heavily represented within the top 100 drugs of the ranking as presented in Table 1 and in Fig. 1. Pharmaceuticals sales increased 10% in the period 2003/2007 [23], and an increase in cardiovascular system and central nervous system has been observed [22].

Given the global increase in population (1 billion/decade), and the current annual increase in drug consumption (3%), a threefold increase in drug consumption is expected, leading to increasing pressures on WWTPs and on the receptive water environment [24]. As their use cannot be avoided, a sound risk assessment of their environmental presence is a key problem. This is the case of antibiotics, given their potential for resistance selection among pathogens, and steroidal hormones, due to their reproductive effects.

2.2   Environmental Pathways: Sources, Fate and Degradation

Due to their high consumption, APIs are continuously introduced into the environment. After administration, the majority of drugs are metabolized through phase-I and/or phase-II reactions.

Both phase-I and phase-II reactions change the physical chemical behaviour of the substance since the metabolization always affords the metabolites more water solubility than the parent compounds. In addition, it was showed that the phase-II metabolites can be even converted back into the original active drug. A significant amount of the original compound reaches the environment unmetabolized or only slightly transformed via urine or faeces [25], being the former more significant [26–30].

Table 1 Consumption of APIs in Portugal [22]

Pharmacotherapeutical group and percentage of consumption (%)	Pharmacotherapeutical subgroup	No. of medicinal products on top of 100 drugs most used	The most representative active ingredients	Position on ranking of 100 most used
Cardiovascular system (33.5)	Anti-hypertensives	24	Indapamide	12
			Losartan	20
	Antilipemics	5	Simvastatin	2
	Cardiotonics	2	Digoxin	75
	Vasodilators	2	Nifedipine	39
Central nervous system (30.5)	Analgesics and antipyretics	5	Paracetamol	1
			Acetylsalicylic acid	11
	Non-steroidal antiinflamatory agents	6	Ibuprofen	8
			Diclofenac	10
	Psychodrugs	16	Alprazolam	4
			Lorazepam	9
			Zolpidem	13
			Fluoxetine	41
Anti-infective products (8.6)	Antibacterial drugs	5	Amoxicilin + Clavulanic acid	5
	Antifungals	1	Fluconazol	96
Endocrine system (10.9)	Oral antidiabetics	4	Metformin	3
	Sex hormones	1	Ethinylestradiol Gestodene	34
Digestive system (7.3)	Antiacids/anti-ulcerous	4	Omeprazol	7
Anticoagulants/ antithrombotics (5.7)		2	Clopidogrel	19
Antineoplasics/ immunomodulators agents (0.24)		2	Tamsulosin	26
Respiratory system (3.3)	Anti-asthmatic agents	2	Montelukast	61
	Bronchodilator	1	Salbutamol	62

Their widespread environmental presence is most likely to occur from WWTPs point source discharges, which incompletely remove these compounds, leading to the contamination of surface waters, seawaters, groundwater and some drinking waters [31].

Hospital wastewater is the other main source of contamination [32–35]. The dilution of hospital effluents by municipal wastewaters will lower the concentration of pharmaceuticals only moderately, because the latter also contain pharmaceuticals from households and veterinary sources [36].

Other anthropogenic contamination sources, such as disposal of unused or expired drugs or pharmaceutical industry discharges, should be assumed [26–30].

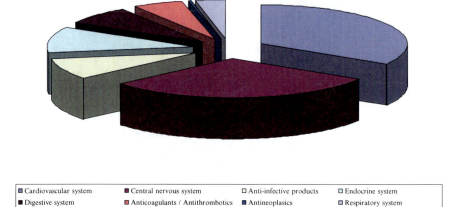

| ☐ Cardiovascular system | ■ Central nervous system | ☐ Anti-infective products | ☐ Endocrine system |
| ■ Digestive system | ☐ Anticoagulants / Antithrombotics | ■ Antineoplasics | ☐ Respiratory system |

Fig. 1 Portuguese consumption percentage of the pharmacotherapeutical groups [22]

Most drugs are designed to be persistent so that they retain their chemical structure long enough to do their therapeutic work and this, coupled with their continual input, may enable them to remain in the environment for a significant period of time [37]. However, when in the environment they may be degraded through three elimination processes: photodegradation, adsorption and biodegradation.

The first depends on factors such as intensity of solar irradiation, latitude, season of the year and presence of photosensitizers (e.g., nitrates, humic acids) [30]. As example, the β-blocker propranolol was reported to be rapidly photodegraded and therefore is not expected to be persistent in surface waters [38]. Photolysis and hydrolysis were also suggested to be rapid ways of removal of amoxicillin in the environment [39].

Adsorption to suspended solids is dependent on the lipophilicity or other ability of binding to soil or sludge, which also depend on pH and/or presence of other constituents (e.g., complexation). Drugs that are mobile in the soil may be a threat to the groundwater or leach to a nearby stream [11]. Acidic pharmaceuticals such as the acetylsalicylic acid, ibuprofen, fenoprofen, ketoprofen, naproxen, diclofenac and indomethacin (pK_a between 4.1 and 4.9) have little tendency of adsorption to the sludge [40]. Whereas basic pharmaceuticals and zwitterions can adsorb to sludge to a significant extent, this is the case of fluoroquinolone antibiotics (FQs) [34]. Low adsorption coefficients make active substances remain in the aqueous phase, favouring their mobility through the aquatic environment.

Biodegradation occur either in aerobic or in anaerobic conditions, and as detailed below, it is the most important elimination process in WWTPs. This is an important degradation process for diclofenac, ofloxacin (OFLO), sulfamethoxazole and tetracyclines (TCs) [40]. Several fungi, such as *Mucor ramannianus*, present in soils, already have been shown to metabolize FQs [41].

2.3 APIs Removal

2.3.1 WWTPs Removal

The scientific published data report the APIs incomplete elimination by WWTPs and that these facilities are regarded as hot spots of aquatic contamination.

In WWTPs, the following events may happen to the drug/metabolites:

1. They could be mineralized by micro-organisms to carbon dioxide and water, for example acetylsalicylic acid [11]. Aerobic/anaerobic bio-conversion occurring either during sewage sludge digestion or during activated sludge treatment seems to be the most efficient process of elimination. Usually, the best biodegradation results are obtained when activated sludge treatment is conducted through an increase in hydraulic retention time and the use of mature sludge [30, 40].
2. Depending on the lipophilicity or other binding possibilities, for example ionic bindings, a part of the compound will be retained in the sludge. Adsorption to suspended solids and subsequent removal by sedimentation, as primary or secondary sludge, is one of the most important elimination processes [42]. If the sludge is used as soil fertilizer, drugs may be dispersed on agricultural fields and bioaccumulation occur [11].
3. If the drug/metabolites are persistent and at the same time very polar and do not bind to solids, they are not retained neither degraded in the WWTPs. Therefore, they can easily reach the aquatic environment [11]. This is the example of the ubiquitous presence of carbamazepine in natural waters due to its persistence and low removal rate (below 20%) [43].

APIs elimination rates in WWTPs depend on their chemical properties and on the technology used in the WWTPs [44, 45]. The suitability of distinct wastewater treatment processes for the elimination of APIs has been studied and, to the best of our knowledge, only for a limited number of pharmaceutical compounds [43].

Conventional wastewater treatment plants (WWTPs) were not designed for an efficient APIs removal. The sludge retention time (SRT) is one of the crucial parameters, which influence on the design, operation and control of WWTPs. APIs can be divided into three major groups: compounds with optimum SRT range for which their removal is the most effective (e.g., antibiotics and anti-inflammatories), compounds on which SRT has no impact (e.g., anticonvulsants, β-blockers and hormones), compounds with visible influence of SRT on their removal rate (e.g., lipid regulators) [43].

Depending on the wastewater treatment, removal rates can differ [46–49]. Moreover, for compounds of the same pharmacotherapeutic group, such as β-blockers, highly variable removal rates were obtained when the same treatment procedure was used. Activated sludge process lead to a significative removal efficiency for betaxolol, bisprolol, carazolol and metoprolol (65–90%) [49, 50], whereas for soltalol and propranolol low removal rates (20–32%) were reported [50–52].

Advanced wastewater treatment techniques, for example oxidation processes, can achieve up to 100% removal for diclofenac [52, 53]. Reverse osmosis, activated carbon and ozonation have been shown to significantly reduce or eliminate antibiotics from wastewater effluents [32]. The efficiency of two tertiary treatments, chlorination and UV disinfection, was compared and chlorination led to lower quantities of antibiotics [54].

The analysis of FQs and TCs in Portuguese wastewater composite samples obtained from influent and effluent of the WWTP located in Coimbra showed that FQs and TCs removal ranged between 53 and 92%, and between 89.5 and 100%, respectively [34, 55]. The removal efficiency of ciprofloxacin (CIPRO) in WWTP depends on the treatment applied and ranges between 22.2 and 100% [56]. In a field study at a full-scale WWTPs, in Switzerland, Golet et al. [57] determined a 49–61% reduction in CIPRO and norfloxacin (NOR) concentrations during biological treatment, 28–35% during mechanical treatment and about 3–4% in the flocculation/filtration step. They observed a combined removal during the wastewater treatment process of 88% for CIPRO and 92% for NOR. The WWTPs removal efficiency of OFLO, based on an activated sludge process and biological nutrient removal, was 77% [32].

However, one should be aware of the fact that if a particular pharmaceutical is not detected in a WWTP effluent, this does not imply that it has been fully removed. On some occasions, it may have been degraded and give rise to unsuspecting or even more toxic compounds that will subsequently contaminate surface waters [30, 58–60]. As an example, anhydrotetracycline (ATC) and epianhydrotetracycline (EATC) are tetracycline (TC) degradation products that cause Fanconi syndrome [61].

Several studies identify WWTPs effluents as the main conveyors of APIs and their metabolites into receiving water sources that are used for drinking-water supply [6, 30, 32, 52, 56, 59, 62–71]. For instance, the presence of ibuprofen, paracetamol and salicylic acid which are readily bio(degradable) is an indicator of non- or poorly treated wastewater [72]

As WWTPs removal rates are lacking, modelling is important; however, there is a need to improve their accuracy for a better evaluation [43].

2.3.2 Drinking-Water Treatments Removal

In this way, APIs and their metabolites, eventually, can reach streams impacted by WWTPs and endure the drinking water treatment processes that typically combine conventional and advanced treatments [52].

Studies on conventional drinking-water treatment processes, such as filtration, coagulation and sedimentation, have shown that coagulation is largely ineffective in removing pharmaceuticals. Free chlorine is able to remove up to approximately 50% of the pharmaceuticals investigated, whereas chloramines have lower removal efficiency. Compounds that showed high removal by free chlorine but low removal by chloramines include antibiotics, such as sulfamethoxazole, trimethroprim and erythromycin [52].

Advanced water treatment processes, such as ozonation, advanced oxidation, activated carbon and membranes (e.g., nanofiltration, reverse osmosis), are able to achieve higher removal rates (above 99%) [52]. For instance, antibiotics can be removed from drinking water by ozonation [73, 74]. Ultraviolet (UV) irradiation at typical disinfection dosages was ineffective for removing most target compounds, even though it can achieve more than 50% removal of sulfamethoxazole, triclosan and diclofenac. However, a combination of higher-dose UV with hydrogen peroxide removed most target compounds [52, 75, 76].

2.4 Occurrence Data

Worldwide monitoring programmes and a literature review have shown the presence of many different classes of pharmaceuticals, confirming a global problem. As shown in Table 2, the presence of these residues was found in a wide range of environmental samples, including surface water, groundwater and drinking water, within concentrations up to several ng L^{-1} as presented in Table 2 [72].

A number of factors can affect the concentrations found in the environment. A monitoring programme requires an accurate sampling collection plan to obtain relevant data on environmental contamination and guarantee the accuracy of the study as a whole. Wastewater composite sampling of 24 hours is essential to be representative of the large variations in concentration between the different sampling periods. A difference of a factor of two between maximum day-time and minimum night-time influent loads has been reported [98, 99].

In addition, seasonal variations may also influence the levels of APIs found. A seasonal influence on the FQs and TCs detection frequency in Coimbra (Portugal) municipal and hospital wastewaters was reported [34, 55]. Seasonal variations of up to a factor of four were also reported in a Finnish river [99, 100]. Seasonal variations can also affect the efficiency of WWTPs and river discharges since in winter the microbial activity is reduced due to low temperatures and the dilution effect is higher. Ultimately, at a wider level, geographical situation and climate may also explain different contamination scenarios between countries in different climatic zones [99, 100].

Contamination profiles of hospital effluents and influents/effluents of WWTPs to water courses are essential [55]. Hospitals are one important point source of contamination due to the presence of higher concentrations of APIs residues and the presence of specific antibiotics, antineoplasic and diagnostic agents; however, little knowledge is available on their contribution. These facilities require a wastewater treatment process more specific before the entrance at municipal WWTPs [58, 101–105]. In Portugal, the presence of FQs and TCs antibiotics has been reported in different hospital wastewaters [34, 55].

Table 2 APIs environmental occurrence data, from 2006 until nowadays

API	Country	Sample	Levels (ng L^{-1})	Reference
Non-steroidal anti-inflammatory drugs				
Acetylsalicylic acid	Romania	Surface water	<30–37.2	[77]
Salicylic acid	Canada	WWTP effluent	2,178.2	[78]
		Surface water	130.4	
		Surface water	286.7	
	Spain	WWTP influent	2,566	[6]
		WWTP effluent	34	
		Surface water	12	
Diclofenac	Spain	WWTP influent	1,500	[79]
		WWTP effluent	900	
	Canada	WWTP effluent	448	[78]
	Spain	WWTP influent	21–148	[80]
	Belgium	WWTP effluent	32–1,420	
	Germany	Surface water	26–72	
	Slovenia	Drinking water	<7	
	Spain	Hospital effluent	60–1,900	[81]
	United Kingdom	WWTP influent	901–1,036	[47]
		WWTP effluent	261–598	
	Spain	WWTP influent	726	[6]
		WWTP effluent	323	
		Surface water	51	
Ibuprofen	Spain	WWTP influent	84,000	[79]
		WWTP effluent	7,100	
	Canada	WWTP effluent	6,718.3	[78]
	Romania	Surface water	<30–115.2	[79]
	Spain	WWTP influent	37–860	[80]
	Belgium	WWTP effluent	18–1,860	
	Germany	Surface water	60–152	
	Slovenia	Drinking water	<12	
	Spain	Hospital effluent	1,500–151,000	[81]
	United Kingdom	WWTP influent	7,741–33,764	[47]
		WWTP effluent	1,979–4,239	
		Surface water	144–2,370	
	Spain	WWTP influent	13,228	[6]
		WWTP effluent	3,090	
		Surface water	46	
Ketoprofen	Canada	WWTP effluent	268	[78]
	Spain	WWTP influent	131	[80]
	Belgium	WWTP effluent	<26	
	Germany	Surface water	<26	
	Slovenia	Drinking water	<26	
	Spain	WWTP influent	1,496	[6]
		WWTP effluent	290	
Ketorolac	Spain	Hospital effluent	200–59,500	[81]
Mefenamic acid	United Kingdom	WWTP influent	136–363	[47]
		WWTP effluent	290–396	

(continued)

Table 2 (continued)

API	Country	Sample	Levels (ng L^{-1})	Reference
Naproxen	Canada	WWTP effluent	7,098.2	[78]
	Spain	WWTP influent	109–455	[80]
	Belgium	WWTP effluent	625	
	Germany	Surface water	70	
	Slovenia	Drinking water	<26	
	Spain	WWTP influent	3,249	[6]
		WWTP effluent	598	
		Surface water	35	
Paracetamol	Spain	WWTP influent	134,000	[79]
		WWTP effluent	220	
	Spain	Hospital effluent	500–29,000	[81]
	United Kingdom	WWTP influent	5,529–69,570	[47]
		WWTP effluent	<20	
	Spain	WWTP influent	10,899	[6]
		WWTP effluent	276	
		Surface water	10	
Antihypertensive				
Enalipril	Spain	WWTP influent	697	[6]
		WWTP effluent	457	
		Surface water	n.d.	
Hydrochlorothiazide	Spain	WWTP influent	1,534	[6]
		WWTP effluent	901	
		Surface water	20	
Blood lipid-lowering agents				
Fibrates				
Clofibric acid	Spain	WWTP influent	25–58	[80]
	Belgium	WWTP effluent	22–107	
	Germany	Surface water	24–35	
	Slovenia	Drinking water	<17	
	United Kingdom	WWTP influent	<20–651	[47]
		WWTP effluent	<20–44	
	Spain	WWTP influent	16	[6]
		WWTP effluent	10	
		Surface water	3	
Benzafibrate	Spain	WWTP influent	351	[6]
		WWTP effluent	167	
		Surface water	6	
Gemfibrozil	Canada	WWTP effluent	403.1	[78]
		Surface water	9.0	
		Surface water	n.d.	
	Spain	WWTP influent	219	[6]
		WWTP effluent	101	
		Surface water	27	
Statins				
Artovastatin	Spain	WWTP influent	86	[6]
		WWTP effluent	122	
		Surface water	2	

(continued)

Table 2 (continued)

API	Country	Sample	Levels (ng L^{-1})	Reference
Antibiotics				
Fluoroquinolones				
Ciprofloxacin	Portugal	Surface water	79.6–119.2	[82]
	USA	WWTP influent	n.d.–1,000	[32]
		WWTP effluent	n.d.	
		Hospital effluent	850–2,000	
		Surface water	n.d.	
	USA	WWTP influent	150	[56]
		WWTP effluent	60	
	Spain	WWTP influent	1,030	[6]
		WWTP effluent	332	
		Surface water	35	
	Italy	WWTP influent	513	[83]
		WWTP effluent	248	
		Surface water	8.8–19	
Enrofloxacin	USA	WWTP influent	250	[56]
		WWTP effluent	270	
	Portugal	Surface water	67.0–102.5	[82]
	Spain	WWTP influent	277	[6]
		WWTP effluent	219	
		Surface water	n.d.	
Norfloxacin	Portugal	Surface water	n.d.	[82]
	France	Surface water	13.2–18.6	[84]
Ofloxacin	USA	WWTP influent	400–1,000	[32]
		WWTP effluent	n.d.–110	
		Hospital effluent	25,500–35,500	
		Surface water	n.d.	
	France	Surface water	10.5	[84]
	Spain	WWTP influent	219	[6]
		WWTP effluent	95	
		Surface water	11	
	Italy	WWTP influent	463	[83]
		WWTP effluent	191	
		Surface water	5-10.9	
β-Lactams				
Amoxicillin	Italy	WWTP influent	18	[83]
		Surface water	<2.08–5.7	
Lincosamide				
Lincomycin	USA	Hospital effluent	300–2,000	[32]
	USA	Groundwater	320	[85]
	Italy	WWTP influent	9.7	[83]
		WWTP effluent	7.2	
		Surface water	5.7–8.1	
Macrolides				
Azithromycin	Spain	WWTP influent	314	[6]
		WWTP effluent	80	
		Surface water	14	

(continued)

Table 2 (continued)

API	Country	Sample	Levels (ng L^{-1})	Reference
Clarithromycin	Spain	WWTP influent	208	[6]
		WWTP effluent	50	
		Surface water	11	
	Germany	Surface water	77	[86]
		Seawater	14	
		WWTP effluent	520	
	Italy	WWTP influent	519	[83]
		WWTP effluent	145	
		Surface water	1.7–25.4	
Erythromycin	Germany	Surface water	22	[86]
		WWTP effluent	173	
	Italy	WWTP influent	12	[83]
		WWTP effluent	72	
		Surface water	2.9–5.4	
Roxithromycin	Germany	Surface water	16	[86]
		WWTP effluent	78	
Spiramycin	Italy	WWTP influent	603	[83]
		WWTP effluent	375	
		Surface water	1.1–7.9	
Sulfonamides				
Sulfadiazine	Italy	Surface water	236	[87]
	Spain	WWTP influent	73	[6]
		WWTP effluent	60	
		Surface water	6	
Sulfadimethoxine	Italy	Surface water	28	[87]
		Surface water	74	
		Drinking water	11	
	USA	Groundwater	46–68	[88]
Sulfamethazine	USA	WWTP influent	160	[56]
		WWTP effluent	n.d.	
	USA	Groundwater	76–215	[88]
	USA	Groundwater	360	[85]
	Spain	WWTP influent	132	[6]
		WWTP effluent	46	
		Surface water	3	
Sulfamethoxazole	USA	WWTP influent	390–1,000	[32]
		WWTP effluent	n.d.–310	
		Hospital effluent	400–2,100	
		Surface water	n.d.–300	
	USA	WWTP influent	300	[56]
		WWTP effluent	200	
	Italy	Surface water	402	[87]
		Drinking water	13–80	
	USA	Groundwater	1,110	[85]
	France	Surface water	12.9–26	[84]
	Spain	WWTP influent	354	[6]
		WWTP effluent	208	
		Surface water	11	

(continued)

Table 2 (continued)

API	Country	Sample	Levels (ng L^{-1})	Reference
	Germany	Surface water	93	[86]
		Seawater	7	
		WWTP effluent	509	
	Belgium	Seawater	n.d.–96	[89]
	Italy	WWTP influent	246	[83]
		WWTP effluent	101	
		Surface water	2.1–5.3	
	Portugal	Surface water	9.14–53.3	[90]
Sulfapyridine	Italy	Surface water	<12–121	[87]
		Surface water	66	
Tetracyclines				
Doxycycline	Portugal	Hospital effluent	8,100	[55]
Minocycline	Portugal	Hospital effluent	531,700–3,177,900	[55]
		WWTP influent	350,000–915,300	
		WWTP effluent	95,800	
Tetracycline	USA	WWTP influent	520	[56]
		WWTP effluent	170	
	Portugal	Hospital effluent	6,000–54,700	[55]
Epi-Tetracycline	Portugal	Hospital effluent	6,000–17,500	[55]
Other				
Metronidazole	Spain	WWTP influent	167	[6]
		WWTP effluent	164	
		Surface water	21	
Trimethoprim	USA	WWTP influent	590–1,000	[32]
		WWTP effluent	n.d.–180	
		Hospital effluent	2,900–5,000	
		Surface water	n.d.	
	USA	WWTP influent	330	[56]
		WWTP effluent	170	
	Spain	WWTP influent	50	[6]
		WWTP effluent	37	
		Surface water	4	
	Belgium	Seawater	n.d.–29	[89]
Sex hormones				
Diethylstilbestrol	China	Surface water	20	[91]
17β-Estradiol	China	Surface water	100	[91]
Estrone	China	Surface water	180	[91]
Antiepileptics				
Carbamazepine	Spain	WWTP influent	150	[79]
		WWTP effluent	130	
	Romania	Surface water	<30–75.1	[77]
	Finland	WWTP influent	290–400	[92]
		WWTP effluent	380–470	
		Surface water	<1.4–66	
		Surface water	23	
	Spain	Hospital effluent	30–70	[81]

(continued)

Table 2 (continued)

API	Country	Sample	Levels (ng L^{-1})	Reference
	Spain	WWTP influent	157	[6]
		WWTP effluent	198	
		Surface water	19	
	Spain	Surface water	82	[93]
	Sweden	Surface water	23.6–26.3	[94]
Carbamazepine-10,11-epoxide	Spain	WWTP influent	350	[79]
		WWTP effluent	160	
Beta-blockers				
Acebutolol	Finland	WWTP influent	390–510	[92]
		WWTP effluent	80–230	
		Surface water	<0.8–8	
		Surface water	8	
Atenolol	Finland	WWTP influent	510–800	[92]
		WWTP effluent	40–440	
		Surface water	<11.8–25	
		Surface water	<11.8	
	Spain	Hospital effluent	400–1,700	[81]
	Spain	WWTP influent	779	[6]
		WWTP effluent	452	
		Surface water	9	
Metoprolol	Finland	WWTP influent	980–1,350	[92]
		WWTP effluent	910–1,070	
		Surface water	<3.8–116	
		Surface water	38	
	Spain	WWTP influent	65	[6]
		WWTP effluent	49	
		Surface water	2	
	Sweden	Surface water	8–18	[94]
Propranolol	Spain	Hospital effluent	200–6,500	[81]
	Spain	WWTP influent	51	[6]
		WWTP effluent	55	
		Surface water	2	
	United Kingdom	WWTP influent	60–119	[47]
		WWTP effluent	195–373	
		Surface water	35–107	
Solatol	Finland	WWTP influent	640–830	[92]
		WWTP effluent	160–300	
		Surface water	<3.9–52	
		Surface water	37	
Antidepressants				
Citalopram	Norway	WWTP influent	62.9–303.6	[95]
		WWTP effluent	21.9–238.4	
	Norway	WWTP influent	13.0–612	[96]
		WWTP effluent	9.2–382	
	Canada	WWTP influent	52.2–52.7	[97]
		WWTP effluent	46.8–57.8	
		Surface water	3.4–11.5	
	Spain	Surface water	43	[93]

(continued)

Table 2 (continued)

API	Country	Sample	Levels (ng L^{-1})	Reference
Fluoxetine	Norway	WWTP influent	1.1–18.7	[95]
		WWTP effluent	0.6–8.4	
	Norway	WWTP influent	0.4–2.4	[96]
		WWTP effluent	<0.12–1.3	
	USA	Groundwater	56	[85]
	Canada	WWTP influent	3.1–3.5	[97]
		WWTP effluent	2.0–3.7	
		Surface water	0.42–1.3	
	Spain	WWTP influent	25	[6]
		WWTP effluent	10	
		Surface water	2	
	Spain	Surface water	14	[93]
Fluvoxamine	Norway	WWTP influent	0.8–1.7	[95]
		WWTP effluent	<0.49–0.8	
		Seawater	0.5–0.8	
	Norway	WWTP influent	0.4–3.9	[96]
		WWTP effluent	<0.15–0.8	
Paroxetine	Norway	WWTP influent	2.9–12.9	[95]
		WWTP effluent	1.0–11.7	
		Seawater	0.6–1.4	
	Norway	WWTP influent	0.6–12.3	[96]
		WWTP effluent	0.5–1.6	
	Canada	WWTP influent	4.6–5.3	[97]
		WWTP effluent	4.3–5.2	
		Surface water	1.3–3.0	
Sertraline	Norway	WWTP influent	8.4–19.8	[95]
		WWTP effluent	3.7–14.6	
		Seawater	<0.16	
	Norway	WWTP influent	1.8–2.5	[96]
		WWTP effluent	0.9–2.0	
	Canada	WWTP influent	6.0–6.1	[97]
		WWTP effluent	5.1–5.8	
		Surface water	0.84–2.4	
Venlafaxine	Spain	Surface water	57	[93]
Anxiolytics				
Nordiazepam	Spain	Surface water	26	[93]
Oxazepam	Spain	Surface water	30	[93]
	Sweden	Surface water	10.4–12.7	[94]
7-aminoflunitrazepam	Spain	Surface water	55	[93]
Lorazepam	Spain	WWTP influent	76	[6]
		WWTP effluent	102	
		Surface water	11	

As above mentioned, in Sect. 2.3.2, none of the wide range of drinking water treatment processes available have been designed to remove pharmaceuticals that may be present in source waters. Therefore, its worldwide presence in drinking water generates a growing concern due to the possible human exposure [52, 106, 107], namely the involuntary and unavoidable exposure of pregnant women and children. APIs reported most frequently in finished drinking water include: carbamazepine, phenytoin, meprobamate, clofibric acid, gemfibrozil, iopromide, iopamidol, ibuprofen and sulfamethoxazole. The six APIs consistently reported to have the highest concentrations are: ibuprofen, triclosan, carbamazepine, phenazone, clofibric acid and acetaminophen. As above mentioned, although ibuprofen is considered to be an indicator of non- or poorly treated wastewater, this API and its methyl ester metabolite were reported as having exceeded a concentration of 1 μg L^{-1} in finished drinking water [107].

Future monitoring studies should close the existing data gaps in the current knowledge and overcome the challenge of data comparability [72]. A European database of results would also be helpful to improve knowledge and management of risk.

2.5   Environmental Risk

2.5.1   Ecotoxicity and Bioaccumulation

Little is known about the biological effects of APIs on the aquatic environment. The sparse research on this emerging problem reflects, in part, the fact that environmental toxicology has traditionally focused on the effects of acute exposure, rather than on low level, chronic exposure [40, 58].

Even though they are found in concentrations well below the therapeutical level, a number of issues suggest the need for more knowledge of the potential health effects of highly diluted pharmaceuticals, ingested over the whole life span, for non-target organisms, as well as for human health.

Many non-target organisms (which possess human- and animal-alike metabolic pathways, similar receptors or biomolecules) are inadvertently exposed to these substances [40, 58]. Since several APIs are known to interact with Cytochrome P-450, there is a potential risk of disruption in the homeostasis of non-target organisms. Moreover, pharmaceuticals that interact with the Glycoprotein-P (P-gp), a multidrug transporter that actively transports xenobiotics out of the cell, increase their sensitivity to environmental pollutants [17].

Ecotoxicological data points out that mixtures of active substances might have different effects than single compounds [108–110], but general knowledge on this issue is still sparse. There are some examples of toxicity studies in literature showing that APIs mixtures may exhibit additive effects leading to toxicity even at low levels [109].

The actual exposure scenario posing the most uncertainty regarding toxicology is long-term exposure to multiple APIs. If many APIs are present and share the same MoA, then the dose is effectively summed accordingly (known as dose additivity). For APIs known to occur in the environment, examples of drug classes whose individual APIs share the same MoA include oestrogens, selected serotonin reuptake inhibitors (SSRIs) antidepressants, non-steroidal anti-inflammatory drugs

(NSAIDs), specific classes of antibiotics (e.g., sulfa drugs) and statin lipid-lowering agents [107].

Consequently, single compound ecotoxicological assays are not sufficient to provide accurate environmental risk assessment of pharmaceuticals [17]. A comprehensive manner to evaluate the toxicity effects on non-target organisms must include the development of specific tests embracing either acute effects (where mortality rates are often registered) or chronic effects (by means of exposure to different concentrations of a chemical compound over a prolonged period of time). In the latter, effects are measured through specific parameters such as growth index or reproduction rates. Unfortunately, studies on acute effects in organisms belonging to different trophic levels (i.e., algae, zooplankton and other invertebrates and fish) predominate relatively to chronic ones. Acute toxicity data is only valuable when accidental discharge of the drugs occurs, since the environmental concentrations usually reported for these compounds are low, typically in a factor of one thousand [111].

As originally proposed, the outcomes that might be expected in the aquatic environment are usually expected to be subtle, such as alteration of behaviour, rather than more obvious effects end points such as growth or survival. One example is a reduction in activity or alteration in behaviour of aquatic organisms when exposed to trace levels of SSRIs or NSAIDs. These might lead to alterations in avoidance or attraction that can change predation and reproductive behaviours, thereby affecting change in ecological community structure. The exposure levels at which these types of effects can be measured can be up to six orders of magnitude lower than the existing no-observed-effects-thresholds for conventional end points.

For highly potent APIs, profound effects can occur at low ng L^{-1} levels, the adverse effect of ethynylestradiol on fish populations is one example [107]. Another example is the development of resistant bacterial strains induced by the release of antibiotics into the environment [112, 113]. Dorne et al. [114] concluded that fluoxetine, ibuprofen, diclofenac, propranolol and metoprolol exhibit relatively high acute toxicity to aquatic species. In addition, due to the inherent properties of these chemicals, pharmacodynamic effects were observed in the heart rate of *Daphnia magna* for the β-blockers propranolol and metoprolol.

The most damning evidence of APIs impact on wildlife comes from studies on fish. Fluoxetine has been detected in tissues of fish species living in a municipal effluent at levels of 0.1 ng g^{-1}. Redox properties of some medicinal products can influence the oxidative metabolism in hepatocytes of rainbow trout leading to oxidative damage [115].

A study showed the collapse of a population of fish in an isolated lake spiked with relatively high levels of the synthetic oestrogen 17α-ethinylestradiol [116]. Other studies on aquatic populations in a waste-impacted stream in Boulder, Colorado, showed reproductive effects from estrogenic wastewater effluent [117].

Compared with aquatic exposure, many more uncertainties surround the potential for outcomes from human exposure. Given the sparse research performed on ultra-low-dose studies, and the complexity introduced by mixed-mode (nonmonotonic) dose–response curves (which effectively prevent extrapolations to lower doses), it might seem unlikely at first, but not improbable, that adverse or even benign effects could occur in humans [107].

Continuous consumption of drugs even at sub-therapeutic concentrations represents a potential threat to public health, although one should bear in mind that it is still impossible to evaluate the effects of exposure on human health [58, 60, 118].

There is little information about the bioaccumulation potential of pharmaceuticals in biota or food webs. Despite low lipid solubilities, pharmaceuticals are being detected in tissues of aquatic organisms in higher concentrations than those in the surrounding environmental waters. This is perhaps partly a result of drugs being designed to take advantage of gaining intracellular access via active transport [40].

There are few monitoring studies on APIs in fish tissues; however, multiple APIs were targeted in fish monitoring. The concentrations across tissues vary by many orders of magnitude as a function of the concentration of exposure, the species and specific tissue, with bile often serving to concentrate the most and the brain the least [107].

Diphenhydramine, diltiazem, carbamazepine and norfluoxetine have been reported simultaneously in the same wild fish [107]. Moreover, diclofenac was found accumulating in vultures [119], fluoxetine, sertraline and the SSRI metabolites norfluoxetine and desmethylsertraline were detected in fish [120]. Diclofenac bioaccumulation factors were 10–2,700 in the liver of fish and 5–1,000 in the kidney, depending on exposure concentrations [40, 121]. Gemfibrozil occurred in blood plasma of goldfish after exposure over 14 days at 113 times higher levels than in water [40].

In summary, probably due to the complex experimental work involved, bioaccumulation and chronic toxicity tests are scarce [40, 58]; however, these studies are very important for a better knowledge of the effects on human health.

2.5.2 Environmental Risk Assessment

The discussion on the presence of pharmaceutical products in environment suggested a need for the development of an ecologically integrated risk assessment for the medicinal products. Organizations at the international level are increasingly and intensively engaged in this debate. In this context, it is recognized that the environmental impact of medicinal products has to be evaluated and more appropriate legislation and regulatory has been implemented in the EU.

The EU Directive 92/18/EEC [122] introduced for the first time the requirement for an ERA, as a prerequisite to obtain marketing authorization for veterinary pharmaceuticals. For this purpose, the European Agency for the Evaluation of Medicinal Products (EMEA) published a "Note for Guidance" where guidelines to assess the environmental risk of veterinary medicinal products are established [123].

The European Commission extended its concerns to pharmaceuticals for human use and, according to Article 8 of the Directive 2001/83/EC [12] (6 November 2001), amended by Directive 2004/27/EC [13] (31 March 2004), an application for a marketing authorization for a medicinal product should be accompanied by an evaluation of its potential environmental risks. An ERA is also required if there is an increase of the environmental exposure (e.g., a new indication).

Any risk relating to the quality, safety or efficacy of the medicinal product regarding human health may influence its marketing authorization. However, given the relevance of a medicine, a negative ERA cannot affect its risk/benefit balance. In support to the estimation of an ERA, the European Medicines Agency/Committee for Medicinal Products for Human use [124] has released a guideline (EMEA/CHMP/SWP/4447/00). The Risk Assessment Strategy for human medicinal products follows the general principle as applied to chemicals. Globally it includes primarily an estimation of exposure of concentration of the active ingredient/excipient in the surface water. Whenever the value of the obtained exposure is above the action limit of 0.01 μg L^{-1} or the active ingredient represents an ecological threat, the risk assessment will be extended.

The Risk Assessment model comprises: (1) prediction of the concentration that will reach the environment (PEC), (2) prediction of the no-effect concentration (PNEC) on a small selection of organisms (3 species), and (3) propose possible minimization measures which will be inserted, in form of standard sentences, in summary of product characteristics (SPC) and/or patient information leaflet of respective Medicinal Product [125].

For the last years, a number of risk assessment and/or prioritization strategies have been implemented in Europe and the USA. Most of them were conducted according to the EMEA guideline. One of the first ERA has been conducted in 2000 in Denmark [126]. United Kingdom, Germany, Sweden, Italy, France and the USA also presented ERA studies of human pharmaceuticals. The 2008 French prioritization strategy gives special emphasis to metabolism and pharmacological data, however, in a first ERA step, based on PEC/PNEC ratios, only a few pharmaceuticals were classified due to lack of ecotoxicological data. Therefore, a pragmatic prioritization approach based on three tiers was implemented [17]. Any assessment of negative environmental potential risk implies precautionary and safety measures. However, it is rather difficult to compare the results of risk assessment and prioritization strategies as methodologies, goals and available information are heterogeneous [17].

In this context, it is recognized that the environmental impact of medicinal products has to be evaluated and more appropriate legislation and regulatory shall be considered for implementation in the EU.

2.6 Environmental Classification of Pharmaceuticals

Prior to implement a monitoring program in the aquatic environment, there is a need to rank APIs according to their environmental relevance since it is not feasible to search all molecules in the environment. Therefore, methodologies need to be developed to select the priority compounds [17].

APIs presenting high prescription and consumption (e.g., β-blockers, antibiotics), poor removal in WWTPs, high persistence (e.g., antiepileptics), high toxicity (e.g., antidepressants, hormones) or producing antibiotic resistance must be considered priority compounds due to their high risk to the aquatic environment. On the contrary, APIs with low prescription and consumption, low risk to aquatic environment, efficiently removed in WWTPs, presenting low persistence (due to biodegradation and photodegradation), and low toxicity are not priority molecules.

The Swedish Classification Scheme initiated in 2005 by the Swedish Association of Pharmacy Industries (LIF), the Swedish Medical Products Agency, Apoteket (National Corporation of Swedish Pharmacies), the Swedish Association of Local Authorities and Regions and the Stockholm County Council, take in account Persistence, Bioaccumulation and Toxicity (PBT) characteristics of pharmaceutical products. This voluntary scheme looks at the environmental hazard and the associated risk of pharmaceutical products. The environmental risk is calculated based on the ratio PEC/PNEC according to the EMEA guideline [17, 124, 127]. The obtained information is only available on the website www.fss.se, since due to European restrictions it is not possible to include warning labels on the packaging of medications [17].

Considerable international interest has been shown towards the Swedish model for environmental classification of pharmaceutical substances, from pharmaceutical companies, from various EU projects, from pharmaceutical authorities and from the various health care providers [128]. Although there are some criticisms regarding this classification model, it must be assumed that this was the very first effort to do an environmental classification of pharmaceuticals.

Nowadays, an FP7 project named PHARMAS (www.pharmas-eu.org) is working on a proposal to implement a common environmental classification system of pharmaceuticals throughout the EU, inducing a dialogue with the proper EU authorities [132].

3   Environmental Concerns and Models of Change

The emphasis placed on environmental concerns led to the concept of ecopharmacostewardship that relates to the environmental activities being pursued by the research pharmaceutical companies to minimize the pharmaceuticals environmental footprint and produce a new generation of green and sustainable products [129].

According to the GlaxoSmithKline "the challenge for the green chemistry community is to develop systematic approaches to introduce greener practices with low cost/benefit" [129]. The pharmaceutical industry has recently achieved numerous successes in greening its synthetic processes, for instance, the active ingredient sertraline had its manufacturing process stream lined from 3 steps to one, using a more selective catalyst and more benign solvent [129].

On the other hand, disposal of leftover drugs into sewers has received considerable attention since few years ago, in Europe, take-back schemes for unused or expired medicines are required by EU legislation since 2004. Therefore, under the provisions of current European Union legislation (Directive 2004/27/EC [13]), all EU Member States must establish collection schemes to recover and safely dispose of unused and expired medicines.

Nowadays, 20 European countries have a drug take-back scheme in place [130]. Some of the schemes, particularly those in France and Sweden, are well established and successful. However, many of the more recent schemes do not appear, as yet, to be very effective [129].

In Portugal, aware of the specificity of the drug as waste, the Portuguese pharmaceutical industry, responsible for the packaging disposal management, joined the other members of the APIs lifecycle, mainly distributors and pharmacies, and created the company VALORMED to yield the management of packaging waste and discarded medicines. In 2010, 838 tonnes of packaging waste and discarded medicines were collected, representing an increase of 17% over the previous year (www.valormed.pt) [131].

In recent years, several initiatives have been launched to establish or strengthen surveillance systems, both in EU member states and at an international level, to monitor the presence of these residues in environmental matrices. When implementing measures, water bodies relevant for drinking water should receive priority. Furthermore, research is required to determine whether observations made from regional sample sets are representative of environmental concentrations nationwide [3], being essential to perform contamination maps and implement surveillance models, needed for the establishment of a sustainable strategy, to minimize environmental impact of medicines.

Pharmaceuticals are greatly increasing in number and kind, with greater likelihood of releases into the environment. According to Daughton and Ternes [58], the enormous array of pharmaceuticals will continue to diversify and grow as the human genome is mapped. The explosion in new drugs will severely exacerbate our limited knowledge of drugs in the environment and possibly increase the exposure/effects risks to non-target organisms.

4 Final Remarks

To the best of current scientific knowledge, pharmaceuticals do not pose an eminent threat to the environment; however there are many more uncertainties than certainties and the scientific interest on the consequences of a lifelong exposure to mixtures of low levels of pharmaceuticals is increasing. Moreover, their pseudopersistent nature makes them an intriguing target for further study.

Almost nothing has been published in the medical literature with sated objective of determining the causes, extent, risks, or solutions to the issues of APIs as pollutants. This multidisciplinary problem highlights the importance of a scientific collaboration between different partners, with different areas of knowledge, to ensure that ecopharmacovigilance delivers its full benefits.

After an in-depth evaluation of the many facets of this issue, one can conclude that human health and environment health are intimately tied and the possible risks should be known in advance so that we can prevent them.

Sustainable strategies for minimizing APIs impact on the environment and prioritizing measures should be established. Many actions could be considered to reduce the APIs environmental footprint by promoting safety approaches along the APIs lifecycle, endorsing their rational use, improving the drug disposal, WWTP treatment process and take-back schemes. It has to be kept in mind that advanced technical solutions are not globally applicable and other options of action like developing green pharmaceuticals should not be neglected. According to the "Precautionary Principle", prevention is the best strategy for reducing contamination.

References

1. Daughton CG (2003) Cradle–to–cradle stewardship of drugs for minimizing their environmental disposition while promoting Green pharmacy & pharmecovigilance: prescribing & the planet human health I Rationale for and avenues toward a green pharmacy. Environ Health Perspect 111:757–774
2. Mompelat S, Le Bot B, Thomas O (2009) Occurrence and fate of pharmaceutical products and by-products, from resource to drinking water. Environ Int 35:803–814
3. Anderson PD, D'Aco VJ, Shanahan P et al (2004) Screening analysis of human pharmaceutical compounds in US surface waters. Environ Sci Technol 38:838–849
4. Ashton D, Hilton M, Thomas KV (2004) Investigating the environmental transport of human pharmaceuticals to streams in the United Kingdom. Sci Total Environ 333:167–184
5. Benoit R (2010) Pharmaceuticals in the environment: current knowledge and need assessment to reduce presence and impact. IWA Publishing, London
6. Gros M, Petrović M, Barceló D (2009) Tracing pharmaceutical residues of different therapeutic classes in environmental waters by using liquid chromatography/quadrupole–linear ion trap mass spectrometry and automated library searching. Anal Chem 81:898–912
7. Kümmerer K (2004) Pharmaceuticals in the environment: sources, fate, effects and risks, 2nd edn. Springer, Berlin
8. Zuccato E, Castiglioni S, Fanelli R et al (2004) Pharmaceuticals in the environment: changes in the presence and concentrations of pharmaceuticals for human use in Italy. In: Kümmerer K (ed) Pharmaceuticals in the environment: sources, fate, effects and risks. Springer, Berlin
9. Focazio MJ, Kolpin DW, Furlong ET (2004) Occurrence of human pharmaceuticals in water resources of the United States: a review. In: Kümmerer K (ed) Pharmaceuticals in the environment: sources, fate, effects and risks. Springer, Berlin
10. Webb SF (2004) Data–based perspective on the environmental risk assessment of human pharmaceuticals II – aquatic risk characterization. In: Kümmerer K (ed) Pharmaceuticals in the environment: sources, fate, effects and risks. Springer, Berlin
11. Jørgensen SE, Halling–Sørensen B (2000) Drugs in the environment. Chemosphere 40(7)
12. Directive 2001/83/EC of the European Parliament and of the Council of 6 November 2001 on the community code relating to human medicinal products. Off J Eur Union L 311:67–128
13. Directive 2004/27/EC of the European Parliament and of the Council of 31 March 2004 on the community code relating to human medicinal products. Off J Eur Union L136:34–57
14. Daughton CG, Ruhoy IS (2008) The afterlife of drugs and the role of pharmecovigilance. Drug Saf 31:1069–1082
15. Velo G, Moretti U (2010) Ecopharmacovigilance for better health. Drug Saf 33:963–968

16. World Health Organization (WHO) (2002) Source: The Importance of Pharmacovigilance. http://apps.who.int/medicinedocs/pdf/s4893e/s4893e.pdf. Accessed 29 Sep 2011
17. Besse JP, Garric J (2010) Environmental risk assessment and prioritization stratagies for human pharmaceuticals review and discussion. In: Benoit R (ed) Pharmaceuticals in the environment: current knowledge and need assessment to reduce presence and impact. IWA Publishing, London
18. Regulation (EC) No 726/2004 of the European Parliament and of the Council of 31 March 2004 laying down Community procedures for the authorisation and supervision of medicinal products for human and veterinary use and establishing a European Medicines Agency. Off J Eur Union L136:1–33
19. European Commission (EC) (2011) Health and Consumers directorate general. Implementing measures in order to harmonise the performance of the pharmacovigilance activities provided for in Directive 2001/83/EC and Regulation (EC) No 726/2004. Brussels, SANCO/D3/FS/cg/ddg1.d.3(2011)1003866
20. Holzer A (2010) Do pharmaceuticals in the environment present an investment risk? In: Kümmerer K, Hemple M (eds) Green and sustainable pharmacy. Springer, Berlin
21. IMS Health (IMAP) (2010) Pharmaceuticals & Biotech Industry Global Report – 2011. www.imap.com. Accessed 29 Sep 2011
22. INFARMED (2009) Monitorization of the market. Available at: http://wwwinfarmedpt/portal/page/portal/INFARMED/PUBLICACOES/TEMATICS/ESTATISTICA_ME-DICAMENTO. Accessed 18 Jan 2011
23. INFARMED (2008) Monitorization of the market. Available in http://wwwinfarmedpt/portal/p age/portal/INFARMED/MONITORIZACAO_DO_MERCADO. Accessed 29 May 2008
24. United Nations (2011) Press release. http://esa.un.org/unpd/wpp/other-information/Press_Release_WPP2010.pdf. Accessed 29 Sep 2011
25. Halling–Sørensen B, Nors-Nielsen S, Lanzky PF et al (1998) Occurrence, fate and effects of pharmaceuticals substances in the environment – a review. Chemosphere 36(2):357–393
26. Braund R, Peake BM, Schieffelbien L (2009) Disposal practises for unused medications in New Zealand. Environ Int 35:952–955
27. Glassmeyer ST, Hinchey EH, Boehme SE et al (2009) Disposal practises for unwanted residential medications in the United States. Environ Int 35:566–572
28. Gros M, Petrović M, Barceló D (2006) Development of a multi–residue analytical methodology based on liquid chromatography–tandem mass spectrometry (LC–MS/MS) for screening and trace level determination of pharmaceuticals in surface and wastewaters. Talanta 70:678–690
29. Persson M, Sabelström E, Gunnarsson B (2009) Handling of unused prescription drugs – knowledge, behaviour and attitude among Swedish people. Environ Int 35:771–774
30. Santos LHMLM, Araújo AN, Fachini A et al (2010) Ecotoxicological aspects related to the presence of pharmaceuticals in the aquatic environment. J Hazard Mater 175:45–95
31. Langford KH, Thomas KV (2009) Determination of pharmaceuticals compounds in hospital effluents and their contribution to wastewater treatment works. Environ Int 35:766–770
32. Brown KD, Kulis J, Thomson B et al (2006) Occurrence of antibiotics in hospital, residential, and dairy effluent, municipal wastewater, and the Rio Grande in New Mexico. Sci Total Environ 366:772–783
33. Lindberg R, Jarnheimer P-A, Olsen B et al (2004) Determination of antibiotic substances in hospital sewage water using solid phase extraction and liquid chromatography/mass spectrometry and group analogue internal standards. Chemosphere 57:1479–1488
34. Seifrtová M, Pena A, Lino CM et al (2008) Determination of fluoroquinolone antibiotics in hospital and municipal wastewaters in Coimbra by liquid chromatography with a monolithic column and fluorescence detection. Anal Bioanal Chem 391:799–805

35. Turiel E, Bordin G, Rodríguez AR (2005) Determination of quinolones and fluoroquinolones in hospital sewage water by off–line and on–line solid–phase extraction procedures coupled to HPLC–UV. J Sep Sci 28:257–267
36. Kümmerer K (2003) Significance of antibiotics in the environment. J Antimicrob Chemother 52:5–7
37. Breton R, Boxall A (2003) Pharmaceuticals and personal care products in the environment: regulatory drivers and research needs. QSAR Comb Sci 22:399–409
38. Qin–Tao L, Williams HE (2007) Kinetics and degradation products for direct photolysis of β–blockers in water. Environ Sci Technol 41(3):803–810
39. Andreozzi R, Caprio V, Ciniglia C et al (2004) Antibiotics in the environment: occurrence in Italian STPs, fate, and preliminary assessment on algal toxicity of amoxicillin. Environ Sci Technol 38:6832–6838
40. Fent K, Weston AA, Caminada D (2006) Ecotoxicology of human pharmaceuticals. Aquat Toxicol 76:122–159
41. Parshikov IA, Freeman JP, Lay JO et al (2000) Microbiological Transformation of enrofloxacin by the Fungus Mucor ramannianus. Appl Environ Microbiol 66:2664–2667
42. Ternes TA, Joss A, Siegrist H (2004) Scrutinizing pharmaceuticals and personal care products in wastewater treatment. Environ Sci Technol 38:392A–399A
43. Zabczyński S, Buntner D, Miksch K et al (2010) Performance of conventional treatment processes of the most resistant PPs. In: Benoit R (ed) Pharmaceuticals in the environment: current knowledge and need assessment to reduce presence and impact. IWA Publishing, London
44. Díaz–Cruz MS, Barceló D (2006) Determination of antimicrobial residues and metabolites in the aquatic environment by liquid chromatography tandem mass spectrometry. Anal Bioanal Chem 386(4):973–985
45. Gulkowska A, Leung HW, So MK et al (2008) Removal of antibiotics from wastewater by sewage treatment facilities in Hong Kong and Shenzhen, China. Water Res 42:395–403
46. Lindqvist N, Tuhkanen T, Kronberg L (2005) Occurrence of acidic pharmaceuticals in raw and treated sewages and in receiving waters. Water Res 39:2219–2228
47. Roberts PH, Thomas KV (2006) The occurrence of selected pharmaceuticals in wastewater effluent and surface waters of the lower Tyne catchment. Sci Total Environ 356:143–153
48. Stumpf M, Ternes TA, Wilken R-D et al (1999) Polar drug residue in sewage and natural waters in the state of Rio de Janeiro, Brazil. Sci Total Environ 225:135–141
49. Ternes TA (1998) Occurrence of drugs in German sewage treatment plants and rivers. Water Res 32:3245–3260
50. Gabet-Giraud V, Miège C, Choubert JM et al (2010) Occurrence and removal of estrogens and beta–blockers by various processes in wastewater treatment plants. Sci Total Environ 408:4257–4269
51. Bendz D (2005) Occurrence and fate of pharmaceutically active compounds in the environment, a case–study: Hoje River in Sweden. J Hazard Mater 122:195–204
52. World Health Organization (WHO) (2011) Pharmaceuticals in drinking water Public health and environment, water, sanitation, hygiene and health. http://www.who.int/water_sanitation_health/publications/2011/pharmaceuticals_20110601.pdf
53. Klavarioti M, Mantzavinos D, Kassinos D (2009) Removal of residual pharmaceuticals from aqueous systems by advanced oxidation processe. Environ Int 35:402–417
54. Renew JE, Huang C-H (2004) Simultaneous determination of fluoroquinolone, sulphonamide, and trimethoprim antibiotics in wastewater using tandem solid phase extraction and liquid chromatography–electrospray mass spectrometry. J Chromatogr A 1042:113–121
55. Pena A, Paulo M, Silva LJG, Seifrtová M, Lino CM, Solich P (2010) Tetracycline antibiotics in hospital and municipal wastewaters: a pilot study in Portugal. Anal Bioanal Chem 396:2929–2936
56. Karthikeyan KG, Meyer MT (2006) Occurrence of antibiotics in wastewater treatment facilities in Wisconsin, USA. Sci Total Environ 361:106–207

57. Golet EM, Xifra I, Siegrist H et al (2003) Environmental exposure assessment of fluoroquinolone antibacterial agents from sewage to soil. Environ Sci Technol 37:3243–3249
58. Daughton CG, Ternes TA (1999) Pharmaceuticals and personal care products in the environment: agents of subtle change? Environ Health Perspect 107(6):907–938
59. Heberer T (2002) Occurrence, fate and removal of pharmaceuticals residues in the aquatic environment: a review of recent research data. Toxicol Lett 131:5–7
60. Zwiener C, Gremm TJ, Frimmel FH (2001) Pharmaceutical residues in the aquatic environment and their significance for drinking water production. In: Kümmerer K (ed) Pharmaceuticals in the environment: sources, fate, effects and risks. Springer, Berlin
61. Sokoloski TD, Mitscher LA, Yuen PH (1977) Rate and proposed mechanism of anhydro epimerization in acid solution. J Pharm Sci 66:1159–1165
62. Xu WH, Zhang G, Zou S-C et al (2007) Determination of selected antibiotics in the Victoria Harbour and the Pearl River, South China using high–performance liquid chromatography—electrospray ionization tandem mass spectrometry. Environ Pollut 145:672–679
63. Zhang YJ, Geissen SU, Gal C (2008) Carbamazepine and diclofenac: removal in wastewater treatment plants and occurrence in water bodies. Chemosphere 73:1151–1161
64. Huerta–Fontela M, Galceran MT, Ventura F (2011) Occurrence and removal of pharmaceuticals and hormones through drinking water treatment. Water Res 45 (3):1432–1442
65. Carballa M, Omil F, Lema JM et al (2004) Behaviour of pharmaceuticals, cosmetics and hormones in a sewage treatment plant. Water Res 38:2918–2926
66. Daughton CG, Jones-Leep T (eds) (2001) Pharmaceuticals and personal care products in the environment: scientiffic and regulatory issues. ACS Symposium Series 791, American Chemical Society, Washington DC
67. Göbel A, McArdell CS, Suter MJ et al (2004) Trace determination of macrolide and sulfonamide antimicrobials, a human sulfonamide metabolite, and trimethoprim in wastewater using liquid chromatography coupled to electrospray tandem mass spectrometry. Anal Chem 76(16):4756–4764
68. Golet EM, Alder AC, Giger W (2002) Environmental exposure and risk assessment of fluoroquinolone antibacterial agents in wastewater and river water of the Glatt Valley Watershed, Switzerland. Environ Sci Technol 36:3645–3651
69. Miao XS, Bishay F, Chen M et al (2004) Occurrence of antimicrobials in the final effluents of wastewater treatment plants in Canada. Environ Sci Technol 38(13):3533–3541
70. Nakata H, Kannan K, Jones PD (2005) Determination of fluoroquinolone antibiotics in wastewater effluents by liquid chromatography and fluorescence detection. Chemosphere 58:759–766
71. Yang S, Cha J, Carlson K (2005) Simultaneous extraction and analysis of 11 tetracycline and sulphonamide antibiotics in influent and effluent domestic wastewater by solid-phase extraction and liquid chromatography-electrospray ionization tandem mass spectrometry. J Chromatogr A 1097:40–53
72. Sadezky A, Löffler D, Schlüsener M et al (2010) Real situation: occurrence of the main investigates PPs in water bodies. In: Benoit R (ed) Pharmaceuticals in the environment: current knowledge and need assessment to reduce presence and impact. IWA Publishing, London
73. Huber MM, Canonica S, Park GY et al (2003) Oxidation of pharmaceuticals during ozonation and advanced oxidation processes. Environ Sci Technol 37:1016–1024
74. Ternes TA, Stuber J, Herrmann N et al (2003) Ozonation: a tool for removal of pharmaceuticals, contrast media and musk fragrances from wastewater. Water Res 37:1976–1982
75. Khiari D (2007) Endocrine disruptors, pharmaceuticals, and personal care products in drinking water: an overview of AwwaRF research to date. Awwa Research Foundation, Denver, CO

76. Rosenfeldt EJ, Linden KG (2004) Degradation of endocrine disrupting chemicals bisphenol A, ethinyl estradiol and estradiol during UV photolysis and advanced oxidation processes. Environ Sci Technol 38:5476–5486

77. Moldovan Z (2006) Occurrence of pharmaceutical and personal care products as micropollutants in rivers from Romania. Chemosphere 641:808–1817

78. Verenitch SS, Lowe CJ, Mazumder A (2006) Determination of acidic drugs and caffeine in municipal wastewaters and receiving waters by gas chromatography–ion trap tandem mass spectrometry. J Chromatogr A 1116:193–203

79. Gómez MJ, Martínez Bueno MJ, Lacorte S et al (2007) Pilot survey monitoring pharmaceuticals and related compounds in a sewage treatment plant located on the Mediterranean coast. Chemosphere 66:993–1002

80. Hernando MD, Heath E, Petrovic M et al (2006) Trace–level determination of pharmaceuticals residues by LC–MS/MS in natural and treated waters. a pilot–survey study. Anal Bioanal Chem 385:985–991

81. Gómez MJ, Petrovíc M, Fernández–Alba AR et al (2006) Determination of pharmaceuticals of various therapeutic classes by solid–phase extraction and liquid chromatography–tandem mass spectrometry analysis in hospital effluent wastewaters. J Chromatogr A 1114 (2):224–233

82. Pena A, Chmielova D, Lino CM, Solich P (2007) Determination of fluoroquinolone antibiotics in surface waters from Mondego River by high performance liquid chromatography using a monolithic column. J Sep Sci 30:2924–2928

83. Zuccato E, Castiglioni R, Bagnati S et al (2010) Source, occurrence and fate of antibiotics in the Italian aquatic environment. J Hazard Mater 179:1042–1048

84. Tamtam F, Mercier F, Eurin J et al (2009) Ultra performance liquid chromatography tandem mass spectrometry performance evaluation for analysis of antibiotics in natural waters. Anal Bioanal Chem 393:1709–1718

85. Barnes KK, Kolpin DW, Furlong ET et al (2008) A national reconnaissance of pharmaceuticals and other organic wastewater contaminants in the United States—I) Groundwater. Sci Total Environ 402:192–200

86. Nödler K, Licha T, Bester K et al (2010) Development of a multi–residue analytical method, based on liquid chromatography–tandem mass spectrometry, for the simultaneous determination of 46 micro–contaminants in aqueous samples. J Chromatogr A 1217:6511–6521

87. Perret D, Gentili A, Marchese S (2006) Sulphonamide residues in Italian surface and drinking waters: a small scale reconnaissance. Chromatographia 63:225–232

88. Batt AL, Bruce IB, Aga DS (2006) Evaluating the vulnerability of surface waters to antibiotic contamination from varying wastewater treatment plant discharges. Environ Pollut 142:295–302

89. Wille K, Noppe H, Verheyden K et al (2010) Validation and application of an LC–MS/MS method for the simultaneous quantification of 13 pharmaceuticals in seawater. Anal Bioanal Chem 397:1797–1808

90. Madureira TV, Barreiro JC, Rocha MJ et al (2010) Spatiotemporal distribution of pharmaceuticals in the Douro River estuary (Portugal). Sci Total Environ 408:5513–5520

91. Yang L, Luan T, Lan C (2006) Solid–phase microextraction with on–fiber silylation for simultaneous determinations of endocrine disrupting chemicals and steroid hormones by gas chromatography–mass spectrometry. J Chromatogr A 1104:23–32

92. Vieno NM, Tuhkanen T, Kronberg L (2006) Analysis of neutral and basic pharmaceuticals in sewage treatment plants and in recipient rivers using solid phase extraction and liquid chromatography–tandem mass spectrometry detection. J Chromatogr A 1134:101–111

93. Alonso SG, Catalá M, Maroto RR et al (2010) Pollution by psychoactive pharmaceuticals in the rivers of Madrid metropolitan area (Spain). Environ Int 36:195–201

94. Magnér J, Filipovic M, Alsberg T (2010) Application of a novel solid–phase–extraction sampler and ultra–performance liquid chromatography quadrupole–time–of–flight mass

spectrometry for determination of pharmaceutical residues in surface sea water. Chemosphere 80:1255–1260

95. Vasskog T, Anderssen T, Pedersen–Bjergaard S (2008) Occurrence of selective serotonin reuptake inhibitors in sewage and receiving waters at Spitsbergen and in Norway. J Chromatogr A 1185(2):194–205

96. Vasskog T, Berger U, Samuelsen PJ et al (2006) Selective serotonin reuptake inhibitors in sewage influents and effluents from Tromsø, Norway. J Chromatogr A 1115(1–2):187–195

97. Lajeunesse A, Gagnon C, Sauvé S (2008) Determination of basic antidepressants and their N–desmethyl metabolites in raw sewage and wastewater using solid phase extraction and liquid chromatography–tandem mass. Anal Chem 80:5325–5333

98. Castiglioni S, Bagnati R, Fanelli R et al (2006) Removal of pharmaceuticals in sewage treatment plants in Italy. Environ Sci Technol 40:357–363

99. Greenwood R, Clark J, Mills G et al (2010) Recommenndations on research and development. In: Benoit R (ed) Pharmaceuticals in the environment: current knowledge and need assessment to reduce presence and impact. IWA Publishing, London

100. Vieno M, Tuhkanen T, Kronberg L (2005) Seasonal variation in the occurrence of pharmaceuticals in effluents from a sewage treatment plant and in the recipient water. Environ Sci Technol 39:8220–8226

101. Kümmerer K (2001) Drugs in the environment: emission of drugs Diagnostic aids and disinfectants into wastewater by hospitals in relation to other sources – a review. Chemosphere 45:957–969

102. Larsson DGJ, Pedro C, Paxeus N (2007) Effluent from drug manufactures contains extremely high levels of pharmaceuticals. J Hazard Mater 148:751–755

103. Li D, Yang M, Hu J et al (2008) Determination of penicillin G and its degradation products in a penicillin production wastewater treatment plant and the receiving river. Water Res 42:307–317

104. Lin AY-C, Tsai Y-T (2009) Occurrence of pharmaceuticals in Taiwan's surface waters: impact of waste streams from hospitals and pharmaceutical production facilities. Sci Total Environ 407:3793–3802

105. Lin AY-C, Yu T-H, Lin C-F (2008) Pharmaceutical contamination in residential, industrial, and agricultural waste streams: risk to aqueous environments in Taiwan. Chemosphere 74:131–141

106. Benotti MJ, Trenholm RA, Vanderford BJ et al (2009) Pharmaceuticals and endocrine disrupting compounds in US drinking water. Environ Sci Technol 43:597–603

107. Daughton CG, Ruhoy IS (2011) Green pharmacy and pharmecovigilance: prescribing and the planet. Exp Rev Clin Pharmacol 4:211–232

108. Cleuvers M (2003) Aquatic ecotoxicity of pharmaceuticals including the assessment of combination effects. Toxicol Lett 142:185–194

109. DeLorenzo ME, Fleming J (2008) Individual and mixture effects of selected pharmaceuticals and personal care products on the marine phytoplankton species Dunaliella tertiolecta. Arch Environ Contam Toxicol 54:203–210

110. Quinn B, Gagné F, Blaise C (2009) Evaluation of the acute, chronic and teratogenic effects of a mixture of eleven pharmaceuticals on the cnidarian, Hydra attenuate. Sci Total Environ 407:1072–1079

111. Crane M, Watts C, Boucard T (2006) Chronic aquatic environmental risks from exposure to human pharmaceuticals. Sci Total Environ 367:23–41

112. Ding C, He J (2010) Effect of antibiotics in the environment on microbial populations. Appl Microbiol Biotechnol 87:925–941

113. Kim S, Aga DS (2007) Potential ecological and human health impacts of antibiotics and antibiotic-resistant bacteria from wastewater treatment plants. J Toxicol Environ Health B Crit Rev 10:559–573

114. Dorne JLCM, Ragas AMJ, Frampton GK et al (2007) Trends in human risk assessment of pharmaceuticals. Anal Bioanal Chem 387:1167–1172

115. Boxall A, Greenwood R (2010) Biological monitoring and endpoints. In: Benoit R (ed) Pharmaceuticals in the environment: current knowledge and need assessment to reduce presence and impact. IWA Publishing, London
116. Kidd KA, Blanchfield PJ, Mills KH et al (2007) Collapse of a fish population after exposure to a synthetic estrogen. Proc Natl Acad Sci USA 104:8897–8901
117. Vajda AM, Barber LB, Gray JL et al (2008) Reproductive disruption in fish downstream from an estrogenic wastewater effluent. Environ Sci Technol 42:3407–3414
118. Stackelberg PE, Furlong ET, Meyer MT et al (2004) Persistence of pharmaceutical compounds and other organic wastewater contaminants in a conventional drinking–water–treatment plant. Sci Total Environ 329:99–113
119. Oaks JL, Gilbert M, Virani MZ et al (2004) Diclofenac residues as the cause of vulture population decline in Pakistan. Nature 427:630–633
120. Brooks BW, Chambliss CK, Stanley JK et al (2005) Determination of select antidepressants in fish from an effluent-dominated stream. Environ Toxicol Chem 24:464–469
121. Schwaiger J, Ferling H, Mallow U et al (2004) Toxic effects of the non–steroidal anti–inflammatory drug diclofenac Part I - Histopathological alterations and bioaccumulation in rainbow trout. Aquat Toxicol 68:141–150
122. Directive 92/18/EEC (1992) Modifying the annex to council directive 81/852/EEC on the approximation of the laws of member states relating to analytical, pharmacotoxicological and clinical standards and protocols in respect of the testing of veterinary medicinal products
123. EMEA (1998) Note for guidance: environmental risk assessment for veterinary medicinal products other than GMO-containing and immunological products, The European Agency for the Evaluation of Medicinal Products: Veterinary Medicines Evaluation Unit, EMEA/CVMP/055/96-FINAL, 1998
124. EMEA/CHMP – European Medicines Evaluation Agency/Committee for Medicinal Products for Human Use (2006) Doc Ref EMEA/CHMP/SWP/4447/00, guideline on the environmental assessment of medicinal products for human use, 12 pp
125. Meisel ML, Costa MC, Pena A (2009) Regulatory approach on environmental risk assessment. Risk management recommendations, reasonable and prudent alternatives. Ecotoxicology 18:1176–1181
126. Stuer–Lauridsen F, Birkved M, Hansen LP et al (2000) Environmental risk assessment of human pharmaceuticals in Denmark after normal therapeutic use. Chemosphere 40 (7):783–793
127. Wennmalm A, Gunnarsson B (2010) Experiences with the Swedish Environmental Classification Scheme. In: Kümmerer K, Hemple M (eds) Green and sustainable pharmacy. Springer, Berlin
128. Kümmerer K (2009) Antibioticis in the aquatic environment – a review – Part II. Chemosphere 75:435–441
129. Taylor D (2010) Ecopharmacostewardship – a pharmaceutical industry perspective. In: Kümmerer K, Hemple M (eds) Green and sustainable pharmacy. Springer, Berlin
130. Vidaurre R, Touraud E, Roig B et al (2010) Recommendations on communication and education. In: Benoit R (ed) Pharmaceuticals in the environment: current knowledge and need assessment to reduce presence and impact. IWA Publishing, London
131. www.valormed.com. Accessed 29 Sep 2011
132. www.pharmas-eu.org/. Accessed 29 Sep 2011

Human Exposure and Health Risks to Emerging Organic Contaminants

Adrian Covaci, Tinne Geens, Laurence Roosens, Nadeem Ali, Nele Van den Eede, Alin C. Ionas, Govindan Malarvannan, and Alin C. Dirtu

Abstract We have reviewed the human exposure to selected emerging organic contaminants, such new brominated flame retardants, organophosphate flame retardants, phthalate substitutes, triclosan, synthetic musks, bisphenol-A, perchlorate, and polycyclic siloxanes. Levels of these emerging contaminants in matrices relevant for human exposure (air, dust, food, water, etc.) and in human matrices (blood, urine, or tissues) have been reviewed, together with some of the relevant health effects reported recently.

Keywords Emerging contaminants, Flame retardants, Human exposure, Human health, Personal care products, Phthalates, Review

Contents

A. Covaci (✉), T. Geens, L. Roosens, N. Ali, N. Van den Eede, A.C. Ionas, G. Malarvannan, and A.C. Dirtu
Toxicological Centre, University of Antwerp, Universiteitspelin 1, 2610 Wilrijk-Antwerp, Belgium
e-mail: adrian.covaci@ua.ac.be

D. Barceló (ed.), *Emerging Organic Contaminants and Human Health*,
Hdb Env Chem (2012) 20: 243–306, DOI 10.1007/698_2011_126,
© Springer-Verlag Berlin Heidelberg 2011, Published online: 20 December 2011

1   Introduction

Emerging contaminants are a broad category of chemicals, previously unknown or unrecognized as being of concern, but that are under increasing scientific scrutiny. In general, most of these contaminants are not regulated in the traditional sense of having allowable levels determined for specific environments. Current understanding of these contaminants often contains significant gaps, including their toxicity (toward humans), bioaccumulation, occurrence, transport, and transformation mechanisms. Many emerging contaminants are families of chemicals over which concern has been raised for at least some of the compounds in the group.

Emerging contaminants generally have "point sources" of release or exposure and are widely used in consumer goods, serving useful day-to-day functions within society. Human exposure and body burdens are generally of key concern for these contaminants precisely due to their broad applications. Several emerging contaminants are persistent, bioaccumulative, and toxic (PBT), while others are associated with endocrine disruption, including estrogenic effects in humans or animals. As advances in the sensitivity of analytical methods are achieved, these compounds will continue to be studied by researchers and regulators. There is yet concern regarding the availability and the comparability of data at European and international levels.

In this chapter, we systematically investigated the levels of selected emerging organic contaminants in matrices relevant for human exposure (air, dust, food, water, etc.) and their levels in human matrices, and reviewed some of the relevant health effects reported recently.

2   Human Exposure to Current-Use Brominated Flame Retardants

Due to their adverse health effects, several restrictions and bans have been imposed on the usage of polybrominated diphenyl ethers (PBDEs), namely the penta-, octa-, and deca-BDE formulations, in Europe, China, North America, and Japan (http://www. bsef.com). The restricted usage of these PBDEs increased the market demand for substitute brominated flame retardants (BFRs) including hexabromocyclododecanes

(HBCDs) [1], tetrabromobisphenol A (TBBPA) [2], and a range of novel BFRs (NBFRs) [3]. Currently, TBBPA is the BFR with the largest production volume worldwide (http://www.bsef.com). Their increased production volume is reflected by an ongoing increase of their environmental detection frequencies [3–5]. However, an extensive review of the existing literature revealed no similar temporal trend for TBBPA which might be due to its shorter half-life in humans [2]. Similar to PBDEs, these chemicals reach the environment through the manufacture, use, and disposal of various BFR-containing consumer products. In addition, various monitoring studies have indicated their presence in wildlife and in humans making them ubiquitous contaminants ([3, 6]). Their accumulation in humans primarily occurs through dietary intake, indoor dust ingestion, and indoor air inhalation (Table 1). Dermal uptake is considered to be less important as the log $K_{o/w}$ of these compounds are >5 [18].

2.1 Human Levels and Profiles

In general, compared to the extensive database of PBDEs in human tissues, less information is available on the levels of substitute BFRs in humans. In recent years, both the levels and the detection frequency of HBCDs in human tissues have increased significantly, which was paralleled by the higher usage of the HBCD technical mixture [19]. While Covaci et al. [6] reported a low detection frequency of HBCDs in serum and human milk (3–50%), this has mounted to 70% in serum samples [20] and 100% in human milk samples [21]. Moreover, the median-detected HBCD levels in serum are now similar to the total median PBDE concentration [20], which used to be the dominant BFR in human tissues [6]. Mean HBCD levels in serum and human milk samples are generally below 5 ng/g lw, but can mount up to 100 ng/g lw in occupationally exposed populations (Table 2). HBCD levels in European samples are in general higher compared to those detected in the USA and Asia, although a lack of regulatory measures in China, combined with a high density of e-recycling facilities, may lead to future higher levels in this region [33].

A note should be made on the dominance of α-HBCD in human samples, which is likely to occur because of a diastereomeric shift due to the preferential metabolism of β-HBCD and γ-HBCD by cytochrome P450 [34]. Several studies have reported the dominance of α-HBCD in human serum [7]. In contrast, some studies reported γ-HBCD to contribute to a higher percentage of the total HBCDs in human tissues [23, 26]. An increase in the percentage of γ-HBCD has been reported in highly exposed population and in occupationally exposed workers, with γ-HBCD making up to 40% of Σ-HBCDs [23, 35]. Although the reasons for the different isomer profiles in human tissues from different studies are not yet clear, it is reasonable to hypothesize that they arise from a combination of differences in external exposures (e.g., α-HBCD predominated in both dust and diet of the present study) and interindividual variations in metabolism.

Table 1 Literature survey regarding human exposure to brominated flame retardants

Compounds	Country	Detection frequency	Population	Exposure estimate (ng/day)	Particularities	References
HBCDs	Belgium			1.2–20	Duplicate diet	Roosens et al. [7]
	Belgium			59.4	Market basket	Goscinny et al. [8]
	The Netherlands			174	Market basket	De Winter-Sorkina et al. [9]
	UK			413	Market basket	Driffield et al. [10]
	Belgium			1.1–15	Indoor dust	Roosens et al. [7]
	UK			32.5	Indoor dust	Abdallah et al. [11]
	USA			2–60	Indoor dust	Abdallah et al. [12]
	UK			3.9	Indoor air	Abdallah et al. [11]
TBBPA	The Netherlands			0.8	Diet	De Winter-Sorkina et al. [9]
	Belgium			0.2	Indoor dust	Geens et al. [13]
	UK			1.3	Indoor dust	Abdallah et al. [11]
	UK			0.3	Indoor air	Abdallah et al. [11]
BTBPE	Belgium and UK	Homes = 85 Class rooms = 86 Office = 100	Adult (working) Adult (non working) Toddler	<0.01 [a,b] <0.01 [a,c] 0.05 [a,d]	Indoor dust (Belgian homes, office, and UK day care centers)	Ali et al. [14]
	New Zealand	44	Adult Toddler	<0.01 [a] 0.01 [a]	Indoor dust (homes)	Ali et al. [15]
	China	97.4	Adult Toddler	1.00 4.00	Indoor dust (homes in e-waste area)	Wang et al. [16]
	China	96.3	Adult Toddler	0.32 1.29	Indoor dust (homes in urban area)	Wang et al. [16]
	UK		Adult Toddler	0.23 [e] 1.00 [f]	Indoor dust (homes, cars, and offices)	Harrad et al. [17]
DBDPE	Belgium and UK	Homes = 100 Class rooms = 75 Office = 100	Adult (working) Adult (non working) Toddler	0.18 [a,b] 0.11 [a,c] 1.89 [a,d]	Indoor dust (Belgian homes, office, and UK day care centers)	Ali et al. [14]

	Country		Population	Exposure[a]		Reference
	New Zealand	88	Adult	0.01[a]	Indoor dust (homes)	Ali et al. [15]
			Toddler	0.16[a]		
	China	100	Adult	3.15	Indoor dust (homes in e-waste area)	Wang et al. [16]
			Toddler	12.6		
	China	100	Adult	137	Indoor dust (homes in urban area)	Wang et al. [16]
			Toddler	547		
	UK		Adult	2.3[e]	Indoor dust (homes, cars and offices)	Harrad et al. [17]
			Toddler	5.5[f]		
TBB	Belgium and UK	Homes = 31 Class rooms = 92 Office = 67	Adult (working) Adult (non working) Toddler	<0.01[a,b] <0.01[a,c] 0.08[a,d]	Indoor dust (Belgian homes, office, and UK day care centers)	Ali et al. [14]
	New Zealand	74	Adult Toddler	<0.01[a] 0.04[a]	Indoor dust (homes)	Ali et al. [15]
TBPH	Belgium and UK	Homes = 97 Class rooms = 97 Office = 100	Adult (working) Adult (non working) Toddler	0.02[a,b] 0.01[a,c] 0.40[a,d]	Indoor dust (Belgian homes, office, and UK day care centers)	Ali et al. [14]
	New Zealand	91	Adult Toddler	0.01[a] 0.19[a]	Indoor dust (homes)	Ali et al. [15]
TBBPA-DBPE	Belgium and UK	Homes = 85 Class rooms = 64 Office = 83	Adult (working) Adult (non working) Toddler	0.08[a] 0.06[b] 1.12[c]	Indoor dust (Belgian homes, office, and UK day care centers)	Ali et al. [14]

[a] Exposure in ng/kg bw/day
[b] Assuming working dust ingestion is 78.9% home and 21.1% office
[c] Assuming nonworking adult dust ingestion is 100% home
[d] Assuming toddler dust ingestion is 79.9% home and 20.1% classroom
[e] Assuming adult dust ingestion is 72% home, 23.8% office, and 4.2% car
[f] Assuming toddler dust ingestion is 95.8% home and 4.2% car

Table 2 Overview of HBCDs, TBBPA, and NBFRs levels in various human matrices

Compounds	Matrix	Country	Mean Levels (ng/g lw)	Particularities	References
HBCDs	Serum	Belgium (16)	1.7		Roosens et al. [7]
		The Netherlands (78)	1.1		Covaci et al. [6]
		Sweden (50)	0.46		Covaci et al. [6]
		Greece (61)	1.32		Kalantzi et al. [20]
		Norway (41)	4.1	Highly exposed	Thomsen et al. [22]
		Norway (10)	101	Occupationally exposed	Thomsen et al. [23]
	Human milk	Japan	4.0		Kakimoto et al. [24]
		UK (34)	5.95		Abdallah and Harrad [21]
		Sweden (64)	0.3		Glynn et al. [19]
		China (24)	<LOD-2.8		Shi et al. [25]
	Adipose tissue	USA (20)	0.02		Johnson-Restrepo et al. [26]
TBBPA	Serum	France (91)	16.1		Cariou et al. [27]
		Sweden (4)	1.1–3.8	Occupationally exposed	Hagmar et al. [28]
		Norway (5)	1.3	Occupationally exposed	Thomsen et al. [29]
		Norway (5)	0.54		Thomsen et al. [29]
		Norway (5)	0.34		Thomsen et al. [29]
	Cord serum	France (91)	54.8		Cariou et al. [27]
	Human milk	UK (34)	0.06		Abdallah and Harrad [21]
		China (24)	<LOD-5.1		Shi et al. [25]
		France (77)	0.48		Cariou et al. [27]
		Norway (9)	0.07		Thomsen et al. [30]
	Adipose tissue	France (44)	<LOD		Cariou et al. [27]
		USA (20)	0.05		Johnson-Restrepo et al. [26]
BTBPE	Serum	Sweden (5)	<1.3		Karlsson et al. [31]
		China (128)	<LOD		Zhu et al. [32]
DBDPE	Serum	Sweden (5)	<1.0		Karlsson et al. [31]
		China (128)	<LOD		Zhu et al. [32]

Levels and detection frequency of TBBPA are generally lower than those of PBDEs and HBCDs in human samples. Consistent with its phenolic structure that can be rapidly conjugated in human liver and subsequently excreted in bile [36], TBBPA has a short human half-life that has been reported to be as low as 2 days in human plasma [28, 37]. In addition, HBCD and PBDEs are additive BFRs, while TBBPA is a reactive BFR, meaning that TBBPA is chemically bound to the polymer structure and, thus, the leaching or release of TBBPA into the environment is limited [38]. Therefore, levels of TBBPA are often lower and detection of this flame retardant is likely to reflect recent rather than past exposure [2, 39], which might explain high TPPBA levels found in certain studies such as by [27] (Table 2). Contrarily, Chinese and Japanese studies often reported higher TBBPA levels in human tissues compared to HBCDs which can be explained as TBBPA was manufactured and used in greater quantity in Asia [40].

TBBPA concentrations in human plasma in the low ng/g lipids range have been reported (Table 2). Concentrations and detection frequency of TBBPA in adipose tissue are even lower due to the relatively low lipophilic properties of TBBPA ($K_{ow} = 5$), and the metabolic fate of this BFR.

Few risk assessments and biomonitoring studies concerning NBFRs have been conducted in humans. Therefore, data concerning human exposure to NBFRs are scarce. Karlsson et al. [31] and Sjödin et al. [39] studied decabromodiphenylethane (DBDPE) and 1,2-bis(2,4,6-tribromophenoxy)ethane (BTBPE) alongside PBDEs in serum samples collected from workers in Swedish electronic recycling plants. BTBPE and DBDPE were not detected in the workers' serum. Analytical methods used by Karlsson et al. [31] and Sjödin et al. [39] were primarily based on PBDE analysis; so the apparent absence of BTBPE and DBDPE could be due to suboptimal analytical conditions for those compounds. Zhu et al. [32] screened a number of BFRs in 128 human serum samples from China. Although PBDEs were present, DBDPE and BTBPE were below detectable levels in all serum samples.

2.2 Health Effects

No information is available on the effects of single exposure to TBBPA in humans. Based on animal studies, it can be concluded that TBBPA is of low acute toxicity by all routes of exposure. It is not a skin or respiratory sensitizer. Chronic exposure was assessed in a 90-day rat study which showed no toxicologically significant effects following oral exposure up to 1,000 mg/kg TBBPA [41]. A decrease in serum T4 levels was observed; however, in the absence of changes in other parameters (TSH, T3, and changes in the liver, thyroid, parathyroid, or pituitary gland) of thyroid homeostasis in a species that is very sensitive to perturbations in thyroid hormone levels, these decreases are not considered to be adverse. There are no studies in humans or animals on the carcinogenic potential of TBBPA. However, there are no indications to raise concerns for carcinogenicity. Neurodevelopmental toxicity of TBBPA was assessed but no convincing evidence of an adverse effect on neurodevelopment at dose levels up to 1,000 mg/kg/day could be provided (http://esis.jrc.ec.europa.eu/doc/existing-chemicals/risk_assessment). The WHO conducted a similar risk assessment and concluded that TBBPA has little potential for bioaccumulation, and the risk for the general population is considered to be insignificant (http://www.icl-ip.com).

The commercial HBCD has shown low acute toxicity [6]. The minimum lethal dose is greater than 20 g/kg for dermal and oral administration [18]. Repeated dose toxicity was assessed by Van der Ven et al. [42] and revealed enhanced endocrine and immune parameters and recorded the liver as target organ with an increased liver weight probably due to enzyme induction as from doses of 100 mg/kg/day. Liver weight increase was slowly reversible upon cessation of exposure. Secondary to the liver effects, disruption of the thyroid hormone system has been noted. Reproductive and developmental toxicity was detected [43, 44].

No particular toxicological studies are reported for NBFRs in human. Based on laboratory animal studies, various NBFRs have no to low acute and chronic toxicities. BTBPE has shown low toxicity for laboratory animals; this might be due to the limited accumulation and absorption and rapid excretion of BTBPE in human, which has been indicated in rats when given orally [45, 46]. However, Hakk et al. [46] identified the hydroxylated metabolites of BTBPE which suggest bio-transformation by cytochrome P450 enzymes. In an in vitro study, BTBPE has shown porphyrinogenic properties on chick embryo liver cultures but only after pretreatment with betanaphtoflavone [47]. Li et al. [48] observed no evidence of skin sensitization properties in 200 professional workers during a repeated application of DBDPE in petroleum during 3 weeks. Lack of toxicity for DBDPE is believed due to the poor bioavailability which is due to its high molecular weight and poor water solubility [49]. However, McKinney et al. [50] using an in vitro system based on liver microsomes from three Arctic marine-feeding mammals and laboratory rat observed oxidative and reductive biotransformation of DBDPE. A depletion of 44–74% of 90 pmol was observed for DBDPE in individuals from all species. In another in vivo and in vitro toxicity study, Nakari and Huhtala [51] revealed that DBDPE is bioavailable, is acutely toxic to water fleas (*Daphina magna*), and have injurious effects on the reproduction physiology of zebrafish (*Danio rerio*). Scientific literature on human health monitoring to tetrabromo-bisphenol A bis(2,3-dibromopropyl ether) (TBBPA-DBPE), 2-ethylhexyl-2,3,4,5-tetrabromobenzoate (TBB), bis(2-ethylhexyl)-3,4,5,6-tetrabromophthalate (TBPH), and hexachlorocyclopentadienyl-dibromocyclooctane (HCDBCO) is scarce [52]. Despite the absence of lethality or overt sign of toxicity, Bearr et al. [53] observed genotoxicity in feathhead minnows over the exposure of Firemaster® BZ54 and 550 mixtures, of which TBB and TBPH are important components. Knudsen et al. [54] reported poor absorption of TBBPA-DBPE through gastrointestinal tract in rats, and the absorbed compound in the liver was slowly metabolized and eliminated in the feces. In mice the oral and dermal LD50 values for TBBPA-DBPE were reported to be >20 g/kg (21 mmol/kg) [55, 56].

3   Human Exposure to Organophosphate Esters

Organophosphate triesters (OPFRs) are used as flame retardants and/or plasticizers in a wide range of polymers, in furniture upholstery, insulation, wall coverings, and floor finishing products. Triethyl phosphate (TEP), tri-*n*-butyl phosphate (T*n*BP), tri-*iso*-butyl phosphate (T*i*BP), tris(2-ethylhexyl) phosphate (TEHP), and tris(2-butoxyethyl) phosphate (TBEP) are commonly used as plasticizers in polymers, such as vinyl resins, cellulose esters and lacquers, and sometimes rubber [57–60]. Another typical use of TBEP is floor polish [60]. Trimethyl phosphate (TMP) and tri-*n*-propyl phosphate (TPrP) are mostly used for industrial applications and are rarely detected in the indoor environment or other matrices [61, 62].

Chloroalkyl phosphates such as tris(1,3-dichloro-2-propyl) phosphate (TDC*i*PP) and tris(chloro-*iso*-propyl) phosphate (TC*i*PP) are more often used in textile backcoatings, in rigid and flexible polyurethane foams, which are used for thermal insulation and for furniture and upholstery, respectively [63]. In contrast, tris(2-chloroethyl) phosphate (TCEP) is now more frequently used in PVC, unsaturated polyester resins, and textile backcoatings [63].

Aryl phosphates such as tricresyl phosphate (TCP), triphenyl phosphate (TPP), 2-ethylhexyl-diphenyl phosphate (EHDPP), butyldiphenyl phosphate (BDPP), and dibutylphenyl phosphate (DBPP), as well as TBP are used in flame retardant hydraulic fluids [64]. TPP and TCP are further used as plasticizers, mainly in vinyl polymers and electric cables [58, 65]. EHDPP, BDPP, and DBPP are also applied in vinyl food packaging. The consumption of OPFRs has increased sharply in the European Union after the restriction in use of PBDEs [296]. Also in the USA and Japan, an increase in the use of OPFRs has been observed as PBDEs have been phased out ([66], [297]). The rise in consumption has resulted in increasing levels of these contaminants in the indoor environment with air and dust being the more important sources of exposure for humans because in this environment people generally spend most of their time. Other plausible routes of exposure are the intake of OPFRs via drinking water and food, though according to recent measurements these matrices seem of minor contribution to human body burdens.

Looking at the extent of occurrence of OPFRs, concerns rise about possible health effects related to constant exposure to these chemicals. Updated safety levels have been suggested for TCP and TDC*i*PP [67].

3.1 Routes of Exposure

Settled dust: Preliminary results on indoor and car dust show that levels of OPFRs in dust collected in public buildings and cars are higher than the levels in dust collected from home environments [61, 68, 69]. There is also a shift in the OPFR profile: levels of TPP and TDC*i*PP increase moderately in the office dust, but a remarkable rise in TDC*i*PP levels is observed in car dust samples. Table 3 gives an overview of the sum of analyzed OPFRs in the countries and the most dominant OPEs in the analyzed samples. Some of the aryl OPFRs namely BDPP, DBPP, EHDPP were so far not detected in dust samples. There seems to be a region-specific consumption of OPFRs as higher levels of, e.g., TPP and TDC*i*PP were observed in house dust from the USA and New Zealand ([15], [297]).

Table 3 Levels of OPFRs in matrices relevant for human exposure

Country	Number of samples	Sum OPFRs (medians)	Dominant OPFRs	Particularities of the study	References
Dust (µg/g)					
USA	50	9.82	TPP	Homes, 3 OPFRs	Stapleton et al. [297]
Belgium	15	16.5	TBEP, TCiPP, TPP	Shops, 9 OPFRs	Van den Eede et al. [298]
Belgium	33	13.1	TiBP, TBEP, TCiPP	Homes, 9 OPFRs	Van den Eede et al. [298]
Germany	216	0.40	TCiPP	Homes, only TCiPP	Ingerowski et al. [70]
Germany	356	0.60	TCEP	Homes, only TCEP	Ingerowski et al. [70]
Germany	10	11	TBEP, TPP	Offices, 9 OPFRs	Brommer et al. [69]
Germany	12	37	TDCiPP, TBEP	Cars, 9 OPFRs	Brommer et al. [69]
Germany	1	3	TCiPP, TBEP	Home, 9 OPFRs	Brommer et al. [69]
Japan	40	5.4	TBEP, TCiPP	Homes, 11 OPFRs	Kanazawa et al. [71]
Japan	8	48.5	TBEP, TCiPP, TCP	Hotel, 11 OPFRs	Takigami et al. [72]
New Zealand	50	4.3	TBEP, TPP	Homes, 8 OPFRs	Ali et al. [15]
Romania	47	7.5	TBEP, TCiPP	Homes 8 OPFRs	Dirtu et al. [73]
Spain	8	21.5	TBEP, TCiPP	Homes, 8 OPFRs	García et al. [74]
Spain	9	24	TBEP, TCiPP	Homes, cars, car wash area, 7 OPFRs	García et al. [75]
Sweden	10	21	TBEP, TCEP	Homes, 10 OPFRs	Bergh et al. [68]
Sweden	10	1600	TBEP (TDCiPP)	Day care center, 10 OPFRs	Bergh et al. [68]
Sweden	10	140	TBEP, TCiPP	Workplace, 10 OPFRs	Bergh et al. [68]
Sweden	15	59	TBEP, TPP	Homes and public locations, 12 OPFRs	Marklund et al. [61]
Air (ng/m)					
Germany	50	10		Homes, only TCEP	Ingerowski et al. [70]
Japan	14	6.6		Offices, TnBP	Saito et al. [66]
Japan	18	4.0		Houses, TnBP	Saito et al. [66]
Japan	40	62.3	TCiPP, TEP, TnBP	Homes	Kanazawa et al. [71]
Sweden	10	78	TCiPP, TnBP	Homes, 10 OPFRs	Staaf et al. (2005)
Sweden	10	40	TiBP, TnBP,	Homes, 11 OPFRs	Bergh et al. [68]
Sweden	10	140	TBEP, TCEP	Day care, 11 OPFRs	Bergh et al. [68]

(continued)

Table 3 (continued)

Country	Number of samples	Sum OPFRs (medians)	Dominant OPFRs	Particularities of the study	References
Sweden	10	160	TCiPP, TDCiPP	Offices, 11 OPFRs	Bergh et al. [68]
Sweden	169	14		Apartments, TCiPP	Bergh et al. [76]
Sweden	169	11		Apartments, TiBP	Bergh et al. [76]
Sweden	5	145	TCiPP, TCEP, TiBP	Schools, daycare center, office, 7 OPFRs	Carlsson et al. [77]
Sweden	12	121.2 (mean)	TBEP, TCEP, TCiPP	Electronics recycling plant, dismantling hall, 8 OPFRs	Sjödin et al. [78]
Sweden	17	160	TCiPP, TnBP	Public locations, 10 OPFRs	Marklund et al. [79]
Switzerland	16	41	TCEP, TEP, TCiPP	Cars, stores, offices, 8 OPFRs	Hartmann et al. [80]
Other matrices					
Sweden		410 ng/g lw	TCiPP, TPP, TCEP	Marine perch	Marklund et al. [81]
Sweden		720 ng/g lw	TCiPP, TPP, EHDPP	Freshwater perch	Marklund et al. [81]
Sweden		110 ng/g lw	TCiPP, TPP, EHDPP	Marine herring	Marklund et al. [81]
Sweden		1900 ng/g lw	TCiPP, TBEP, EHDPP	Freshwater perch (STP)	Marklund et al. [81]
USA		15 ng/L	TBP	Finished drinking water	Stackelberg et al. [82]
USA		4 ng/L	TCEP	Finished drinking water	Stackelberg et al. [82]
USA		12 ng/L	TDCiPP	Finished drinking water	Stackelberg et al. [82]
Germany	1	6.25 ng/L	TCiPP, TDCiPP, TCEP	Finished drinking water, 8 OPFRs	Andresen and Bester [83]

Indoor air: The most abundant OPFRs in indoor air samples from homes were TEP, TiBP, TnBP, TCEP, and TCiPP, which are outlined in Table 3. Heavier OPFRs such as TBEP, TPP, TCP, TDCiPP, and TEHP are usually present in low concentrations or even below the detection limit. However, in some work environments and cars high levels of TBEP and TDCiPP were observed [66, 68, 84]. The higher air concentrations were associated with a higher dust concentration though it is hard to establish a significant correlation due to the high particle-adsorbed fraction of the OPFR concentrations in air [68, 79].

Drinking water: OPFRs have been detected at the ng/L level in drinking water, resulting from an incomplete removal from wastewater [85] or from groundwater polluted with surface water leachate [299]. Chlorinated OPFRs such as TCEP and TC*i*PP in particular are more persistent to biodegradation and are not easily removed by bank filtration due to their low soil adsorption coefficient (Regnery 2011). However, activated carbon filtration did remove TCEP and TCPP according to Andresen and Bester [83]. This also appears in the maximum concentrations of TBEP (350 ng/L) and TDC*i*PP (250 ng/L) reported in finished drinking water by Stackelberg et al. [86] in contrast to the mean concentrations displayed in Table 3 [83, 86].

Food: So far only a handful of studies have investigated the presence of OPFRs in food samples. In a market basket study performed by the US FDA [87], most of the OPFRs (EHDPP, TCEP, TC*i*PP, TnBP, TPP, TCP) were found only at the ng/g level. More than 91% of the results were below the quantification limit in most types of samples. Higher detection frequencies were observed for TPP in margarine and caramels (mean 45 ng/g).

Three studies investigated the presence of OPFRs in fish samples. Campone et al. [88] developed and applied a method to 24 samples, but no OPFRs were detected (detection limits of 0.2–9 ng/g). Marklund et al. [81] investigated fish from Swedish rivers and the Baltic Sea (Table 3). Remarkable observations were the higher levels of TBP, TBEP, BDPP, and DBPP near point sources such as sewage treatment plants and airports. Brandsma et al. [89] analyzed different samples representative for the pelagic and the benthic food web in the western Scheldt. For most OPFRs, concentrations in sediment were much higher compared to lipid-normalized levels in invertebrates and fish, indicating trophic dilution. TBEP and TCEP and TC*i*PP were found to be more bioaccumulative as levels in flounder and herring exceeded those in sculpin and poulting. These data, however, are not generalized enough to make a conclusion on the contribution of dietary ingestion to general human exposure.

Dermal exposure to OPFRs: Dermal contact to OPFR-treated fabrics or dust containing OPFRs could lead to additional exposure. Furthermore, Weschler and Nazaroff [300] suggested that air concentrations of OPFRs may cause a significant exposure through dermal absorption. Since most OPFRs possess a log K_{ow} between 2 and 5, the hypothesis of dermal absorption seems acceptable and indeed some of them, namely TBP, TCP, TDC*i*PP, were reported to be absorbed in animal and in vitro studies ([57, 63, 65]). Yet, low toxicity has been evidenced following dermal exposure to TEP, TPP, and TC*i*PP. TBEP was poorly absorbed from skin, but caused skin irritation [60]. No data are currently available on EHDPP, DBPP, or BDPP. Only a few studies have shown the extent of dermal exposure by taking hand washing samples or hand wipes. Makinen et al. [301] found by average a total of 3.5 μg and 34 μg for people working in a circuit board factory and furniture workshop, respectively. TPP and TCP were the dominating compounds in the samples: Cooper and Stapleton [90] detected up to 200 ng TPP and up to 2,000 ng of TDC*i*PP in hand wipes after leisure activities.

3.2   Levels and Profiles in Human Matrices

OPFRs appear to be readily metabolized and so the parent compounds are not frequently detected in human samples. TBP and TDC*i*PP have been detected in a few adipose tissue samples at the ng/g level [57, 58, 63]. TDC*i*PP has been detected in semen as well [91]. Some studies could detect TPP in blood, but it originated from the PVC packaging [92]. Marklund et al. [81] detected OPFRs in pools of human milk samples collected from the 1990s to now. In Table 4, the most important compounds are mentioned, namely TBEP, TBP, and TCiPP. Other OPFRs were determined as 5 ng/g lw or lower.

In urine, OPFR metabolites have been detected, but the median concentrations of the diaryl and dialkyl phosphates remained mostly below the quantification limit. Ranges of concentrations of dialkyl and diaryl phosphates are shown in Table 4. In the study of Schindler et al. [302, 303], the detection frequency was highest for BCEP (50%), followed by DPP (30%), BCPP (12%), and DBP (3%). Di-*m*-cresyl and di-*p*-cresyl phosphate were not found in any sample, probably due to lower exposure in the indoor environment. Reemtsma et al. [304] found also monoaryl and alkyl phosphates in human urine, of which the monobutyl phosphate was the

Table 4 Concentrations of OPFRs and OPFR metabolites in human samples

Compounds	Country	Number of samples	Concentration range	Matrix	References
DBP	Germany	25	ND–0.26 µg/L	Urine	Schindler et al. [303]
DPP	Germany	30	ND–4.1 µg/L	Urine	Schindler et al. [303]
BCEP	Germany	30	ND–27.5 µg/L	Urine	Schindler et al. [303]
BC*i*PP	Germany	25	ND–0.85 µg/L	Urine	Schindler et al. [303]
BDC*i*PP	USA	9	0.05–1.7 µg/L	Urine	Cooper et al. [93]
DPP	USA	9	0.3–7.4 µg/L	Urine	Cooper et al. [93]
MBP	Germany		ND–158 µg/L	Urine	Reemtsma et al. [304]
DBP	Germany		ND–0.52 µg/L	Urine	Reemtsma et al. [304]
DPP	Germany		ND–28.6 µg/L	Urine	Reemtsma et al. [304]
TC*i*PP	Sweden	6 Pooled samples	22–82 ng/g lw	Milk	Marklund et al. [81]
TBP	Sweden	6 Pooled samples	11–57 ng/g lw	Milk	Marklund et al. [81]
TPP	Sweden	6 Pooled samples	3.2–11 ng/g lw	Milk	Marklund et al. [81]
TDC*i*PP	USA		5–50 ppb	Semen	Hudec [91]

dominant congener (Table 4). Cooper et al. [93] found higher concentrations of DPP and was able to detect BDC*i*PP in urine, probably because of a higher indoor exposure in the USA for TDC*i*PP and TPP as mentioned for dust samples.

3.3 Health Effects

Suspected and observed effects in animals: Acute toxicity of OPFRs is associated with typical cholinergic symptoms, such as salivation, diarrhea, piloerection, tremor, ataxia, and respiratory depression. The LD_{50} values were in the range of 1–5 g/kg bodyweight, and as a result the OPFRs were classified as "safe." Concerning subchronic exposure to OPFRs, adverse effects included an absolute or relative increase in liver and kidney masses, and a smaller increase in body mass for young animals compared to controls. No teratogenicity was observed for OPFRs in rodents. However, exposure before and during mating resulted in a decrease in the number of litters and live pups per litter, which is an indication for reproductive toxicity. The safety of OPFRs following chronic exposure can also be questioned: TDC*i*PP caused a development of carcinomas in the liver, kidney, testes, and other organs [63]. Neurotoxic and carcinogenic properties were also observed for TCEP [63].

Possible adverse health effects in humans: So far only a limited number of studies found associations between adverse health effects and exposure of humans to OPFRs, but few have been completely proven.

There were some cases of T*o*CP poisoning in the late nineteenth and early twentieth century. This compound causes delayed neuropathy, but is now only used as a minor component in TCP isomer mixtures [65].

Camarasa and Serra-Baldrich [94] reported allergic contact dermatitis after repeated contact with TPP-treated plastics. Meeker and Stapleton [95] indicated endocrine disruptive properties for TPP and TDC*i*PP, through a negative correlation with semen quality and thyroid hormone levels, respectively. Kanazawa et al. [71] associated mucosal symptoms of the sick building syndrome with high indoor exposure to TBP. These symptoms include irritation to the eyes, nose, and throat symptoms such as flushing, and mucosal symptoms such as irritation to the eyes, nose, and throat; the latter symptoms were strongly associated with TBP levels in air and dust.

4 Human Exposure to Phthalates and Substitutes

The extensive use of phthalates as plasticizers in various materials (furniture, plastics, electronics equipment, textiles, etc.) has led to the widespread and substantial contamination of the indoor environment, e.g., air and dust [3, 96, 97]. Indoor environment and dietary intake are of special concern for the increasing human exposure to such chemicals for which various adverse health effects have been already reported [97]. For example, several phthalic acid esters were classified as potential carcinogens as well as endocrine disruptors [98–100].

As plasticizers, phthalates are imparting optimum overall performance properties in PVC at low costs [101], but since they are not chemically bound to the end products, they gradually migrate into the environment. Since 1949, the most intense used phthalate was di-(2-ethylhexyl) phthalate (DEHP), but due to relatively recent reports on environmental impact, their use starts to be restricted: in July 2005, DEHP, dibutyl phthalate (DBP), and benzylbutyl phthalate (BBP) were banned from all children's articles within the European Union [102]. The DEHP are based mainly on long-chain phthalates, like di-*iso*-nonyl phthalate (DINP) and di-*iso*-decyl phthalate (DIDP) [103]. As a consequence, by 2008, the percentage of DINP and DIDP in plasticizers used in Western Europe reached up to 67% of the overall plasticizer consumption, while the percentage of DEHP decreased from 42% in 1999 to 17.5% in 2008 [104]. Due to the increase and for precautionary reasons, the use of these substitutes in toys which might be taken into the children's mouth was also banned in the EU [102]. Currently, there are two different types of DINP commercialized on the market, even if their chemical composition is not sufficiently described in the scientific literature [105]. DINP 1 (CAS 68515-48-0) is a mixture of esters of *o*-phthalic acid with C8–C10 alkyl alcohols of different chain lengths and branching distributions; DINP 2 (CAS 28553-12-0) consists solely of isomeric C9 alcohols in the ester chain, while a third DINP type has in the meantime banned from the market [106, 107].

Since DINP may not provide the required flexibility properties as well as low temperature viscosity [108], it is not possible to replace DEHP by DINP in all applications (e.g., in medical devices). During the last few years, many plasticizer producers promoted several alternatives to substitute DEHP as a general purpose plasticizer for the use in PVC applications. One of such nonphthalates which share similar technical properties as DEHP is di-*iso*-nonyl 1,2-cyclohexanedicarboxylic acid (DINCH) [109]. It seems that DINCH has no effects on fertility up to the highest administrated dose of 1,000 mg/kg bw/day [1, 110], and moreover, its specific migration rate is considerably lower when compared to DEHP [111, 112]. Consequently, this substitute was initially provided for use in applications that are particularly sensitive, such as medical products and toys [112–114]. However, since 2007, the use of DINCH was extended for food contact applications, like cling film, sealants, cap closures, artificial wine corks, and gloves [1], as well as a wide range of applications, like printing inks, can coatings, childcare articles, high solid paints, marine finishes, coil coatings, protective coatings, textile printings, and sport products [112, 115].

Other DEHP substitutes currently in use, important in terms of produced volumes, are also terephthalates, e.g., di(2-ethylhexyl) terephthalate (DEHT). DEHT is produced in volumes up to 50,000 tones [116], and the main applications are coatings, vinyl floorings, electric connectors, vinyl water stops, coating for clothes, bottle caps, toys, and medical devices [117].

4.1 Levels and Profiles in Humans and Matrices Relevant for Human Exposure

To evaluate human exposure to phthalates and their substitutes, the main approaches investigate either the levels of chemicals in matrices relevant for human exposure (indoor air, dust, food and packages, etc.) or the levels of parent and metabolite compounds in human samples (serum, urine, or breast milk). An overview of phthalate and nonphthalate plasticizers together with their metabolites commonly reported in literature is presented in Table 5. The half-lives for the most of these compounds are already established and therefore, by evaluating the levels of their metabolites in human urine, the levels of their parent compounds may be

Table 5 Overview of phthalate and nonphthalate plasticizers together with their main metabolites commonly reported in literature

Plasticizer	Abbreviation	Metabolite	Abbreviation
Dimethyl phthalate	DMP	Mono-methyl phthalate	MMP
Diethyl phthalate	DEP	Mono-ethyl phthalate	MEP
Di-n-propyl phthalate	DPP		
Di-n-butyl phthalate	DnBP	Mono-n-butyl phthalate	MnBP
Di-isobutyl phthalate	DiBP	Mono-isobutyl phthalate	MiBP
Di-n-pentyl phthalate (Di-n-amyl phthalate)	DPP (DAMP)	Mono-(4-hydroxypentyl) phthalate	MHPP
Butylbenzyl phthalate	BBzP	Mono-benzyl phthalate	MBzP
Di-2-ethylhexyl phthalate	DEHP	Mono-(2-ethylhexyl) phthalate	MEHP
		Mono-(2-ethyl-5-oxohexyl) phthalate	5oxo-MEHP (MEOHP)
		Mono-(2-ethyl-5-hydroxyhexyl) phthalate	5OH-MEHP (MEHHP)
		Mono-(2-carboxymethylhexyl) phthalate	2cx-MMHP (MCMHP)
		Mono-(2-ethyl-5-carboxypentyl) phthalate	5cx-MEPP (MECPP)
Dicyclohexyl phthalate	DcHP		
Diphenyl phthalate	DPhP		
Di-isononyl phthalate	DiNP	MiNP metabolites with a keto functional group	oxo-MiNP (MOINP)
		MiNP metabolites with a hydroxy functional group	OH-MiNP (MHINP)
		MiNP metabolites with a carboxy functional group	cx- MiNP
Di-isodecyl phthalate	DiDP		
Di-(2-propylheptyl) phthalate	DPHP		
Di-n-octyl phthalate	DNOP		
Di(2-ethylhexyl)adipate	DEHA		
Di(2-ethylhexyl) terephthalate	DEHT	Terephthalic acid	TPA

easily back calculated later. Additionally, since the excretion time is relatively short (<6 h), it is more relevant to evaluate the extent of human exposure to phthalates by measuring the levels of their metabolites in urine samples, while the levels of parent compounds are to be investigated in matrices like serum or even breast milk [118].

Therefore, following the models shown to be relevant for human exposure to other organic contaminants, namely flame retardants, indoor exposure to phthalates and their substitutes played also an important role for various research groups worldwide, several publications reporting that they are present in the highest concentrations indoors, especially in house dust samples (Table 6). As expected from production volumes, dust levels are consistent among literature, and the most important plasticizer measured in this matrix being DEHP with concentrations up to few hundred mg/kg of dust. Since DEHP, together with DBP and BBP, serve nearly exclusively as plasticizer, the high concentrations measured in house dust seem to be explained by the building characteristics. Therefore, PVC floorings, plastic-coated wall papers, plastic ceilings and wall coverings, plastic furniture, and plastic (covered) doors and windows were supposed to be possible sources for phthalates [122]. However, interesting results have been also shown for nonphthalate plasticizers, DINCH and DEHT, when following their P95 levels in dust samples over time (2001–2009) reported for Germany: while they were not detected in samples collected in 2001/2002, their P95 levels seems to be increased by few times from 2003/2006 until 2009, suggesting a rise in the use of phthalate substitutes indoors. This result accounts in particular for DINCH and being in agreement with the fact that this plasticizer was launched on the market in 2002 [112].

After ingestion, the oxidative pathway of phthalate metabolism seems to be more effective in children than in adults [125]. Neonates show a further deviation in oxidative DEHP metabolism with 5cx-MEPP being by far the predominating metabolite [126]. Age-dependent metabolism of phthalates may also have relevance to health: the oxidation products are longer in the human body than the simple monoesters; they might be more toxic [127] and young children are a more vulnerable population group. Therefore, the EC has prohibited the sale of toys and childcare articles intended to be placed in the mouth by children <3 years of age made of soft PVC containing >0.1% by weight of six phthalates (Decision 1999/815/EC). Studies performed on materials and plasticizers used in soft children's products (samples collected from The Netherlands) shown that DINP and DEHP were the predominant plasticizers and were usually found in concentrations between 30% and 45% by weight [128]; occasionally some products being found to contain more than 45% by weight of DINP, which is not in compliance with Decision 1999/815/EC.

Considering the migration possibilities of such chemicals from their original products, complex studies were performed for "classical" plasticizers such as DEHP or DBP to evaluate the relevance of each pathway to human exposure assessments. Therefore, the total exposure of adults of such plasticizers was evaluated by quantifying the target substances in duplicate diet portions (collected daily over 7 consecutive days), and also by measuring indoor air and dust concentrations [129]. The results indicate that dietary exposure is the dominant

Table 6 Literature survey on levels of most commonly used phthalates and their substitutes in house dust (median concentrations and in brackets P95 concentrations, results expressed in μg/g) and in indoor air samples (median concentrations, results expressed in ng/m³)

Location	Compounds							References
	BBP	DBP	DEHP	DIDP	DINP	DINCH	DEHT	
Germany, 2003 N = 65 (*62), indoor dust	19	47	600	31*	72*	–	–	Kersten and Reich [119]
Germany, 2004 N = 30, dust indoor	29.7	47	703.4	–	–	–	–	Fromme et al. [120]
Germany, 2005 N = 278, indoor dust	13	29	480	60	80	–	–	Nagorka et al. [121]
Germany, 2008 N = 29, indoor dust	28	51	970	–	–	–	–	Butte et al. [122]
Germany, 2009 N = 30, indoor dust	15.2	87.7	604	33.6	129	–	–	Abb et al. [123]
Sweden, 2010 N = 16, indoor air	11	140	380	–	–	–	–	Bergh et al. [124]
Germany, 2001–2002 N = 274, indoor dust	–	–	–	–	–	<0.3 (<0.3)	<0.8 (<0.8)	Nagorka et al. [112]
Germany, 2003–2004 N = 593, indoor dust	–	–	–	–	–	<0.3 (<0.3)	<0.8 (4.6)	
Germany, 2004 N = 593, indoor dust	–	–	–	–	–	<0.3 (0.7)		
Germany, 2004–2005 N = 593, indoor dust	–	–	–	–	–	<0.3 (3.4)		
Germany, 2005–2006 N = 593, indoor dust	–	–	–	–	–	<0.3 (5.3)		
Germany, 2009 N = 36, indoor dust	–	–	–	–	–	2.2 (76)	1.4 (24)	

BBP benzyl-butyl phthalate, *DBP* di-butyl phthalate, *DEHP* di-(2-ethylhexyl) phthalate, *DEHT* di-(2-ethylhexyl) terephthalate, *DIDP* di-*iso*-decyl phthalate, *DINP* di-*iso*-nonyl phthalate, *DINCH* di-iso-nonyl 1,2-cyclohexane di-carboxylic acid

*DINP and DIDP were measured only in 62 dust samples

intake pathway. The estimated 95 percentile of total daily intake from diet, dust ingestion, and inhalation contributed to 11.1% (DEHP), 17.3% (DnBP), and 23.9% (DiBP) of the tolerable daily intake (TDI). It was therefore suggested that the recent exposure of an adult population to DEHP is unlikely to pose a significant health risk. However, the higher share of DiBP and DnBp in the TDI indicates that other important sources are present which should be reduced under precautionary principles [129].

The monitoring of metabolites in urine is more convenient since it avoids external contamination with parent compounds, making possible to obtain easier reliable results from biological samples [105]. Table 7 shows a summary of the literature available data on the concentrations reported for the most important phthalate and nonphthalate plasticizers as well as for some of their most important metabolites from human samples, especially urine and breast milk. Since it was evidenced that the metabolism of such plasticizers is an age-dependent process [125], it became also important to evidence their presence in diet of newborns besides other sources for infant's intake, like dust (Table 6). Therefore, breast milk analysis for such chemicals shows that they are present in this matrix, sometimes at elevated levels (Table 7). Moreover, when addressing the relationship between metabolite concentration and age of the sampled individuals, a recent study showed that two important metabolites, namely mono-carboxy-*iso*-octyl phthalate (MCOP), a metabolite of DINP, and mono-carboxy-*iso*-nonyl phthalate (MCNP), a metabolite of DIDP, proven to be found at significantly higher level in urine samples collected from children of 6–11-year-old age group compared to any other age group included in an American survey at national level including more than 2,500 individuals with age over 60 [140]. Interesting also is that the literature reported results seem to show a different profile as a function of the continent sampled, as follows: relatively higher values for DEHP were reported for the North American continent, while higher values for DnBP seem to characterize the samples collected from Asia (Table 7).

Some studies showed already that the levels of phthalate substitute's metabolites measured in human's urine are usually lower when compared to DEHP metabolites (Table 7) [105, 135]. However, such comparison should be carefully addressed since it was shown through rats exposure to such plasticizers that a considerable portion (about 50%) of the orally dosed DINP is excreted via the feces while it is known that DEHP is mainly extracted in humans via urine [105, 141].

4.2 Health Effects

Phthalates in general, as well as their substitutes, became large volume workplace chemicals and further ubiquitous environmental contaminants. Their impact on organism's health was relatively intense tested through several exposure experiments ([142, 305]). However, it seems that in general they display low toxicity [305], their effects being recorded only for high dose exposure levels

Table 7 Levels (I: median/range concentrations expressed in μg/L of sample; II: median/range creatinine adjusted concentrations expressed in μg/g) of phthalates, substitutes, and their metabolites from human fluids samples

Phthalate plasticizers

Location	Compounds							References
	DEP	DiBP	DnBP	DEHP	DcHP	BzBP	DAP	
Germany, 2007–2008 N = 78, milk (ng/g)	<3	1.2/<0.1–5.3	0.8/<0.1–7.4	3.9/<0.5–23.5	<4/<4–9.1	<5	<10/<10–20	Fromme et al. [130]
Sweden, 2001 N = 42, milk	0.2/0.2–1.5	–	1.5/1.5–20	9/0.5–305	–	0.5/0.1–4.4	–	Hogberg et al. [131]
Germany, 1997 N = 5, milk	–	–	–/10–50	–/10–110	–	–	–	Bruns-Weller and Pfordt [132]
Canada, 2003–2004 N = 86, milk	<0.2/0.2–8.1	–	0.5/0.1–11	116/1.2–2,920	–	–	–	Zhu et al. [133]
China, 2008 N = 40, milk	<0.05	–	53.5/0.6–174	–	–	–	–	Chen et al. [134]
Sweden, 2001 N = 42, blood–serum	0.24/0.07–1.1	–	0.8/0.2–9.1	0.5/0.5–129	–	0.25/0.05–1.4	–	Hogberg et al. [131]

Table 7 (continued)

Phthalate metabolites

Location	Compounds	MEP	MnBP	MiBP	MBzP	MEHP	5-oxo-MEHP	5-HO-MEHP	5-cx-MEPP	2-cx-MMHP	Oxo-MiNP	OH-MiNP	References
Germany, 2005 N = 27, urine	I	–	48/17–307	36/16–164	8/2.6–150	4/0.1–16.3	14/3.6–61.5	17.5/5–66	26/4–92.4	9/1.6–30	3/0.5–8	6/1–12	Fromme et al. [135]
	II	–	47/31–173	45/17–93	8/3–84	4/0.2–12.4	15/5.4–47	18/8–63	25/7–80	9/4.5–32	3/1–11	5/1–21	
Germany, 2005 N = 23, urine	I	–	50/25–114	47/23–120	6.4/3–38	5/2–12	16/8–27	23/10–40	24/15–54	10/4–24	3/1.4–20	5.5/2–49	Fromme et al. [135]
	II	–	41/23–87	50/21–113	5/3–21	4/1.4–7.5	12/7–20	18.5/9–24	21/13–35	8/3–12	3/1–18	5/2–37	
Denmark, 2006 N = 60, urine	I	54/6–4115	37/2–159	47/3.5–343	36/4–509	5/0.2–59	20/0.7–265	26/<LOD-424	17/1.5–126	–	2/<LOD-89	3/<LOD-142	Frederiksen et al. [118]
	II	–	–	–	–	–	–	–	–	–	–	–	
Sweden, 2001 N = 38, urine	I	35/0.5–761	46/5–198	16/0.5–130	13/2–38	9/3–57	11/0.5–83	15/1.4–126	–	–	–	–	Hogberg et al. [131]
	II	39/5.5–862	50/18–191	15/1–110	17/4.5–63	15/4.6–74	15/3–57	24/5–86	–	–	–	–	
USA, 2004–2005 N = 33, urine[a]	I	73/9.2–6892	14/1.3–53.5	4/0.7–12	9.6/2–76	3/1.5–86	12/2.6–224	18.6/4–336	27.3/4–364	–	–	–	Hines et al. [136]
	II	145/29–3156	18.3/10–41	5/2.6–14	14.4/6–70	7.6/2–47	18/7.3–122	24.5/10–183	37/17–248	–	–	–	
USA, 2004–2005 N = 30, urine[a]	I	74/13–1662	7/<LOD-156	2/0.6–19	11/1–141	<LOD/<LOD-21	8.4/1.3–78	10.9/1.4–182	24.8/3–237	–	–	–	Hines et al. [136]
	II	113/29–2444	18.2/4.3–133	3.4/1–12.2	13.4/5–63	3.6/1.3–58	13.5/4.3–45	17/4.5–81	32.5/13–135	–	–	–	
USA, 1998–2002 N = 295, urine[b]	I	386/151–1025	36/17–73	6/2.6–12.2	24/9–51	6/3–14.5	18/8–37	20/10–41	36/16–72	–	–	–	Engel et al. [137]
	II	–	–	–	–	–	–	–	–	–	–	–	
USA, 2000–2005 N = 246, urine[c]	I	35.5/7–175	35.5/7–175	10/2–36	17/1.7–147	5/0.5–47	17.5/3–108	20/2.5–150	37/6–232	–	–	–	Adibi et al. [138]
	II	–	–	–	–	–	–	–	–	–	–	–	
Israel, 2006 N = 19, urine	I	–	30.8/3.5–131	15.6/<LOD-67	5.3/0.5–71	6.8/<LOD-21	17.5/1.7–64	21.5/1.8–71	27/6–107	–	–	–	Berman et al. [139]
	II	–	46/4–90	28/<LOD-53	9.6/0.5–109	7/<LOD-41	17.5/5–129	24/5–134	41/14–193	–	–	–	
Germany, 2011 N = 74, breast milk	I	–	2.1/0.4–18	11.8/4.4–44	–	2.3/<1–27.4	<LOD	<LOD	–	–	–	–	Fromme et al. [130]
	II	–	–	–	–	–	–	–	–	–	–	–	

[a] 10th and 90th percentile

[b] 1st and 3rd quartile

[c] 5th and 95th percentile

(1,000 mg/kg bw/day) and influencing mainly the reproductive system of developing organisms. As a consequence, at European level regulatory decisions were taken for the moment by stopping the use of "classical" plasticizers such as DEHP, but also of some substitutes like DINP or DIDP on children's articles [102]. When addressing the toxicity of the other substitutes like DINCH or DEHT, it was shown that they have no effects on fertility as reported for phthalate plasticizers up to the same high exposure doses [1, 110]. As far as currently known, DEHT has no carcinogenic, genotoxic, or developmental effects (NOAEL 800–1,000 mg/kg/day for reproductive toxicity, NOEL 666 mg/kg/day for carcinogenicity) [103, 143].

Most of the effects recorded through exposure experiments for the above-mentioned plasticizers occurred at doses well above the estimated intake of the general population. However, some recent experimental results shown that biological changes may also be induced at low, human relevant doses being suggested that different active phthalates can have cumulative effects [144–146]. However, uncertainties in the epidemiological database, difficulties in animal to human extrapolations, and the lack of knowledge on the significance of low-dose effects for human health preclude a better understanding of the real risks for humans.

5 Human Exposure to Personal Care Products

5.1 Triclosan

Triclosan (TCS) is a broad spectrum antimicrobial used in personal care products (PCPs), such as toothpaste, soap, shampoo, deodorant, mouthwash, cosmetics, typically in a concentration range of 0.1–0.3%. TCS can also be found in kitchen utensils, toys, and textiles [147]. Due to the widespread use of TCS in these applications, human exposure can occur through oral or dermal contact with these PCPs or through the consumption of contaminated food or drinking water [148]. Absorption of TCS after oral administration of TCS-containing products is fairly rapid and complete [148, 149]. TCS can also reach the systemic circulation after dermal application with absorption up to 9% of the administered amount [149]. Once absorbed, TCS is readily metabolized to its glucuronide and sulfate conjugates and primarily eliminated through urinary excretion. TCS has a low potential for bioaccumulation with a calculated half-life of approximately 11 h [150].

5.1.1 Levels in Human Matrices

TCS has been identified in numerous studies in urine, serum, plasma, and human breast milk (Table 8). All levels are expressed as total TCS (unconjugated and conjugated).

Table 8 Levels of TCS and musks in human matrices

Compound	Matrix	Country	Population	Concentrations/exposure	Detection frequency	Particularities	References
TCS	Urine	USA	30 adults	Mean 127 ng/mL, median 12.5 ng/mL	80%		Ye et al. [151]
TCS	Urine	USA	90 (girls 6–8 years)	Range <1.6–956 ng/mL; GM 10.9 ng/mL; median 7.2 ng/mL	68%		Wolff et al. [152]
TCS	Urine	USA	2,517 (≥6 years)	Range <2.3–3,790 ng/mL; GM 13.0 ng/mL; median 9.2 ng/mL	75%		Calafat et al. [153]
TCS	Urine	Belgium	193 (14–16 years)	Range 0.1–706 ng/mL; GM 2.2 ng/mL; median 1.3 ng/mL	100%		Milieu en Gezondheid [154]
TCS	Urine	Spain	120 pregnant women	Median 6.1 ng/mL	60%		Casas et al. [155]
TCS	Urine	Spain	30 (boys 4 years)	Median 1.2 ng/mL	37%		Casas et al. [155]
TCS	Urine	China	287 (3–24 years)	Range < LOQ–558 μg/g creatinine, GM 3.8 ng/mL; median 3.4 ng/mL	93%		Li et al. [156]
TCS	Plasma	Sweden	51	Range 0.01–19 ng/g ww	100%	No use of TCS containing PCPs	Allmyr et al. [157]
TCS	Plasma	Sweden	18	Range 0.40–38 ng/g ww	100%	Use of TCS containing PCPs	Allmyr et al. [157]
TCS	Serum		15	Range <1.1–13.7 ng/mL	67%	Commercial serum samples	Ye et al. [158]
TCS	Serum	Belgium	7 adults	Median 0.5 ng/mL	100%		Dirtu et al. [159]
TCS	Serum	Belgium	14 pools (women 55–60 years)	Median 1.1 ng/mL	100%		Dirtu et al. [159]
TCS	Serum	Belgium	20 adolescents	Range <0.1–9.2 ng/mL	55%		Geens et al. [160]
TCS	Human milk	Sweden	51	Range <0.02–0.35 ng/g ww	43%	No use of TCS containing PCPs	Allmyr et al. [157]
TCS	Human milk	Sweden	18	Range 0.02–0.95 ng/g ww	100%	Use of TCS containing PCPs	Allmyr et al. [157]

(continued)

Table 8 (continued)

Compound	Matrix	Country	Population	Concentrations/exposure	Detection frequency	Particularities	References
TCS	Human milk	USA (California, Texas)	62	Range <5–2,100 ng/g lw	82%		Dayan [161]
TCS	Human milk	Australia	151 (16–45 years)	Range <0.02–19 ng/g ww	93%		NICNAS [162]
HHCB	Human milk	China	10	Range 12–68 ng/g lw	100%		Wang et al. [163]
HHCB	Human milk	China	100	Median 63 ng/g lw	99%		Zhang et al. [164]
HHCB	Human milk	Switzerland	54	Range 6.1–310 median 36 ng/g lw	83%		Schlumpf et al. [165]
HHCB	Human milk	South Korea	20	Range 0.06–0.52	100%		Kang et al. [166]
HHCB	Human milk	USA	31	Median 136 ng/g lw	97%		Reiner et al. [308]
AHTN	Human milk	China	10	Range 23–118 ng/g lw	100%		Wang et al. [163]
AHTN	Human milk	China	100	Median 5 ng/g lw	75%		Zhang et al. [164]
AHTN	Human milk	Switzerland	54	Range 4.8–28.8 median 10.2 ng/g lw	13%		Schlumpf et al. [165]
AHTN	Human milk	South Korea	20	Range 0.02–0.09 ng/g lw	65%		Kang et al. [166]
AHTN	Human milk	USA	31	Median 53 ng/g lw	56%		Reiner et al. (2007)
MK	Human milk	China	10	<LOQ	60%		Wang et al. [163]
MK	Human milk	China	100	Median 4 ng/g lw	60%		Zhang et al. [164]

	Matrix	Country	n	Value	%	Note	Reference
MK	Human milk	Switzerland	54	Range 0.25–12.0 median 0.64 ng/g lw	63%		Schlumpf et al. [165]
MK	Human milk	South Korea	20	Range 0.02–0.22 ng/g lw	53%		Kang et al. [166]
MK	Human milk	USA	31	Median 58 ng/g lw	85%		Reiner et al. (2007)
MX	Human milk	China	100	Median 17 ng/g lw	83%		Zhang et al. [164]
MX	Human milk	Switzerland	54	Range 0.25–31.6 median 1.34 ng/g lw	87%		Schlumpf et al. [165]
MX	Human milk	South Korea	20	Range 0.02–0.15 ng/g lw	65%		Kang et al. [166]
MX	Human milk	USA	31	Median 17 ng/g lw	36%		Reiner et al. [308]
HHCB	Plasma	China	204	Median 0.85 µg/L	98%		Hu et al. [167]
HHCB	Plasma	Austria	53	Max 6.9 µg/L	89%		Hutter et al. [168]
HHCB	Serum	Austria	114	Median 0.42 µg/L	91%		Hutter et al. [169]
HHCB	Serum	South Korea	20	Range 0.17–1.4 ng/g lw	90%	Maternal serum	Kang et al. [166]
HHCB	Serum	South Korea	20	Range 0.67–2.7 ng/g lw	70%	Cord serum	Kang et al. [166]
AHTN	Plasma	China	204	Median 0.53 µg/L	85%		Hu et al. [167]
AHTN	Plasma	Austria	53	Max 0.29 µg/L	19%		Hutter et al. [168]
AHTN	Serum	Austria	114	Median < LOD	17%		Hutter et al. [169]
AHTN	Serum	South Korea	20	Range <0.17–1.4 ng/g lw	35%	Maternal serum	Kang et al. [166]
AHTN	Serum	South Korea	20	Range <0.67–2.7 ng/g lw	15%	Cord serum	Kang et al. [166]
MK	Plasma	Austria	53	Max 0.19 µg/L	43%		Hutter et al. [168]
MK	Serum	South Korea	20	<LOQ	0%	Maternal serum	Kang et al. [166]
MK	Serum	South Korea	20	<LOQ	0%	Cord serum	Kang et al. [166]
MX	Plasma	Austria	53	Max 0.28 µg/L	62%		Hutter et al. [168]
MX	Serum	South Korea	20	Range 0.17–0.51 ng/g lw	5%	Maternal serum	Kang et al. [166]
MX	Serum	South Korea	20	<LOQ	0%	Cord serum	Kang et al. [166]

Urine: Urine is the most appropriate matrix for exposure assessment since urinary excretion is the major route of elimination with the major fraction excreted within 24 h [148, 150]. Urinary concentrations cover a wide range with a difference of approximately three orders of magnitude. The largest study was performed as part of the US National Health and Nutrition Examination Survey (NHANES) >2,500 urine samples of a population ≥6 years. Concentrations differed by age, with the highest concentrations found during the third decade of life. Furthermore, levels of TCS were also influenced by socioeconomic status with higher levels detected in people with higher household incomes. Race/ethnicity or sex had no influence on TCS levels [153]. Wolff et al. [152] determined TCS in the urine of 90 girls aged between 6 and 8 years. Mean and median values were comparable with the general population measured by Calafat et al. [153].

Similar biomonitoring studies in Belgium and China conducted with adolescents (14–16 years) and children and students (3–24 years), respectively, showed a high detection frequency of more than 90% [154, 156]. However, mean and median values were lower compared to the US data. In the Chinese study, females had a statistically higher least square geometric mean concentration than males. They also observed a decreasing tendency with age in the 7–24 age group [156].

Serum and plasma: TCS could be measured in all plasma samples from 69 nursing mothers in Sweden, consisting of two groups, one with a known use of TCS-containing PCPs and the other a control group without known usage of products containing TCS. The first group had significant higher median levels of TCS compared to the control group. Yet, the presence of TCS in all samples of the control group is an indication for the exposure to TCS through other sources or can be due to the incomplete labeling of PCPs [157].

Another study of Allmyr et al. [170] investigated the influence of age, gender, and region of residence on the concentration of TCS in pooled Australian serum samples. No influence of residence place was observed, while a small but significant influence was observed for age and gender. Higher levels were observed for males, while the 31–45-year age group had higher levels of TCS compared to the other age groups. The same increase in TCS exposure during the third decade of life was observed in the urine samples of the NHANES study in the USA [153]. The concentration of TCS in the Australian samples was approximately a factor two higher than the previously mentioned Swedish plasma samples [170].

Human milk: With a log K_{ow} of 4.76, TCS susceptible for transfer to breast milk. In the earlier-mentioned study with nursing mothers from Sweden performed by Allmyr et al. [157] the same trend was observed as for plasma. Mothers who used TCS-containing PCPs had significantly higher levels of TCS compared to milk samples from mothers who did not use TCS-containing PCPs. However, participants had significantly lower TCS levels in milk compared to their plasma levels [157]. A risk assessment of TCS in human breast milk was made by Dayan [161]. A conservative exposure of 7.4 µg/kg/day was calculated from the levels of TCS measured in 62 human milk samples (range <5–2,100 ng/g lw). Based on a calculated margin of safety (MOS) of approximately 6,760 between the exposure and NOAEL for pups, the authors conclude that the levels of TCS measured in breast milk do not pose a risk to breastfeeding infants [161].

5.1.2 Health Effects

The tolerability and safety of TCS have been evaluated in human volunteers with little indication of toxicity or sensitization. Clinical data demonstrated that TCS was not a dermal or oral mucosal irritant and was shown to have very low sensitization/allergenic potential at concentrations found in PCPs ($\leq 1\%$) [149].

Based on mammalian studies conducted in mice, rats, rabbits, and dogs, it was demonstrated that TCS has a low acute toxicity with a $LD_{50} > 1$ g/kg following oral administration [148, 150, 171]. Subacute and chronic toxicity data from mammalian species, including mice, rats, hamsters, and baboons, were primarily limited to changes in liver and kidneys. In the liver, changes were observed in liver weight, liver enzymes, liver hypertrophy, and an increase in peroxisome size and number. Renal toxicity was evidenced by inflammation and tubular regeneration [149, 150]. Based on a number of in vitro and in vivo studies using classical assay systems, TCS did not show genotoxic, mutagenic, or carcinogenic effects. Also no reproductive or developmental toxicity was observed in several mammalian studies, even not in the highest administered dose of 3,000 mg/kg [149, 150].

Concern has been raised about the potential disruption of the thyroid homeostasis. This concern is based on the structural similarity of TCS to thyroid hormones and the detected decreases in serum T_4 and total serum T_3 in rats [172–174]. However, these decreases in thyroid hormones are not observed in a study conducted in humans under real exposure conditions. During 2 weeks, 12 adults were instructed to brush their teeth with a 0.3% TCS-containing toothpaste twice a day. While a significant increase in plasma TCS levels was observed, no significant changes in thyroid hormones could be detected [175]. Other endocrine-disrupting effects have been allocated to TCS in a large number of in vitro and in vivo studies in animals, although it remains unclear whether TCS has (anti)estrogenic effects, (anti)androgenic effects, or both [148, 150, 176]. MOS calculated by Rodricks et al. [149] were based on product-based estimates and biomonitoring-based estimates. The product-based estimates revealed MOS of approximately 1,000, 730, and 630 for men, women, and children, respectively, while the MOS of biomonitoring-based estimates were 5,200, 6,700, and 11,750 for men, women, and children, respectively. Based on these estimations, Dann and Hontela [150] could conclude that hypothetical exposures to TCS in consumer products are not expected to cause adverse health effects in children or adults who use these products as intended, even in sensitive individuals.

5.2 Musks

Musks are pleasant smelling fragrances found in various personal and home care products. Since the 1950s, the usage of natural musks has declined due to the relatively easy and inexpensive production of their synthetic counterparts. There are three major types of synthetic musks, nitro musks, polycyclic musks, and macro cyclic musks. Due to the bioaccumulative properties and health adverse reactions of nitro musks [177], some compounds (musk tibetene, musk moskene, and musk ambrette) were banned during the 1990s [178, 179], while musk ketone (MK) and musk xylene (MX) could still be used in cosmetics but with restrictions [180, 181]. In 2011, the European Commission announced its decision to ban MX under the new European Chemicals Legislation – Registration, Evaluation, Authorisation and Restriction of Chemicals (REACH), bringing EU regulations in line with the global International Fragrance Association (IFRA) Standards. This ban was paralleled by an increasing use of polycyclic musks, a second group of synthetic musks which comprises several high volume use products, such as tonalide® (AHTN) and galaxolide® (HHCB). Although polycyclic musks have been tested in the past and showed no toxicological and dermatological effects [181], their high levels of use, chemical stability, and low biodegradability [182] led to their gradual replacement by a third group of fragrances, the macro cyclic musks [183]. Synthetic musks are readily released into the environment through household water and wastewater, and they are found in various human tissues (Table 8). The most likely exposure pathway is dermal exposure and absorption through the skin [184]. However, recent studies based on the excretion in urine and feces indicate that this absorption might be less than 4% of dermally applied musks [306]. Recently, the focus has shifted toward indoor air inhalation and indoor dust ingestion as synthetic musks in indoor air have been reported to be approximately tenfold higher than the concentrations detected in outdoor air [185] due to the use of musks in diverse household products, and their high particle-binding affinities [186]. Although the overall impact of synthetic musks on the environment and on humans is currently unknown, musks are an active area of international research.

5.2.1 Levels in Human Matrices

Synthetic musks have been detected in human tissues (Table 8) due to their lipophilic nature and their low biodegradability. The occurrence of these fragrance-related chemicals is subjected to a variable pattern with substantial interindividual differences, opposed to other environmental contaminants such as polychlorinated biphenyls (PCBs) or pesticides [165].

Human milk: In general, the existing literature reports that levels of polycyclic musks in European human milk are about one order of magnitude higher than those of nitro musks ([307, 165]). An identical trend is observed for US milk samples [308]. Overall, 1,3,4,6,7,8-hexahydro-4,6,6,7,8,8-hexamethylcyclopenta-c-2-benzopyran (HHCB) shows the highest median concentration in human milk, followed by 7-acetyl-1,1,3,4,4,6-hexamethyl-1,2,3,4-tetrahydronaphthalene (AHTN) and MX. Concentrations of MK were lowest and often below the quantification limit (Table 8). This profile corresponds with the greater production and usage of HHCB, compared to other synthetic musks [187]. Lignell et al. (2008) suggested a significant decline in AHTN and MX concentrations in breast milk between 1996 and 2003, but no temporal trend was detected for HHCB.

A downward trend regarding MX is likely, since the industry has voluntarily replaced the nitro musks with polycyclic musks [180]. This decrease has also been seen in human serum studies (Table 8). Kang et al. [166] observed lower concentrations in breast milk than in serum, and a weak correlation between the levels in umbilical serum, maternal serum, and breast milk. However, women with a high use of perfume during pregnancy had elevated milk concentrations of HHCB, and elevated concentrations of AHTN were observed among women reporting use of perfumed laundry detergent, which strongly suggests that perfumed products are important sources of musk exposure both among the mothers and the nursed infants [307].

Serum and plasma: Serum and plasma samples partially mimic the trend observed in human milk samples. Overall, polycyclic musks can be detected more frequently and at higher concentrations compared to the nitro musk compounds (Table 8). HHCB is by far the most common of the polycyclic musks as production and use of this compound increased at the same time as production and use of nitro musks decreased [168]. However, MX still seems to be a common contaminant in human serum, with a considerably high detection frequency compared to AHTN. A high percentage of the population is still exposed to nitro musk compounds [168], although a moderate decline in MX and a strong decline in MK are observed.

The change of MX in the European countries in blood samples had been studied intensively during the 1990s. A decreasing trend of MX was detected in serum from 1994 [188] to 1997 [189]. Kafferlein and Angerer [309] reported a remarkable decrease in MX concentration by comparing levels between 1992/1993 and 1998. The change was consistent with the decreased use of MX in European countries. Similarly, these low concentrations of MX and MK have been reported in Asian studies, finding which is in accordance with the regulations presented in the 10th Five-Year Plan of China Light industry in 2001 restricting the use of nitro musk in fragrances [167]. Considering the trend of increasing fragrances' use (perfumes, deodorants, lotions, creams, hair products, and soaps), it can be assumed that the exposure to products containing nitro musks will persist.

Significant associations between HHCB and MX concentrations in serum and the frequent use of perfumes, deodorants, and shampoos have been detected. Hutter et al. [168] reported higher plasma concentrations in older persons. This finding could be due to the higher use of lotions and crèmes on face and hands and a more frequent use of skin care products. In addition, physiological aging-related changes might be responsible for higher dermal absorption of synthetic musks.

5.2.2 Health Effects

The health effects are discussed for the most relevant synthetic musks present in human tissues (MK, MX, HHCB, and AHTN). It should be stressed that all synthetic musks undergo metabolism and health effects are thus influenced [190–193].

MX and MK are not acutely toxic, but they do cause an increased absolute and relative liver weight [194]. Mouse and rat studies have confirmed that MX and MK are nongenotoxic substance but due to their enzyme-inducing properties, they can exhibit cogenotoxic activity. Especially the significant induction of liver enzymes (cytochromes P450 1A1, 1A2 and 2B and cytochrome b5) might be responsible for the carcinogenic properties of the substance seen in mouse liver [195]. MK and MX are quite similar with respect to enzyme induction properties, but MX is the more potent one. Given its resemblance to phenobarbital, MX is classified as a carcinogen category 3 (R40), although it is a borderline case. There are no carcinogenicity data for MK, but as MK is quite comparable to MX with respect to physicochemical and toxicokinetic properties, and in particular both MK and MX are phenobarbital-like inducers of liver enzymes in both rats and mice [196], there is a concern that MK may be hepatocarcinogenic in mice as well [190, 191]. In mice, MK also caused histological changes in the liver. In a dermal toxicity study with rats, effects included a decreased body weight gain without a decrease in food consumption, decreases in red blood cell parameters, and an increase in absolute and relative liver weight without a histopathological correlation [191].

Very few toxic reports have been made concerning HHCB and AHTN. They are not acutely toxic but AHTN is listed as a photosensitizer. This effect was only in animal studies but the negative human data do not overrule the positive animal data. Photosensitivity of AHTN might be due to AHTN itself, or to sensitizing effects from photodegradation products arising from the interaction of AHTN and UV light [193]. Some dermal subchronic studies revealed slight liver weight increases and body weight decreases due to HHCB exposure, but the magnitude of these effects was not reported and they were not confirmed by other studies. There are no carcinogenicity data available for both compounds. HHCB and AHTN are not genotoxic [197]. They do have a very weak estrogenic potency in vitro but such effects were not seen in vivo [198]. In an oral developmental study, there were signs of maternal toxicity for HHCB with an increased incidence of skeletal malformations and decreased ossification in fetuses at the highest dose of 500 mg/kg bw/day [192]. AHTN exposure has led to clear mild hematological effects with a dark discoloration of the liver and mesenteric lymph nodes, but all effects were reversible.

6 Human Exposure to Bisphenol-A

Bisphenol-A (BPA) or 2,2-bis(4-hydroxyphenyl)propane is one of the highest volume chemicals produced worldwide, with a production volume of approximately 3.8 million tons in 2006 [199]. BPA is mainly used as monomer in the production of polycarbonate and epoxy resins. Polycarbonate is a transparent polymer with a high impact resistance which can be used in different consumer goods such as reusable drinking bottles and food containers. Epoxy resins on the other hand are used as inner coating of food and drink cans. Release and exposure of

BPA can occur due to an incomplete polymerization, followed by migration into the food or environment [200]. Furthermore, BPA monomer is used as color developer in thermal paper and is consequently also present in recycled paper and paperboard which can be used as food packaging material [201]. Human exposure to BPA is believed to occur mainly through contaminated food, although recently dermal exposure and absorption have gained attention [202–204].

After oral administration, BPA is rapidly and efficiently absorbed from the gastrointestinal tract and undergoes extensive first-pass metabolism in the gut wall and liver. BPA is mainly transformed into BPA-glucuronide and to a lesser extent into BPA-sulfate. These highly water-soluble metabolites are rapidly cleared from blood by the kidneys and excreted with urine [205]. Völkel et al. determined that the terminal half-life of BPA is less than 6 h in humans [206].

6.1 Levels in Human Matrices

Due to the fast excretion of BPA in urine, this is the most appropriate matrix for biomonitoring and subsequent exposure assessment to BPA. Therefore, only urine will be discussed further. In Table 9, only studies with an adequate number of samples are included.

Two recent large-scale studies were performed in the USA and Canada including 2,514 and 5,476 participants, respectively. Exposure to BPA was ubiquitous with a detection frequency of more than 90% in both studies. Highest urinary concentrations were detected in adolescents (12–19 years) followed by children (6–11 years) and adults (>19 years), and unadjusted BPA concentrations were higher among men than women. After adjusting BPA levels for creatinine concentration, children had the highest BPA concentrations, followed by adolescents and adults. Women had higher adjusted urine concentrations than men. The difference between adults and children can be due to a difference in exposure sources or due to a difference in metabolism and excretion. Despite the similarity between the different age categories in both the Canadian and American study, it is interesting to note that for each age group, concentrations found among Canadians were about half of those found in Americans [207, 209]. Concentrations of 193 Belgian adolescents were between those for Canadian and American adolescents [154].

Several studies were conducted in Asian countries. He et al. [214] determined BPA in 922 individuals. BPA levels were influenced by gender and smoking status which resulted in higher unadjusted concentrations for the male population and smokers. The geometric mean of 0.87 ng/mL and a detection frequency of 50% in this study are much lower than most other studies. The authors suggest this may be due to a different usage of BPA and BPA-contained productions between China and developed countries which results in varied exposure routines. Therefore, they believe that the Chinese population might have more chances to be exposed to BPA by inhalation, but fewer chances by ingestion than people in the developed countries [214]. However, other recent studies in different Asian countries resulted

Table 9 Concentrations of BPA in humans

Compound	Matrix	Country	Population	Concentrations/exposure	Detection frequency	References
BPA	Urine	USA	2,514 (≥6 to ≥60 years)	GM 2.6 ng/mL (2.6 µg/g cr)	93%	Calafat et al. [207]
			314 (6–11 years)	GM 3.6 ng/mL (4.3 µg/g cr)		
			713 (12–19 years)	GM 3.7 ng/mL (2.8 µg/g cr)		
			950 (20–59 years)	GM 2.6 ng/mL (2.4 µg/g cr)		
			537 (≥60 years)	GM 1.9 ng/mL (2.3 µg/g cr)		
BPA	Urine	USA	394 adults	GM 1.33 ng/mL (1.36 µg/g cr)	95%	Calafat et al. [208]
BPA	Urine	Canada	5,476 (6–79 years)	GM 1.16 ng/mL (1.40 µg/g cr)	91%	Bushnik et al. [209]
			6–11 years	GM 1.30 ng/mL (2.00 µg/g cr)	93%	
			12–19 years	GM 1.50 ng/mL (1.31 µg/g cr)	94%	
			20–39 years	GM 1.33 ng/mL (1.49 µg/g cr)	91%	
			40–59 years	GM 1.04 ng/mL (1.33 µg/g cr)	88%	
			60–79 years	GM 0.90 ng/mL (1.26 µg/g cr)	88%	
BPA	Urine	Germany	599 (3–14 years)	GM 2.66 ng/mL; median 2.74 ng/mL	99%	Becker et al. [210]
			137 (3–5 years)	GM 3.55 ng/mL; median 3.53 ng/mL	99%	
			145 (6–8 years)	GM 2.72 ng/mL; median 2.81 ng/mL	99%	
			149 (9–11 years)	GM 2.22 ng/mL; median 2.13 ng/mL	99%	
			168 (12–14 years)	GM 2.42 ng/mL; median 2.60 ng/mL	98%	
BPA	Urine	Germany	147	<0.3–9.3 ng/mL		Völkel et al. [211]
BPA	Urine	Belgium	193 (14–16 years)	0.1–53.4 ng/mL (0.18–32.4 µg/g cr)	99%	Milieu en Gezondheid [154]
				2.22 ng/mL (1.66 µg/g cr)		
BPA	Urine[a]	Italy	715 (20–74 years)	GM 3.59 ng/mL		Galloway et al. [212]
			111 (20–40 years)	GM 4.31 ng/mL, median 4.4 ng/mL		
			157 (41–65 years)	GM 3.95 ng/mL, median 3.7 ng/mL		
			452 (66–74 years)	GM 3.32 ng/mL, median 3.2 ng/mL		
BPA	Urine	Korea	516	Mean 2.74 ng/mL, median 0.64 ng/mL	76%	Hong et al. [213]
BPA	Urine	China	419 males	GM 1.41 ng/mL (0.72 µg/g cr)	58%	He et al. [214]
			503 females	GM 0.58 ng/mL (0.23 µg/g cr)	44%	

	Matrix	Country	Sample	Concentration	Detection	Reference
BPA	Urine	China	287 (3–24 years)	GM: 3 ng/mL (2.75 µg/g cr) 0.41–198.05 µg/g cr	100%	Li et al. [156]
BPA	Urine	China	116	GM 1.10 ng/mL (1.03 µg/g cr)	90%	Zhang et al. [215]
BPA	Urine	Vietnam	30	GM 1.42 ng/mL (1.27 µg/g cr)	100%	Zhang et al. [215]
BPA	Urine	Malaysia	29	GM 1.00 ng/mL (1.93 µg/g cr)	97%	Zhang et al. [215]
BPA	Urine	India	21	GM 1.59 ng/mL (2.51 µg/g cr)	100%	Zhang et al. [215]
BPA	Urine	Kuwait	32	GM 1.24 ng/mL (1.09 µg/g cr)	81%	Zhang et al. [215]
BPA	Urine	Japan	36	GM 0.84 ng/mL (0.67 µg/g cr)	100%	Zhang et al. [215]
BPA	Urine	Korea	32	GM 2.00 ng/mL (2.53 µg/g cr)	97%	Zhang et al. [215]
BPA	Urine	Germany	91 samples from 47 infants (1–5 months)	<0.45–17.85 ng/mL	42%	Völkel et al. [216]
BPA	Urine	Spain	30 (boys 4 years)	Median 4.2 ng/mL	97%	Casas et al. [155]
BPA	Urine	USA	90 (girls 6–8 years)	GM 2.0 ng/mL (3.0 µg/g cr) median 1.8 ng/mL <0.3–54.3 ng/mL	94.4%	Wolff et al. [152]
BPA	Urine	USA	195 samples from 35 children (6–10 years)	GM 3.4 ng/mL (3.4 µg/g cr) median 3.6 ng/mL (3.5 µg/g cr) <0.36–40 ng/mL (0.2–36.3 µg/g cr)	95%	Teitelbaum et al. [217]
BPA	Urine[b]	USA	81 (23–64 months)	GM 4.8 ng/mL (6.6 µg/g cr) 0.4–211 ng/mL (0.5–334 µg/g cr)	100%	Morgan et al. [218]
BPA	Urine	USA	404 pregnant women	Median 1.3 ng/mL <0.36–35.2 ng/mL	91%	Wolff et al. [219]
BPA	Urine	The Netherlands	100 pregnant women	GM 1.5 ng/mL (1.7 µg/g cr) median 1.2 ng/mL (1.6 µg/g cr) <0.26–46 ng/mL (0.1–22.7 µg/g cr)	82%	Ye et al. [220]
BPA	Urine	Spain	120 pregnant women	Median 2.2 ng/mL	91%	Casas et al. [155]

[a]24 h Urine samples
[b]48 h Sampling period
[c]3rd Trimester of pregnancy

in detection frequencies ranging from 76% to 100% [156, 164, 213, 215]. Zhang et al. analyzed the urine from 296 individuals in seven Asian countries. Overall, a detection frequency of 94% was obtained with a geometric mean concentration of 1.20 ng/mL. Highest concentration of BPA was found in samples from Kuwait followed by Korea, India, Vietnam, China, Malaysia, and Japan. As for other studies, the highest median concentration was observed for the age group ≤ 19 years, while in this study, no gender difference was observed [164, 215].

Different studies have focused on the levels of BPA in infants and young children. Völkel et al. [216] found BPA in the urine of 42% (LOQ 0.45 ng/mL) of 91 samples from 47 infants, aged between 1 and 5 months. The highest concentration observed (17.9 ng/mL) was 18-fold below the TDI of 50 µg/kg body weight per day. Infants who were fed using baby bottles showed approximately twofold higher median levels of total BPA. Important to note is that the sampling of the infants was performed before the ban of polycarbonate baby bottles in the European Union was declared (EU Directive 2011/8/EU).

In a study of Casas et al. [155], 120 pregnant women and 30 of 4-year-old boys were involved. The median urine concentration of the 4-year-old boys (4.2 ng/mL) was approximately double of the median concentration of the pregnant women (2.2 ng/mL). Other different studies conducted in the USA have focused on children exposure [152, 217, 218]. In all three studies, detection frequency was more than 94% which demonstrates the ubiquitous exposure of the complete population.

In Italy, the InCHIANTI study tries to identify the risk factors for mid- and late-life morbidity. As part of this study, BPA was determined in 24 h urine samples of 715 adults aged 20–74 years. A geometric mean of 3.59 ng/mL was detected while a geometric mean excretion of 5.63 µg BPA/day was obtained. Lower excretion rates were detected with advancing age, while higher excretion was seen among men and in those with increasing waist circumference and weight. No associations were obtained between daily BPA excretion and years of education or smoking status [212].

Mahalingaiah et al. [221] investigated the temporal variability and predictors of urinary BPA concentrations in 45 women and 37 men. In total 217 urine samples were collected with a median concentration of 1.30 ng/mL. Age, body mass index, and sex were not significant predictors of urinary BPA concentrations. BPA urinary concentrations among pregnant women were higher (not significantly) than those among the same women when not pregnant. In the study population, 24 women and men were partners. The urinary BPA concentrations of the female and male partner on the same day were correlated [221]. Several other studies determined BPA concentrations in urine of pregnant women [155, 219, 220]. Levels were not remarkably different from other populations. In the study of Wolff et al. [219], urinary concentrations of pregnant women in their third trimester were associated (not significant) with offspring birth weight.

Overall it can be concluded that BPA exposure is a global concern and that almost the entire population is exposed. In some studies, a gender difference was observed. When such a difference was seen, men usually had higher unadjusted BPA concentrations than female while female had higher creatinine adjusted BPA concentrations than men. All age categories are affected, where children and adolescent had higher levels than adults.

6.2   Health Effects

Even though the toxicity of BPA has been extensively investigated compared to many other compounds, there is still no consensus regarding at what exposure levels BPA poses a health risk [222]. Already for a long time both the in vitro and in vivo estrogenic activity of BPA is known [223]. The current TDI of 0.05 mg/kg body weight per day, derived by the European Food Safety Authority (EFSA), is mainly based on body weight changes and changes in livers and kidneys weights in two- and three-generation studies in mice and rats [224]. BPA is not genotoxic and not carcinogenic, and the guideline conforming repeated dose toxicity studies, including studies on reproductive and developmental toxicity covering a wide dose range, showed adverse effects only at doses >50 mg/kg bw/day [205]. However, more than 150 studies described effects of BPA at doses lower than 50 mg/kg bw/day. These effects included altered development of the male and female reproductive tracts, organization of sexually dimorphic circuits in the hypothalamus, onset of estrus cyclicity and earlier puberty, altered body weight, altered organization of the mammary gland, and cancers of the mammary gland and prostate [225, 226]. The relevance of the low dose effects of BPA in animals for human hazard assessment remains unclear. However, it has been suggested that the exposure to xenoestrogens such as BPA during early development may be a major contributing factor to the increased incidence of infertility, obesity, genital tract abnormalities, and prostate and breast cancer observed in European and US human populations over the last 50 years [226].

The biochemical mechanism through which BPA acts as endocrine disruptor remains not completely clear. In biochemical assays, BPA binds to both ERα and ERβ, with an approximately ten times higher affinity to ERβ; however, the affinity of BPA is still 10,000-fold weaker than that of estradiol [223]. Recent studies have revealed a variety of pathways through which BPA stimulates cellular responses at very low concentrations, below the levels where BPA is expected to bind to the classical nuclear ERs [227]. Several membrane steroid receptors have been described, including a membrane bound form of ERα and a transmembrane ER called G protein-coupled receptor [223]. Moreover, BPA interacts with the androgen receptor, thyroid hormone receptor, pregnane X-receptor, and estrogen-related receptor-γ (ERR γ), which may play a role in differentiation and maturation of the fetal brain [222, 224]. However, effects of BPA mediated by binding to androgen and thyroid hormone receptors appear to require higher doses than those required to elicit estrogenic or antiestrogenic responses [225].

Despite the uncertainty at which concentration BPA shows toxic effects, it is important to conclude that current human exposure, based on biomonitoring studies, is still far below the current TDI for the general population [224].

7 Human Exposure to Perchlorate and Cyclic Siloxanes

7.1 Perchlorate

Perchlorate is produced and used in the form of salts of ammonium, sodium, and potassium as an oxidizer in solid rocket propellants, missiles, flares, and fireworks [228]. Perchlorate salts are also used as components or additives in nuclear reactors, vehicle airbag inflators, electronic tubes, lubricating oils, leather tanning and finishing processes, electroplating and aluminum refining, and paint and enamel production [228]. Large-scale production of perchlorate began in the 1940s. Natural sources of perchlorate have also been reported, such as in sodium nitrate deposits in Chile [229]. In addition, it is thought that atmospheric processes can produce perchlorate [230]. Perchlorate is readily soluble in water (solubility in the range of tens to hundreds of gram per liter), mobile in aquatic systems, and does persist for many decades under typical groundwater and surface water conditions [228, 231].

Perchlorate is incorporated into foodstuffs such as cow milk and leafy vegetables, through food chain transfer and accumulation [232–234]. Drinking of water is a route of exposure of humans to perchlorate. Since 1997, when it was found in groundwater and in some surface waters across the USA, perchlorate has been listed as a contaminant in drinking water monitoring programs [235]. The widespread occurrence of perchlorate in drinking water in the USA [236] led to concerns regarding perchlorate-induced iodine deficiency, and thereby thyroid hormone level alterations [237]. Thyroid hormone (viz., T3 and T4) deficiency during pregnancy along with low iodine intake can adversely affect neurodevelopmental outcome in the fetus [238, 239]. In 2005, the United States Environmental Protection Agency (EPA) established a reference dose for perchlorate of 0.7 µg/kg/day [240].

Exposure of humans to perchlorate via foodstuffs and drinking water has been documented [241]. Urine, breast milk, amniotic fluid, saliva, and blood have been used as matrices in biomonitoring of human exposures to perchlorate [233, 242–253] (Table 10). Assessment of human exposures to perchlorate is important, since this compound blocks iodine uptake in the thyroid gland, which can lead to a decrease in the production of thyroid hormones (T3 and T4) essential for neurodevelopment [260].

Table 10 Literature survey on levels (I: mean or median/range concentrations expressed in μg/L of sample; II: mean or median/range creatinine adjusted concentrations expressed in μg/g) of perchlorate in various human body fluids

Compound	Matrix	Country	Population	Concentrations/exposure	Detection frequency	Particularities of the study	References
Perchlorate	Urine	Chile	30	II: Median 180 μg/g cr		Collected from children in Chile	Gibbs et al. [254]
Perchlorate	Urine	Chile	297	Median 35 μg/L	–	3 Cities in Northern Chile	Tellez et al. [255]
Perchlorate	Urine	USA	61	I: Median 3.2 μg/L; II: Median 7.8 μg/g		Living in Georgia, USA	Valentin-Blasini et al. [256]
Perchlorate	Urine	USA	2,820 2,818	I: GM 3.54 / 3.29–3.81 μg/L; II: GM 3.56/ 3.34–3.80 μg/g cr	100%	30 Locations throughout the USA	Blount et al. [242]
Perchlorate	Urine	USA	13	Median 5.5 μg/L		Southern California	Braverman et al. [257]
Perchlorate	Maternal urine	USA	34	a0.900 μg/L; b7.71 μg/L	100%		Blount et al. [248]
Perchlorate	Urine	USA	273	I: Median 2.14 μg/L; II: Median 2.19 μg/g cr	100%	An urban community in New Jersey	Borjian et al. [253]
Perchlorate	Breast milk	USA	36	Mean 10.5/0.60–92.2 μg/L		18 Different states	Kirk et al. [233]
Perchlorate	Breast milk	USA	110	Median 3.8/0.5–40 μg/L			Kirk et al. [244]
Perchlorate	Breast milk	USA	49	Median 9.1/1.3–410 μg/L		Boston area	Pearce et al. [245]
Perchlorate	Breast milk	USA	457	Median 7.3/0.01–48 μg/L		Donors were 24–34 years old	Dasgupta et al. [258]
Perchlorate	Breast milk	USA	276	Median 4.38 μg/L/ 0.30–99.5 μg/L	100%	Urban community in New Jersey	Borjian et al. [253]
Perchlorate	Amniotic fluid	USA	48	GM 0.18/0.057–0.71 μg/L			Blount and Valentin-Blasini [259]
Perchlorate	Fetal Amniotic Fluid	USA	130	a0.056 μg/L b0.380 μg/L	97%		Blount et al. [248]
Perchlorate	Saliva	India	74	Median 0.91/0.2–4.7 μg/L	94%	Adult volunteers in several cities	Kannan et al. [251]

(continued)

Table 10 (continued)

Compound	Matrix	Country	Population	Concentrations/ exposure	Detection frequency	Particularities of the study	References
Perchlorate	Saliva	USA	85	Mean 5.3/ 0.4–37 µg/L	99%	Albany, New York	Oldi and Kannan [250]
Perchlorate	Serum	Israel	1,156	[c]Mean 1.97 µg/L (exposed) [d]0.44 µg/L (control)		Newborns who were born in hospitals and were screened	Amitai et al. [246]
Perchlorate	Cord blood serum	USA	126	[a]0.050 µg/L [b]0.480 µg/L	67%		Blount et al. [248]
Perchlorate	Maternal blood serum	USA	132	[a]0.050 µg/L [b]0.893 µg/L	94%		Blount et al. [248]
Perchlorate	Serum and plasma	USA	82	Mean 0.20/0.05–7.7 µg/L	99%	Adults from 3 cities	Oldi and Kannan [249]
Perchlorate	Whole blood	China	131	Mean 2.68/0.51–10.5 µg/L	100%	Infants, children, and adults	Zhang et al. [252]

GM Geometric mean

[a]5th Percentile

[b]95th Percentile

[c]Exposed populations

[d]Control populations

7.1.1 Levels in Human Matrices

Urine: Urine is the principal route by which nonlactating humans excrete perchlorate [261, 262]. Urinary perchlorate provides a reasonable measure of human exposure because 70–95% of perchlorate dose is excreted unchanged in the urine with a half-life of ~8 h [261–263]. Creatinine (CR) adjustment is typically used to minimize the effects of variation of analyte concentration in urine either among samples produced by different individuals or among samples produced by the same individual.

Blount et al. [242] determined perchlorate in urine samples collected from a representative sample of 2,820 persons, aged 6 years and older, as part of the 2001–2002 National Health and Nutrition Examination Survey (NHANES). Perchlorate in all 2,820 urine samples tested from NHANES 2001–2002, with levels ranging from 0.19 to 160 µg/L (mean 3.54 µg/L). Children had higher median urinary perchlorate levels (5.2 µg/L; 5.79 µg/g creatinine) compared with adults (3.5 µg/L; 3.25 µg/g creatinine). Significantly higher levels of urinary perchlorate were found in populations in northern Chile consuming tap water with perchlorate levels as high as 114 µg/L [255]. As expected, urinary perchlorate levels in these highly exposed Chilean populations (median 35 µg/L) were significantly higher than the levels found in this study. Estimation of perchlorate dose in adults revealed a median of 0.066 µg/kg/day and a 95th percentile of 0.234 µg/kg/day. These estimated perchlorate dose levels are lower than the current EPA reference dose of 0.7 µg/kg/day.

Several small studies have also found measurable perchlorate levels in human urine. For 61 adults living in Georgia, USA, all urine samples contained measurable levels of perchlorate, with a median of 3.2 µg/L and a log-normal distribution [256]. Similar background levels of perchlorate (median 5.5 µg/L) were detected in urine from 13 subjects in a Southern California study [257]. Similarly, detectable levels of perchlorate were also found in all urine samples provided ($n = 273$) in an urban community in New Jersey [253]. The range, mean ± SD, and median for all urine perchlorate samples were 0.18–18.3 µg/L, 3.19 ± 3.64 µg/L, and 2.14 µg/L, respectively.

Breast milk: During lactation human mammary tissue expresses the sodium iodide symporter [260], and thus significant transfer of perchlorate into human milk is likely. The presence of micrograms per liter concentrations of perchlorate in milk collected from US women [233] confirms lactation as a relevant perchlorate excretion path. If lactating women are secreting perchlorate in milk, then urine-based estimates of total perchlorate exposure for these individuals are likely to be lower than actual [242].

Borjian et al. [253] determined perchlorate in breast milk samples provided from 276 lactating mothers. The range, mean ± SD, and median for all breast milk perchlorate samples were 0.30–99.5 ng/mL, 6.80 ± 8.76 ng/mL, and 4.38 ng/mL, respectively, and the levels are comparable to perchlorate levels detected in breast milk in other studies. In a study by Pearce et al. [245] involving lactating Boston area women, perchlorate was detected in all breast milk samples ($n = 49$) and levels ranged from 1.3 to 411 ng/mL (mean 33 ± 77). Kirk et al. [233] obtained 36 breast milk samples from women in 18 different states. Perchlorate was detected in all 36 samples, ranging from 0.60 to 92.2 ng/mL (mean 10.5). These results included three samples from New Jersey, with levels of 50.7 ± 2.2, 92.2 ± 5.8, and 3.2 ± 0.3 ng/mL. In a smaller study of ten lactating women, perchlorate levels ranged from 0.5 to 39.5 ng/mL (mean 5.8 ± 6.2) [244]. Similarly, detectable levels of perchlorate were also found in 457 breast milk samples [258]. The range, mean ± SD, and median for all breast milk perchlorate samples were 0.01–48 ng/mL, 9.3 ± 7.5 ng/mL, and 7.3 ng/mL, respectively.

The range of perchlorate concentrations in breast milk in this study and the aforementioned studies varies greatly in range. It cannot be presumed that during lactation the mother's diet remains consistent. Intake changes can result from cultural beliefs, recommendations from the subject's doctor, and the mother's change in appetite and energy needs due to the changing needs of the breastfed infant [242, 264]. Detectable levels of perchlorate were found in all urine and breast milk samples collected, which indicates that the general population including infants is exposed to perchlorate and that breast milk can be an exposure route.

Amniotic fluid: Evaluation of the health effects of in utero perchlorate exposure requires effective assessment of fetal exposure. In utero exposure is commonly estimated based on maternal diet or toxicant levels in fluids (e.g., umbilical cord blood) collected at birth. Alternatively amniotic fluid (AF) is sampled during amniocentesis procedures in the second and third trimesters, with the residual AF from this procedure providing a matrix for characterizing fetal exposure. Amniotic fluid is a complex and dynamic milieu that changes as pregnancy progresses. During the first 20 weeks of gestation, AF composition is similar to fetal plasma. After 20 weeks of gestation, fetal kidneys produce enough urine to contribute significantly to the composition of AF [265].

Blount and Valentin-Blasini [259] detected perchlorate in all amniotic fluid samples ($n = 48$) tested, ranging from 0.057 to 0.71 µg/L with a geometric mean of 0.18 µg/L. No comparison data for perchlorate in AF were available in the scientific literature. The perchlorate levels previously reported for human urine and milk are an order of magnitude higher than the levels found in this group of 48 AF samples [233, 256]. Lower levels of perchlorate in human AF compared with human milk could result from low NIS expression in the placenta compared to the lactating breast [265].

Saliva: The use of saliva as a diagnostic fluid has been studied for many years [266]. While the ease and noninvasiveness with which a sample can be obtained make this matrix attractive to the medical community, the use of saliva to detect exposures of persons to environmental contaminants has not been investigated in many studies. However, it has been established that the measurement of cotinine, an indicator of exposure to environmental tobacco smoke, in saliva is correlated with concentrations of cotinine in serum [267].

Perchlorate was found in saliva samples of Indian adults at concentrations above 0.2 µg/L [251]. Perchlorate concentrations as high as 4.7 µg/L were found in saliva, with an average value of 1.3 ± 1.3 μg/L (median: 0.91 μg/L). When saliva samples were stratified by city, gender, or age, no notable differences in perchlorate concentrations were found among the groups. A recent study reported perchlorate concentrations in saliva samples collected from subjects in Albany, New York [250]. The mean concentration of perchlorate in the saliva samples from Albany was 5.3 μg/L, and the maximum value was 37 μg/L. The mean concentration of perchlorate in saliva samples from the Indian donors was approximately fivefold lower than the concentrations reported for Albany.

Blood: Very few studies have yet reported perchlorate levels in human blood. Zhang et al. [252] determined perchlorate in 131 blood samples from Nanchang, China. The mean concentration of perchlorate (4.07 ng/mL) in blood from infants was twofold greater than the concentration in adults. Oldi and Kannan [249] detected perchlorate in 82 human serum and plasma samples from adults in the USA and found a mean perchlorate concentration of 0.20 ng/mL. Blount et al. [248] reported a 50th percentile value of 0.22 ng/mL for perchlorate in 132 maternal serum samples collected in the USA. Perchlorate levels in human serum from Israel were reported to be 5.99, 1.19, and 0.44 ng/mL for high, medium, and low exposure groups, respectively [246]. The mean perchlorate concentration in the blood samples from Nanchang adults was similar to that reported for Israel, but was tenfold greater than the concentration previously reported for adult human serum from the USA. The sampling site (Jiangxi Province) has the second largest fireworks manufacturing operations in China. The manufacture and exhibition of fireworks in China are notable sources of perchlorate in the environment. A recent study found high levels of perchlorate in drinking water from Nanchang [268].

There is relatively little information in the scientific literature assessing concurrent exposure to chemicals that can affect the thyroid gland, but the results of recent reviews and scientific studies suggest that exposure to nitrate and thiocyanate from drinking water or food accounts for a more significant proportion of iodine uptake inhibition in comparison to perchlorate [234, 269]. As information accumulates to assess human exposure to perchlorate and other chemicals with similar mechanisms of toxicity, these types of data will be useful for scientists in assessing cumulative risks associated with environmental chemicals that can have adverse effects on human thyroid function.

7.1.2 Health Effects

Perchlorate pharmacology and effects on the thyroid gland: Perchlorate is one of the several environmental chemicals capable of affecting the thyroid gland in humans. When thyroid hormone function is abnormally high, this can result in the clinical diagnosis of hyperthyroidism. When thyroid hormone function is abnormally low, this can result in the clinical diagnosis of hypothyroidism. The developing fetus, neonate, and pregnant women are considered to be sensitive populations at higher risk of adverse outcomes from deficient thyroid gland function (hypothyroidism). Perchlorate continues to have important medical uses for treating hyperthyroidism caused by the antiarrhythmic drug amiodarone [260, 270]. Agranulocytosis is a rare and dose-dependent side effect that has been reported to occur with other pharmaceutical drugs (in addition to perchlorate) used in treating hyperthyroidism [271]. Perchlorate has a relatively short half-life in serum (6–8 h), and is rapidly eliminated from the body by urinary excretion [263, 265].

Epidemiological studies of human exposure to perchlorate (drinking water): There have been several epidemiological studies that have assessed whether there are adverse effects on thyroid function in populations living in communities where perchlorate has been detected in drinking water. These studies have included assessments of neonates, children, and adults. The majority of these epidemiological studies have not found evidence of an association between perchlorate in drinking water and adverse effects on thyroid function. [238, 239, 272, 273]. To strengthen the validity of exposure assumptions in these epidemiological studies, it would be helpful to confirm exposure at the individual level by using biomarkers of exposure (which reflect internal dose). These types of measurements were conducted in the epidemiological study in Chile, and the results confirmed that higher levels of perchlorate were found in urine samples obtained from school-age children living in Taltal, where drinking water had the highest levels of perchlorate [254].

Epidemiological studies of human exposure to perchlorate (occupational studies): Occupational exposure to perchlorate has been studied to assess the possibility of adverse effects on thyroid function among workers with high levels of inhalation exposure. One study assessed a cohort of workers with long-term exposure to airborne ammonium perchlorate dust (at an average single-shift dose of 36 µg/kg), and no significant changes or adverse effects on thyroid hormone or functions were observed [274]. A more recent publication studied workers with long-term, intermittent occupational exposure to perchlorate at a manufacturing facility [275]. Transient inhibitory effects on the uptake of iodide by the thyroid gland were observed among these workers, but there was no evidence of adverse effects on TSH or other thyroid function parameters in comparison to control subjects without a history of occupational exposure. The doses described in these occupational studies are generally much higher than doses that might be encountered from dietary or drinking water exposure, and several orders of magnitude lower than doses that have been used in the pharmacological treatment of hyperthyroidism.

Studies of low-dose perchlorate exposure in healthy human subjects: A small number of studies have been published investigating the effects of low doses of perchlorate in thyroid function in healthy adults (without thyroid disease). One study was conducted in healthy male volunteers, involving the administration of 10 mg of perchlorate in drinking water for 14 days. A significant decrease in the uptake of iodine by the thyroid was observed at this dose, but there was no evidence of adverse effects on thyroid hormones or TSH concentrations [262]. Another recent study was conducted in healthy adults to determine the highest dose of perchlorate at which there is no effect on the uptake of iodine by the thyroid gland [263].

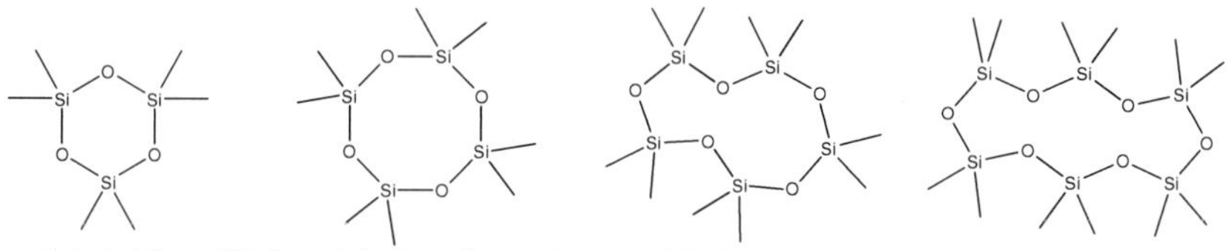

Hexamethylcyclotrisiloxane (D3) Octamethylcyclotetrasiloxane (D4) Decamethylcyclopentasiloxane (D5) Dodecamethylcyclohexasiloxane (D6)

Fig. 1 The structures of the most used cyclic siloxanes

Table 11 Chemical and physical properties of the most common cyclic siloxanes [277–280]

Name	Molecular weight (g/mol)	Boiling point (°C)	Vapor pressure (Pa) at 25°C	Water solubility (mg/L) at 25°C	Log K_{ow}	Henry's law constant (Pa m^3/mol)
D3	222.5	135	1,147	1.56	3.85	6.4×10^3
D4	296.6	175	132	0.056	6.49	1.2×10^6
D5	370.8	211	23	0.017	8.03	3.3×10^6
D6	444.9	245	4	0.005	9.06	4.9×10^6

7.2 Cyclic Siloxanes

Cyclic siloxanes are a group of chemicals more and more present in our daily lives. They are found in shower gels, shampoos, hair care products, deodorants or antiperspirants, skin cleansers or foundations, and even in children's products: baby lotions, diaper creams, and even pacifiers [276]. Cyclic siloxanes have a backbone structure of alternating $-Si(CH_3)_2-O-$ units that are singly bonded and form a ring. The simplest compound in this class is hexamethylcyclotrisiloxane (D3) and it is composed of three of these units forming a ring. Similarly, the compound with four $-Si(CH_3)_2-O-$ units is called D4, the one with five is called D5, and so on (Fig. 1). Cyclic siloxanes are ideal solvents for the active ingredients used in cosmetic and PCPs. They are volatile, have high thermal stability and low surface tension, and are transparent, hydrophobic, and odorless (Table 11) [277]. Products containing cyclic siloxanes have a smooth and silky texture. D4, D5, and D6 have already been classified by the Organization for Economic Cooperation and Development and by the US Environmental Protection Agency as high production volume chemicals [277]. Until almost two decades ago, they were believed to be inert.

7.2.1 Levels in Matrices Relevant for Human Exposure

The main source of human exposure to cyclic siloxanes is PCPs. Other less significant sources are rubber products, sealants, cookware, silicone grease, pharmaceuticals, medical devices, electronics, dust, and even indoor and outdoor air [277, 281]. The highest concentrations of cyclic siloxanes reported by Lu et al. [282] in PCPs from China were 72.9 µg/g D4 and 1,110 µg/g D5 in shampoos and hair conditioners. The highest concentrations of D6 and D7 found were in make-up products and especially liquid foundations (367 and 78.6 μg/g, respectively).

In a survey on PCPs from the US market, Horii and Kannan [276] found the highest concentrations of D5 and D6 in a liquid foundation cosmetic sample: 81,800 and 43,100 μg/g, respectively. Household sanitation products (furniture polish) were found to contain up to 9,380 μg/g D4, while the highest concentrations of D7 (846 μg/g) were found, surprisingly, in a baby pacifier.

In the study done by Wang et al. [277] on PCPs from Canada, the compound with the highest detection frequency was D5, with a top concentration of 683,000 μg/g in an antiperspirant sample. In 12 of the 13 tested antiperspirants, considerable concentrations of D5 were found. The other cyclic siloxanes were detected in lower levels: 98,000 μg/g for D6 (in a baby diaper cream sample) and 11,000 μg/g for D4 (body lotion sample). The mean levels of D5 concentrations were at least two orders of magnitude higher than mean concentrations for D3, D4, or D6.

The cyclic siloxanes have also been found in water from wastewater treatment plants in concentrations up to 710 μg/L in the influent and 13 μg/L in the effluent [283]. In fish, cyclic siloxane levels are usually in the low ng/g range but can go up to several hundreds of ng/g. Kierkegaard et al. [284] found a top D5 concentration of 300 ng/g wet weight in flounder from the Humber estuary in England.

Lu et al. [281] also found cyclic siloxanes in dust. Of the 88 dust samples analyzed, all of them contained at least one of these compounds. The concentration of siloxanes in dust (cyclic and linear) ranged from 21.5 to 21,000 ng/g, and the mean concentration for the cyclic siloxanes was 2,850 ng/g.

Generally, electronic components are coated in siloxane-containing materials to insulate them and to increase stability against electrical shock and mechanical damage. Due to the heat generated by these components, siloxanes can volatilize and accumulate in dust. Indeed, significantly higher concentrations of D4 and a few linear siloxanes were found in dust samples from rooms containing a high number of electronic devices [281].

7.2.2 Potential Routes of Human Exposure

The principal exposure routes to cyclic siloxanes are dermal exposure, inhalation, or ingestion. This latter pathway of exposure is less important than the first two, because, as presented above, the concentrations of siloxanes in water, food, and dust are typically much lower than those in PCPs.

Exposure through dermal contact: The PCPs, main source of contamination with cyclic siloxanes, are all applied to or come in contact with the skin. Therefore, dermal contact would be the main route of human exposure if these siloxanes would not be so volatile. When a product like a lotion or an antiperspirant is applied to the skin, most of the cyclic siloxanes evaporate shortly after application.

Jovanovic et al. [285] studied the dermal absorption of D4 and D5 by applying both neat and antiperspirant formulations containing ^{14}C-D4 and ^{14}C-D5 to human skin (in vitro) and to the skin of a number or rats (in vivo study). By using liquid scintillation counting, it was determined that most (more than 90%) of the D4 and D5 applied to the human skin volatilized before being absorbed into the skin. The actual quantity absorbed was only 0.5% and 0.04% for D4 and D5, respectively, of the total applied amount. Of these absorbed quantities, more than 90% remained in the skin. As for the in vivo study, less than 1% and 0.2% for D4 and D5, respectively, was absorbed, and of these absorbed amounts only 60% and 30%, respectively, reached the systemic compartments.

Exposure through inhalation: As most of the quantity of the cyclic siloxanes evaporate shortly after application of the PCPs or volatilize from electronics, adhesives and sealants, polishes and surface cleaners, etc., the main pathway of exposure is through inhalation. Plotzke et al. [286] studied the inhalation exposure to D4 by exposing rats to ^{14}C-D4. Using liquid scintillation counting, it was determined that the retention of inhaled D4 in the body of the rats was 5–6%. The radioactivity reached maximum concentrations in the fat 24 h after exposure, but in the plasma and other tissues (except for fat) in only 3 h. The fat tissue acted as a depot because the elimination of the radioactivity from it was slower than from other tissues.

7.2.3 Health Effects

The largest number of studies was conducted on the health effects of D4: the compound causes changes in organ weights in rats [287, 288], has weak estrogenic effects [288, 289] and adverse effects on reproductive health and function [290]: even a single 6-h exposure in the day before mating leads to a significant reduction in fertility.

D5 has slightly different properties than D4, and it does not have any estrogenic activity [289]. It does, however, also have adverse effects on the reproductive system, much like D4, but also on the adipose tissue, bile production, and even immune system due to D5's effect of reducing the prolactin levels [291]. In addition, it was determined that D5 causes a significant increase in uterine tumors in rats after a 160 ppm exposure. However, it is proposed that the tumors occur in rats through a mechanism that would not affect humans [291]. D5 also acts as a dopamine agonist and it can cause adverse effects on the nervous system in humans [291]. For exposures to D6 in rats, an increase in liver and thyroid mass and reproductive effects were observed [292].

Potential for bioaccumulation: Due to their high Log K_{ow} values and high fat: blood partition coefficient, the cyclic siloxanes are likely to be stored into the lipid tissue. However, bioaccumulation is not dependent just on the lipophilicity of the compound, but also in how fast it leaves the contaminated organism. Other indicators of bioaccumulation are the bioconcentration factor (BCF) and bioaccumulation factor (BAF). Values over 5,000 are usually characteristic for the bioaccumulative compounds. D4 has a BCF of 12,400 L/kg [293], D5 of 7,060 L/kg [279], and D6 of 1,160 L/kg [280], values calculated for fish.

Most compounds with such high BCF and Log K_{ow} values are highly bioaccumulative. The cyclic siloxanes, however, also have a high rate of clearance from the body through exhaled breath and a fast metabolism, so they might not be as bioaccumulative as the BCF and Log K_{ow} values indicate [285, 294, 295]. Kierkegaard et al. [284] compared the bioaccumulation potential of D4, D5, and D6 with the one of PCB 180, a compound known to be very bioaccumulative. D5 was found to be more bioaccumulative than PCB 180 and D4 even more so. D6, however, was found to have a lower bioaccumulation potential than PCB 180.

Although the cyclic siloxanes are more bioaccumulative in fish because they cannot eliminate them through breath as fast as land mammals, other factors such as metabolism, kinetic constraints, or reduced dietary absorption limit these compounds' potential for bioaccumulation [284].

8   Concluding Remarks and Future Perspectives

From the information presented in this chapter, it appears that a wealth of data exists on the various exposure pathways for these chemicals to humans or on their levels in human tissues. However, there is missing information regarding the relative importance of various exposure pathways and the multiple factors which influence the magnitude of these exposures. Moreover, more studies are needed to evaluate the significance of measured body burdens for each chemical or group of chemicals in relation with the existing toxicological data. With other words, what does this exposure means in terms of toxicological relevance? By far, there is a lack of standardized testing of endpoint toxicity for different groups of chemicals which makes the comparison of toxic effects and health issues for various chemicals very difficult, if not impossible. There is also a lack of concerted action regarding epidemiological studies, the chemicals investigated, and the health endpoints (including clinical parameters) which are tested.

The list of described emerging contaminants of concern for human exposure is not exhaustive. A range of other contaminants have been described in the literature as potentially dangerous for human exposure, but there is much less information known than for those described above. Also for other (emerging) contaminants, further studies should unravel the relative importance of various exposure pathways, but also the possibility of reducing this exposure.

References

1. The European Food Safety Authority (EFSA) (2006) Advice of the scientific panel on contaminants in the food chain on a request from the Commission related to relevant chemical compounds in the group of brominated flame retardants for monitoring in feed and food. EFSA J 328:1–4. http://www.efsa.europe.eu

2. Covaci A, Voorspoels S, Abdallah MA, Geens T, Harrad S, Law RJ (2009) Analytical and environmental aspects of the flame retardant tetrabromobisphenol-A and its derivatives. J Chromatogr A 1216:346–363
3. Covaci A, Harrad S, Abdallah MAE, Ali N, Law RJ, Herzke D, de Wit CA (2011) Novel brominated flame retardants: a review of their analysis, environmental fate and behaviour. Environ Int 37:532–556
4. Law K, Palace VP, Halldorson T, Danell R, Wautier K, Evans B, Alaee M, Marvin C, Tomy GT (2006) Dietary accumulation of Hexabromocyclododecane diastereoisomers in juvenile rainbow trout (*Oncorhynchus mykiss*) I: biomaccumulation parameters and evidence of bioisomerization. Environ Toxicol Chem 25:1757–1761
5. Vorkamp K, Rigét FR, Bossi R, Dietz R (2011) Temporal trends of hexabromocyclo-dodecane, polybrominated diphenyl ethers and polychlorinated biphenyls in ringed seals from East Greenland. Environ Sci Technol 45:1243–1249
6. Covaci A, Gerecke AC, Law RJ, Voorspoels S, Kohler M, Heeb NV, Leslie H, Allchin CR, De Boer J (2006) Hexabromocyclododecanes (HBCDs) in the environment and humans: a review. Environ Sci Technol 40:3679–3688
7. Roosens L, Abdallah MAE, Harrad S, Neels H, Covaci A (2009) Exposure to hexabromo-cyclododecanes (HBCDs) via dust ingestion, but not diet, correlates with concentrations in human serum: preliminary results. Environ Health Perspect 117:1707–1712
8. Goscinny S, Vandevijvere S, Maleki M, Van Overmeire I, Windal I, Hanot V, Blaude MN, Vleminckx C, Van Loco J (2011) Dietary intake of hexabromocyclododecane diastereoisomers in the Belgian adult population. Chemosphere 84:279–288
9. De Winter-Sorkina R, Bakker MI, van Donkersgoed G, van Klaveren JD (2003) Dietary intake of brominated flame retardants by the Dutch population. RIVM Report 310305001. http://rivm.openrepository.com/rivm/bitstream/10029/8876/1/310305001.pdf. Accessed Sept 2009
10. Driffield M, Harmer N, Bradley E, Fernandes AR, Rose M, Mortimer D, Dicks P (2008) Determination of brominated flame retardants in food by LC-MS/MS: diastereoisomer-specific hexabromocyclododecane and tetrabromobisphenol A. Food Addit Contamin 25:895–903
11. Abdallah MAE, Harrad S, Covaci A (2008) Hexabromocyclododecanes and tetrabromo-bisphenol-A in indoor air and dust in Birmingham, UK: implications for human exposure. Environ Sci Technol 42:6855–6861
12. Abdallah MAE, Harrad S, Ibarra C, Diamond M, Melmymuk L, Robson M, Covaci A (2008) Hexabromocyclododecanes in indoor dust from Canada, the United Kingdom, and the United States. Environ Sci Technol 42:459–464
13. Geens T, Roosens L, Neels H, Covaci A (2009) Assessment of human exposure to Bisphenol-A, Triclosan and Tetrabromobisphenol-A through indoor dust intake in Belgium. Chemosphere 76:755–760
14. Ali N, Harrad S, Goosey E, Neels H, Covaci A (2011) "Novel" brominated flame retardants in Belgian and UK indoor dust: implications for human exposure. Chemosphere 83:1360–1365
15. Ali N, Harrad S, Dirtu AC, Van den Eede N, 't Mannetje A, Coackley J, Douwes J, Neels H, Covaci A. (2011) Assessment of human exposure to alternative flame retardants in New Zealand via indoor dust ingestion. Organohalog Compd 73
16. Wang J, Maa YJ, Chena SJ, Tiana M, Luoa XJ, Maia BX (2010) Brominated flame retardants in house dust from e-waste recycling and urban areas in South China: implications on human exposure. Environ Int 36:535–541
17. Harrad S, Ibarra C, Abdallah MA, Boon R, Neels H, Covaci A (2008) Concentrations of brominated flame retardants in dust from United Kingdom cars, homes and offices: causes of variability and implications for human exposure. Environ Int 34:1170–1175
18. ESIS (2006) Summary risk assessment report on 2,2',6,6'-tetrabromo-4,4'-isopropylidene diphenol (tetrabromobisphenol-A or TBBP-A) Part II – Human Health 2006. http://esis.jrc.ec.europa.eu

19. Glynn A, Lignell S, Darnerud PO, Aune M, Ankarberg EA, Bergdahl IA, Barregård L, Bensryd I (2011) Regional differences in levels of chlorinated and brominated pollutants in mother's milk from primiparous women in Sweden. Environ Int 37:71–79

20. Kalantzi OI, Geens T, Covaci A, Siskos PA (2011) Distribution of polybrominated diphenyl ethers (PBDEs) and other persistent organic pollutants in human serum from Greece. Environ Int 37:349–353

21. Abdallah MAE, Harrad S (2011) Tetrabromobisphenol-A, hexabromocyclododecane and its degradation products in UK human milk: relationship to external exposure. Environ Int 37:443–448

22. Thomsen C, Knutsen HK, Liane VH, Frøshaug M, Kvalem HE, Haugen M, Meltzer HM, Alexander J, Becher G (2008) Consumption of fish from a contaminated lake strongly affects the concentrations of polybrominated diphenyl ethers and hexabromocyclododecane in serum. Mol Nutr Food Res 52:228–237

23. Thomsen C, Molander P, Daae HL, Janaak K, Froshaug M, Liane V, Thorud S, Becher G, Dybing A (2007) Occupational exposure to hexabromocyclododecane at an industrial plant. Environ Sci Technol 41:5210–5216

24. Kakimoto K, Akutsu K, Konishi Y, Tanaka Y (2008) Time trend of hexabromocyclododecane in the breast milk of Japanese women. Chemosphere 71:1110–1114

25. Shi ZX, Wu YN, Li JG, Zhao YF, Feng JF (2009) Dietary exposure assessment of chinese adults and nursing infants to tetrabromobisphenol-A and hexabromocyclododecanes: occurrence measurements in foods and human milk. Environ Sci Technol 43:4314–4319

26. Johnson-Restrepo B, Adams DH, Kannan K (2008) Tetrabromobisphenol A (TBBPA) and hexabromocyclododecanes (HBCDs) in tissues of humans, dolphins, and sharks from the United States. Chemosphere 70:1935–1944

27. Cariou R, Antignac JP, Zalko D, Berrebi A, Cravedi JP, Maume D, Marchand P, Monteau F, Riu A, Andre F, Le bizec B (2008) Exposure assessment of French women and their newborns to tetrabromobisphenol-A: occurrence measurements in maternal adipose tissue, serum, breast milk and cord serum. Chemosphere 73:1036–1041

28. Hagmar L, Sjodin A, Hoglund P, Thuresson K, Rylander L, Bergman A (2000) Biological halflives of polybrominated diphenyl ethers and tetrabromobisphenol-A in exposed workers. Organohalog Compd 47:198–201

29. Thomsen C, Lundanes E, Becher G (2001) Brominated flame retardants in plasma samples from three different occupational groups in Norway. J Environ Monit 3:366–370

30. Thomsen C, Lundanes E, Becher G (2002) Brominated flame retardants in archived serum samples from Norway: a study on temporal trends and the role of age. Environ Sci Techn 36:1414–1418

31. Karlsson M, Julander A, van Bavel B, Hardell L (2007) Levels of brominated flame retardants in blood in relation to levels in household air and dust. Environ Int 33:62–69

32. Zhu L, Ma B, Hites RA (2009) Brominated flame retardants in serum from the general population in Northern China. Environ Sci Technol 43:6963–6968

33. Gao S, Wang J, Yu Z, Guo Q, Sheng G, Fu G (2011) Hexabromocyclododecanes in surface soils from e-waste recycling areas and industrial areas in South China: concentrations, diastereoisomer- and enantiomer-specific profiles, and inventory. Environ Sci Technol 45:2093–2099

34. Zegers BN, Mets A, Van Bommel R, Minkenberg C, Hamers T, Kamstra J, Pierce GJ, Boon JP (2005) Levels of HBCD in harbour porpoises and common dolphin from western European seas with evidence for stereospecific biotransformation by cytochrome P450. Environ Sci Technol 39:2095–2100

35. Eljarrat E, Guerra P, Martínez E, Farré M, Alvarez JG, López-Teijón M, Barcelo D (2009) Hexabromocyclododecane in human breast milk: levels and enantiomeric patterns. Environ Sci Technol 43:1940–1946

36. Schauer UM, Volkel W, Dekant W (2006) Toxicokinetics of tetrabromobisphenol a in humans and rats after oral administration. Toxicol Sci 91:49–58

37. Hakk H, Larsen G, Bergman A, Orn U (2000) Metabolism, excretion and distribution of the flame retardant tetrabromobisphenol-A in conventional and bile-duct cannulated rats. Xenobiotica 30:881–890
38. de Wit CA (2002) An overview of brominated flame retardants in the environment. Chemosphere 46:583–624
39. Sjödin A, Bergman A, Pattersson DG Jr (2009) A review on human exposure to brominated flame retardants – particularly polybrominated diphenyl ethers. Environ Int 29:829–839
40. Kawashiro Y, Fukata H, Omori-Inoue M, Kubunoya K, Jotaki T, Takigami H, Sakai S, Mori C (2008) Perinatal exposure to brominated flame retardants and polychlorinated biphenyls in Japan. Endocr J 55:1071–1084
41. Saegusa Y, Fujimoto H, Woo GH, Inoue K, Takahashi M, Mitsumori K, Hirose M, Nishikawa A, Shibutani M (2009) Developmental toxicity of brominated flame retardants, tetrabromo-bisphenol A and 1,2,5,6,9,10-hexabromocyclododecane, in rat offspring after maternal exposure from mid-gestation through lactation. Reprod Toxicol 28:456–467
42. Van der Ven LTM, Verhoef A, Van de Kuil T, Slob W, Leonards PEG (2006) A 28-day oral dose toxicity study of HBCD in Wistar rats. Toxicol Sci 94:281–292
43. Eriksson P, Fisher C, Wallin M, Jakobsson E, Fredriksson A (2006) Impaired behaviour, learning and memory in adult mice neonatally exposed to HBCD. Environ Toxicol Pharmacol 21:317–322
44. Ema M (2008) Two generation reproductive toxicity study of the flame retardant HBCD in rats. Reprod Toxicol 25:335–351
45. Nomeir AA, Markham PM, Ghanayem BI, Chadwick M (1993) Disposition of the flame retardant 1,2- bis(2,4,6-tribromophenoxy)ethane in rats following administration in the diet. Drug Metab Dispos 21:209–214
46. Hakk H, Larsen G, Bowers J (2004) Metabolism, tissue disposition, and excretion of 1,2-bis (2,4,6-tribromophenoxy)ethane (BTBPE) in male Sprague–Dawley rats. Chemosphere 54:1367–1374
47. Koster P, Debets FMH, Strik JJTW (1980) Porphyrinogenic action of fire retardants. Bull Environ Contamin Toxicol 25:313–315
48. Li X, Yang L, Liu E, Xu W (2004) Research on the risk characteristics of decabrominated diphenyl ethane. Prog Saf Sci Technol, Proc Int Symp B:2164–2166
49. Hardy ML, Margitich D, Ackerman L, Smith RL (2002) The subchronic oral toxicity of ethane, 1,2- bis(pentabromophenyl) (Saytex 8010) in rats. Int J Toxicol 21:165–170
50. McKinney MA, Dietz R, Sonne C, De Guise S, Skirnisson K, Karlsson K, Steingrímsson E, Letcher RJ (2011) Comparative hepatic microsomal biotransformation of selected PBDEs, including decabromodiphenyl ether, and decabromodiphenyl ethane flame retardants in Arctic marine-feeding mammals. Environ Toxicol Chem 30:1506–1514
51. Nakari T, Huhtala S (2010) In vivo and in vitro toxicity of decabromodiphenyl ethane, a flame retardant. Environ Toxicol 25:333–338
52. Harju M, Heimstad ES, Herzke D, Sandanger T, Posner S, Wania F (2009) Emerging "new" brominated flame retardants in flame retarded products and the environment. Report 2462, Norwegian Pollution Control Authority, Oslo, Norway
53. Bearr JS, Stapleton HM, Mitchelmore CL (2010) Accumulation and DNA damage in fathead minnows (pimephales promelas) exposed to 2 brominated flame-retardant mixtures, Firemaster 550 and Firemaster BZ-54. Environ Toxicol Chem 29:722–729
54. Knudsen GA, Jacobs LM, Kuester RK, Sipe IG (2007) Absorption, distribution, metabolism and excretion of intravenously and orally administered tetrabromobisphenol A (2,3-dibromopropyl ether) in male Fischer-344 rats. Toxicology 237:158–167
55. Great Lakes Chem. Corp. (1987) Summaries of toxicity data. PE-68, Bis(2,3-dibromopropyl ether) of tetrabromobisphenol A. West Lafayette, IN
56. International Programme on Chemical Safety (IPCS) (1995) Environmental Health Criteria 172. Tetrabromobisphenol A and derivatives. World Health Organization, Geneva, Switzerland. http://www.inchem.org/documents/ehc/ehc/ehc172.htm

57. World Health Organisation (WHO) (1991a) Tri-*n*-butyl phosphate. Environmental Health Criteria 112. WHO, Geneva
58. World Health Organisation (WHO) (1991b) Triphenyl phosphate. Environmental Health Criteria 111. WHO, Geneva
59. Organisation for Economic Cooperation and Development (1998) Screening information data set, Triethyl Phosphate (CAS 78-40-0)
60. World Health Organisation (WHO) (2000) Flame retardants: Tris(2-butoxyethyl)phosphate, tris(2-ethylhexyl) phosphate and tetrakis(hydroxymethyl)phosphonium salts. Environmental Health Criteria 218, WHO, Geneva
61. Marklund A, Andersson B, Haglund P (2003) Screening of organophosphorus compounds and their distribution in various indoor environments. Chemosphere 53:1137–1146
62. Quintana JB, Rodil R, Reemstma T, Garcia-Lopez M, Rodriguez I (2008) Organophosphorus flame retardants and plasticizers in water and air II. Anal Methodol Trac Trend Anal Chem 27:904–915
63. World Health Organisation (WHO) (1998) Flame retardants: Tris(chloropropyl) phosphate and tris(2-chloroethyl) phosphate. Environmental Health Criteria 209, WHO, Geneva
64. World Health Organisation (WHO) (1997) Flame retardants: a general introduction. Environmental Health Criteria 192, WHO, Geneva
65. World Health Organisation (WHO) (1990) Tricresyl Phosphate. Environmental Health Criteria 110. WHO, Geneva
66. Saito I, Onuki A, Seto H (2007) Indoor organophosphate and polybrominated flame retardants in Tokyo. Indoor Air 17:28–36
67. National Research Council (2000) Toxicological risks of selected flame-retardant chemicals. National Academy Press, Washington DC, pp 338–416
68. Bergh C, Torgrip R, Emenius G, Östman C (2011) Organophosphate and phthalate esters in air and settled dust – a multi-location indoor study. Indoor Air 21:67–76
69. Brommer S, Harrad S, Van den Eede N, Covaci A (2011) Concentrations of organophosphate and brominated flame retardants in german indoor dust samples. Organohalog Compd 73
70. Ingerowski G, Friedle A, Thumulla J (2001) Chlorinated ethyl and isopropyl phosphoric acid triesters in the indoor environment – an inter-laboratory exposure study. Indoor Air Int J Indoor Air Qual Clim 11:145–149
71. Kanazawa A, Saito I, Araki A, Takeda M, Ma M, Saijo Y, Kishi R (2010) Association between indoor exposure to semi-volatile organic compounds and building-related symptoms among the occupants of residential dwellings. Indoor Air 20:72–84
72. Takigami H, Suzuki G, Hirai Y, Ishikawa Y, Sunami M, Sakai S (2009) Flame retardants in indoor dust and air of a hotel in Japan. Environ Int 35:688–693
73. Dirtu AC, Ali N, Van den Eede, Neels H, Covaci A (2011) Profile for chlorinated and brominated organic contaminants in indoor dust. Case study for Iasi, Eastern Romania. Organohalog Compd 73
74. García M, Rodríguez I, Cela R (2007) Microwave-assisted extraction of organophosphate flame retardants and plasticizers from indoor dust samples. J Chromatogr A 1152:280–286
75. García M, Rodríguez I, Cela R (2007) Optimisation of a matrix solid-phase dispersion method for the determination of organophosphate compounds in dust samples. Anal Chim Acta 590:17–25
76. Bergh C, Aberg KM, Svartengren M, Emenius G, Östman C (2011) Organophosphate and phthalate esters in indoor air: a compoarison between multi-storey buildings with high and low prevalence of sick building symptoms. J Environ Monit 13:2001–2009
77. Carlsson H, Nilsson U, Becker G, Östman C (1997) Organophosphate ester flame retardants and plasticizers in the indoor environment: analytical methodology and occurrence. Environ Sci Technol 31:2931–2396
78. Sjödin A, Carlsson H, Thuresson K, Sjölin S, Bergman Å, Östman C (2001) Flame retardants in indoor air at an electronics recycling plant and at other work environments. Environ Sci Technol 35:448–454

79. Marklund A, Andersson B, Haglund P (2005) Organophosphorus flame retardants and plasticizers in air from various indoor environments. J Environ Monit 7:814–819
80. Hartmann PC, Bürgi D, Giger W (2004) Organophosphate flame retardants and plasticizers in indoor air. Chemosphere 57:781–787
81. Marklund A, Olofsson U, Haglund P (2010) Organophosphorus flame retardants and plasticizers in marine and fresh water biota and human milk. J Environ Monit 12:943–951
82. Stackelberg PA, Gibs J, Furlong ET, Meyer MT, Zaugg SD, Lippincott RL (2007) Efficiency of conventional drinking-water-treatment processes in removal of pharmaceuticals and other organic compounds. Sci Total Environ 377:255–272
83. Andresen J, Bester K (2006) Elimination of organophosphate ester flame retardants and plasticizers in drinking water purification. Water Res 40:621–629
84. Staaf T, Östman C (2005) Organophosphate triesters in indoor environments. J Environ Monit 7:883–887
85. Meyer J, Bester K (2004) Organophosphate flame retardants and plasticisers in wastewater treatment plants. J Environ Monit 6:599–605
86. Stackelberg PE, Furlong ET, Meyer MT, Zaugg SD, Henderson AK, Reissman DB (2004) Persistence of pharmaceutical compounds and other organic wastewater contaminants in a conventional drinking water treatment plant. Sci Total Environ 329:99–113
87. US Food and Drug Administration (2006) Total Diet Study – Pesticides and industrial contaminants, Analytical results, Revision 3. http://www.fda.gov/Food/FoodSafety/FoodContaminantsAdulteration/TotalDietStudy/ucm184293.htm
88. Campone L, Piccinelli AL, Östman C, Rastrelli L (2010) Determination of organophosphorous flame retardants in fish tissues by matrix solid-phase dispersion and gas chromatography. Anal Bioanal Chem 397:799–806
89. Brandsma S, de Boer J, Leonards P (2011) Determination of organophosphorous flame retardants in the food web of the western Scheldt – including *in vitro* biotransformation. Organohalog Compd 73
90. Cooper E, Stapleton HM (2011) Flame retardants tris(1,3-dichloroisopropyl) phosphate and triphenyl phosphate in recreational equipment: a mini case study. Organohalog Compd 73
91. Hudec T, Thean J, Kuehl D, Dougherty RC (1981) Tris(dichloropropyl)phosphate, a mutagenic flame-retardant – frequent occurence in human seminal plasma. Science 211:951–952
92. Shah M, Meija J, Cabovska B, Caruso JA (2006) Determination of phosphoric acid triesters in human plasma using solid-phase microextraction and gas chromatography coupled to inductively coupled plasma mass spectrometry. J Chromatogr A 1103:329–336
93. Cooper E, Covaci A, van Nuijs ALN, Webster TF, Stapleton HM (2011) Analysis of the flame retardant metabolites bis(1,3-dichloro-2-propyl) phosphate (BDCPP) and diphenyl phosphate (DPP) in urine using liquid chromatography–tandem mass spectrometry. Anal Bioanal Chem. doi:10.1007/s00216-011-5294-7
94. Camarasa JG, Serra-Baldrich E (1992) Allergic contact dermatitis from triphenyl phosphate. Contact Dermatitis 26:264–265
95. Meeker JD, Stapleton HM (2010) House dust concentrations of organophosphate flame retardants in relation to hormone levels and semen quality parameters. Environ Health Persp 118:318–323
96. Rudel RA, Perovich LJ (2009) Endocrine disrupting chemicals in indoor and outdoor air. Atmos Environ 43:170–181
97. Harrad S, de Wit CA, Abdallah MAE, Bergh C, Bjorklund JA, Covaci A, Darnerud PO, de Boer J, Diamond M, Huber S, Leonards P, Mandalakis M, Oestman C, Haug LS, Thomsen C, Webster TF (2010) Indoor contamination with hexabromocyclododecanes, polybrominated diphenyl ethers, and perfluoroalkyl compounds: an important exposure pathway for people? Environ Sci Technol 44:3221–3231
98. ECHA (2008a) Member State Committee–Support document for identification of benzyl butyl phthalate (BBP) as a substance of very high concern. European Chemicals Agency, 1

October. http://echa.europa.eu/doc/candidatelist/svhc supdoc bbp publication.pdf. Accessed April 2010

99. ECHA (2008b) Member State Committee–Support document for identification of dibutyl phthalate (DBP) as a substance of very high concern. European Chemicals Agency, 1 October. http://echa.europa.eu/doc/consultations/recommendations/prioritisations/prioritisationdbp.pdf. Accessed April 2010

100. Oomen AG, Janssen PJCM, Dusseldorp A, Noorlander CW (2008) Exposure to chemicals via house dust. RIVM Report 609021064/2008, National Institute for Public Health and Environment. http://rivm.nl/bibliotheek/609021064.html. Accessed April 2010

101. Krauskopf LG, Godwin A (2005) Plasticizers. In: Wilkens CE, Daniels CA, Summers JW (eds) PVC Handbook. Carl Hanser, Munich

102. European Parliament and the Council (2005) Directive 2005/84/EC of the European Parliament and of the Council of the 14. December 2005 amending for the 22nd time CouncilDirective 76/769/EEC on the approximation of the laws, regulations and administrative provisions of the Member States relating to the restrictions on the marketing and use of certain dangerous substances and preparations (phthalates in toys and childcare articles). OJ L344/40:27. December. European Union. Commission Directive 2011/8/EU of 28 January 2011 amending Directive 2002/72/EC as regards the restriction of use of bisphenol A in plastic infant feeding bottle

103. ECHA (2009) Data on manufacture, import, export, uses and releases of Bis(2-ethylhexyl) phthalate (DEHP) as well as information on potential alternatives to its use. European Chemicals Agency, Revised version of 29 January. http://echa.europa.eu/doc/consultations/recommendations/techreports/techrepdehp.pdf. Accessed April 2010

104. ECPI (2009) European plasticizer consumption – trends. European Council for Plasticizers and Intermediates

105. Koch HM, Muller J, Angerer J (2007) Determination of secondary, oxidised di-isononylphthalate (DINP) metabolites in human urine representative for the exposure to commercial DINP plasticizers. J Chromatogr B 847:114–125

106. European Union Risk Assessment Report (2003) 1,2-Benzenedicarboxylic acid, di-C8-10-branched alkyl esters, C9-rich and di-"isononyl" phthalate (DINP), European commission Joint Research Centre, EUR 20784 EN, Office for Official Publications of the European Communities

107. Kavlock R, Boekelheide K, Chapin R, Cunningham M, Faustman E, Foster P, Golub M, Henderson R, Hinberg I, Little R, Seed J, Shea K, Tabacova S, Tyl R, Williams P, Zacharewski T (2002) NTP Center for the evaluation of risks to human reproduction: phthalates expert panel report on the reproductive and developmental toxicity of di-isononyl phthalate. Reprod Toxicol 16:679–708

108. Gärtner E (2009) Risikobewertung–Verunsicherung über PVC-Weichmacher in der EU. September. http://www.firmenpresse.de/pressinfo22630.html. Accessed April 2010

109. Crespo JE, Balart R, Sanchez L, Lopez J (2007) Substitution of di(2-ethylhexyl) phthalate by di(isononyl) cyclohexane-1,2-dicarboxylate as a plasticizer for industrial vinyl plastisol formulations. J Appl Polym Sci 104:1215–1220

110. NICNAS (2008) Full Public Report: 1,2-Cyclohexanedicarboxylic acid, 1,2-diisononyl ester ('Hexamoll DINCH'). National Industrial Chemicals Notification and Assessment Scheme (NICNAS), File No: STD/1259. http://www.nicnas.gov.au/publications/car/new/std/stdfullr/std1000fr/std1259fr.pdf. Accessed April 2010

111. Welle F, Wolz G, Franz R (2005) Migration von Weichmachern aus PVC Schläuchen in enterale Nahrungslösungen. Pharma International 3 (in German)

112. Nagorka R, Conrad A, Scheller C, Süßenbach B, Moriske H-J (2011) Diisononyl 1,2-cyclohexanedicarboxylic acid (DINCH) and di(2-ethylhexyl) terephthalate (DEHT) in indoor dust samples: concentration and analytical problems. Int J Hyg Environ Health 214:26–35

113. Pfrimmer Nutricia (2005) Neu von Pfrimmer Nutricia: DEHP-freie Überleitungsgeräte. Pressrelation Pfrimmer Nutricia PN-05-001 03, 21. January (in German)

114. Premiumpresse (2002) Innovativer Weichmacher für sensible Anwendungen. Premiumpresse.de, pressrelations GmbH, pressrelation 135447, 03. July (in German) [online]. http://www.pressrelations.de/new/standard/dereferrer.cfm?r=98832. Accessed April 2010

115. SpecialChem (2009) SpecialChem4 coatings: Hexamoll Applications. Specialchem 2009. http://www.specialchem4coatings.com/tc/plasticizers/index.aspx?id=application. Accessed April 2010

116. Scientific Committee on Emerging and Newly-Identified Health Risks (SCENIHR) (2007) Preliminary report on the safety of medical devices containing DEHP-plasticized PVC or other plasticizers on neonates and other groups possibly at risk. European Commission, Health and Consumer Protection

117. Eastman Chemical Company (2009) Eastman 168 Plasticizer. Product Information, Eastman website. http://www.eastman.com/Brands/Eastmanplasticizers/Pages/ProductHome.aspx?product=71045700. Accessed April 2010

118. Frederiksen H, Jorgensen N, Andersson AM (2010) Correlations between phthalate metabolites in urine, serum, and seminal plasma from young Danish men determined by isotope dilution liquid chromatography tandem mass spectrometry. J Anal Toxicol 34:400–410

119. Kersten W, Reich T (2003) Non-volatile organic substances in Hamburg indoor dust. Gefahrst Reinhalt L 63:85–91

120. Fromme H, Lahrz T, Piloty M, Gebhart H, Oddoy A, Ruden H (2004) Occurrence of phthalates and musk fragrances in indoor air and dust from apartments and kindergartens in Berlin (Germany). Indoor Air 14:188–195

121. Nagorka R, Scheller C, Ullrich D (2005) Plasticizer in house dust. Gefahrst Reinhalt L 65:99105

122. Butte W, Hostrup O, Walker G (2008) Phthalates in house dust and air: associations and potential sources indoors. Gefahrst Reinhalt L 68:79–81

123. Abb M, Heinrich T, Sorkau E, Lorenz W (2009) Phthalates in house dust. Environ Int 35:965–970

124. Bergh C, Torgrip R, Östman C (2010) Simultaneous selective detection of organophosphate and phthalate esters using gas chromatography with positive ion chemical ionization tandem mass spectrometry and its application to indoor air and dust. Rapid Commun Mass Spectrom 24:2859–2867

125. Wittassek M, Angerer J (2008) Phthalates: metabolism and exposure. Int J Androl 31:131–138

126. Koch HM, Preuss R, Angerer J (2006) Di(2-ethyl-hexyl)phthalate (DEHP): human metabolism and internal exposure – an update and latest results. Int J Androl 29:155–165

127. Stroheker T, Cabaton N, Nourdin G, Regnier JF, Lhuguenot JC, Chagnon MC (2005) Evaluation of anti-androgenic activity of di-(2-ethylhexyl)phthalate. Toxicology 208:115–121

128. Bouma K, Schakel DJ (2002) Migration of phthalates from PVC toys into saliva simulant by dynamic extraction. Food Addit Contamin 19:602–610

129. Fromme H, Korner W, Gruber L, Heitmann D, Schlummer M, Volkel W, Bolte G (2010) Exposure of the population to phthalates – results from the INES study. Gefahrst Reinhalt L 70:77–81

130. Fromme H, Gruber L, Seckin E, Raab U, Zimmermann S, Kiranoglu M, Schlummer M, Schwegler U, Smolic S, Volkel W (2011) Phthalates and their metabolites in breast milk – results from the Bavarian Monitoring of Breast Milk (BAMBI). Environ Int 37:715–722

131. Hogberg J, Hanberg A, Berglund M, Skerfving S, Remberger M, Calafat AM, Filipsson AF, Jansson B, Johansson N, Appelgren M, Hakansson H (2008) Phthalate diesters and their metabolites in human breast milk, blood or serum, and urine as biomarkers of exposure in vulnerable populations. Environ Health Perspect 116:334–339

132. Bruns-Weller E, Pfordt J (2000) Bestimmung von phthalsäureestern in lebensmitteln, frauenmilch, hausstaub und textilien. UWSF-Z Umweltchem Ökotox 12:125–130
133. Zhu J, Phillips SP, Feng YL, Yang X (2006) Phthalate esters in human milk: concentration variations over a 6-month postpartumtime. Environ Sci Technol 40:5276–5281
134. Chen J, Liu H, Qiu Z, Shu W (2008) Analysis of di-*n*-butyl phthalate and other organic pollutants in Chongqing women undergoing parturition. Environ Pollut 156:849–853
135. Fromme H, Bolte G, Koch HM, Angerer J, Boehmer S, Drexler H, Mayer R, Liebl B (2007) Occurrence and daily variation of phthalate metabolites in the urine of an adult population. Int J Hyg Environ Health 210:21–33
136. Hines EP, Calafat AM, Silva MJ, Mendola P, Fenton SE (2009) Concentrations of phthalate metabolites in milk, urine, saliva, and serum of lactating North Carolina women. Environ Health Perspect 117:86–92
137. Engel SM, Zhu C, Berkowitz GS, Calafat AM, Silva MJ, Miodovnik A, Wolff MS (2009) Prenatal phthalate exposure and performance on the Neonatal Behavioral Assessment Scale in a multiethnic birth cohort. Neurotoxicology 30:522–528
138. Adibi JJ, Whyatt RM, Williams PL, Calafat AM, Camann D, Herrick R, Nelson H, Bhat HK, Perera FA, Silva MJ, Hauser R (2008) Characterization of phthalate exposure among pregnant women assessed by repeat air and urine samples. Environ Health Perspect 116:467–473
139. Berman T, Hochner-Celnikier D, Calafat AM, Larry L. Needham LL, Amitai Y, Wormser U, Richter E (2009) Phthalate exposure among pregnant women in Jerusalem, Israel: results of a pilot study. Environ Int 35:353–357
140. Calafat AM, Wong L-Y, Silva MJ, Samandar E, Preau JL Jr, Jia LT, Needham LL (2011) Selecting adequate exposure biomarkers of diisononyl and diisodecyl phthalates: data from the 2005–2006 National Health and Nutrition Examination survey. Environ Health Perspect 119:50–55
141. Koch HM, Bolt HM, Angerer J (2004) Di(2-ethylhexyl)phthalate (DEHP) metabolites in human urine and serum after a single oral dose of deuterium-labelled DEHP. Arch Toxicol 78:123–130
142. Hellwing J, Freudenberger H, Jackh R (1997) Differential prenatal toxicity of branched phthalate esters in rats. Food Chem Toxicol 35:501–512
143. Faber WD, Deyo JA, Stump DG, Navarro L, Ruble K, Knapp J (2007) Developmental toxicity and uterotrophic studies with di-2-ethylhexyl therephthalate. Birth Defects Res B Dev Reprod Toxicol 80:396–405
144. Howdeshell KL, Furr J, Lambright CR, Rider CV, Wilson VS, Gray LE (2007) Cumulative effects of dibutyl phthalate and diethyl hexyl phthalate on male rat reproductive tract development: altered fetal steroid hormones and genes. Toxicol Sci 99:190–202
145. Howdeshell KL, Wilson VS, Furr J, Lambright CR, Rider CV, Blystone CR, Hotchkiss AK, Gray LE (2008) A mixture of five phthalate esters inhibits fetal testicular testosterone production in the Sprague Dawley rat in a cumulative, dose additive manner. Toxicol Sci 105:153–165
146. Martino-Andrade AJ, Morais RN, Botelho GG, Muller G, Grande SW, Carpentieri GB, Leao GMC, Dalsenter PR (2009) Coadministration of active phthalates results in disruption of foetal testicular function in rats. Int J Androl 32:704–712
147. Scientific Committee on Consumer Safety, SCCS (2010) Opinion on triclosan (antimicrobial resistance). http://ec.europa.eu/health/scientific_committees/consumer_safety/docs/sccs_o_023.pdf
148. Fang JL, Stingley RL, Beland FA, Harrouk W, Lumpkins DL, Howard P (2010) Occurrence, efficacy, metabolism, and toxicity of triclosan. J Environ Sci Health C Environ Carcinog Ecotoxicol Rev 28:147–171
149. Rodricks JV, Swenberg JA, Borzelleca JF, Maronpot RR, Shipp AM (2010) Triclosan: a critical review of the experimental data and development of margins of safety for consumer products. Crit Rev Toxicol 40:422–484
150. Dann AB, Hontela A (2011) Triclosan: environmental exposure, toxicity and mechanisms of action. J Appl Toxicol 31:285–311

151. Ye X, Kuklenyik Z, Needham LL, Calafat AM (2005) Automated on-line column-switching HPLC-MS/MS method with peak focusing for the determination of nine environmental phenols in urine. Anal Chem 77:5407–5413
152. Wolff MS, Teitelbaum SL, Windham G, Pinney SM, Britton JA, Chelimo C, Godbold J, Biro F, Kushi LH, Pfeiffer CM, Calafat AM (2007) Pilot study of urinary biomarkers of phytoestrogens, phthalates, and phenols in girls. Environ Health Perspect 115:116–121
153. Calafat AM, Ye X, Wong LY, Reidy JA, Needham LL (2008) Urinary concentrations of triclosan in the U.S. population: 2003–2004. Eniviron Health Perspect 116:303–307
154. Milieu en Gezondheid (2010) Vlaams Humaan Biomonitoringsprogramma 2007–2011. Resultatenrapport: deel referentiebiomonitoring
155. Casas L, Fernández MF, Llop S, Guxens M, Ballester F, Olea N, Iruzun MB, Rodríguez LSM, Riaño I, Tardón A, Vrijheid M, Calafat AM, Sunyer J (2011) Urinary concentrations of phthalates and phenols in a population of Spanish pregnant women and children. Environ Int 37:858–866
156. Li X, Ying GG, Zhao JL, Chen ZF, Lai HJ, Su HC (2011) 4-nonylphenol, bisphenol-A and triclosan levels in human urine of children and students in China and the effects of drinking these bottles materials on levels. Environ Int (in press)
157. Allmyr M, Adolfsson-Erici M, McLachlan MS, Sandborgh-Englund G (2006) Triclosan in plasma and milk from Swedish nursing mothers and their exposure via personal care products. Sci Total Environ 372:87–93
158. Ye X, Tao LJ, Needham LL, Calafat AM (2008) Automated on-line column-switching HPLC–MS/MS method for measuring environmental phenols and parabens in serum. Talanta 76:865–871
159. Dirtu AC, Roosens L, Geens T, Gheorghe A, Neels H, Covaci A (2008) Simultaneous determination of bisphenol A, triclosan, and tetrabromobisphenol A in human serum using solid-phase extraction and gas chromatography-electron capture negative-ionization mass spectrometry. Anal Bioanal Chem 391:1175–1181
160. Geens T, Neels H, Covaci A (2009) Sensitive and selective method for the determination of bisphenol-A and triclosan in serum and urine as pentafluorobenzoate-derivatives using GC-ECNI/MS. J Chromatogr B Anal Technol Biomed Life Sci 877:4042–4046
161. Dayan AD (2007) Risk assessment of triclosan [Irgasan®] in human breast milk. Food Chem Toxicol 45:125–129
162. NICNAS (2009) Priority existing chemical assessment. Report no. 30: triclosan. http://www.nicnas.gov.au/publications/car/pec/pec30/pec_30_full_report_pdf.pdf
163. Wang H, Zhang J, Gao F, Yang Y, Duan H, Wu Y, Berset JD, Shao B (2011) Simultaneous analysis of synthetic musks and triclosan in human breast milk by gas chromatography tandem mass spectrometry. J Chromatogr B 879:1861–1869
164. Zhang X, Liang G, Zeng X, Zhou J, Sheng G, Fu J (2011) Levels of synthetic musk fragrances in human milk from three cities in the Yangtze River Delta in Eastern China. J Environ Sci 23:983–990
165. Schlumpf M, Kypke K, Wittassek M, Angerer J, Mascher H, Mascher D, Vökt C, Birchler M, Lichtensteiger W (2010) Exposure patterns of UV filters, fragrances, parabens, phthalates, organochlor pesticides, PBDEs, and PCBs in human milk: correlation of UV filters with use of cosmetics. Chemosphere 81:1171–1183
166. Kang CS, Lee JH, Kim SK, Lee KT, Lee JS, Park PS, Yun SH, Kannan K, Yoo YW, Ha JY, Lee SW (2010) Polybrominated diphenyl ethers and synthetic musks in umbilical cord Serum, maternal serum, and breast milk from Seoul, South Korea. Chemosphere 80:116–122
167. Hu Z, Shi Y, Niu H, Cai Y, Jia G, Wu Y (2010) Occurence of synthetic musk fragrances in human blood from 11 cities in China. Environ Toxicol Chem 29:1877–1882
168. Hutter HP, Wallner P, Hartl W, Uhl M, Lorbeer G, Gminski R, Mersch-Sundermann V, Kundi M (2010) Higher blood concentrations of synthetic musks in women above fifty years than in younger women. Int J Hyg Environ Health 213:124–130

169. Hutter HP, Wallner P, Moshammer H, Hartl W, Sattelberger R, Lorbeer G, Kundi M (2005) Blood concentrations of polycyclic musks in healthy young adults. Chemosphere 59:487–492
170. Allmyr M, Harden F, Toms LML, Mueller JF, McLachlan MS, Adolfsson-Erici M, Sandborgh-Englund G (2008) The influence of age and gender on triclosan concentrations in Australian human blood serum. Sci Total Environ 393:162–167
171. Bhargava HN, Leonard PA (1996) Triclosan: applications and safety. Am J Infect Control 24:209–218
172. Crofton KM, Paul KB, De Vito MJ, Hedge JM (2007) Short-term *in vivo* exposure to the water contaminant triclosan: evidence for disruption of thyroxine. Environ Toxicol Pharmacol 24:194–197
173. Zorilla LM, Gibson EK, Jeffay SC, Crofton KM, Setzer WR, Cooper RL, Stoker TE (2009) The effects of TCS on puberty and thyroid hormones in male Wistar rats. Toxicol Sci 107:56–64
174. Paul KB, Hedge GM, De Vito MJ, Crofton KM (2010) Short-term exposure to triclosan decreases thyroxine *in vivo* via upregulation of hepatic catabolism in young Long-Evans rats. Toxicol Sci 113:367–379
175. Allmyr M, Panagiotidis G, Sparve E, Diczfalusy U, Sandborgh-Englund G (2009) Human exposure to triclosan via toothpaste does not change CYP3A4 activity or plasma concentrations of thyroid hormones. Basic Clin Pharmacol Toxicol 105:339–344
176. Witorsch RJ, Thomas JA (2010) Personal care products and endocrine disruption: a critical review of the literature. Crit Rev Toxicol 40:1–3
177. Van der Burg B, Schreurs R, Van der Linden S, Seinen W, Brouwer A, Sonneveld E (2008) Endocrine effects of polycyclic musks: do we smell a rat? Int J Androl 31:188–193
178. European Commission (1995) Eightieth Commission Directive 95/34/EC of 10 July 1995 adapting to technical progress Annexes II, III, VI and VII to Council Directive 76/768/ EEC on the approximation of the laws of the Member States relating to cosmetic products. Offic J Eur Commun L 167:19–21
179. European Commission (1998) Twenty-third Commission Directive 98/ 62/EC of 3 September 1998 adapting to technical progress Annexes II, III, VI and VII to Council Directive 76/ 768/ EEC on the approximation of the laws of the Member states relating to cosmetic products. Offic J Eur Commun L 253:20–23
180. Rimkus GG (1999) Polycyclic musk fragrances in the aquatic environment. Toxicol Lett 111:37–56
181. European Commission (2002) Twenty-sixth Commission Directive 2002/34/EC of 15 April 2002 adapting to technical progress Annexes II, III, VII to Council Directive 76/768/ EEC on the approximation of the laws of the Member States relating to cosmetic products. Offic J Eur Commun L 015:19–31
182. Luckenbach T, Epel D (2005) Nitromusk and polycyclic musk compounds as long-term inhibitors of cellular xenobiotic defense systems mediated by multidrug transporters. Environ Health Perspect 113:17–24
183. Greenpeace (2005) TNO report: Phtalates and artificial musks in perfumes. http://www.greenpeace.org/raw/content/international/press/reports/phthalates-and-artificial-musk.pdf
184. Reiner JL, Kannan K (2006) A survey of polycyclic musks in selected household commodities from the United States. Chemosphere 62:867–873
185. Kallenborn R, Gatermann R (2004) Synthetic musks in ambient and indoor air. In: Rimkus G (ed) The handbook of environmental chemistry. Spring, Heidelberg, pp 85–104
186. Werner B (2004) Synthetic musks in house dust. In: Rimkus G (ed) The handbook of environmental chemistry. Springer, Heidelberg, pp 105–121
187. Heberer T (2002) Occurrence, fate, and assessment of polycyclic musk residues in the aquatic environment of urban areas–a review. Acta Hydrochem Hydrobiol 30:227–243
188. Helbing KS, Schnid P, Schlatter C (1994) The trace analysis of musk xylene in biological samples: problems associated with its ubiquitous occurrence. Chemosphere 29:477–484

189. Angerer J, Käfferlein HU (1997) Gas chromatographic method using electron-capture detection for the determination of musk xylene in human blood samples. J Chromatogr B 693:71–78

190. European Commission (2005a) Summary risk assessment report (5-tert-butyl-2,4,6-trinitro-m-xylene) musk xylene. http://esis.jrc.ec.europa.eu

191. European Commission (2005b) Summary risk assessment report (4'-tert-butyl-2',6'-dimethyl-3',5'-dinitroacetophenone) musk ketone. http://esis.jrc.ec.europa.eu

192. European Commission (2008a) Summary risk assessment report (1,3,4,6,7,8-hexahydro-4,6,6,7,8,8-hexamethylcyclopenta-γ-2-benzopyran) HHCB. http://esis.jrc.ec.europa.eu

193. European Commission (2008b) Summary risk assessment report 1-(5,6,7,8-tetrahydro-3,5,5,6,8,8-hexamethyl-2-naphtyl ethane-1-one) AHTN. http://esis.jrc.ec.europa.eu

194. Suter-Eichenberger R, Boelsterli UA, Conscience-Egli M, Lichtensteiger W, Schlumpf M (1999) CYP 450 enzyme induction by chronic oral musk xylene in adult and developing rats. Toxicol Lett 111:117–132

195. Apostolidis S, Chandra T, Demirhan I, Cinatl J, Doerr HW, Chandra A (2002) Evaluation of carcinogenic potential of two nitro-musk derivatives, musk xylene and musk tibetene in a host-mediated *in vivo*/*in vitro* assay system. Anticancer Res 22:2657–2662

196. Lehman-McKeeman LD, Caudill D, Vassallo JD, Pearce RE, Madan A, Parkinson A (1999) Effects of musk xylene and musk ketone on rat hepatic cytochrome P450 enzymes. Toxicol Lett 111:105–115

197. Api AM, San RHC (1999) Genotoxicity tests with 6-acetyl-1,1,2,4,4,7-hexamethyltetraline and 1,3,4,6,7,8-hexahydro-4,6,6,7,8,8-hexamethylcyclopenta-g-2- benzopyran. Mutat Res Genet Toxicol Environ Mutagenesis 446:67–81

198. Seinen W, Lemmen JG, Pieters RHH, Verbruggen EMJ, van der Burg B (1999) AHTN and HHCB show weak estrogenic – but no uterotrophic activity. Toxicol Lett 111:161–168

199. Plastics Europe (2007) BPA applications. http://www.bisphenol-a-europe.org/uploads/BPA%20applications.pdf. Accessed August 2011

200. Geens T, Goeyens L, Covaci A (2011) Are potential sources for human exposure to bisphenol-A overlooked? Int J Hyg Environ Health 214:339–347

201. Ozaki A, Yamaguchi Y, Fujita T, Kuroda K, Endo G (2004) Chemical analysis and genotoxicological safety assessment of paper and paperboard used for food packaging. Food Chem Toxicol 42:1323–1337

202. Biedermann S, Tschudin P, Grob K (2010) Transfer of bisphenol A from thermal printer paper to the skin. Anal Bioanal Chem 398:571–576

203. Mielke H, Partisch F, Gundert-Remy U (2011) The contribution of dermal exposure to the internal exposure of bisphenol A in man. Toxicol Lett 204:190–198

204. Zalko D, Jacques C, Duplan H, Bruel S, Perdu E (2011) Viable skin efficiently absorbs and metabolizes bisphenol A. Chemosphere 82:424–430

205. Dekant W, Völkel W (2008) Human exposure to bisphenol A by biomonitoring: methods, results and assessment of environmental exposures. Toxicol Appl Pharmacol 228:114–134

206. Völkel W, Colnot T, Csanady GA, Filser JG, Dekant W (2002) Metabolism and kinetics of bisphenol-A in humans at low doses following oral administration. Chem Res Toxicol 15:1281–1287

207. Calafat AM, Ye XY, Wong LY, Reidy JA, Needham LL (2008) Exposure of the US population to bisphenol A and 4-tertiary-octylphenol: 2003–2004. Environ Health Perspect 116:39–44

208. Calafat AM, Kuklenyik Z, Reidy JA, Caudill SP, Needham LL (2005) Urinary concentrations of bisphenol A and 4-nonylphenol in a human reference population. Environ Health Perspect 113:391–395

209. Bushnik T, Haines D, Levallois P, Levesque J (2010) Lead and bisphenol A concentrations in the Canadian populations. Stat Can Health Rep 21:7–18

210. Becker K, Göen T, Seiwert M, Conrad A, Pick-Fuβ H, Müller J, Wittassek M, Schulz C, Kolossa-Gehring M (2009) GerES IV: Phthalate metabolites and bisphenol A in urine of German children. Int J Hyg Environ Health 212:685–692

211. Völkel W, Kiranoglu M, Fromme H (2008) Determination of free and total bisphenol A in human urine to assess daily uptake as a basis for a valid risk assessment. Toxicol Lett 179:155–162

212. Galloway T, Cipelli R, Guralnik J, Ferrucci L, Bandinelli S, Corsi AM, Money C, McCormack P, Melzer D (2010) Daily bisphenol A excretion and associations with sex hormone concentrations: results from the InCHIANTI adult population study. Environ Health Perspect 118:1603–1608

213. Hong YC, Park EY, Park MS, Ko JA, Oh SY, Kim H, Lee KH, Leem JH, Ha EH (2009) Community level exposure to chemicals and oxidative stress in adult population. Toxicol Lett 184:139–144

214. He Y, Miao M, Herrinton LJ, Wu C, Yuan W, Zhou Z, Li DK (2009) Bisphenol A levels in blood and urine in a Chinese population and the personal factores affecting the levels. Environ Res 109:629–633

215. Zhang Z, Alomirah H, Cho HS, Li YF, Liao C, Minh TB, Mohd MA, Nakata H, Ren N, Kannan K (2011) Urinary bisphenol A concentrations and their implications for human exposure in several Asian countries. Environ Sci Technol 45:7044–7050

216. Völkel W, Kiranoglu M, Fromme H (2011) Determination of free and total bisphenol A in urine of infants. Environ Res 111:143–148

217. Teitelbaum SL, Britton JA, Calafat AM, Ye X, Silva MJ, Reidy JA, Galvez MP, Brenner BL, Wolff MS (2008) Temporal variability in urinary concentrations of phthalate metabolites, phytoestrogens and phenols among minority children in the United States. Environ Res 106:257–269

218. Morgan MK, Jones PA, Calafat AM, Ye X, Croghan CW, Chuang JC, Wilson NK, Clifton MS, Figueroa Z, Sheldon LS (2011) Assessing the quantitative relationship between pre-school children's exposure to bisphenol A by route and urinary biomonitoring. Environ Sci Technol 45:5309–5316

219. Wolff MS, Engel SM, Berkowitz GS, Ye X, Silva MJ, Zhu C, Wetmur J, Calafat AM (2008) Prenatal phenol and phthalate exposures and birth outcomes. Environ Health Perspect 116:1092–1097

220. Ye X, Pierik FH, Hauser R, Duty S, Angerer J, Park MM, Burdorf A, Hofman A, Jaddoe VWV, Mackenbach JP, Steegers EAP, Tiemeier H, Longnecker MP (2008) Urinary metabolite concentrations of organophosphorous pesticides, bisphenol A, and phthalates among pregnant women in Rotterdam, the Netherlands: the generation R study. Environ Res 108:260–267

221. Mahalingaiah S, Meeker JD, Pearson KR, Calafat AM, Ye X, Petrozza J, Hauser R (2008) Temporal variability of urinary bisphenol A concentrations among men and women. Environ Health Perspect 116:173–178

222. Bondesson M, Jönsson J, Pongratz I, Olea N, Cravedi JP, Zalko D, Hakansson H, Halldin K, Di Lorenzo D, Behl C, Manthey D, Balaguer P, Demeneix B, Fini JB, Laudet V, Gustafsson JA (2009) A cascade of effects of bisphenol A. Reprod Toxicol 28:563–567

223. Vandenberg LN, Maffini MV, Sonneschein C, Rubin BS, Soto AM (2009) Bisphenol-A and the great divide: a review of controversies in the field of endocrine disruption. Endocr Rev 30:75–95

224. Hengstler JG, Foth H, Gebel T, Kramer PJ, Lilienblum W, Schweinfurth H, Völkel W, Wollin KM, Gundert-Remy U (2011) Critical evaluation of key evidence on the human health hazards of exposure to bisphenol A. Crit Rev Toxicol 41:263–291

225. Richter CA, Birnbaum LS, Farabollini F, Newbold RR, Rubin BS, Talsness CE, Vandenbergh JG, Walser-Kuntz DR, vom Saal FS (2007) In vivo effects of bisphenol A in laboratory rodent studies. Reprod Toxicol 24:199–224

226. Vandenberg LN, Chahoud I, Heindel JJ, Padmanabhan V, Paumgartten FJR, Schoenfelder G (2010) Urinary, circulating, and tissue biomonitoring studies indicate widespread exposure to bisphenol A. Environ Health Perspect 118:1055–1070

227. Welshons WV, Thayer KA, Judy BM, Taylor JA, Curran EM, vom Saal FS (2003) Large effects from small exposures I: mechanisms for endocrine-disrupting chemicals with estrogenic activity. Environ Health Perspect 111:994–1006

228. Urbansky ET (1998) Perchlorate chemistry: implications for analysis and remediation. Bioremed J 2:81–95

229. Urbansky ET, Brown SK, Magnuson ML, Kelty SK (2001) Perchlorate levels in samples of sodium nitrate fertilizer derived from Chilean caliche. Environ Pollut 112:299–302

230. Dasgupta PK, Martinelango PK, Jackson WA, Anderson TA, Tian K, Tock RW, Rajagopalan S (2005) The origin of naturally occurring perchlorate: the role of atmospheric processes. Environ Sci Technol 39:1569–1575

231. Gal H, Ronen Z, Weisbrod N, Dahan O, Nativ R (2008) Perchlorate biodegradation in contaminated soils and the deep unsaturated zone. Soil Biol Biochem 40:1751–1757

232. Kirk AB, Smith EE, Tian K, Anderson TA, Dasgupta PK (2003) Perchlorate in milk. Environ Sci Technol 37:4979–4981

233. Kirk AB, Martinelango PK, Tian K, Dutta A, Smith EE, Dasgupta PK (2005) Perchlorate and iodide in dairy and breast milk. Environ Sci Technol 39:2011–2017

234. Sanchez CA, Crump KS, Krieger RI, Khandaker NR, Gibbs JP (2005) Perchlorate and nitrate in leafy vegetables of North America. Environ Sci Technol 39:9391–9397

235. United States Environmental Protection Agency (USEPA) (1998) Drinking water contaminant list. EPA document No. 815-F-98-002, GPO, Washington, DC

236. Urbansky ET (2002) Perchlorate as an environmental contaminant. Environ Sci Pollut Res 9:187–192

237. Crump KS, Gibbs JP (2005) Benchmark calculations for perchlorate from three human cohorts. Environ Health Perspect 113:1001–1008

238. Li FX, Byrd DM, Deyhle GM, Sesser DE, Skeels MR, Katkowsky SR, Lamm SH (2000) Neonatal thyroid-stimulating hormone level and perchlorate in drinking water. Teratology 62:429–431

239. Li Z, Li FX, Byrd D, Deyhle GM, Sesser DE, Skeels MR, Lamm SH (2000) Neonatal thyroxine level and perchlorate in drinking water. J Occup Environ Med 42:200–205

240. United States Environmental Protection Agency (USEPA) (2005) http://yosemite.epa.gov/opa/admpress.nsf/b1ab9f485b098972852562e7004dc686/c1a57d2077c4bfda85256fac005b8b32!opendocument. Accessed October 2011

241. Murray CW, Egan SK, Kim H, Beru N, Bolger PM (2008) US Food and Drug Administration's total diet study: dietary intake of perchlorate and iodine. J Exp Sci Environ Epidemiol 18:571–580

242. Blount BC, Pirkle JL, Osterloh JD, Valentin-Blasini L, Caldwell KL (2006) Urinary perchlorate and thyroid hormone levels in adolescent and adult men and women living in the United States. Environ Health Perspect 114:1865–1871

243. Blount BC, Valentin-Blasini L, Osterloh JD, Mauldin JP, Pirkle JL (2007) Perchlorate exposure of the US Population, 2001–2002. J Exp Sci Environ Epidemiol 17:400–407

244. Kirk AB, Dyke JV, Martin CF, Dasgupta PK (2007) Temporal patterns in perchlorate, thiocyanate, and iodide excretion in human milk. Environ Health Perspect 115:182–186

245. Pearce EN, Leung AM, Blount BC, Bazrafshan HR, He X, Pino S, Valentin-Blasini L, Braverman LE (2007) Breast milk iodine and perchlorate concentrations in lactating Boston-area women. J Clin Endocrinol Metab 92:1673–1677

246. Amitai Y, Winston G, Sack J, Wasser J, Lewis M, Blount BC, Valentin-Blasini L, Fisher N, Israeli A, Leventhal A (2007) Gestational exposure to high perchlorate concentrations in drinking water and neonatal thyroxine levels. Thyroid 17:843–850

247. Ohira S, Kirk AB, Dyke JV, Dasgupta PK (2008) Creatinine adjustment of spot urine samples and 24 h excretion of iodine, selenium, perchlorate, and thiocyanate. Environ Sci Technol 42:9419–9423
248. Blount BC, Rich DD, Valentin-Blasini L, Lashley S, Ananth CV, Murphy E, Smulian JC, Spain BJ, Barr DB, Ledoux T (2009) Perinatal exposure to perchlorate, thiocyanate, and nitrate in New Jersey mothers and newborns. Environ Sci Technol 43:7543–7549
249. Oldi JF, Kannan K (2009) Perchlorate in human blood serum and plasma: relationship to concentrations in saliva. Chemosphere 77:43–47
250. Oldi JF, Kannan K (2009) Analysis of perchlorate in human saliva by liquid chromatography–tandem mass spectrometry. Environ Sci Technol 43:142–147
251. Kannan K, Praamsma M, Oldi JF, Kunisue T, Sinha RK (2009) Occurrence of perchlorate in drinking water, groundwater, surface water and human saliva from India. Chemosphere 76:22–26
252. Zhang T, Wu Q, Sun WH, Rao J, Kannan K (2010) Perchlorate and iodide in whole blood samples from infants, children, and adults in Nanchang, China. Environ Sci Technol 44:6947–6953
253. Borjian M, Marcella S, Blount B, Greenberg M, Zhang J, Murphy E, Blasini VL, Robson M (2011) Perchlorate exposure in lactating women in an urban community in New Jersey. Sci Total Environ 409:460–464
254. Gibbs JP, Narayanan L, Mattie DR (2004) Study among school children in Chile: subsequent urine and serum perchlorate levels are consistent with perchlorate in water in Taltal. J Occup Environ Med 46:516–517
255. Tellez R, Chacon PM, Crump KS, Blount BC, Gibbs JP (2005) Chronic environmental exposure to perchlorate through drinking water and thyroid function during pregnancy and the neonatal period. Thyroid 15:963–975
256. Valentin-Blasini L, Mauldin JP, Maple D, Blount BC (2005) Analysis of perchlorate in human urine using ion chromatography and electrospray tandem mass spectrometry. Anal Chem 77:2475–2481
257. Braverman LE, Pearce EN, He X, Pino S, Seeley M, Beck B (2006) Effects of six months of daily low-dose perchlorate exposure on thyroid funct ion in healthy volunt eers. J Clin Endocrinol Metab 91:2721–2724
258. Dasgupta PK, Kirk AB, Dyke JV, Ohira S (2008) Intake of iodine and perchlorate and excretion in human milk. Environ Sci Technol 42:8115–8121
259. Blount BC, Valentin-Blasini L (2006) Analysis of perchlorate, thiocynate, nitrate and iodide in human amniotic fluid using ion chromatography and electrospray tendem mass spectrometry. Anal Chim Acta 567:87–93
260. Wolff J (1998) Perchlorate and the thyroid gland. Pharmacol Rev 50:89–105
261. Anbar M, Guttmann S, Lweitus Z (1959) The mode of action of perchlorate ions on the iodine uptake of the thyroid gland. Int J Appl Radiat Isot 7:87–96
262. Lawrence JE, Lamm SH, Pino S, Richman K, Braverman LE (2000) The effect of short-term low-dose perchlorate on various aspects of thyroid function. Thyroid 10:659–663
263. Greer MA, Goodman G, Pleus RC, Greer SE (2002) Health effects assessment for environmental perchlorate contamination: the dose response for inhibition of thyroidal radioiodine uptake in humans. Environ Health Perspect 110:927–937
264. Institute of Medicine (IOM) (1991) Subcommittee on Nutrition during Lactation. Nutrition during lactation. National Academy Press, Washington, DC
265. NAS (2005) Health implications of perchlorate ingestion. National Research Council, National Academy Press, Washington DC, http://www.nap.edu/books/0309095689/html. Accessed October 2011
266. Malamud D (1992) Malamud D, Tabak L (eds) Saliva as a diagnostic fluid. Ann NY Acad Sci 694:36–47
267. Bernert JT, McGuffey JE, Morrison MA, Pirkle JL (2000) Comparison of serum and salivary cotinine measurements by a sensitive high-performance liquid chromatography-tandem mass

spectrometry method as an indicator of exposure to tobacco smoke among smokers and nonsmokers. J Anal Toxicol 24:333–339

268. Wu Q, Zhang T, Sun H, Kannan K (2010) Perchlorate in drinking water, groundwater, surface waters and bottled water from China, and its association with other inorganic anions and with disinfection byproducts. Arch Environ Contam Toxicol 58:543–550

269. De Groef B, Decallonne BR, Van der Geyten S, Darras VM, Bouillon R (2006) Perchlorate versus other environmental sodium/iodide symporter inhibitors: potential thyroid-related health effects. Eur J Endocrinol 155:17–25

270. Erdogan MF, Gulec S, Tutar E, Baskal N, Erdogan G (2003) A stepwise approach to the treatment of amiodarone-induced thyrotoxicosis. Thyroid 13:205–209

271. Cooper DS (2005) Antithyroid drugs. N Engl J Med 352:905–917

272. Crump C, Michaud P, Tellez R, Reyes C, Gonzalez G, Montgomery EL, Crump KS, Lobo G, Becerra C, Gibbs JP (2000) Does perchlorate in drinking water affect thyroid function in newborns or school-age children? J Occup Environ Med 42:603–612

273. Li FX, Squartsoff L, Lamm SH (2001) Prevalence of thyroid diseases in Nevada counties with respect to perchlorate in drinking water. J Occup Environ Med 43:630–634

274. Gibbs JP, Ahmad R, Crump KS, Houck DP, Leveille TS, Findley JE, Francis M (1998) Evaluation of a population with occupational exposure to airborne ammonium perchlorate for possible acute or chronic effects on thyroid function. J Occup Environ Med 40:1072–1082

275. Braverman LE, He X, Pino S, Cross M, Magnani B, Lamm SH, Kruse MB, Engel A, Crump KS, Gibbs JP (2005) The effect of perchlorate, thiocyanate, and nitrate on thyroid function in workers exposed to perchlorate long-term. J Clin Endocrinol Metab 90:700–706

276. Horii Y, Kannan K (2008) Survey of organosilicone compounds, including cyclic and linear siloxanes, in personal-care and household products. Arch Environ Contamin Toxicol 55:701–710

277. Wang R, Moody RP, Koniecki D, Zhu JP (2009) Low molecular weight cyclic volatile methylsiloxanes in cosmetic products sold in Canada: implication for dermal exposure. Environ Int 35:900–904

278. Brooke DN, Crookes MJ, Gray D, Robertson S (2009) Environmental Risk Assessment Report: Octamethylcyclotetrasiloxane. Environment Agency of England and Wales, Bristol, UK. http://publications.environment-agency.gov.uk/pdf/SCHO0309BPQZ-e-e.pdf

279. Brooke DN, Crookes MJ, Gray D, Robertson S (2009) Environmental risk assessment report: decamethylcyclopentasiloxane. Environment Agency of England and Wales, Bristol, UK. http://publications.environment-agency.gov.uk/pdf/SCHO0309BPQX-e-e.pdf

280. Brooke DN, Crookes MJ, Gray D, Robertson S (2009) Environmental risk assessment report: dodecamethylcyclohexasiloxane. Environment Agency of England and Wales, Bristol, UK. http://publications.environment-agency.gov.uk/pdf/SCHO0309BPQY-e-e.pdf

281. Lu Y, Yuan T, Yun SH, Wang W, Wu Q, Kannan K (2010) Occurrence of cyclic and linear siloxanes in indoor dust from China, and implications for human exposures. Environ Sci Technol 44:6081–6087

282. Lu Y, Yuan T, Wang W, Kannan K (2011) Concentrations and assessment of exposure to siloxanes and synthetic musks in personal care products from China. Environ Pollut 12:3522–3528

283. Maguire RJ (2001) Preliminary environmental assessment of organosilicon substances. Environment Canada, National Water Research Institute, Burlington/Saskatoon, NWRI Contribution No. 01-037

284. Kierkegaard A, van Egmond R, McLachlan MS (2011) Cyclic volatile methylsiloxane bioaccumulation in Flounder and Ragworm in the Humber Estuary. Environ Sci Technol 45:5936–5942

285. Jovanovic ML, McMahon JM, McNett DA, Tobin JM, Plotzke KP (2008) In vitro and in vivo percutaneous absorption of ^{14}C-octamethylcyclotetrasiloxane (^{14}C-D4) and ^{14}C-decamethyl cyclopentasiloxane (^{14}C-D5). Regul Toxicol Pharmacol 50:239–248

286. Plotzke KP, Crofoot SD, Ferdinandi ES, Beattie ES, Reitz JG, McNett RH, Meeks DA (2000) Disposition of radioactivity in Fischer 344 rats after single and multiple inhalation exposure to [C-14]-octamethylcyclotetrasiloxane ([C-14]D4). Drug Metab Dispos 28:192–204

287. McKim JM, Wilga PC, Breslin WJ, Plotzke KP, Gallavan RH, Meeks RG (2001) Potential estrogenic and antiestrogenic activity of the cyclic siloxane octamethylcyclotetrasiloxane (D4) and the linear siloxane hexamethyldisiloxane (HMDS) in immature rats using the uterotrophic assay. Toxicol Sci 63:37–46

288. He B, Rhodes-Brower S, Miller MR, Munson AE, Germolec DR, Walker VR, Korach KS, Meade BJ (2003) Octamethylcy-clotetrasiloxane exhibits estrogenic activity in mice via ER alpha. Toxicol Appl Pharmacol 192:254–261

289. Quinn AL, Regan JM, Tobin JM, Marinik BJ, McMahon JM, McNett DA, Sushynski CM, Crofoot SD, Jean PA, Plotzke KP (2007) In vitro and in vivo evaluation of the estrogenic, androgenic, and progestagenic potential of two cyclic siloxanes. Toxicol Sci 96:145–153

290. Meeks RG, Stump DG, Siddiqui WH, Holson JF, Plotzke KP, Reynolds VL (2007) An inhalation reproductive toxicity study of octamethylcyclotetrasiloxane (D-4) in female rats using multiple and single day exposure regimens. Reprod Toxicol 23:192–201

291. OEHHA (2007) Toxicity data review: decamethylcyclopentasiloxane (D5). September, 2007. http://www.arb.ca.gov/toxics/dryclean/oehhad5review.pdf

292. Environment Canada (2008) Screening assessment for the challenge. Dodecamethylcyclo-hexasiloxane (D6). Chemical Abstracts Service Registry Number: 540-97-6. http://www.ec.gc.ca/substances/ese/eng/challenge/batch2/batch2_540-97-6_en.pdf

293. Environment Canada (2007) Existing substances evaluation. Substance profile for the challenge. Decamethylcyclopentasiloxane (D5). Chemical Abstracts Service Registry Number: 541-02-6. http://www.ec.gc.ca/substances/ese/eng/challenge/batch2/batch2_541-02-6.cfm

294. Reddy MB, Dobrev ID, Plotzke KP, Andersen ME, Reitz RH, Morrow P, Utell M (2003) A physiologically based pharma-cokinetic model for inhalation of octamethylcyclotetra-siloxane (D4) in Human during rest and exercise. Toxicol Sci 72:3–18

295. Andersen ME, Reddy MB, Plotzke KP (2008) Are highly lipophilic volatile compounds expected to bioaccumulate with repeated exposures? Toxicol Lett 179:85–92

296. EFRA (2007) Market Statistics. http://www.cefic-efra.com/Objects/2/Files/European%20FR%20Market.pdf

297. Stapleton HM, Klosterhaus S, Eagle S, Fuh J, Meeker JD, Blum A, Webster TF (2009) Detection of organophosphate flame retardants in furniture foam and US house dust. Environ Sci Technol 43:7490–7495

298. Van den Eede N, Dirtu AC, Neels H, Covaci A (2011) Analytical developments and preliminary assessment of human exposure to organophosphate flame retardants from indoor dust. Environ Int 37: 454–461

299. Regnery J, Puttmann W, Merz, C, Berthold G (2011) Occurrence and distribution of organo-phosphorus flame retardants and plasticizers in anthropogenically affected groundwater. J Environ Monit 13: 347–354

300. Weschler CJ, Nazaroff WW (2008) Semivolatile organic compounds in indoor environments. Atm Environ 42: 9018–9040

301. Makinen MSE, Makinen MRA, Koistinen JTB, Pasanen AL, Pasanen PO, Kalliokoski PI, Korpi AM (2009) Respiratory and Dermal Exposure to Organophosphorus Flame Retardants and Tetrabromobisphenol A at Five Work Environments. Environ Sci Technol 43:941–947

302. Schindler BK, Foerster K, Angerer J (2009a) Determination of human urinary organophos-phate flame retardant metabolites by solid-phase extraction and gas chromatography-tandem mass spectrometry. J Chromatogr B 877:375–381

303. Schindler BK, Foerster K, Angerer J (2009b) Quantification of two urinary metabolites of organophosphorus flame retardants by solid-phase extraction and gas chromatography-tan-dem mass spectrometry. Anal Bioanal Chem 395: 1167–1171

304. Reemtsma T, Lingott J, Roegler S (2011) Determination of 14 monoalkyl phosphates, dialkyl phosphates and dialkyl thiophosphates by LC-MS/MS in human urinary samples. Sci Total Environ 409:1990–1993

305. Martino-Andrade AJ, Chahoud I (2010) Reproductive toxicity of phthalate esters. Mol Nutr Food Res 54:148–157

306. Hawkins DR, Elsom LF, Kirkpatrick D, Ford RA, Api AM (2002) Dermal absorption and disposition of musk ambrette, musk ketone and musk xylene in human subjects. Toxicol Lett 131:147–151

307. Lignell S, Darnerud PO, Aune M, Cnattingius S, Hajslova J, Setkova L, Glynn A (2008) Temporal trends of synthetic musk compounds in mother's milk and associations with personal use of perfumed products. Environ Sci Technol 42:6743–6748

308. Reiner JL, Wong CM, Arcaro KF, Kannan K (2007) Synthetic musk fragrances in human milk from the United States. Environ Sci Technol 41:3815–3820

309. Kafferlein HU, Angerer J (2001) Trends in the musk xylene concentrations in plasma samples from the general population from 1992/1993 to 1998 and the relevance of dermal uptake. Intern Arch Occup Environ Health 74:470-476

Occurrence of Phthalates and Their Metabolites in the Environment and Human Health Implications

Mario Antonio Fernández, Belén Gómara, and María José González

Abstract Phthalates are chemicals that have been used for over 80 years in large quantities due to their wide range of applications, mainly in the plastic industry. For many years, these compounds were not considered dangerous for humans due to their low toxicity shown in the preliminary studies and their low persistence. However, research conducted in recent years has evidenced their activity as endocrine disruptors, and they are now considered as emerging contaminants and included in the priority list of dangerous substances in the legislation of many countries. This chapter provides an overview on the properties, major uses, emission sources, environmental and human levels, current legislation, behavior and fate of phthalates, and their metabolites, with special emphasis on their toxicity and human exposure.

Keywords Environmental levels, Human exposure, Phthalates, Properties, Toxicity

Contents

M.A. Fernández, B. Gómara, and M.J. González (✉)
Instrumental Analysis and Environmental Chemistry Department, General Organic Chemistry
Institute, CSIC, Juan de la Cierva 3, 28006 Madrid, Spain
e-mail: mariche@iqog.csic.es

D. Barceló (ed.), *Emerging Organic Contaminants and Human Health*,
Hdb Env Chem (2012) 20: 307–336, DOI 10.1007/698_2011_127,
© Springer-Verlag Berlin Heidelberg 2011, Published online: 20 December 2011

1 General Overview

Phthalates are well known chemical compounds. In fact, they began to be used in the plastics industry more than 80 years ago. During these years it has been confirmed that their presence in the environment is in large quantities, but the results of different studies about their toxicity and persistence concluded that phthalates had not been considered dangerous for human health. However, during the last years, their activity as endocrine disruptors, their impact on the normal development of living organisms, as well as their teratogenic activity in both humans and animal studies has made them to be considered as new emerging contaminants. As a result, a high amount of researches is now being conducted in order to know their levels and behavior in the environment and in humans, their main pollution sources as well as the way of decreasing their levels in our nearby. They are now in the priority list of dangerous substances in the legislation of most of the industrialized countries, where the use of plastic materials containing food or for whatever children use is limited, and maximum permitted levels for water environment and ambient workplaces are established.

Phthalates, also known as phthalic acid esters (PAEs), are esters of phthalic acid and are mainly used as plasticizers. The structure of the most commonly used phthalates is shown in Fig. 1. The differences among various phthalates are in the structures of the two hydrocarbon chains (R1 y R2) linked to the two carboxylic acid functional groups. The various esters used in industry have alkyl side chains containing from 1 to 13 carbon atoms. The smallest chain in phthalates is the methyl group forming dimethyl phthalate (R1=R2=CH_3; DMP), and the longest chain is the tridecyl group forming ditridecyl phthalate (R=R2=$C_{13}H_{27}$; DTDP). The large differences in the length of the side chains provide very different physical properties such as vapor pressure (VP), partition coefficients K_{AW} (air–water), K_{OW} (octanol–water), K_{OA} (octanol–air) [1], and its behavior in the environment may differ from each other, producing very different effects on the living organisms. The most important properties of the most commonly used phthalates are summarized in Table 1.

As previously mentioned, phthalates are mainly used as plasticizers, but they are also used as solubilizing or stabilizing agents in other applications. They are components of many plastics products such as food wraps, detergents, building

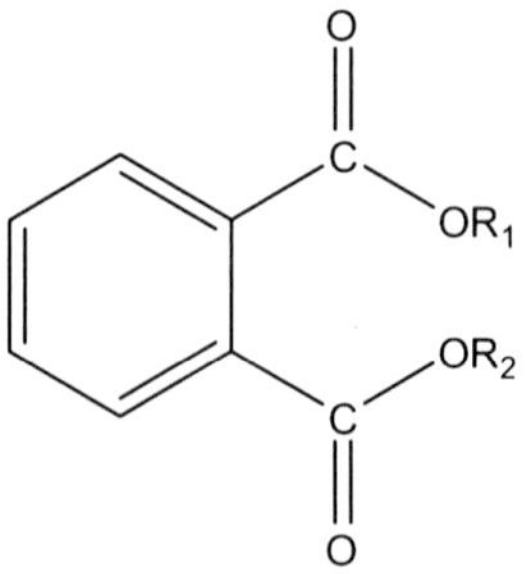

Fig. 1 General structure
of PAEs

Table 1 Physical–chemical properties of the most commonly monitored phthalates [1]

Phthalate ester	Abbreviation	CAs number	Molar mass (g/mol)	Molar volume (cm³/mol)	Melting point (°C)	VP^a (Pa) (25°C)	log K_{OW}^{b} (25°C)	log K_{OA}^{c} (25°C)	log K_{AW}^{d} (25°C)
Dimethyl phthalate	DMP	131-11-3	194.2	206.6	5.5	0.263	1.61	7.01	−5.40
Diethyl phthalate	DEP	84-66-2	222.2	254.0	−40	0.65×10^{-3}	2.54	7.55	−5.01
Dipropyl phthalate	DPP	131-16-8	250.3	298.4	–	1.75×10^{-2}	3.40	8.04	−4.64
Di-n-butyl phthalate	DBP	84-74-2	278.4	342.8	−35	4.73×10^{-3}	4.27	8.54	−4.27
Diisobutyl phthalate	DiBP	84-69-5	278.4	342.8	−58	4.73×10^{-3}	4.27	8.54	−4.27
Butylbenzyl phthalate	BBP	85-68-7	312.4	364.8	−35	2.49×10^{-3}	4.70	8.78	−4.08
Di(2-ethylhexyl) phthalate	DEHP	117-81-7	390.6	520.4	−46	2.5×10^{-5}	7.73	10.5	−2.80
Di-n-octyl phthalate	DnOP	117-84-0	390.6	520.4	−25	2.5×10^{-5}	7.73	10.53	−2.80
Diisononyl phthalate	DiNP	68515-48-0 28553-12-0 (C = 8–10)	418.6	564.8	−48	6.81×10^{-6}	8.60	11.03	−2.43
Diisodecyl phthalate	DiDP	68515-49-1 26761-40-0 (C = 9–11)	446.7	609.2	−46	1.84×10^{-6}	9.46	11.52	−2.06

^{a}VP vapour pressure
^{b}K_{OW} the octanol–water partition coefficient
^{c}K_{OA} the octanol–air partition coefficient
^{d}K_{AW} the air–water partition coefficient

products (flooring, sheeting, and films), lubricating oils, carriers in pesticide formulations, PCB substitutes, solvents, personal care products, cosmetics, toys, some medical devices, etc. For example, di(2-ethylhexyl)phthalate (DEHP) is added to polyvinyl chloride (PVC), a thermoplastic polymer, to make it softer and more flexible. DEHP is also used in food packaging materials, medical products such as intravenous tubing, plastic toys, vinyl upholstery, shower curtains, adhesives, and coatings. Phthalates with smaller side chains such as diethyl phthalates (DEP) and di-n-butyl phthalate (DBP) are used as solvents and perfumes [2].

Various phthalate esters have been reported to be present in the environment, including outdoor air, water, and soil [3–6], consumer products [7, 8], medical devices [9], marine ecosystems [10], and indoor air and dust [5, 11, 12]. They are also widely present in dairy products [13] and food in general [14].

Phthalates are easily released into the environment because there is no covalent bond between them and plastics in which they are mixed. The major portion of phthalates that are found in the environment comes from the slow releases of phthalates from plastics and other phthalate containing articles due to weathering. At natural conditions, phthalates are hydrolyzed to some extent yielding their corresponding monoesters, which are also environmental pollutants [15]. They show poor mobility in soil but aqueous leachates from landfills may contain trace amounts of more soluble products of phthalate degradation [11, 16].

For the general population, the oral route of exposure has been considered the major route, including inhalation of air (indoors and outdoors), ingestion of food, incidental ingestion of soil, and ingestion of dust (indoors), as well as direct contact with products that contain phthalates. Some studies suggested [17, 18] that food represents the most important source of exposure to DMP, DEP, DBP, butyl benzyl phthalate (BBP), and DEHP. A few studies showed that air inhalation could be also an important route of exposure [19–21], while others did not find any significant correlation between urinary levels in children and home dust measurements of phthalates [22].

Generally, phthalates are metabolized and excreted quickly and do not accumulate in the body [23]. Ingested phthalate diesters are initially hydrolyzed in the intestine to their corresponding monoester, which is then absorbed [24], and could be further oxidized in the body.

For example, in the intestinal tract and liver of both humans and animals DEHP is rapidly hydrolyzed by esterases to yield mono-(2-ethylhexyl) phthalate (MEHP) and 2-ethylhexanol [25]. The latter metabolite is subsequently oxidized enzymatically to 2-ethyl hexanoic acid (2-EHXA) [26]. MEHP, 2-hethylhexanol, and/or their metabolites are the immediate inducers of the majority of enzymes known to be affected by exposure of DEHP [27]. Due to the high importance of the primary and secondary PAE metabolites in the human exposure studies, during the last years a big number of studies have been conducted to prove that some of them are appropriate biomarkers to calculate human PAE intake [28–30] and that their determination is easier than calculate it through food intake, which are more time consuming and subjects to several error sources.

Studies on health effects of PAEs in humans have remained controversial due to limitations of the study design. Some findings in human populations are consistent with animal data, suggesting that PAEs and their metabolites produce toxic effects in the reproductive system. Some studies associate monoesters PAEs with semen parameters, sperm DNA damage, and hormones in human population, but none of them are statistically significant. Urinary monomethyl phthalate (MMP), monobenzyl phthalate (MBzP), mono-*n*-butyl phthalate (MBP), MEHP, and monoethyl phthalate (MEP) were associated with poor sperm morphology and vigor, and with low sperm concentration, motility, and linearity [31–35]. However, it is not yet possible to conclude whether phthalate exposure is harmful for human reproduction.

2 Production, Properties, and Uses

Phthalates were first introduced in the 1920s and quickly replaced the volatile and odorous camphor. In the earliest 1930s, the commercial availability of PVC and the development of DEHP caused the boom of the plasticizer PVC industry.

The global production of DEHP in 1994 was estimated to be between 1 and 4 million tons per year. The production volume of DEHP in Western Europe was 505,000 tons per year in 1997. In 2000, the European Union (EU) estimated a production of phthalates around 1 million ton per year in Western Europe (worldwide approximately 7 million tons), being DEHP the 60% of the production [36]. More recent information from industry shows that the use of DEPH in the EU has decreased to 221,000 in 2004, whilst the use of diisononyl phthalate (DiNP) and diisodecyl phthalate (DiDP) has increased during the same period. In fact, the annual production of DiNP in the EU in 2005 was estimated around 500,000 tons [37]. DiNP and DiDP are replacing DEHP as plasticizer because the use of DEHP has been limited due to it has been classified as Category 1A reprotoxin and it is included in the Annex XIV of the EU REACH legislation [38]. Between 1999 and 2004 the proportion of DEHP to total phthalate usage decreased from 42% to 22% and the proportion of DiNP and DiDP rose from 35% to 58% [37].

According to their physical chemical characteristics (Table 1), phthalate esters have a wide range of properties that extends up to six orders of magnitude. Especially important is the marked decreases in volatility and solubility in water with increased molar volume or alkyl chain length. The low solubility in water and the high K_{OW} values of the high-molecular weight phthalates could result in their strong sorption to dissolve organic carbon and their availability to suffer bioconcentration process. The lower molecular weight phthalate esters are quite volatile, but owing to their very low K_{AW} values they will volatilize very slowly from aqueous solution. The log K_{OW} values vary from 1.61 to 9.46; thus, the high-molecular weight esters are very hydrophobic and will sorb strongly to organic matter and surfaces. The high values of K_{OA} indicate that the higher molecular weight esters present in the atmosphere are weakly adsorbed to aerosol particles and to soil and vegetation. Air–water partition coefficients increase with increasing molecular weight; thus, the higher molecular weight phthalate esters will potentially evaporate more rapidly from water, but this will be mitigated by adsorption to suspended matter in the water column. The phthalate esters also show significant and systematic differences in reactivity or half-life, with the primary biodegradation half-life increasing with increasing alkyl chain length, showing the opposite trend the photo-oxidation half-life.

PAEs are mainly used as plasticizers but they are also used in a large variety of products, from entering coating of pharmaceutical pills and nutritional supplements to viscosity control agents, film formers, stabilizers, dispersants, lubricants, binders, emulsifying, and suspending agents. End-applications include adhesives and glues, agricultural adjuvants, building materials, personal-care products, medical devices, detergents and surfactants, packaging children's toys, modeling clay, waxes, paints, printing inks and coatings, pharmaceuticals, food products, and textiles. Most of the high molecular weight phthalates esters are used in the manufacture of a wide variety of vinyl goods, both industrial and consumers. The lower molecular weight phthalates have a very broad use which includes consumer products and pharmaceutical [8].

The uses of PAEs are directly related with their physical–chemical properties (Table 1), which are of very broad spectrum. So, DMP is used to manufacture solid rocket propellant and some consumer products such as insect repellents and plastics [39, 40]. DEP is an industrial solvent used in many consumer products, particularly those containing fragrances. Products that may contain DEP include perfume, cologne, deodorant, soap, shampoo, and hand lotion [39, 40]. DBP and diisobutyl phthalate (DiBP) are industrial solvents or additives used in many consumer products such as nail polish, cosmetics, some printing inks, pharmaceutical coatings, and insecticides [41]. About 76% of DBP goes to plasticizing of PVC or other polymers, 14% is used in adhesives, 7% in printing inks and the remaining 3% of DBP is used in other miscellaneous application. BBP is an industrial solvent and additive used in products such as adhesives, vinyl-flooring products, sealants, car-care products, and to a lesser extent, some personal-care products [39, 40]. More than 90% of BBP goes to plasticizing of PVC or other polymers. DEHP is primarily used (80%) to produce flexible plastics, mainly polyvinyl chloride, which is used for many home and garden products, toys, packaging film, and blood-product storage and intravenous delivery systems [42]. Concentrations in plastic materials may reach 40% by weight. DEHP has been removed from or replaced in most children's toys and food packaging in the United States (US) and the EU. Other sources of exposure include foods and foods in contact with plastic containing DEHP.

Di-*n*-octylphthalate (DnOP) is used primarily to produce flexible plastics. Diisononyl phthalate (DiNP) is actually a mixture of phthalates with branched alkyl side chains of varying length (C8, C9, and C10). DiNP is primarily used (more than 95%) to produce flexible plastics and has been used to replace DEHP in some plastics. DiNP is now widely used in products such as children's toys, flooring, gloves, drinking straws, and garden hoses. The remaining 5% of DiNP is used in non-PVC applications [43], such as rubber, inks and pigments, adhesives, sealant, paints and lacquers, and lubricants [44]. Diisodecyl phthalates (DiDP) is a mixture of esters of *o*-phthalic acid with C9–C11 (C10 enrichment) alkyl alcohols. These alcohols can be obtained by different processes, yielding different ratios of chain length and branching distribution, which result in different DiDP types. Presently, there are two different DiDP types being used. Many of the constituents are common in both DiDP mixtures and only differ by isomeric distribution curves. Both DiNP and DiDP are mixtures that overlap chemically with each other and cannot analytically be distinguished clearly if both are present in a mixture. Almost 100% of DiDP goes to plasticizing of PVC or other polymers.

3   Legislation

Regulations on phthalate esters cover all aspects of their production, transportation, use, and disposal. Phthalates are regulated under the Clean Water Act, so that at certain manufacturing facilities in the US, wastewater to be treated in municipal sewage treatment plants may be required to undergo pretreatment prior to leaving the facility (Pretreatment Standards).When they become waste products, certain phthalates are subject to Resource Conservation and Recovery Act [45] requirements. Releases to the environment of several phthalate esters are required to be publicly reported in the US, Canada, and Japan.

Phthalate esters have undergone comprehensive risk assessments regarding virtually all aspects of environmental and human health under "existing substances" regulations in the US, Canada, the EU, and at the Organization for Economic Cooperation and Development (OECD) level. The results of the various risk assessments completed to date have led to varying conclusions ranging from no further information needed and/or no needed for further restrictions on use, to proposed requirements for some use-specific risk reduction measures. In this way, up to 12 PAEs, including DBP, BBP, and DEHP are in the list of the proposed substances suspected to produce endocrine alterations published by the EU [46]. According to Section 307 of the US Clean Water Act, DEP, DMP, DEHP, BBP, DBP, and DnOP are considered Priority Toxic Pollutants [45]. On the other hand, DEHP, BBP, and DBP are listed in Annex XIV of the REACH regulation [38]. In 2006, The Australian Government declared DEHP, DiDP, DMP, DiNP, DBP, BBP, DnOP, DEP, and bis(2-methylethyl) phthalate (DMEP) as priority existing chemicals and initiated public risk assessments for these phthalates. Australia concluded the PAE public health hazard assessments in 2008 [47].

DEHP is the most prevalent phthalate used and, thus, the most regulated. The EU has included it in the list of 33 substances of priority or possibly priority substances in the field of water policy and in the Water Framework Directive 2001/2455/EC [48], with the aim to reduce uses and emissions of DEHP to surface waters. A limit

value for surface waters for DEHP of 1.3 ng/mL was set by EU Directive 2008/105/
EC [49]. The World Health Organization (WHO) has established a guideline value
of 8 ng/mL for DEHP for fresh and drinking water [50], which is similar to the
maximum contaminant level (MCL) for DEHP set by the Environmental Protection
Agency [51] (6 ng/mL). This agency recommends the closely monitoring of
concentrations above 0.6 ng/mL [52]. Other institutions as the Netherlands National
Institute of Public Health and Environment [53] and the Danish Environmental
Protection Agency [54] have also established similar limits.

Concerning occupational exposure, the Occupational Safety and Health Administration [55] sets a maximum average of 5 mg of DEHP per cubic meter of air
(5 mg/m^3) in the workplace during an 8 hour shift. The short-term (15 minutes)
exposure limit is 10 mg/m^3. The National Institute for Occupational Safety and
Health [56] and the American Conference of Governmental Industrial Hygienists
[57] recommended a maximum concentration of 5 mg/m^{3}of DEHP in workplace air
for an 8 to 10 hour workday, 40 hour workweek.

The first warning against the use of phthalates in toys was the Recommendation
adopted by the European Commission on 1 July 1998 concerning toys and childcare
articles intended to be placed in the mouth by children under three years of age,
made of soft PVC and containing phthalates. The Commission Decision of 7
December 1999 (1999/815/EC) [58] made it possible to prohibit the use of certain
phthalates on the basis of the legislation on general product [59]. Since 1999, The
Commission Decision 1999/815/EC [58] was extended more than 20 times in the
name of the precautionary principle until the adoption of Directive 2005/84/EC
[60]. This Directive restricted the use of DEHP, DBP, and BBP in the manufacture
of toys and childcare articles intended for children; and DiNP, DiDP, and DnOP are
limited only in toys and childcare articles which can be placed in the mouth. The
restriction states that the amount of phthalates may not be greater than 0.1% by
mass of the plasticized material part of the toys. The member states of the EU
applied this directive from 16 January 2007.

Regarding plastic materials and articles intended to come into contact to
foodstuffs, some restrictions for phthalates were stated in the Commission Directive
2002/72/EC [61]. In 2005, the opinion on certain phthalates (DBP, BBP, DEPH,
DiNP, and DnOP) by the European Food Safety Agency (EFSA) authorities was
published [62]. They set tolerable daily intakes (TDI) for certain phthalates and
estimated that the exposure to humans of particular phthalates was in the same
range as the TDI. In 2007, the Commission Directive 2007/19/EC of 30 March 2007
[63] restricted the use of BBP, DEHP, DBP, DiNP, and DiDP to be employed only
as plasticizer in single use materials and articles containing non-fatty foods except
for infant formulae and in technical support agent in concentrations up to 0.1%
(0.05% in the case of DBP) in the final products. The same directive also fixed a
maximum of the specific migration limits (SML) between 0.3 (for DBP) and 30 (for
BBP) mg/kg food stimulant.

Similar restrictions for materials in contact with foodstuffs were adopted by the
US [64–66], Canada [67], Japan [68], and Australia [47].

4   Levels in the Environment

The migration of PAEs from the polymers leads emissions to the environment during their production, transport, storage, manufacture, use, and disposal [8, 15, 40, 69]. Once in the different environmental compartments phthalates are subject to photo degradation, biodegradation, aerobic and anaerobic degradation and, thus, generally do not persist in the outdoor environment [8, 70].

Some of the most recent data in the literature regarding air PAE levels (Table 2) indicate that phthalates are ubiquitous in the air compartment, with indoor air concentrations generally higher than outdoor air concentrations [5, 11]. Regarding PAEs determined in outdoor air samples, concentrations in urban traffic were higher than in an industrial site in Thessaloniki (Greece). Concentrations of DEHP were significantly higher ($p < 0.05$) at the urban-traffic site (ranging from 4.63 to 45 ng/m^3) than at the urban-industrial site (ranging from ND to 6.5 ng/m^3, what implies an input from vehicular emissions). DEHP was the dominant phthalate with maximum concentration of 45 ng/m^3 [6]. DBP and BBP exhibited lower abundances, whereas DMP, DEP, and DnOP were not found at detectable levels. DEHP was also the prevailing PAE in air particles from Paris (up to 10.4 ng/m^3) followed by DBP (up to 4.6 ng/m^3) [71]. The same trend was observed by Guidotti and coworkers [72] in the breathable fraction of aerosols in the urban area of Rieti (Italy) with high levels of DEHP (up to 1,439 ng/m^3) and DBP (up to 17.6 ng/m^3). Rudel and coworkers [5] reported total concentrations of DEHP, DBP, and BBP up to 230, 32, and 8.5 ng/m^3, respectively, in outdoor air of California. In general, concentrations in urban and suburban areas are usually higher than in rural and remote areas [5]. Higher concentrations of PAEs at urban compared to suburban area were found at Nanjing (China), and were attributed to both continuing releases from many point-sources and environmental recycling [73]. Emissions from waste-water treatment plants and from waste combustion are also considered as possible sources of PAEs in the atmosphere. Phthalates (predominantly butyl, isobutyl, and 2-ethylhexyl derivatives) have been detected in aerosols emitted from the aeration tank of a sewage treatment plant at concentrations ranging from 71.1 to 228.1 ng/m^3 [76, 77]. Those studies reported the occurrence of DEHP at significant concentrations in emissions from uncontrolled burning of domestic wastes since this compound showed the highest emission rate (336.3 mg/kg) in open-air barrel burning.

Because of volatilization and leaching from their application in consumer and personal care products, phthalate esters are ubiquitous contaminants in indoor environment, and the levels found in dust from homes in different countries (Table 2), showed that the less volatile phthalates such as DEHP and BBP, are the predominant in dust samples [11, 16, 74], and that the percentage of both carpet and plastic materials (furniture, decoration, and home electronics) could be associated with higher concentrations of BBP and DEHP in house dust [12, 74, 75]. The proportion of DiNP in house dust from Germany in 2009 [12], indicates that the

Table 2 Phthalate range (or median) concentrations in indoor and outdoor air (expressed in ng/m³) and home dust (expressed in μg/g dry weight) samples from different sites and countries

Sample (site)	City, country	Year	DMP	DEP	DBP	DiBP	BBP	DEHP	DnOP	DiNP	Reference
Outdoor air (urban)	Thessaloniki, Greece	2011	<DL	<DL	0.43–2.4	NA	0.04–0.98	4.63–45	ND–0.11	NA	[6]
Outdoor air (industrial)	Thessaloniki, Greece	2011	<DL	<DL	1.2–3.36	NA	0.11–0.80	ND–6.5	ND–0.11	NA	[6]
Outdoor air (urban)	Paris, France	2006	ND–0.3	0.1–0.9	0.6–4.6	NA	0.1–0.5	2.4–10.4	ND–0.6	NA	[71]
Outdoor air (industrial)	California, USA	2010	NA	90–200	6–32	5–18	0.8–8.5	8–230	ND	NA	[11]
Indoor air (inside home)	California, USA	2010	NA	110–2,500	50–1,200	17–1,700	1–180	40–220	ND	NA	[11]
Outdoor air (urban)	Rieti, Italy	1998	NA	NA	17.6	NA	NA	1,439	NA	NA	[72]
Outdoor air (urban, city)	Nanjing, China	2008	3.9–21.8	1.1–9.3	28.1–66.7	NA	1.0–7.1	6.8–28.7	0.3–2.4	NA	[73]
Outdoor air (suburban)	Nanjing, China	2008	1.9–7.3	0.08–3.7	4.1–20	NA	0.01–2.1	3.4–12	ND	NA	[73]
Indoor air (inside home)	USA	2003	NA	130–4,300	52–1,100	NA	<31–480	<59–1,000	NA	NA	[5]
Indoor dust (inside home)	Beijing, China	2011	ND–1.6	0.1–0.6	7–31.5	7.2–83.2	0.1–1.1	47.6–883	ND–0.5	NA	[16]
Indoor dust (inside home)	Albany, USA	2011	ND–3.3	0.7–11.8	4.5–94.5	0.7–33.4	3.6–393	37.2–9,650	ND–14.1	NA	[16]
Indoor dust (inside home)	Fyn island, Denmark	2010	NA	1.7	15	27	3.7	210	NA	NA	[74]
Indoor dust (inside home)	Sweden	2005	NA	NA	150	47	135	770	NA	NA	[75]
Indoor dust (inside home)	Germany	2009	NA	NA	87	NA	15	604	NA	129	[12]

ND non detected, *NA* non analyzed, *DL* detection limit

substitution of BBP and DEHP by DiNP in various products, is increasing their levels in our nearby.

The presence and persistence of PAEs in aquatic environments directly depend on their chemical–physical properties, and degradation/biodegradation processes. They are practically insoluble in pure water (Table 1), but they are soluble by fulvic and humic acids as complexes and/or adsorbed onto particulate matter, with sediments being the final sink [78, 79]. Chemical–physical degradation processes (e.g., photolysis and hydrolysis) are not significantly effective for environmental degradation [80, 81]. However, microorganism degradation (anaerobic, aerobic, and facultative) is thought to be the principle mechanism for PAE degradation in soil and sediments [15]. According to the results from Yuan and coworkers [78], the average aerobic biodegradation half-life range in sediments were from 2.5 (DEP) to 14.8 (DEHP) by day while the average anaerobic biodegradation range were from 14.4 (DBP) to 34.7 (DEHP). In relation to these numbers, the aerobic degradation half-lives are smaller than anaerobic degradation, being DEHP the most difficult to degrade in any case.

In accordance with previous considerations, and taking into account that DEHP is the most used PAEs, it is logical that DEHP dominates the phthalate concentrations in the environment, with levels found in sediments higher than in surface water.

Thus, in a study conducted in Germany in 2002 [3], BBP, DBP, and DEHP were measured in various compartments (surface water, sediments, sewage treatment plants effluents, sewage sludge, dump water, and liquid manure). DEHP dominated the phthalate concentrations, which ranged from 0.3 to 97.8 μg/L (surface water), 1.74 to 182 μg/L (sewage effluents), 27.9 to 154 μg/g dry weight (d.w.) (sewage sludge), and 0.21 to 8.44 μg/g (sediment). DBP was found only in minor concentrations and BBP only in a few samples and in low amounts. Very high concentrations of phthalates were confirmed in waste dump water and compost water samples as well as in the liquid manure samples.

In another study conducted in river waters and sediments of central Italy, a direct relationship between PAEs concentration levels in water samples from rivers and lakes with input of urban or industrial treated wastewaters near the sampling point were also found [82]. They also found an accumulation factor in sediment samples ranging from 10 to 100, showing the trend of PAEs to be absorbed in sediments. DEHP and DBP were found in higher concentration levels than the other seven PAEs investigated. The presence of PAEs in the studied freshwaters was closely related to the input of urban and industrial treated wastewaters. DEPH concentrations in freshwater and sediment samples ranged from 0.3 to 31.2 μg/L and from 0.003 to 0.49 μg/g, respectively.

Eight PAEs were measured in 14 surface waters and 6 sediments taken from various rivers in Taiwan [78]. Like in the previous mentioned studies [3, 82], the concentration levels of DEHP and DBP found were higher than the rest of PAEs investigated. DEHP concentrations in the water and sediment samples ranged from ND to 18.5 μg/L and from 0.5 to 23.9 μg/g, respectively. DBP concentrations in the water and sediment samples ranged from 1.0 to 13.5 μg/L and from 0.3 to 30.3 μg/g, respectively. They found that aerobic degradation rates were up to ten times faster than anaerobic degradation rates, being DEHP difficult to degrade in both, aerobic and anaerobic degradation.

5 Toxicity of PAEs and Their Metabolites

All phthalates have low acute toxicity, but some of them are suspected of having chronic effects in humans. Most of the toxicological studies have been conducted in animals (mainly rodents), showing adverse effects on liver, kidney, and the reproductive system, and acting as endocrine disrupting agents [83], but they have not been classified with respect to carcinogenicity by the International Agency for Research on Cancer (IARC), the Environmental Protection Agency (EPA), and the National Toxicology Program (NTP), because of peroxisomal proliferation, which may be a pathway to the development of liver cancers in animals, may be a less relevant pathway in humans [84].

Phthalates do not seem to act via direct hormonal mimicking. However, in rodents, some phthalates (BBP, DiBP, DBP, DEP, DEHP, and DiNP) can modulate the endogenous production of fetal testicular testosterone [85–87], resulting in functional and structural impairment of male reproduction and development [85, 88–90], but these effects have not been proved when tested in non-human primates.

Although PAEs act as endocrine disruptors, their way of action is not very clear since the results of different studies are sometimes contradictory, although not all metabolites have been tested. Thus, *in vitro* studies showed that some phthalates have weak or no estrogenic, antiestrogenic, or androgenic activity [91–94], while *in vivo* studies showed no estrogenic [95, 96] and anti-androgenic activities, indicating that exposure to high doses of DEHP, DBP, and BBP during the fetal period have produced lowered testosterone levels, testicular atrophy, and Sertoli cell abnormalities in male animals and ovarian abnormalities in female animals [97, 98]. Additional studies have shown that DEHP and MEHP cause adverse effects on the male reproductive system in rodents [85], and anti-androgenic effects in castrated rats [99, 100], and DBP and DEHP affects sperm parameters [101–103].

Phthalates are suspected of acting as endocrine disrupters also in humans, affecting male and female reproductive tract development. Exposure to PAEs in adult men has been associated with semen quality and alterations in sexual behavior [104], and with endometriosis and intrauterine inflammation (which is a risk factor for prematurity) in adult women [105, 106], as well as other effects. These studies suggest that DEHP may play a role in inducing the intrauterine inflammatory process. Besides the reproductive effects of PAEs, recent studies have also shown the genotoxicity of DEHP, DBP, and DiBP in human lymphocytes and mucosal cells [107, 108].

Regarding toxicity of primary metabolites, some of them (MMP, MBzP, MBP, MEHP and MEP) were associated with poor sperm morphology and vigor, and with low sperm concentration, motility, and linearity [31–35]. Urinary MEP level was associated with lower serum luteinizing hormone (LH) values [34] and sperm DNA damages [32, 109] however, these results regarding to MEP are not supported by animal studies [110, 111]. Urinary MBP and MEHP levels were negatively correlated with free testosterone in exposed workers [112]. Additionally, MBzP exposure was significantly associated with a decrease in serum follicle-stimulating hormone (FSH) level, a marker of spermatogenesis for infertile males in clinical evaluation [113].

Some studies have been conducted to know the relation among PAEs and human infants, which are the subgroups of humans that could be most affected by the presence of PAEs in their parents. The urinary concentration of four phthalate metabolites (MEP, MBP, MBzP, and MiBP) of mother's during pregnancy were negatively correlated with the anogenital index (AGI) which is a weight normalized index of anogenital distance (AGD) [AGD/weight (mm/kg)] in 134 boys of 2–36 months of age [114]. Although some studies of human infants indicate possible associations between PAE exposure and the development of the human reproductive system, other studies of adolescents exposed to DEHP from medical devices as neonates showed no significant adverse effects on their maturity or sexual activity [115].

Studies on health effects of PAEs in humans have remained controversial due to limitations of the study designs. Some findings in human populations are consistent with animal data suggesting that PAEs and their metabolites produce toxic effects in the reproductive system. However, it is not yet possible to conclude whether phthalate exposure is harmful for human reproduction [116].

It is sometimes claimed that the use of animal data for estimating human risk does not provide strong scientific support. However, because it is difficult to find alternative methods to test the direct toxic effects of chemicals, continuance of studies in animals is required for risk assessment of chemicals including PAEs.

6   Estimation of PAEs Exposure in Human Populations

As it was stated earlier, many studies have suggested that PAEs and their metabolites produce reproductive and developmental toxicities in laboratory animals. Although the most of these animals were exposed to PAEs at relatively high level to exam toxicological effects, some studies showed that relatively low doses of PAEs caused toxic effects [88, 117, 118]. Thus, there is a question of whether humans are exposed to PAEs at a severe enough level to generate human health effects.

Human exposure to PAEs may occur via ingestion, inhalation, and dermal pathways. Dermal exposition is usually very low, being via ingestion (food) and inhalation (indoor/outdoor ambient) the most important ones. An estimation study concerning sources of phthalate exposure among Europeans showed that dermal application of consumer products dominated the sources of DMP, DEP, and BBP, whereas dietary intake was the major source of exposure to DBP, DiBP, and DEHP [119]. In a study of general population in Japan, dietary intake and inhalation accounted for less than 50% of total daily exposure to DMP, DEP, and DBP, whereas dietary intake was the dominant source of exposure to BBP and DEHP [120]. The fact that the application of exposure models for the evaluation of sources of phthalate exposures yielded variable results means that the extent of exposure via different pathways depends on the individual phthalate, the geographic area, the age, and the lifestyle of the consumers [121].

Food intake could be an important way of PAE exposure in human population. Some authors [17, 18, 122] estimated that contaminated food makes up around 90% of the DEHP intake. PAEs (DEHP, BBP, DBP, DiDP, and DiNP) may be present in foodstuffs, either due to migration from food contact materials containing PAEs or due to its widespread presence as an environmental contaminant which can be found in air, water, soil, and food. Contamination of food by PAEs can occur during processing, handling, transportation, and packaging of food and via secondary food storage articles. During processing, food may be contaminated from PVC tubing and other process equipment containing DEHP [123], from lubricants used in the food processing industry, via polymer and non-polymer components of food packaging [124] where high amount of DBP was found in the food wrap or food packaging material.

In 2005, the EFSA [62] made an estimation of PAE exposure in human populations based on the limited available literature on DEHP, DBP, BBP, DiNP, and DiDP concentration in foods and diets. Some studies have been conducted in two different populations in United Kingdom (UK) and Denmark from 1996 to 2003 [124–129]. Based on the information obtained from the mentioned studies, the EFSA estimated the daily oral intake and the maximum dietary exposure (calculated in the 95th percentile) (MDE) for the most used PAEs (Table 3).

In general, the total daily oral intakes estimated for the PAEs studied were higher in Denmark studies [128, 129] than those estimated in UK studies [124–127]. In the study carried out at the regional levels in Denmark in 2003 [129], the total daily oral intakes estimated for DBP, BBP, DEHP, DiNP, and DiDP for children aged from 1 to 6 years were always higher than those estimated for children aged from 7 to 14 years, being the adult values always the lowest ones. In the case of DiNP and DiDP the values for children aged from 6 to 12 months were very high. In the case of DiNP, the difference between the estimated oral daily intake in Denmark for infants aged 6-12 months (216 μg/kg b.w./day) and adults (5 μg/kg b.w./day) was of 211 μg/kg b.w./day, which was the highest one (Table 3).

Based on the calculated Maximum Dietary Exposure (MDE) of PAEs, and the Non Observed Adverse Effect Level (NOAEL) calculated from the available toxicology evidence, mainly hepatic, renal changes and reproductive toxicity in animals [88, 89, 130–133], and making an uncertainty factor between 100 and 200, the EFSA panel calculated the Tolerable Daily Intake (TDI) for DBP, BBP, DEHP,

Table 3 Estimation of maximum dietary exposure (MDE) (95th percentile), non observed adverse effect level (NOAEL) and tolerable daily intake (TDI) of the most used PAEs according to EFSA [62]

Phthalate	MDE (μg/kg b.w./day)	NOAEL (mg/kg b.w./day)	TDI (μg/kg b.w./day)
DBP	0.5 (adults)[a]	2 (LOAEL)	10
	10.2 (adults)[b]		
	1.6 (adults)[c]		
	3.5 (7–14 years)[c]		
	8 (1–6 years)[c]		
BBP	3 (adults)[a]	50	0.5
	4.5 (adults)[b]		
	1 (adults)[c]		
	2.4 (7–14 years)[c]		
	5.9 (1–6 years)[c]		
DEHP	5 (adults 60 kg)[a]	5	50
	15.7 (adults 70 kg)[b]		
	4.5 (adults)[c]		
	11 (7–14 years)[c]		
	26 (1–6 years)[c]		
DiNP	<0.17 (adults 60 kg)[d]	15	150
	2.4 (0–6 months)[e]		
	1.8 (>6 months)[f]		
	5 (adults)[c]		
	10 (7–14 years)[c]		
	63 (1–6 years)[c]		
	216 (6–12 months)[c]		
DiDP	<0.17 (adults 60 kg)[d]	15	150
	2.4 (0–6 months)[e]		
	1.8 (>6 months)[f]		
	3 (adults)[c]		
	7 (7–14 years)[c]		
	53 (1–6 years)[c]		
	210 (6–12 months)[c]		

[a][124]
[b][128]
[c][129]
[d][125]
[e][126]
[f][127]

DiNP, and DiDP (Table 3). The EFSA Panel concluded that exposure to DEHP and BPD from food consumption was in the range of the TDIs, and that of DiNP and DiDP were well below the TDI values.

The TDI values settle by the EFSA in 2005 (Table 3) were different from those calculated by the Scientific Committee for Toxicity, Ecotoxicity and the Environment (CSTEE) in 1998 [134] based on the phthalate migration from soft PVC toys and child care articles, and the available toxicity studies on animals at that time [117, 122, 135], with the following values: DBP (TDI of 100 μg/kg b.w./day), BBP (TDI of 200 μg/kg b.w./day), and DiDP (TDI of 200 μg/kg b.w./day), but they

Table 4 Estimated daily intakes (EDI) of phthalates based on the geometric mean values for urinary metabolites estimated by David [137] for CDC data measured in 289 US individuals [138] and the tolerable daily intake (TDI) values calculated by EFSA [62], CSTEE [134], and MHLW [68], as well as the reference dose of phthalates (RfD) calculated by EPA [136] (in μg/kg b.w./day)

Phthalate	EDI [138]		TDI (EU) [134]	TDI (EU) [62]	RfD (US) [136]	TDI (Japan) [68]
	Geometric mean	95th percentile				
DEP	12.34	93.3				
BBP	0.73	3.34	200	500	200	–[a]
DBP	1.56	6.87	100	10	100	–[a]
DEHP	0.60	3.05	37	20	20	4.0–140
DnOP	<LOD	–[a]	370	–[a]	–[a]	–[a]
DiNP	0.21	1.08	150	150	–[a]	150

[a]Non established

were similar for DEHP and DiNP (Table 3). The TDI values calculated by the EFSA (2005) were similar to the reference dose (RfD) calculated by the EPA (BBP: 200 μg/kg b.w./day, DBP: 100 μg/kg b.w./day, and DEHP: 20 μg/kg b.w./day) established in 2006 [136] and those TDI values established in 2002 by the Japanese Government (DEHP: 4–140 μg/kg b.w./day and DiNP: 150 μg/kg b.w./day [68]; Table 4).

Nevertheless, to estimate human exposure to PAEs measuring the chemicals in foodstuffs, collecting survey/questionnaire data on personal lifestyle and food consumption are not very satisfactory because there are other sources which contributed to the overall human exposure to PAEs (e.g., dermal contact and environmental media). Because of that, since the late 1990s many studies have been conducted with the target to prove that the urinary concentration of PAE metabolites could be used as biomarkers to estimate dose in risk human assessment of PAEs.

Blood and urine are the most common matrices for biomonitoring the most common PAEs metabolites, shown in Table 5, but urine is the matrix of choice for biomonitoring of non-persistent chemicals because urinary concentrations of these compounds or their metabolites are higher than blood concentrations. People exposed to DMP, DEP, and DBP will excrete MMP, MEP, and MBP in their urine. The amount of MMP, MEP, and MBP is an indicator of how much contact with the parental PAEs occurred. Small amounts of 3-carboxy-monopropyl phthalate (MCPP) are also produced from DBP. In addition, people exposed to BBP will excrete MBzP and small amounts of MBP in their urine [23]. DEHP is metabolized into various metabolites [20, 24, 139, 140]. Three of these metabolites are MEHP, 5oxo-mono(2-ethylhexyl) phthalate (5oxo-MEHP), and 5OH-mono(2-ethylhexyl) phthalate (5OH-MEHP). MEHP is primarily formed by the hydrolysis of DEHP in the gastrointestinal tract and then absorbed. 5OH-MEHP and 5oxo-MEHP are produced by the oxidative metabolism of MEHP and are present at roughly three- to tenfold higher concentrations than MEHP in urine [139, 140]. People exposed to DnOP will excrete small amounts of mono-*n*-octyl phthalate (MnOP) in their urine. People exposed to DiNP will excrete small amounts of monoisononyl phthalate (MiNP) in their urine.

Table 5 The most common parent phthalates and their corresponding monoester (primary) and oxidized (secondary) metabolites

Parent phthalate	Monoester metabolite	Secondary oxidized metabolite
Dimethyl phthalate (DMP)	Monomethyl phthalate (MMP)	
Diethyl phthalate (DEP)	Monoethyl phthalate (MEP)	
Dicyclohexyl phthalate (DCHP)	Monocyclohexyl phthalate (MCHP)	
Di-*n*-pentyl phthalate (D*n*PeP)	Mono-*n*-pentyl phthalate (M*n*PeP)	
Butylbenzyl phthalate (BBP)	Monobenzyl phthalate (MBzP)	
Diisobutyl phthalate (DiBP)	Monoisobutyl phthalate (MiBP)	
Di-*n*-butyl phthalate (DBP)	Mono-*n*-butyl phthalate (MBP)	
		3-Carboxy-mono-propyl phthalate (MCPP)
Di(2-ethylhexyl) phthalate (DEHP)	Mono(2-ethylhexyl) phthalate (MEHP)	
		5OH-mono (2-ethylhexyl) phthalate (5OH-MEHP)
		5oxo-mono (2-ethylhexyl) phthalate (5oxo-MEHP)
		5carboxy-mono (2-ethylhexyl) phthalate (5cx-MEPP)
		2carboxy-mono (2-ethylhexyl) phthalate (2cx-MMHP)
Di-*n*-octyl phthalate (DnOP)	Mono-*n*-octyl phthalate (MnOP)	
Diisononyl phthalate (DiNP)	Monoisonyl phthalate (MiNP)	
		7OH-monomethyloctyl phthalate (OH-MiNP)
		7oxo-monomethyloctyl phthalate (oxo-MiNP)
		7carboxy-monomethylheptyl phthalate (cx-MiNP)
Diisodecyl phthalate (DiDP)	Monoisodecyl phthalate (MiDP)	
		OH-monoisodecyl phthalate (OH-MiDP)
		oxo-monoisodecyl phthalate (oxo-MiDP)
		Carboxymonoisodecyl phthalate (cx-MiDP)

As with DEHP, other oxidative metabolites of DiNP are probably the most abundant urinary metabolites [141]. Because DiNP is a complex mixture, MiNP may not reflect total exposure to all DiNP components. Thus, low molecular weight phthalates (e.g., DEP and DBP) mostly metabolize to their hydrolytic monoesters and high molecular weight phthalates (e.g., DEHP and DiNP) metabolize to their hydrolytic monoesters, which are transformed to oxidative products.

Since Blount and coworkers [138] published in 2000 the levels of seven urinary monoesters phthalates metabolites: MEP, MBP, MBzP, monocyclohexyl phthalate (MCHP), DEHP, MnOP, and MiNP found in 289 US adults the number of PAE metabolites have increased to 22, although only 11 have been used as biomarkers [142]. The results obtained by Blount and coworkers [138] were used by David [137] to estimate the daily intake levels (Table 4) by using the following equation:

Daily intake (mg/kg/day) = urinary concentration (mg/g creatine) × creatinine excretion (g/kg/day) × (monoester in urine (mol)/diester ingested (mol)) × (molecular weight of diester (g/mol)/molecular weight of monoester (g/mol)).

As shown in Table 4, all estimated PAE intakes in the US population in the year 2000 were lower than the TDI values settle by the EU Scientific Committee for Toxicity, Ecotoxicity and the Environment [134], the US Reference Dose [136], the TDI values established by Japanese Government [68], and the EFSA [62]. Among these PAEs, the estimated daily intake levels of DEHP were not as high as expected, taking into account that DEHP is the most commonly used plasticizer for flexible PVC formulations at that time and it is a widespread environmental contaminant [18]. Later studies showed the low sensitivity of the biomarker MEHP for assessing DEHP exposure, exploring two secondary metabolites of DEHP, the 5oxo-MEHP and the 5OH-MEHP [139, 140, 143]. They found that the major metabolites of DEHP in urine were 5oxo-MEHP and 5OH-MEHP, which were fourfold higher than MEHP.

Recently, other secondary oxidized metabolites of DEHP have been recognized [144]. Although 5OH-MEHP and 5oxo-MEHP in the urine reflect short-term exposure levels of DEHP, other secondary oxidized metabolites of DEHP such as 5carboxy-mono (2-ethylhexyl) phthalate (5cx-MEPP) and 2carboxy-mono (2-ethylhexyl) phthalate (2cx-MMHP) are considered excellent parameters for measurement of the time-weighted body burden of DEHP due to their long half-times of elimination. Biological monitoring in a German population ($n = 19$) indicated that 5cx-MEPP is the major urinary metabolite of DEHP. Median concentrations of the metabolites of DEHP were 85.5 µg/L (5cx-MEPP), 47.5 µg/L (5OH-MEHP), 39.7 µg/L (5oxo-MEHP), 9.8 µg/L (MEHP), and 36.6 µg/L (2cx-MMHP) [145]. Furthermore, oxidized metabolites of DiNP have been recently introduced as new biomarkers for measurement of DiNP exposure [146, 147].

Table 6 shows a brief overview of biomonitoring urinary data on phthalate exposure in Germany, US, Taiwan, and China populations. In some of those studies only primary metabolites have been analyzed (i.e., Taiwan), in others also the oxidized metabolites of DEHP were included in the study, even the DiNP metabolites. It is clear the need for analyzing the secondary metabolites in the case of DEHP and DiNP, because as it can be seen in all the studies gathered in Table 6, the MEHP represents a percentage lower than 12% of the total DEHP metabolites. In the case of DiNP the need for calculating the oxidized metabolites is still important, due to the fact that in three studies [28, 29, 146, 148, 150] the levels

Table 6 Selected biomonitoring data on phthalate metabolites exposure in different countries (median, range or 95th percentile), expressed in µg/L, found in different population

		Germany	US	Taiwan	China[a]	Germany	Germany
Country							
Sample number		25	2605/129 (DiNP, DiDP)	60	99	85	399
Age		Adults	>6 years, DiDP and DiNP adults	Adults	Adult women	7–64 years	14–60 years
Sampling year		Unknown	2003–2006	<2008	<2010	2002	2005
Reference		[28]	[142, 146, 148]	[149]	[16]	[140]	[150]
Parent phthalate	Metabolite						
DMP	MMP		1.2 (16.3)	32.3 (115)	12		
DEP	MEP		174 (2,660)	ND (193)	21.5	90.2 (2.4–2,111)	
BBP	MBzP	5.6 (1.7–27.4)	14.3 (101)	2.2 (15.1)	0.6	21 (1.2–268)	7.2 (<0.3–538)
DiBP	MiBP	48.4 (8.1–133)	2.6 (17.9)		56.7		44.9 (3.1–525)
DBP	MnBP	50.8 (12.6–367)	23.2 (121)	36.5 (142)	61.2	181 (14.5–1,256)	49.6 (2.3–1,509)
	MCPP		3.1 (15.2)				
DnOP	MnOP		<LOQ (<LOQ)			<LOQ (<LOQ–5.9)	
DEHP	MEHP	4.7 (<LOQ–142)	1.9 (31)	15.9 (56.2)	2.1	10.3 (<LOQ–177)	4.9 (<0.3–207)
	5OH-MEHP	14.7 (3–258)	21.1 (266)		7	46.8 (0.5–818)	19.2 (0.6–682)
	5oxo-MEHP	13.5 (1.3–174)	14.4 (157)		11.3	36.5 (0.5 544)	14.7 (0.5–447)
	5cx-MEPP	25.1 (3.3–358)	33 (337)		30		26.2 (0.8–947)
	2cx-MMHP	6.5 (1.4–98)			18.5		8.3/0.4–301)
DiNP	MiNP		<LOQ (<LOQ)				
	OH-MiNP	2.5 (<LOQ–287)	13.2 (43.7)				5.5 (<LOD–698)
	oxo-MiNP	1.3 (<LOQ–174)	1.2 (6.6)				3 (<LOD–304)
	cx-MiNP	5 (<LOQ–260)	8.4 (46.2)				
DiDP/DPHP	MiDP		<LOQ (<LOQ)				
	OH-MiDP		4.9 (70.6)				
	oxo-MiDP		1.2 (15)				
	cx-MiDP		4.4 (104)				

LOD limit of detection, *LOQ* limit of quantification, *ND* non detected

[a]Median

of MiNP are lower than LOQ, and only calculating the secondary metabolites, it would be possible to asses the exposure levels of DiNP.

The comparison of the PAE metabolite concentrations obtained in different populations (Table 6) should be carried out cautiously due to differences in study design (e.g., different way, type and time of sampling, range of ages, number of samples, etc.). Besides, it is known that there are variations from person to person in the proportions or amounts of the metabolite excreted after people received similar doses [23] as well as variation in the same person during repetitive monitoring [116, 151]. In addition, the proportion of each metabolite for a given phthalate may vary also by differing routes of exposure [19, 152].

Taking into account the previous considerations, the data reported in Table 6 suggest that PAE levels are within the same order of magnitude in the four countries (US, Germany, China, and Taiwan), but some differences in the exposure of PAE metabolites could be appreciated.

The median concentrations of MEP in urine samples reported by China and Taiwan are much lower than the median concentration reported in the US population [148]. Concentrations of MiBP reported by the US [148] are about 16-fold lower than those reported by Germany [28] and China [16]. The exposure to DEHP appears nearly similar in the US, Germany, and China, while the exposure to DiNP is higher in US [146] than in German populations [28, 150].

The EDI of phthalates in China, Germany, Taiwan, and US populations are shown in Table 7. The calculation was based on phthalate metabolite (primary and secondary) concentrations, the model of David [137] and the excretion fractions according to various authors [23, 28, 143, 144]. DEHP median values are very close or clearly exceed the TDIs and RfD values (Table 4). The median values for the rest of PAEs are below levels determined to be safe for daily exposures estimated by the US (RfD), the EU and Japan (TDI) (Table 4). However, the upper percentiles of DBP and DEHP urinary metabolite concentrations suggested that for some people, these daily phthalate intakes might be substantially higher than previously assumed and exceed the RfD and TDIs.

Furthermore, special situations including using DBP containing medications [41], platelet donation [144], or intensive medical interventions [154–156] can result in daily intakes that exceed the RfD or TDI for long periods of time and/or are close to levels where first toxic effects have been observed in animals. The toxicological significance of reaching the RfD and/or TDI of multiple phthalate exposures among susceptible subpopulations (e.g., children and pregnant women), of the time when these high exposures occurred (e.g., prenatal exposures) and of coexposures to other endocrine disruptor chemicals (EDCs) remains unclear and warrants further investigation.

With the exception of MEP, the NHANES 2003–2004 [148] subsamples showed that children aged 6–11 years excreted higher concentrations of metabolites than did older age groups, a finding that has been noted in other studies of German adults and children for DEHP metabolites [22, 157]. The 2003–2004 NHANES [148] subsamples also showed other differences in concentrations of specific phthalate metabolites by age, gender, and race/ethnicity. Finding a detectable amount of one

Table 7 Estimated daily intakes of phthalates in different countries, median (range or 95th percentile), based in urinary metabolite concentrations, expressed in μg/kg b.w./day

| | DMP | DEP | DBP | DiBP | BBP | DEHP | | | | DiNP | |
						MEHP	5OH-MEHP	5oxo-MEHP	5cx-MEPP	MiNP	OH-MiNP
China n = 99 Adult women <2010 [16]	0.6	1.1	8.5			8.7	1.5	2.2			
Germany n = 399 14–60 years 2005 [150]		1.7 (4.2)		1.7 (5.2)	0.2 (1.2)	2.2 (7.2)	2 (6.5)	2.3 (7.2)			0.7 (3.5)
Taiwan n = 60 Adults <2008 [149]			ND (ND–27.9)	2.2 (ND–23.5)	0.2 (ND–1.6)	33.9 (0.1–310)					
Germany n = 85 7–64 years 2002 [140]		2.32 (0.33–69)		5.2 (1.8–22.6)	0.6 (0.16–4.5)	10.3 (<LOQ–165)	13.5 (2.6–166)	14.2 (2.9–147)			
US n = 214 Pregnant women 1991–2002 [153]		6.64 (112)	0.84 (2.33)	0.12 (0.41)	0.5 (2.47)	1.32 (9.32)	1.7 (10.7)				

ND non detected, *LOQ* limit of quantification

or more phthalate metabolites in urine does not mean that they cause a negative health effect. More research is needed to know if these levels of phthalate metabolites could cause negative effects in human health. The PAE metabolite levels found in urine from different populations only provide a reference range and, at the moment, it is only possible to determine whether people have been exposed to levels above those affecting the general population. These data will also help scientists to design new research studies on phthalate exposure and health effects.

Acknowledgments Authors thank MICINN (AGL2009-09733) and CM (P2009/AGR-1464) for financial support.

References

1. Cousins IT, Mackay D, Pakerton TF (2003) Physical–chemical properties and evaluative fate modeling of phthalates esters. The handbook of environmental chemistry, vol. 3, Part Q. Springer, Berlin
2. Heudorf U, Mersch-Sundermann V, Angerer J (2007) Phthalates: toxicology and exposure. Int J Hyg Environ Health 210:623–634
3. Fromme H, Küchler T, Otto T, Pilz K, Müller J, Wenzel A (2002) Occurrence of phthalates and bisphenol A and F in the environment. Water Res 36:1429–1438
4. Polo M, Llompart M, Garcia-Jares C, Cela R (2005) Multivariate optimization of a solid-phase microextraction method for the analysis of phthalate esters in environmental waters. J Chromatogr A 1072:63–72
5. Rudel RA, Dodson RE, Perivich LJ, Morrelo-Frosch R, Camann DE, Zuniga MM, Yau AY, Just AC, Brody JG (2010) Semivolatile endocrine disrupting compounds in paired indoor and outdoor air in two Northern California communities. Environ Sci Technol 44:6583–6590
6. Salapasidou M, Samara C, Voutsa D (2011) Endocrine disrupting compounds in the atmosphere of the urban area of Thessaloniki, Greece. Atmos Environ 45:3720–3729
7. Uhde E, Bednarek M, Fuhrmann F, Salthammer T (2001) Phthalic esters in the indoor environment-test chamber studies on PVC-coated wallcoverings. Indoor Air 11:150–155
8. Stanley MK, Robillard KA, Staples CA (2003) Phthalate esters. In: Staples CA (ed) The handbook of environmental chemistry, vol 3, Part Q. Springer, Berlin
9. Inoue K, Higuchi T, Okada F, Iguchi H, Yoshimura Y, Sato A, Nakazawa H (2003) The validation of column-switching LC/MS as a high-throughput approach for direct analysis pf di(2-ethylhexyl) phthalate released from PVC medical devices in intravenous solution. J Pharm Biomed Anal 31:1145–1152
10. Lin Z-P, Ikonomou MG, Jing H, Mackintosh Ch, Gobas FAPC (2003) Determination of phthalate ester congeners and mixtures by LC/ESI-MS in sediments and biota of an urbanized marine inlet. Environ Sci Technol 37:2100–2108
11. Rudel RA, Camann DE, Spengler JD, Korn LR, Brody JG (2003) Phthalates, alkylphenols, pesticides, polybrominated diphenyl ethers, and other endocrine-disrupting compounds in indoor air and dust. Environ Sci Technol 37:4543–4553
12. Abb M, Heinrich T, Sorkau E, Lorenz W (2009) Phthalates in house dust. Environ Int 35:965–970
13. Zhu P, Phillips SP, Cao X-L (2010) Chemical contaminants: phthalates. In: Nollet LML, Toldrá F (eds) Handbook of dairy foods analysis. CRC Press, Boca Raton
14. Cao X-L (2010) Phthalate esters in foods: sources, occurrence, and analytical methods. Compr Rev Food Sci Food Saf 9:21–43
15. Staples CA, Peterson DR, Parkerton TF, Adams WJ (1997) The environmental fate of phthalate esters: a literature review. Chemosphere 35:667–749

16. Guo Y, Wu Q, Kannan K (2011) Phthalate metabolites in urine from China, and implications for human exposures. Environ Int 37:893–898

17. Meek ME, Chan PK (1994) Bis(2-ethylhexyl) phthalate: evaluation of risks to health from environmental exposure in Canada. J Environ Sci Health Part C 12:179–194

18. Kavlock R, Boekelheide K, Chapin R, Cunningham M, Faustman E, Foster P, Golub M, Henderson R, Hinberg I, Little R, Seed J, Shea K, Tabacova S, Tyl R, Williams P, Zacharewski T (2002) NTP center for the evaluation of risks to human reproduction: phthalates expert panel report on the reproductive and developmental toxicity of di(2-ethylhexyl) phthalate. Reprod Toxicol 16:529–653

19. Liss GM, Albro PW, Hartle RW, Stringer WT (1985) Urine phthalate determinations as an index of occupational exposure to phthalic anhydride and di(2-ethylhexyl) phthalate. Scand J Work Environ Health 11:381–387

20. Dirven HA, van der Broek PH, Jongeneelen FJ (1993) Determination of four metabolites of the plasticizer di(2-ethylhexyl) phthalate in human urine samples. Int Arch Occup Environ Health 64:555–560

21. Adibi JJ, Perera FP, Jedrychowski W, Camann DE, Barr D, Jacek R, Whyatt RB (2003) Prenatal exposures to phthalates among women in New York City and Krakow, Poland. Environ Health Perspect 111:1719–1722

22. Becker K, Seiwert M, Angerer J, Heger W, Koch HM, Nagorka R, Roßkamp E, Schlüter C, Seifert B, Ullrich D (2004) DEHP metabolites in urine of children and DEHP in house dust. Int J Hyg Environ Health 207:409–417

23. Anderson WAC, Castle L, Scotter MJ, Massey RC, Springall C (2001) A biomarker approach to measuring human dietary exposure to certain phthalate diesters. Food Addit Contam 18:1068–1074

24. Albro PW, Corbett JT, Schroeder JL, Jordan S, Matthews HB (1982) Pharmacokinetics, interactions with macromolecules and species differences in metabolism of DEHP. Environ Health Perspect 45:19–25

25. Rowland IR (1974) Metabolism of di-(2-ethyl-hexyl) phthalate by the contents of the alimentary tract of rat. Food Cosmetic Toxicol 12:293–303

26. Collins MD, William JS, Scott JM, David AE, Heinz N (1992) Murine teratology and pharmacokinetics of the enantiomers of sodium 2-ethylhexanoate. Toxicol Appl Pharmacol 112:257–265

27. Pollack GM, Shenand DD, Dorr MB (1989) Contribution of metabolites to the route- and time-dependent hepatic effects of di-(2-ethylhexyl) phthalate in rats. J Pharmacol Exp Ther 248:176–181

28. Koch HM, Müller J, Angerer J (2007) Determination of secondary, di-iso-nonylphthalate (DINP) metabolites in human urine representative for the exposure to commercial DINP plasticizers. J Chromatogr B 847:114–125

29. Silva MJ, Reidy JA, Kato K, Preau JL, Needham LL, Calafat AM (2007) Assessment of human exposure to di-isodecyl phthalate using oxidative metabolites as biomarkers. Biomarkers 12:133–144

30. Koch HM, Calafat AM (2009) Review. Biomonitoring chemicals used in plastics. Philos Trans R Soc London Ser B 364:2063–2078

31. Duty SM, Silva MJ, Barr DB, Brock JW, Ryan L, Chen ZY, Herrick RF, Christiani DC, Hauser R (2003) Phthalate exposure and human semen parameters. Epidemiology 14:269–277

32. Duty SM, Singh NP, Silva MJ, Barr DB, Brock JW, Ryan L, Herrick RF, Christiani DC, Hauser R (2003) The relationship between environmental exposures to phthalates and DNA damage in human sperm using the neutral comet assay. Environ Health Perspect 111:1164–1169

33. Duty SM, Calafat AM, Silva MJ, Brock JW, Ryan L, Chen Z, Overstreet J, Hauser R (2004) The relationship between environmental exposure to phthalates and computer-aided sperm analysis motion parameters. J Androl 25:293–302

34. Jonsson BA, Richthoff J, Rylander L, Giwercman A, Hagmar L (2005) Urinary phthalate metabolites and biomarkers of reproductive function in young men. Epidemiology 16:487–493

35. Hauser R, Meeker JD, Duty S, Silva MJ, Calafat AM (2006) Altered semen quality in relation to urinary concentrations of phthalate monoester and oxidative metabolites. Epidemiology 17:682–691

36. European Commission (EC) (ed) (2000) Green paper on environmental issues of PVC. COM (2000) 406 final, European Commission, Brussels

37. ECPI (2006) European Council for Plasticizers and Intermediates. The phthalates information center. http://www.plasticisers.org/plasticisers/high-phthalates

38. ECHA (2009) Recommendation of the European Chemical Agency for the inclusion of substances in Annex XIV. European Chemical Agency. 01.06.2009. http://echa.europa.eu/doc/authorisation/annex_xiv_rec/annex_xiv_subst_inclusion.pdf

39. David RM, McKee RH, Butala JH, Barter RA, Kayser M (2001) Esters of aromatic mono-, di-, and tricarboxylic acids, aromatic diacids, and di-, tri-, or polyalcohols. In: Bingham E, Cohrssen B, Powell Patty's CH (eds) Toxicology, vol 6, 5th edn. Wiley, New York

40. EC (2008) Risk assessment reports on phthalates. European Commission. European Chemicals Bureau. http://ecb.rc.it/home.php?CONTENU=/DOCUMENTS/Existing-Chemicals/RISK_ASSESSMENT/

41. Hauser R, Duty S, Godfrey-Bailey L, Calafat AM (2004) Medications as a source of human exposure to phthalates. Environ Health Perspect 112:751–753

42. US FDA (2001) Safety assessment of di(2-ethylhexyl) phthalate (DEHP) released from PVC medical devices. Center for Devices and Radiological Health, U.S. Food and Drug Administration, Rockville. http://www.fda.gov/cdrh/ost/dehp-pvc.pdf

43. ECPI (1997) European Council for Plasticizers and Intermediates. The phthalates information center. http://www.dinp-facts.com/upload/documents/webpage/document2.pdf

44. Legrand P (1996) Plasticizer consumption and end-use patterns in Western Europe. ECPI document, 18.03.1996

45. US EPA (2002) Federal Water Pollution Control Act (Clean Water Act). http://epw.senate.gov/water.pdf

46. EC (2001) Commission communication to the council and the European parliament on the implementation of the community strategy for endocrine disrupters—a range of substances suspected of interfering with the hormone systems of humans and wildlife, COM (1999) 706; COM (2001) 262 final, Brussels

47. Australian Government (2008) Phthalates hazard compendium. A summary of physicochemical and human health hazard data for 24 ortho-phthalate chemicals. Department of Health and Ageing, National Industrial Chemicals Notification and Assessment Scheme (NICNAS). Sydney, Australia

48. EC (2001) Commission Decision 2001/2455/EC of the European Parliament and of the Council of 20 November 2001 establishing the list of priority substances in the field of water policy and amending Directive 2000/60/EC, Off J Eur Commun L 331

49. EC (2008) Commission Directive 2008/105/EC of the European Parliament and of the Council of 16 December 2008 on environmental quality standards in the field of water policy, amending and subsequently repealing Council Directives 82/176/EEC, 83/513/EEC, 84/156/EEC, 84/491/EEC, 86/280/EEC and amending Directive 2000/60/EC of the European Parliament and of the Council. Off J Eur Commun L 348, 24.12.2008, pp 84–97

50. WHO (2003) Chapter 8: Chemical aspects, Guidelines for drinking-water quality, 3rd edn (Draft), World Health Organization, Geneva. http://www.who.int/docstore/watersanitationhealth/GDWQ/Updating/draftguidel/2003gdwq8.pdf

51. US EPA (1993) Integrated Risk Information System (IRIS). Di(2-ethylhexyl) phthalate (DEHP) (CASRN 117-81-7). U.S. Environmental Protection Agency. http://www.epa.gov/IRIS/subst/0014.htm. Accessed 12 Jan 2008

52. US EPA (1991) National Primary Drinking Water Regulations Federal Register, Part 12, 40 CFR Part 141, US EPA, Washington, 1 July
53. NIPHE (1997) Resultaten meetprogramma drinkwater 1997. National Institute of Public Health and the Environment. RIVM rapport 703713 007. October 1998. http://www.rivm. nl/bibliotheek/rapporten/703713007.pdf
54. DEPA (2002) Guidelines on remediation of contaminated sites. Environmental Guidelines No. 7, Danish Environmental Protection Agency. http://www.mst.dk/udgiv/publications/ 2002/87-7972-280-6/pdf/87-7972-281-4.pdf
55. OSHA (2006) Occupational Safety and Health Administration. Permissible Exposure Limits (PELs) Establishing PELs. Regulations (Standards - 29 CFR) Table Z-1 Limits for Air Contaminants. 1910.1000 Table Z-1. http://www.oshagov/SLTC/pel/recognition.html
56. NIOSH (1992) NIOSH recommendation for occupational safety and health. U.S. Department of Health and Human Services, National Institute for Occupational Safety and Health, Cincinnati, OH. Compendium of policy documents and statements, 92-100. http://www. atsdr.cdc.gov/toxprofiles/tp73.pdf
57. ACGIH (2001) Threshold limit values for chemical substances and physical agents and biological exposure indices. American Conference of Governmental Industrial Hygienists, Cincinnati, OH
58. EC (1999) Commission decision 1999/815/EC of 7 December 1999 adopting measures prohibiting the placing on the market of toys and childcare articles intended to be placed in the mouth by children under three years of age made of soft PVC containing one or more of the substances diiso-nonyl phthalate (DINP), di(2-ethylhexyl) phthalate (DEHP), dibutyl phthalate (DBP), di-isodecyl phthalate (DIDP), di-n-octyl phthalate (DnOP), and butylbenzyl phthalate (BBP). Off J Eur Commun L 315
59. EC (1976) Council directive 76/769/EEC on the approximation of the laws, regulations and administrative provisions of the member states relating to restrictions on the marketing and use of certain dangerous substances and preparations. Off J Eur Commun L 262, pp 201–203
60. EC (2005) Commission Directive 2005/84/ EC of the European Parliament and of the council of 14 December 2005 amending for the 22nd time council directive 76/769/EEC on the approximation of the laws, regulations and administrative provisions of the member states relating to restrictions on the marketing and use of certain dangerous substances and preparations (phthalates in toys and childcare articles). Off J Eur Commun L 344
61. EC (2002) Commission Directive 2002/72/EC of 6 August 2002 relating to plastic materials and articles intended to come into contact with foodstuffs. Off J Eur Commun L 220, 15.8.2002, pp 18–58
62. EFSA (2005) Opinion of the Scientific Panel on food additives, flavourings, processing aids and materials in contact with food (AFC). European Food Safety Authority. http://www.efsa. europa.eu/EFSA/ScientificPanels/efsa_locale-1178620753812_AFC.htm
63. EC (2007) Commission Directive 2007/19/EC of 30th March 2007 amending Directive 2002/ 72/EC relating to plastics materials and articles to come into contact with food and Council Directive 85/572/EEC laying down the list of stimulants to be used for testing migration of constituents of plastic materials and articles intended to come into contact with foodstuffs. Off J Eur Commun L 91
64. CRS (2008) Congressional Research Service. CRS report to congress phthalates in plastics and possible human health effects, updated July 29, 2008. http://www.policyarchive.org/bitstream/ handle/10207/19121/RL34572_20080729.pdf?sequence=2. Accessed 28 Dec 2009)
65. US CPSC (2008) Consumer Product Safety Improvement Act. Section 108. Products containing certain phthalates. Consumer Product Safety Commission. http://www.cpsc.gov/ ABOUT/Cpsia/ICPHSO2009.pdf
66. US FDA (2008) Phthalates and Cosmetic Products. U.S. Food and Drug Administration. 7.02.2008. http://www.cfsan.fda.gov/~dms/cos-phth.html. Accessed 31 Aug 2008
67. Ministry of Health of Canada (2009) Canada Consumer Product Safety Act. Bill C-6. 15.12.2009. http://www.parl.gc.ca/HousePublications/Publication.aspx?Docid=3986468&file=4

68. MHLW (2002) Notification of Pharmaceutical and Food Safety Bureau. Yakushoku-hatsu No. 0611001 Amendments to standards for devices, containers package and toys

69. Rudel RA, Perovich LJ (2009) Endocrine disrupting chemicals in indoor and outdoor air. Atmos Environ 43:170–181

70. Xie Z, Ebinghaus R, Temme C, Lohmann R, Caba A, Ruck W (2007) Occurrence and air–sea exchange of phthalates in the Arctic. Environ Sci Technol 41:4555–4560

71. Teil MJ, Blanchard M, Chevreuil M (2006) Atmospheric fate of phthalate esters in an urban area (Paris, France). Sci Total Environ 354:212–223

72. Guidotti M, Giovinazzo R, Cedrone O, Vitali M (1998) Investigation on the presence of aromatic hydrocarbons, persistent organochlorine compounds, phthalates and the breathable fraction of atmospheric particulate in the air of Rieti urban area. Ann Chim-Rome 88:419–427

73. Wang P, Wang SL, Fan CQ (2008) Atmospheric distribution of particulate and gas phase phthalic esters (PAEs) in a Metropolitan City, Nanjing, East China. Chemosphere 72:1567–1572

74. Langer S, Weschler CJ, Fischer A, Bekö G, Toftum J, Clausen G (2010) Phthalate and PAH concentrations in dust collected from Danish homes and daycare centers. Atmos Environ 44:2294–2301

75. Bornehag CG, Lundgren B, Weschler CJ, Sigsgaard T, Hägerhed-Engman L, Sundell J (2005) Phthalates in indoor dust and their associations with building characteristics. Environ Health Perspect 113:1399–1404

76. Lepri L, Del Bubba M, Masi F, Udisti R, Cini R (2000) Particle size distributions of organic compounds in aqueous aerosols collected from above sewage aeration tanks. Aerosol Sci Technol 32:404–420

77. Sidhu S, Gullet B, Striebich R, Klosterman JR, Contreras J, DeVito M (2005) Endocrine disrupting chemical emissions from combustion sources: diesel particulate emissions and domestic waste open burn emissions. Atmos Environ 39:801–811

78. Yuan SY, Liu C, Liao CS, Chang BV (2002) Occurrence and microbial degradation of phthalate esters in Taiwan river sediments. Chemosphere 49:1295–1299

79. Huang P-C, Tien C-J, Sun Y-M, Hsieh C-Y, Lee C-C (2008) Occurrence of phthalates in sediment and biota: relationship to aquatic factors and the biota-sediment accumulation factor. Chemosphere 73:539–544

80. Wolfe NL, Steen WC, Burns LA (1980) Phthalate ester hydrolysis: linear free energy relationships. Chemosphere 9:403–408

81. Peterson DR, Staples CA (2003) Degradation of phthalate esters in the environment. The handbook of environmental chemistry, vol. 3, Part Q. Springer, Berlin

82. Vitali M, Guidotti M, Macilenti G, Cresmini C (1997) Phthalate esters in freshwaters as markers of contamination sources. A site of study in Italy. Environ Int 23:337–347

83. CERHR (2003) National Toxicology Program—Centre for the Evaluation of risks to human reproduction. Monograph on the Potential Human Reproductive and Developmental Effects of Di-Isodecyl Phthalate (DIDP). NIH Pub No. 03-4485. US Department of Health and Human Services, April 2003. http://ntp.niehs.nih.gov/ntp/ohat/phthalates/didp/DIDP_Monograph_Final.pdf

84. Melnick RL (2001) Is peroxisome proliferation an obligatory precursor step in the carcinogenicity of di(2-ethylhexyl) phthalate (DEHP)? Environ Health Perspect 109:437–442

85. Gray LE, Ostby J, Furr J, Price M, Veeramachaneni DNR, Parks L (2000) Perinatal exposure to the phthalates DEHP, BBP, and DINP, but not DEP, DMP, or DOTP, alters sexual differentiation of the male rat. Toxicol Sci 58:350–365

86. Borch J, Axelstad M, Vinggaard AM, Dalgaard M (2006) Diisobutyl phthalate has comparable antiandrogenic effects to di-n-butyl phthalate in fetal rat testis. Toxicol Lett 163:183–190

87. Howdeshell KL, Furr J, Lambright CR, Rider CV, Wilson VS, Gray LE (2007) Cumulative effects of dibutyl phthalate and diethylhexyl phthalate on male rat reproductive tract development: altered fetal steroid hormones and genes. Toxicol Sci 99:190–202

88. Lee KY, Shibutani M, Takagi H, Kato N, Shu T, Unemaya C, Hirose M (2004) Diverse developmental toxicity of di-*n*-butyl phthalate in both sexes of rat offspring after maternal exposure during the period from late gestation through lactation. Toxicology 203:221–238
89. Tyl RW, Myers CB, Marr MC (2004) Reproductive toxicity evaluation of butyl benzyl phthalate in rats. Reproduc Toxicol 18:241–264
90. Foster PM (2006) Disruption of reproductive development in male rat offspring following in utero exposure to phthalate esters. Int J Androl 29:140–147
91. Coldham NG, Dave M, Silvapathasundaram S, McDonnell DP, Connor C, Sauer MJ (1997) Evaluation of a recombinant yeast cell estrogen screening assay. Environ Health Perspect 105:734–742
92. Harris CA, Henttu P, Park MG, Sumpter JP (1997) The estrogenic activity of phthalate esters in vitro. Environ Health Perspect 105:105–108
93. Parks LG, Ostby JS, Lambright CR, Abbott BD, Klinefelter GR, Barlow NJ, Gray LE Jr (2000) The plasticizer diethylhexyl phthalate induces malformations by decreasing fetal testosterone synthesis during sexual differentiation in the male rat. Toxicol Sci 58:339–349
94. Okubo T, Suzuki T, Yokoyama Y, Kano K, Kano I (2003) Estimation of estrogenic and anti-estrogenic activities of some phthalate diesters and monoesters by MCF-7 cell proliferation assay in vitro. Biol Pharm Bull 26:1219–1224
95. Zacharewski TR, Meek MD, Clemons JH, Wu ZF, Fielden MR, Matthews JB (1998) Examination of the in vitro and in vivo estrogenic activities of eight commercial phthalate esters. Toxicol Sci 46:282–293
96. Milligan SR, Balasubramanian AV, Kalita JC (1998) Relative potency of xenobiotic estrogens in an acute in vivo mammalian assay. Environ Health Perspect 106:23–26
97. McKee RH, Butala JH, David RM, Gans G (2004) NTP center for the evaluation of risks to human reproduction reports on phthalates: addressing the data gaps [review]. Reprod Toxicol 18:1–22
98. Jarfelt K, Dalgaard M, Hass U, Borch J, Jacobsen H, Ladefoged O (2005) Antiandrogenic effects in male rats perinatally exposed to a mixture of di(2-ethylhexyl) phthalate and di(2-ethylhexyl) adipate. Reprod Toxicol 19:505–515
99. Stroheker T, Cabaton N, Nourdin G, Regnier JF, Lhuguenot JC, Chagnon MC (2005) Evaluation of anti-androgenic activity of di-(2-ethylhexyl) phthalate. Toxicology 208:115–121
100. Lee B, Koo H (2007) Hershberger assay for antiandrogenic effects of phthalates. J Toxicol Environ Health Part A 70:1365–1370
101. Agarwal DK, Eustis S, Lamb JC IV, Reel JR, Kluwe WM (1986) Effects of di(2-ethylhexyl) phthalate on the gonadal pathophysiology, sperm morphology, and reproductive performance of male rats. Environ Health Perspect 65:343–350
102. Higuchi TT, Palmer JS, Gray LE Jr, Veeramachaneni DN (2003) Effects of dibutyl phthalate in male rabbits following in utero, adolescent, or postpubertal exposure. Toxicol Sci 72:301–313
103. Akingbemi BT, Youker RT, Sottas CM, Ge R, Katz E, Klinefelter GR, Zirkin BR, Hardy MP (2001) Modulation of rat Leydig cell steroidogenic function by di(2-ethylhexyl) phthalate. Biol Reprod 65:1252–1259
104. Swan SH (2008) Environmental phthalate exposure in relation to reproductive outcomes and other health endpoints in humans. Environ Res 108:177–184
105. Cobellis L, Latini G, De Felice C, Razzi S, Paris I, Ruggieri F, Mazzeo P, Petraglia F (2003) High plasma concentrations of di-(2-ethylhexyl) phthalate in women with endometriosis. Hum Reprod 18:1512–1515
106. Reddy BS, Rozati R, Reddy BV, Raman NV (2006) Association of phthalate esters with endometriosis in Indian women. BJOG 113:515–520
107. Kleinsasser NH, Wallner BC, Kastenbauer ER, Weissacher H, Harréus UA (2001) Genotoxicity of di-butyl-phthalate and di-iso-butyl-phthalate in human lymphocytes and mucosal cells. Teratogenesis Carcinog Mutagen 21:189–196

334   M.A. Fernández et al.

M.A. Fernández et al.

108. Kleinsasser NH, Harréus UA, Kastenbauer ER, Wallner BC, Sassen AW, Staudenmaier R, Rettenmeier AW (2004) Mono(2-ethylhexyl) phthalate exhibits genotoxic effects in human lymphocytes and mucosal cells of the upper aerodigestive tract in the comet assay. Toxicol Lett 148:83–90
109. Hauser R, Meeker JD, Singh NP, Silva MJ, Ryan L, Duty S, Calafat AM (2007) DNA damage in human sperm is related to urinary levels of phthalate monoester and oxidative metabolites. Hum Reprod 22:688–695
110. Foster PM, Thomas LV, Cook MW, Gangolli SD (1980) Study of the testicular effects and changes in zinc excretion produced by some n-alkyl phthalates in the rat. Toxicol Appl Pharmacol 54:392–398
111. Lamb JC, Chapin RE, Teague J, Lawton AD, Reel JR (1987) Reproductive effects of four phthalic acid esters in the mouse. Toxicol Appl Pharmacol 88:255–269
112. Pan G, Hanaoka T, Yoshimura M, Zhang S, Wang P, Tsukino H, Inoue K, Nakazawa H, Tsugane S, Takahashi K (2006) Decreased serum free testosterone in workers exposed to high levels of di-n-butyl phthalate (DBP) and di-2-ethylhexyl phthalate (DEHP): a cross-sectional study in China. Environ Health Perspect 114:1643–1648
113. Subhan F, Tahir F, Ahmad R, Khan ZD (1995) Oligospermia and its relation with hormonal profile. J Pak Med Assoc 45:246–247
114. Swan SH, Main KM, Liu F, Stewart SL, Kruse RL, Calafat AM, Mao CS, Redmon JB, Ternand CL, Sullivan S, Teague JL (2005) Decrease in anogenital distance among male infants with prenatal phthalate exposure. Environ Health Perspect 113:1056–1061
115. Main KM, Mortensen GK, Kaleva MM, Boisen KA, Damgaard IN, Chellakooty M, Schmidt IM, Suomi AM, Virtanen HE, Petersen DV, Andersson AM, Toppari J, Skakkebaek NE (2006) Human breast milk contamination with phthalates and alterations of endogenous reproductive hormones in infants three months of age. Environ Health Perspect 114:270–276
116. Hoppin JA, Brock JW, Davis BJ, Baird DD (2002) Reproducibility of urinary phthalate metabolites in first morning urine samples. Environ Health Perspect 110:515–518
117. Arcadi FA, Costa C, Imperatore C, Marchese A, Rapidisarda A, Salemi M, Trimarchi GR, Costa G (1998) Oral toxicity of bis(2-ethylhexyl) phthalate during pregnancy and suckling in Long-Evans rat. Food Chem Toxicol 36:963–970
118. Wormuth M, Scheringer M, Vollenweider M, Hungerbuhler K (2006) What are the sources of exposure to eight frequently used phthalic acid esters in Europeans? Risk Anal 26:803–824
119. Itoh H, Yoshida K, Masunaga S (2007) Quantitative identification of unknown exposure pathways of phthalates based on measuring their metabolites in human urine. Environ Sci Technol 41:4542–4547
120. Franco A, Prevedouros K, Alli R, Cousins IT (2007) Comparison and analysis of different approaches for estimating the human exposure to phthalate esters. Environ Int 33:283–291
121. CERHR (2000) Center for the Evaluation of Risks to Human Reproduction. National Toxicology Program. National Institute of Environmental Health Sciences [online]. http://cerhr.niehs.nih.gov/news/phthalates/monographs.html, 4/7/05
122. Poon R, Lecavalier P, Mueller R, Valli VE, Procter BB, Chu I (1997) Subchronic oral toxicity of di-n-octyl phthalate and di(2-ethylhexyl) phthalate in the rat. Food Chem Toxicol 35:225–239
123. Castle L, Gilbert J, Eklund T (1990) Migration of plasticizer from poly(vinyl chloride) milk tubing. Food Addit Contam 7:591–596
124. MAFF (1996) Ministry of Agriculture, Fisheries and Food. Survey of plasticizer levels in food contact materials and in foods. Food Surveillance Papers No 21, UK
125. MAFF (1996) Ministry of Agriculture, Fisheries and Food. Phthalates in Food. Food Surveillance Information Sheet No 82, UK
126. MAFF (1996) Ministry of Agriculture, Fisheries and Food. Phthalates in Infant Formulae Food Surveillance Information Sheet No 83, UK
127. MAFF (1998) Ministry of Agriculture, Fisheries and Food. Phthalates in Infant Formulae—Follow-up Survey Food Surveillance Information Sheet No 168, UK

128. Petersen JH, Breindahl T (2000) Plasticizers in total diet samples, baby food and infant formulae. Food Addit Contam 17:133–141
129. Müller AM, Nielsen A, Ladefoged O (2003) Ministeriet for Fødevarer, Landbrug og Fiskeri Veterinær- og Fødevaredirektoratet. Human exposure to selected phthalates in Denmark, rapport 15
130. Hazleton Laboratories (1968) 13-Week Dietary Administration—Dogs Plasticiser (DIDP) submitted to WR Grace and Company
131. Exxon Biomedical Sciences (1986) Chronic toxicity/oncogenicity study in F-344 Rats. Test Material: MRD-83-260. Project No 326075 performed at Exxon Biomedical Sciences, Inc., Unpublished Laboratory Report
132. Wolfe GW, Layton KA (2003) Multigeneration reproduction toxicity study in rats (unaudited draft): diethylhexylphthalate: multigenerational reproductive assessment when administered to Sprague-Dawley rats in the diet. The Immune Research Corporation (Gaithersburg, Maryland), TRC Study no 7244-200
133. Tyl RW, Myers CB, Marr MC (2001) Two-generation reproductive toxicity evaluation of butyl benzyl phthalate administered in the feed to CD (Sprague-Dawley) rats. RTI Project No. 65C-0726-200, RTI Protocol No. RTI-761
134. CSTEE (1998) Phthalate migration from soft PVC toys and child-care articles: opinion expressed ate the CSTEE third plenary meeting Brussels (EU Scientific Committee on Toxicity, Ecotoxicity and the Environment) 24 April 1998
135. Lington AW, Bird MG, Plutnick RT, Stubblefield WA, Scala RA (1997) Chronic toxicity and carcinogenic evaluation of diisononyl phthalate in rats. Fundam Appl Toxicol 36:79–89
136. US EPA (2006) Integrated Risk Information System. http://www.epa.gov.iris/index.html
137. David RM (2000) Exposure to phthalate esters. Environ Health Perspect 108:A440
138. Blount BC, Silva MJ, Caudill SP, Needham LL, Pirkle JL, Sampson EJ, Lucier GW, Jackson RJ, Brock JW (2000) Levels of seven urinary phthalate metabolites in a human reference population. Environ Health Perspect 108:979–982
139. Barr DB, Silva MJ, Kato K, Reidy JA, Malek NA, Hurtz D, Sadowski M, Needham LL, Calafat AM (2003) Assessing human exposure to phthalates using monoesters and their oxidized metabolites as biomarkers. Environ Health Perspect 111:1148–1151
140. Koch HM, Rossbach B, Drexler H, Angerer J (2003) Internal exposure of the general population to DEHP and other phthalates—determination of secondary and primary phthalate monoester metabolites in urine. Environ Res 93:177–185
141. McKee RH, El Hawari M, Stoltz M, Pallas F, Lington AW (2002) Absorption, disposition and metabolism of di-isononyl phthalate (DINP) in F-344 rats. J Appl Toxicol 22:293–302
142. Silva MJ, Samandar E, Preau JL Jr, Reidy JA, Needham LL, Calafat AM (2007) Quantification of 22 phthalate metabolites in human urine. J Chromatogr B 860:106–112
143. Koch HM, Bolt HM, Angerer J (2004) Di(2-ethylhexyl) phthalate (DEHP) metabolites in human urine and serum after a single oral dose of deuterium-labelled DEHP. Arch Toxicol 78:123–130
144. Koch HM, Bolt HM, Preuss R, Angerer J (2005) New metabolites of di(2-ethylhexyl) phthalate (DEHP) in human urine and serum after single oral doses of deuterium-labelled DEHP. Arch Toxicol 79:367–376
145. Preuss R, Koch HM, Angerer J (2005) Biological monitoring of the five major metabolites of di-(2-ethylhexyl) phthalate (DEHP) in human urine using column-switching liquid chromatography-tandem mass spectrometry. J Chromatogr B 816:269–280
146. Silva MJ, Reidy JA, Preau JL Jr, Needham LL, Calafat AM (2006) Oxidative metabolites of diisononyl phthalate as biomarkers for human exposure assessment. Environ Health Perspect 114:1158–1161
147. Koch HM, Angerer J (2007) Di-iso-nonylphthalate (DINP) metabolites in human urine after a single oral dose of deuterium-labelled DINP. Int J Hyg Environ Health 210:9–19
148. NHANES. The National Health and Nutrition Examination Survey. http://www.cdc.gov/nchs/nhanes.htm

149. Chen M-L, Chen J-S, Tang C-L, Mao IF (2008) The internal exposure of Taiwanese to phthalate—an evidence of intensive use of plastic materials. Environ Int 34:79–85

150. Fromme H, Bolte G, Koch HM, Angerer J, Boehmer S, Drexler H, Mayer R, Liebl B (2007) Occurrence and daily variation of phthalate metabolites in the urine of an adult population. Int J Hyg Environ Health 210:21–33

151. Hauser R, Meeker JD, Park S, Silva MJ, Calafat AM (2004) Temporal variability of urinary phthalate metabolite levels in men of reproductive age. Environ Health Perspect 112:1734–1740

152. Peck CC, Albro PW (1982) Toxic potential of the plasticizer di (2-ethylhexyl) phthalate in the context of its disposition and metabolism in primates and man. Environ Health Perspect 45:11–17

153. Marsee K, Woodruff TJ, Axelrad DA, Calafat AM, Swan SH (2006) Estimated daily phthalate exposure in a population in mothers of male infant exhibiting reduced anogenital distance. Environ Health Perspect 114:805–809

154. Calafat AM, Needham LL, Silva MJ, Lambert G (2004) Exposure to di-(2-ethylhexyl) phthalate among premature neonates in a neonatal intensive care unit. Pediatrics 113: e429–e434

155. Koch HM, Preuss R, Angerer J (2006) Di(2-ethylhexyl) phthalate (DEHP): human metabolism and internal exposure—an update and latest results. Int J Androl 29:155–165

156. Weuve J, Sanchez BN, Calafat AM, Schettler T, Green RA, Hu H, Hauser R (2006) Exposure to phthalates in neonatal intensive care unit infants: urinary concentrations of monoesters and oxidative metabolites. Environ Health Perspect 114:1424–1431

157. Koch HM, Drexler H, Angerer J (2004) Internal exposure of nursery-school children and their parents and teachers to di(2-ethylhexyl) phthalate (DEHP). Int J Hyg Environ Health 207:15–22

Perfluorinated Compounds in Drinking Water, Food and Human Samples

Francisca Pérez, Marta Llorca, Marinella Farré, and Damià Barceló

Abstract Perfluorinated compounds are industrial chemicals widely used for more than 60 years. However, during the last decade, due to their high resistance to degradation, bioaccumulation attached to proteins, biomagnification to the food chain and their relation to toxicological effects, especially during early stages of life, these compounds have gained scientific and regulatory attention.

In addition, the difficulty associated with their analysis in complex matrices, such as biota, food and human fluids and tissues samples, should be mentioned.

This chapter provides a comprehensive examination of the current knowledge of drinking water and food contamination by PFCs and their bioaccumulation in humans, with special attention given to the fundamental role chemical analysis played in the evaluation of these compounds' sources, levels, exposure and risk assessment.

Keywords Dietary intake, Food analysis, Liquid chromatography, Mass spectrometry, Perfluorinated compounds, Perfluorinated compounds, Risk assessment

F. Pérez, M. Llorca, and M. Farré (✉)
Department of Environmental Chemistry, Institute of Environmental Assessment and Water Studies, IDAEA-CSIC, C/Jordi Girona 18-26, 08034 Barcelona, Spain
e-mail: mfuqam@cid.csic.es

D. Barceló
Department of Environmental Chemistry, Institute of Environmental Assessment and Water Studies, IDAEA-CSIC, C/Jordi Girona 18-26, 08034 Barcelona, Spain

Catalan Institute of Water Research (ICRA), C/Emili Grahit, 101, 17003 Girona, Spain

D. Barceló (ed.), *Emerging Organic Contaminants and Human Health*,
Hdb Env Chem (2012) 20: 337–374, DOI 10.1007/698_2011_136,
© Springer-Verlag Berlin Heidelberg 2012, Published online: 2 March 2012

Contents

1 Introduction

Perfluorinated compounds (PFCs) comprise a large group of compounds widely used in industrial applications that are characterised by a fully fluorinated hydrophobic linear carbon chain attached to one or more hydrophilic head. PFCs repel both water and oil, and are therefore ideal chemicals for surface treatments [1]. These compounds have been used for many industrial applications [2] and have been at the centre of an increasing number of environmental monitoring studies mainly because of their persistence, bioaccumulative potential [3] and global distribution [4, 5], particularly in aquatic biological samples [6]. These compounds have been detected worldwide in sediments and biota [3, 7]. In recent years, an increasing number of papers report high levels of PFCs in blood, tissues and breast milk from both occupationally and nonoccupationally exposed human populations [4, 8–10]. The most important exposure pathways of perfluorinated compounds for humans are thought to be intake of drinking water, food and inhalation of dust [11–13].

Because of their bioaccumulation [14–16] and potential health concerns including toxicity [17–21], and their possible contribution to cancer promotion, nongovernmental organisations, national and international authorities have addressed the PFCs problem by several pressure and legislative actions. The total production of perfluorooctanesulphonate (PFOS) has been significantly reduced from 2000 to 2005. One major fire-fighting foam manufacturer, 3M, abandoned production of PFOS in 2000. In February 2006, E.E.U.U. regulators reached a voluntary agreement with eight companies to phase out the use of perfluorooctane acid (PFOA).

Under the agreement, companies will reduce emissions of these compounds from their facilities and consumer products by 95 per cent by 2010, and work towards eliminating the sources of PFOA by no later than 2015. Furthermore, PFOS and PFOA as well as other perfluorocarboxylic acids (PFCAs) are stable degradation products and/or metabolites of neutral PFCs such as fluorotelomer alcohols (PFTOHs), perfluorinated sulphonamides (PFASAs) and perfluorinated sulphonamide ethanols (PFASEs) [22]. Therefore, the largest global manufacturer and supplier of fluorotelomers such as Capstone, DuPont have adapted its entire product line to utilise short-chain chemistry because short-chain molecules cannot break down to PFOA in the environment.

Canada was the first country to take precautionary action on PFCs by banning in July 2004, for a two-year period, four fluorinated polymers that contain telomer alcohols, and then recommending their permanent ban. In Europe, the hazard assessment of the Organisation for Economic Co-operation and Development (OECD) from the year 2002 identified PFOS as a PBT-chemical (persistent, bioaccumulative and toxic). As a result, PFOS was proposed as a candidate in the Stockholm convention on persistent organic pollutants (POPs). In December 2006, the European Parliament and the Council decided to restrict marketing and use of PFOS, with a few exceptions, by amending Council Directive 76/769/EC on dangerous substances for PFOS. The directive will be later transposed into the EU's forthcoming REACH chemical legislation. It is currently being discussed if PFOA should be incorporated into this Directive. In addition, PFOS and PFOA are considered to be included in the so-called priority substances regulated by the Water Framework Directive (WFD). Although these restrictive actions provided additional security, in January 2005, OECD published a report with results from a survey on the production and use of PFOS and related substances in the OECD area. One conclusion from the survey is that PFOS is still manufactured by Germany (20–60 tonnes in 2003) and Italy (<22 tonnes in 2003).

PFCs are now included in different health programs in EEUU, Canada and Europe. The EU to provide a better assessment about the distribution, toxicity and persistence of these compounds is supporting several projects of the VII European Research Framework Programme targeted on PFCs. During the last years, several reviews have been published on PFCs that summarise the analytical strategies [23], biological monitoring data [24] and recent advances in toxicology and their mode of action [25]. However, data on levels of PFCs in the human diet are rather scarce [26, 27]. In addition, PFOS remains the predominant PFC found in all aquatic species, tissues and locations analysed around the world. PFOA and PFOS accumulation in marine mammals (e.g., >1,200 ng/g wwt in liver) is common in the Arctic [28, 29], Europe [30, 31] and Asia [32], while levels are much lower in the southern hemisphere. These contamination trends are likely due to continued use of PFOA and PFOS precursors, such as fluorotelomer alcohols and polyfluoroalkyl phosphate and/or continued oceanic and atmospheric inputs of sources resulting in exposure and bioavailability of these PFCs in the northern hemisphere [6].

This chapter provides a comprehensive examination of the current knowledge of drinking water and food contamination by PFCs and their bioaccumulation in humans, with special attention given to the fundamental role chemical analysis played in the evaluation of these compounds' sources, levels, and exposure and risk assessment.

2 Overview of Analytical Methods for the Analysis of Perfluorinated Compounds

2.1 Sample Storage and Conservation

Storage and conservation of samples for PFCs analysis presents some critical steps because losses or contamination of the samples can easily occur. In order to avoid or reduce sources of contamination, different protocols have been suggested. For liquid samples is very important the pre-cleaning of bottles and recipients with a rinsing with semi-polar solvents before sampling [33]. However, less attention has been paid to the potential losses during the storage. Main causes of losses are the adsorption to sample containers, the volatilisation of some PFCs, or transformations due to an inappropriate conservation. There have been controversies about whether and which PFCs can absorb to glass surfaces [34, 35]. The partial adsorption to glass containers of high concentrations standard solutions has been reported [36], but it is expected that this will not happen in real samples with more complex matrices [37]. On the other hand, some authors reported that polymeric container, such as polypropylene (PP) and high density ethylene (HDPE), can also partially adsorb long-chain compounds, such as PFOSA, PFOS and PFOA [38]; however, these are in general more used materials. Another cause of losses is volatilisation that can affect some volatile compounds, such as fluorotelomer alcohols (FTOHs), during sampling, storage and sample pre-treatment. In order to minimise these losses, it has been recommended avoiding headspace in sampling bottles [39]. Long-term conservation of the samples is also a critical point. Most of the authors report freezing, refrigeration, solvents addition or acidification combined with refrigeration to preserve the samples [40]. However, it has been shown that when pH decreases, PFCs become increasingly associated with the available protons, and then PFCs can be more easily adsorbed to the container's surface [41]. Szostek et al. [42] investigated the stability of FTOHs in water and water samples mixed with acetonitrile during the storage. They concluded that aqueous samples could safely be stored in the freezer in a glass vial, sealed with a septum lined with alumina foil. Finally, biodegradation and biotransformations should be prevented. Whereas good results were obtained when conservations were conducted in the freezer or using combinations of solvent (such as acetonitrile) and freezing [43], the use of biological inhibitors, such as formalin, was found to suppress the MS responses during the analysis [44].

2.2 Food, Biota and Aqueous Samples Including Drinking Water: Extraction and Clean-Up

2.2.1 PFAS Analysis in Aqueous Matrices

Different extraction procedures have been proposed including the use of solid-phase micro extraction (SPME), liquid–liquid extraction (LLE), large-volume direct injection and solid-phase extraction (SPE). For example, Alzaga et al. [45] proposed a method based on ion-pair SPME combined with in-port derivatisation and gas chromatography–negative chemical ionisation–tandem mass spectrometry (GC–NCI-MS-MS) determination of various PFCAs in STP effluent and seawater [45]. Gonzalez-Barreiro et al. [46] have compared a method based on LLE of various PFAS from STP effluents in comparison with SPE at two different pH values. Schultz et al. [44] used large-volume direct injection liquid chromatography-tandem mass spectrometry (LC/MS/MS) after centrifugation of STP influent and effluent samples without prior analyte enrichment. This rapid, straightforward method covered a broad range of compounds. Sample filtration was considered but rejected since the filters either retained parts of the analytes or resulted in sample contamination with certain PFAS. Large-volume direct injection LC/MS/MS offers the advantage of low contamination potential and quantitative recoveries (apart from the particle bound PFAS fraction). Additionally, the injected sample volume could be increased employing a column-switch system with a pre-concentration column.

However, most of the authors use SPE for the extraction and clean-up of water and liquid samples. Moody and Field [47] described the first SPE method for PFCAs from contaminated groundwater originating from a fire-training site. The compounds were derivatised and analysed by GC/EI-MS with a limit of quantification (LOQ) of 36,000 ng/L (PFOA). Since then, due to the different polarities of PFCs different extraction SPE cartridges have been explored. Broadly, good recoveries were reported using Oasis WAX-SPE cartridges including short-chain (C_4–C_6) compounds, and have been applied in many monitoring studies [13, 48, 49]. For longer chain PFCs, less polar phases (C_{18} and Oasis HLB) may be applied. When an ion-pairing agent is used, that decreases the polarity of the ion-pair complex, a non-polar solvent (MTBE) may be used. Non-ionic PFCs may be extracted from the matrix by non-polar media (C_{18} SPE or hexane). Moderate polar media (Oasis HLB and Oasis WAX-SPE, a hexane–acetone mixture or acetonitrile) have also been applied for extraction of non-ionic PFCs. However, one of the critical points in PFCs analysis is background contamination in the analytical blanks [24, 50]. One known source of procedural contamination is contact with laboratory materials made of, or containing, fluoropolymers [44]. Water samples may be filtered [44, 51] to separate solids from the liquid phase. However, filtration can result in losses by adsorption of PFCs on the filters, or on the contrary levels can increase by contamination from the filters, as was found by

Schultz et al. for fibre, nylon, cellulose acetate and poly ether sulphone filters [44]. They applied centrifugation as alternative for separation the liquid from the solids.

Controversial studies reported the cross contamination of samples during PFCs analysis using different SPE cartridges. Yamashita et al. [52] examined the source of blank contamination at various different steps, including sample collection, extraction and treatment of samples. PFOS and PFOA contamination in the SPE cartridges, OASIS HLB and Sep-Pak (C_{18}), was evaluated. Both SPE cartridges were a cause of contamination by PFOS and PFOA. However, higher concentrations of PFOS and PFOA were reported for Sep-Pak cartridges. In the case of the Oasis HLB, PFOS, PFOA, PFHS and PFBS were detected, but at lower concentrations than those found in the Sep-Pak cartridges. On the other hand, Taniyashu et al. [3] evaluated Oasis HLB and Oasis-WAX columns for the extraction of PFCs. In this study, few target perfluorinated compounds were detected in procedural blanks at a few pg/L in the final extract. However, PFOA, PFDA and PFUnDA were still found at relatively high concentrations. In general, the performance of these columns was comparable. Recoveries were good (70–100%) for most compounds, but for short-chain PFCAs recoveries using Oasis WAX-SPE cartridges were higher. Losses due to evaporation during analysis, and adsorption to the polypropylene sample container surface as discussed earlier were suggested causes for the lower recoveries.

2.2.2 Food and Biota Samples

Main methodologies for sample treatment and extraction for the analysis of PFAS in food and biota are summarised in Table 1. Main sample preparation and extraction procedures have been based on:

- Ion-pair extraction
- Solid liquid extraction (SLE)
- Alkaline digestion
- Pressurised liquid extraction (PLE)

Ylinen et al. [53] developed an ion-pair extraction procedure employing tetrabutylamonium (TBA) counter ions for determination of PFOA in plasma and urine in combination with gas chromatography (GC) and flame ionisation detection (FID). Later on, Hansen et al. [35] improved the sensitivity of the ion-pair extraction approach using methyl tertiary butyl ether (MTBE) and by the inclusion of a filtration step to remove solids from the extract making it amenable to liquid chromatography coupled to tandem mass spectrometry (LC-MS/MS) determination. Ion-pair extraction procedure has been the basis of several procedures for biota [49, 54–58] and food samples [50, 59, 60]. However, this method has shown to have some limitations, such as (1) co-extraction of lipids and other matrix constituents and the absence of a clean-up step to overcome the effects of matrix compounds and (2) the wide variety of recoveries observed, typically ranging.

Table 1 Main methodologies for sample treatment and extraction for the analysis of PFAS in food and biota

Sample	Compounds	Extraction and clean-up	Stationary Phase	Mobile phase	Detection	LOD-LOQ	Ref.
Seafood	PFOS PFHS PFUnDA PFDA PFNA PFOA PFHpAPFHxA	Ion-pairing liquid extraction method	Betasil C18	2 mM ammonium acetate/methano	LC-(QqQ)-MS/MS	LOQ 250 ng/Kg in fish	[133]
Pelican tissue	PFOS, PFOA, PFHxS, and PFOSA	Ion-pairing liquid extraction method	Betasil C18	2 mM ammonium acetate/methano	LC-(QqQ)-MS/MS	LOQs were 0.1, 1, 30, and 0.5 ng/g, wet weight, for PFOS, PFOA and PFHxS,	[155]
Fish and oysters	PFOA, PFOS and PFDA	C18-SPE	Betasil C18	1 mM ammonium acetate (pH 6.0) and a methanol	LC-(QqQ)-MS/MS	LOQs were between 0.5 and 6 ng/l in 250 ml of water sample, while 5–50 ng/g (dry weight) for biological tissue sample	[156]
Mussels and Oysters	PFOS, PFHS, PFBS, PFOSA, PFDoDA, PFUnDA, PFDA, PFNA, PFOA, PFHpA and PFHxA,	Alkaline digestion and SPE with Oasis HLB	Betasil C18	2 mM ammonium acetate and methanol	LC-(QqQ)-MS/MS		[157]
Biota samples	PFOS, PFHS, PFBS, PFOSA, PFOcDA, PFHxDA, PFTeDA PFDoDA, PFDA, PFNA, PFOA, PFHpA, PFHxA, PFPeA, PFBA 7:1 FTOH PDFO, 8:2 FTCA, 8:2 FTUCA	*KOH digestion–Oasis WAX*	Betasil C18	2 mM ammonium acetate aqueous solution and methanol	LC-(QqQ)-MS/MS	LOD 0.01–1 ng/L	[3]

(continued)

Table 1 (continued)

Sample	Compounds	Extraction and clean-up	Stationary Phase	Mobile phase	Detection	LOD-LOQ	Ref.
Fish, mussel, amphipods, and algae	PFOS PFOSA PFOA PFHS	Ion-pairing liquid extraction method	Betasil C18 column	2 mM ammonium acetate aqueous solution and methanol	LC-(QqQ)-MS/MS	LOQ were in the range of 0.2–2 ng/g	[136]
Fish, birds and marine mammals	PFOS, PFOSA, PFOA, PFHxA,	Ion-pairing liquid extraction method	C8 Luna 5 mm, 150! 2.00 mm analyticalcolumn	2 mM ammonium acetate aqueous solution and methanol	LC-(QqQ)-MS/MS	LOD were in the range 3–7 (ng/g wet weight) LOQs were in the range of 4–12 (ng/g wet weight)	[151]
Human milk	PFOSA, Me-PFOSA-AcOH, Et-PFOSA-AcOH, PFHxS, PFOS, PFPeA, PFHxAPFHpA, PFNA, PFDeA, PFU/A, PFDoA	Oasis HLP- SPE	Betasil C8 column	20 mM ammonium acetate (pH 4) in water and methanol	LC-(QqQ)-MS/MS	LOD 0.1 –3.2 (mg/mL of milk)	[71]
Human milk and serum	PFOS, PFHxS, PFOSA, PFOA, PFNA	Formic acid digestion–Oasis WAX	C18 analytical column	2 mM ammonium acetate aqueous solution and methanol	LC-(QqQ)-MS/MS	LOD 0.1–1.1(ng/mL of milk) and 0.005–0.209 (ng/mL for serum)	[9]
Treated tap water	PFOA, PFOSA	Sep-Pak plus PS-2 cartridges SPE (adjusted pH to 2)	RP-Amide column	5 mM ammonium acetate/acetonitrile	LC-(QqQ)-MS/MS	LOQ 0.1 ng/L for PFOS and 1.0 ng/L for PFOA.	[158]
Human breast milk	PFOS, PFOA	Acetonitrile digestion and vortexed and centrifuged	Online with a Reprosil-Pur C18 AQ and separate by Reprosil-Purvasic C8	2 mM ammonium acetate aqueous solution and acetonitrile	LC-(QTRAP)-MS/MS	LOQ 20 ng/L for PFOS and 200 ng/L for PFOA	[149]

Chicken egg	PFOS, PFHxS, PFBS, PFOSA, PFNA, PFOA, PFHxA, PFHpA, PFDA, PFUnDA, PFDoDA	Ion-pairing liquid method following by Oasis WAX SPE cartridges extraction	Betasil C18 column	2 mM ammonium acetate aqueous solution and methanol	LC-(QqQ)-MS/MS	LOQ 0.01–0.08 ng/g	[159]
Meat, fish, fast food	PFHpA, PFOA, PFNA, PFDA, PFUnA, PFDoA, PFTeDA	MeOH digestion	Genesis C18 analytical column	5 mM ammonium formate in water and acetonitrile/methanol	LC-(QqQ)-MS/MS	LOD between 0.5 and 1 ng/g range. LOQ estimated as three times the LOD value.	[27]
Lake trout	PFHxS, PFOS, PFOSA, PFHpA, PFOA, PFNA, PFDA, PFUnA, PFDoA, PFTeA, PFDS,	Ion-pairing liquid extraction and filtered	Genesis C18 column	0.01 mM ammonium acetate aqueous solution and methanol	LC-(QTRAP)-MS/MS	LOD between 1 and 16 pg/g	[125]
Surface water, fish and birds	PFOS, PFHS, PFBS, PFOA, PFOSA	Ion-pairing liquid extraction (fish and bird) Oasis HLB SPE cartridges (water)	Betasil C18 column	2 mM ammonium acetate aqueous solution and methanol	LC-(QqQ)-MS/MS	LOQ 0.8 ng/L (water), 1.5 ng/g (fish livers) and 7.5 ng/g (bird livers)	[160]

Perfluorobutanesulfonate (PFBS), perfluorohexanesulfonate (PFHS), PFOS, perfluorohexanoic acid (PFHxA), perfluoroheptanoicacid (PFHpA), PFOA, perfluorononanoicacid (PFNA), perfluorodecanoicacid (PFDA), and perfluoroundecanoicacid (PFUnDA),perfluoroundecanoate (PFUnDA), perfluorodecanoate (PFDA), PFNA, PFOA, perfluoroheptanoate (PFHpA) and PFHxA,perfluorotetradecanoate (PFTeDA)

PFOS were the isotope-labelled IS used in this study. Fromme et al. [61] reported an SLE procedure using sonication with methanol. Extracts were cleaned up using SPE with an anionic exchange cartridge.

Sample preparation by alkaline digestion has also been widely applied for the analysis of PFCs in food. This procedure is based on digestion with sodium or potassium hydroxide in methanol followed by SPE. This procedure combined with SPE using Oasis-WAX cartridges have been applied for diverse foodstuff analysis. Vegetables, cheese, margarine, milk, bread, strawberry jam, pork, beef, chicken, egg, fish, canned mackerel, salmon, cod, cod liver were also analysed using alkaline digestion followed by SPE with Oasis-WAX by Haug et al. [62]. In another study, Jogsten et al. [63] used the alkaline digestion followed by SPE with Oasis-WAX for the analysis of a wide variety of foodstuff including raw, grilled and fried veal, pork and chicken, lamb liver, pate of pork liver, foie gras of duck, Frankfurt sausages, marinated salmon, lettuce and common salt. Llorca et al. used this extraction procedure to study the PFCs content in fish [49] and commercial baby food [10].

Modern extraction and clean-up techniques, such as pressurised liquid extraction and microwave-assisted extraction, have almost not applied to the analysis of PFCs yet. Llorca et al. [49] reported the development and application of a PLE method for PFCs determination in fish. This technique provided rapid and accurately clean extracts for sensitive analysis.

2.3   Pre-treatments, Extraction and Clean-Up for Human Biological Samples

PFCs have been investigated in humans mainly through the analysis of whole blood, serum, plasma and breast milk. Typical extraction procedures and clean-up procedures based on LLE in combination with ion-pair extraction using MTBE [64, 65] can also be used for extraction of FTOHs from plasma. Szostek and Pricket extracted 8:2 FTOH from rat plasma with MTBE [66]. The extract was analysed without further clean-up. Recoveries were 86–113% and the LOD was estimated at 5 ng/mL. Different approaches based on online and offline SPE. In these cases, protein precipitation is in general required blood [37, 67], or in serum and plasma [36, 68–70] in order to prevent clogging of the SPE columns, using trichloroacetic acid, formic acid or acetonitrile. This step is typically followed by centrifugation to separate the precipitates from the liquid phase. Kukyenlik et al. [71] designed an online SPE-LC-MS/MS method in which no precipitation step was required, thereby greatly reducing sample handling time. Only the addition of formic acid to the sample in the vial prior to analysis was required [72, 73].

SPE strategies have been carried out using a variety of media according to the different polarities exhibited by PFCs. Ionic perfluorocarboxylates (PFCAs) and perfluorosulphonates (PFSAs) require moderately polar media (Oasis WAX-SPE or methanol and acetonitrile) for efficiently trapping of water-soluble short-chain

(C4–C6) compounds. For longer chains, less polar or non-polar SPE phases (C18 and Oasis HLB) may be applied. When an ion-pairing agent is used, that decreases the polarity of the ion-pair complex, a non-polar solvent (MTBE) may be used. Non-ionic PFASs may be extracted from the matrix by non-polar media (C18 SPE or hexane). Moderate polar media (Oasis HLB and Oasis WAX-SPE, a hexane–acetone mixture or acetonitrile) have also been applied for extraction of non-ionic PFASs.

For certain types of samples with sample size limitations, such as cord blood, methods based on line SPE present a series of advantages in addition to the reduction of sample manipulation and therefore increase in robustness, based on the need of low sample volumes. The group of Calafat [74, 75] presented a method for the analysis of eighteen PFCs in human serum consisting in dilution with 0.1 M formic acid and 20 min sonication followed by SPE-LC-MS/MS showing LOD in the range between 50 and 800 ng/L. More recently, Gosetti et al. [60] presented another online SPE method. In this approach, 1 mL of blood is required, and the method consisted of the blood centrifugation for 10 min, then dilution of the supernatant in acetonitrile, centrifugation 10 min. more, dilution of the supernatant with water and methanol acidified with formic acid followed by online SPE and ultra-high performance liquid chromatography (UHPLC)–MS/MS analysis. This method has presented good LOD for eight PFCs, in the range between 0.003 and 0.015 ng/mL. Another method based on UPLC-MS/MS was presented by Lien et al. [67]; in this case, the samples were processed with protein precipitation with formic acid and methanol, followed by sonication and centrifugation, and the supernatants were analysed by UPLC-MS/MS. The method has shown LOD between 0.05 and 1 ng/mL for the twelve PFCs.

Recently, turbulent flow chromatography (TFC) has shown a great potential for online sample pre-treatment in the analysis of PFCs. Up to now, the use of this technique in food and environmental analysis is scarce, but some successful applications have been developed. Among them, the analysis of PFCs has been carried out in cord blood and also in less invasive human samples, hair and urine. In these works, the main advantages presented were the simplified sample preparation, robustness and sensitivity. In addition in the case of cord blood, a low volume of sample was required.

2.4 Qualitative and Quantitative Aspects of the Determination

Liquid chromatography-mass spectrometry (LC-MS) or liquid chromatography-tandem-mass spectrometry (LC-MS/MS) has been in general the technique of choice for the analysis of PFCs. Therein detailed information about the main experimental conditions used for analysis, such as LC-MS/MS precursor-product ion transitions, were reported.

LC separation of PFCs has been mainly carried out with C_{18} and C_8 columns. In spite of the wide use of RP-C_{18} columns for PFCs analysis, the interference producing the enhancement of spectral signal has been reported. RP columns with shorter alkyl chain bonded phases (e.g., C_8, C_6, phenyl, phenylhexyl) also separated the branch isomers, but to a lesser extent. To minimise the separation of branched isomers, the authors increased the LC column temperature to 35°C or 40°C [71, 75, 76]. Taniyasu et al. [77] explored the chromatographic properties and separation of short-chain PFAs on RP-C_{18} and ion-exchange columns. The results showed that the peaks of PFPrA and PFEtS were broad and not adequately resolved whereas that of TFA was not retained in the analytical column eluting with the solvent front. This suggested that RP columns are not suitable for the analysis of short-chain PFAs, especially TFA. As a proper alternative, ion-exchange columns have superior retention properties for more hydrophilic substances enabling the analysis of short-chain PFAs, TFA, PFPrA, PFBA, PFEtS, PFPrS and PFBS together with several long-chain PFAs, in water samples.

Due to the complexity of the food samples, it is possible that the presence of some compounds in the matrix interferes with analyte determination. To date, this problem has been partially solved using LC–MS/MS. However, even when working in LC–MS/MS, certain compounds present in the sample can affect the initial ionisation of the analyte through what is often called ion suppression or matrix effects.

ESI operating in the negative ion (NI) mode has been the interface most widely used for the analysis of anionic perfluorinated surfactants. In addition, ESI has also been optimised for the determination of neutral compounds such as the sulphonamides PFOSA, Et-PFOSA and t-Bu-PFOS. The use of atmospheric pressure photoionisation (APPI) has been explored in few works [78–80]. Takino et al. [78] found as the main advantage of this technology, the absence of matrix effects, but the limits of detection were considerably higher than those obtained by LC–ESI-MS/MS.

LC-MS/MS performed using triple quadrupole mass spectrometer (QqQ) combined with multiple reaction monitoring (MRM) is one of the more widely applied detector, as well as, to be one of the better suited for quantification of PFCs. Nowadays the performance of ion trap (IT) and time of flight (TOF) have been also considered for trace quantification of PFCs. PFCs contain carboxylic, sulphonic, hydroxy or sulphonamide group. They have acidic properties and can therefore dissociate. Therefore, electrospray ionisation in the negative mode [ESI (-)] is best suited. LC-(ESI)-MS/MS is the technique most widely used in food analysis, allowing limits of detection in the pg–ng/g range. Pseudo molecular ions are formed such as $[M-K]^-$ for PFOS (m/z 499), $[M-H]^-$ for PFOA (m/z 413) and PFOSA (m/z 498, which are generally selected as precursor ions for MS^2 experiments using ion trap and a triple quadrupole instruments. Berger et al. [81] has presented a comparison between IT, QqQ and TOF instruments. Tandem mass spectrometry showed excellent specificity, but the matrix background is eliminated by the instrument, thus it cannot be visualised. Applying TOF-MS gives an estimation of the amount of matrix left in the extract, which could impair the ionisation performance and the high mass resolution of the TOF-MS instrument offers excellent specificity for PFCs identification after a crude sample injection. Recently, the analytical suitability of three different LC-MS/MS systems: QqQ), conventional 3D ion trap (IT) and quadrupole-linear IT (QqLIT) to determine trace levels of PFCs in fish and shellfish was compared [82]. In this study, the accuracy was similar in the three systems, with recoveries always over 70%. Precision was better for the QqLIT and QqQ systems (7–15%) than for the IT system (10–17%). The QqLIT (working in SRM mode) and QqQ systems offered a linear dynamic range of at least three orders of magnitude, whereas that of the IT system was two orders of magnitude. The main advantage of QqLIT system is the high sensitivity, at least 20-fold higher than the QqQ system. Another advantage of QqLIT systems is the possibility to use enhanced product ion (EPI) mode and MS^3 modes in combination of MRM node for confirmatory purposes of target analytes in complex matrices. These modes were applied to assess the content of PFCs in a breast milk samples and commercial baby food by Llorca et al. [10].

3 Sources of Food Contamination

Two main sources of food contamination can be distinguished:

- Direct environmental exposure of plants and animals and/or bioaccumulation through the food chain
- Indirect contamination: Cooking, food packaging and food processes

3.1 Direct Environmental Exposure of Plants and Animals and/or Bioaccumulation Through the Food Chain

One of the main inputs of PFCs in the environment is produced during their production process. Then a series of transformations have been identified reverting sometimes in PFOS and PFOA.

PFCs are produced via two major synthesis processes: electrochemical fluorination (ECF) and telomerisation. The ECF process was used by 3M Company from the 1950s to 2001 to obtain perfluorooctanesulfonamide ethanols (FOSEs), perfluorooctane sulphonamide (PFOSA) and PFOS. The FOSEs and PFOSA have been shown to degrade abiotically [83, 84] to PFOS and PFOA as well as via biotic degradation to PFOS. It should be pointed out that the degradation of these compounds would only yield PFOA and not other PFCAs. The ECF process was also used to manufacture PFOA from 1947 to 2002. In 2001, the 3M Company announced the voluntary phase-out of its production in favour of shorter chain length compounds. However, it has been reported that PFOS has been directly produced in China since 2003, which may influence global emission patterns. In contrast, the telomerisation process has been used since the 1970s for the production of fluorotelomer alcohols (FTOHs), fluorotelomer olefins, fluorotelomer acrylates, fluorotelomer iodides and PFCAs. FTOHs have been shown to degrade via abiotic [85], and biotic [86] mechanisms to PFCAs. More recent research has shown that the fluorotelomer olefins [87], iodides [88] and acrylates [89] may form PFCAs via atmospheric oxidation. Compounds manufactured via the ECF process have been shown to contain both the linear and several branched isomers [90, 91]. In contrast, compounds produced by the telomerisation process contain only the linear isomer [92].

Following their release into the environment, PFCs can enter plants and animals at the bottom of the food chain that are then consumed by animals higher up. Therefore, one of the main inputs of PFCs in the food chain is the exposure of food-producing animals or plants to these compounds via environmental routes. Specially, contamination of the water cycle has been identified as one of the major causes of PFCs in food. Several studies report PFCs contamination in drinking water [33, 93–96], and the removal efficiency of ionic PFCs has been shown often very limited [44]. Non-ionic PFCs transform into the stable end products PFOS and PFOA. Due to this reason, wastewater is one of the main influences of PFCs in the water cycle. On the other hand, the use of sewage sludge as fertiliser and subsequent run-off was also found to contribute significantly to the contamination of surface, food and drinking water [33]. In addition, bioaccumulation in food chains will lead to increased levels of PFCs in animal-derived foods. Bioaccumulation of fish has been shown to be the main influences of PFCs in dietary exposure. Since 2006, more than 40 studies have reported concentrations of PFCs in aquatic biological samples and a clear picture of the global PFC contamination in aquatic biota can be drawn from these concentration data for invertebrates, fish, reptiles, aquatic birds and marine mammals. The majority of recent PFC biological assessment studies were located in Asia, Europe, North America and the Arctic, although other regions of the world, such as South America, Antarctica and Russia, are now being investigated.

Fish, bird and marine mammal have been studied at various locations around the world while most invertebrate work has been limited to East Asia. A few extensive monitoring studies have been conducted such as the assessment of PFCs in fish from 59 lakes and the Mississippi River in Minnesota, USA [97] and PFCs and precursors in herring gull eggs from 15 colonies in the Laurentian Great Lakes, Canada/USA [98, 99]. These extensive monitoring studies yielded highly valuable information on the distribution and the potential sources of these chemicals over large but specific territories. In addition, a high number of studies have reported PFCs concentrations for regions or in species for which very little or no data existed before, providing additional information on the environmental distribution of these chemicals [49, 100–103].

Additional information on the exposure of aquatic wildlife to PFCs within aquatic organisms was provided by studies of tissue distribution. For example, analyses of ten harbour seal organs showed that PFCs tend to accumulate primarily in blood (38% of the total PFC burden) > liver (36%) > muscle (13%) > lung (8%) > kidney (2%) > blubber (2%) > heart (1%) > brain (1%) > thymus (<0.01%) and thyroid (<0.01%) [104] (1982, 8.8 ng/g wwt; 2006, 36.1 ng/g wwt).

Some of these studies have related the fish consumption and human health risk [72, 105–107]. This issue will be discussed in Sects. 4 and 5, but for example,

Zhao et al. [105] have been calculated the hazard ratio (HR) of PFOS for fish consumption and the risks and potential effects of PFCs to health of coastal population in the Pearl River Delta. Due to the contamination levels of more consumed species (mandarin fish, bighead carp, grass carp and tilapia), the authors have concluded that the levels of PFCs in these fish species might pose an unacceptable risk to human health.

In a market basket study in Sweden Berger et al. found that PFOS and PFOA concentrations were below the quantification limits in composite samples of foods of animal origin. However, predatory fish from the largest lake in Sweden had substantially elevated levels of several PFCs. In another work, Ericson et al. [12] studied the dietary exposure to PFCs in Spain. In this study, the dietary intake of PFCs was estimated for different age and gender groups and was found to be on average between 0.9 and 1.1 ng/kg bw/day for the adult male population. Fish, followed by dairy products and meats, were the main contributors to PFOS intake due to their bioaccumulation and biomagnification through the food chain. Similar conclusions were reported by Berger et al. [108]. In this work, fish consumption was identified as one of the main sources of human exposure in Sweden. Ostertag et al. [109] estimated the dietary exposure to PFCs from traditional food among Inuit in northern Canada. In this study, the bioaccumulation of PFCs through the food chain and their contribution to the Inuit dietary exposure was revealed. Recently, Haug et al. [110] explored the possible associations between concentrations of PFCs in serum and seafood consumption. Concluding that, a significant relationship between estimated dietary intakes and serum concentrations exist.

3.2   Indirect Contamination: Cooking, Food Packaging and Food Processes

Food preparation is another source of contamination [27], but preliminary data on the influence of domestic cookware on levels of PFCs in the preparation of food indicated no elevated levels for a limited number of experiments [111]. In addition, Del Gobbo et al. [112] reported the cooking decreases of PFCs concentrations in fish.

Packaging may also introduce chemicals into food, for example PFCs used in greaseproof packaging for fast foods and special packaging. In these situations, PFCs entry into food via migration from food package [27]. Fluorochemical-treated paper was tested to determine the amount of migration that occurs into foods and food-simulating liquids and the characteristics of the migration [113]. Additionally, microwave popcorn and chocolate spread were used to investigate migration. Results indicate that fluorochemicals paper additives migrate to food during actual package use. For example, we found that microwave popcorn contained 3.2 fluorochemical mg/kg popcorn after popping and butter contained 0.1 mg/kg after 40 days at 4ºC. Tests also indicate that common food-simulating liquids for migration testing and package material evaluation might not provide an accurate indication of the amount of fluorochemical that actually migrates to food. Tests show that oil containing small amounts of an emulsifier can significantly enhance migration of a fluorochemical from paper.

4 Food and Drinking Water Contamination: Daily Intakes and Safety Limits

The characterisation of health hazards of food contaminants, the assessment of the occurrence of undesirable compounds in food and the estimation of the dietary intake are key issues in the risk assessment. In 2000, the European Commission published a White Paper on Food Safety, which underlined the importance of ensuring the highest possible standards of food safety and proposed a new approach to achieve them. Recently, PFCs have gained increased scientific and socioeconomic interest as emerging environmental contaminants due to the unique combination of persistence, toxicity and environmental prevalence. Risk assessment of the dietary exposure to PFCs, however, is hampered by the lack of sufficient data about the occurrence of these contaminants in food.

A growing number of studies report on the occurrence of PFCs in food. The outcome of these studies has been related to potential dietary intake and exposure levels (mainly by the estimation of the daily intake). Most selected examples from the literature can be seen in Table 2. It is important to remark that PFOS and PFOA tend to bind to certain proteins rather than bioconcentrate in fat, but they have also some potential to bioaccumulate in the food chain.

In the next sections, data published about the presence of PFCs in drinking water and food will be revised. Special attention will be paid to fish contamination since it has been well documented that PFCs may accumulate in fish and this accumulation tends to increase with increasing chain length [114–116]. Therefore, fish are an important dietary source of PFCs for humans. In addition, a revision of daily intakes and safety limits are reported.

4.1 Drinking Water Contamination

During the last decade, several studies have reported the concentrations of PFCs found in drinking water worldwide. Table 3 presents a summary of detected compounds and the ranges of concentrations reported during the last years.

In Asia, Saito et al. [117] reported concentrations of PFCs in drinking water from areas with known PFC sources; their results ranged between 5.4 and 40.0 ng/L for PFOS. For PFOA, the concentrations ranged between 1.1 and 1.6 ng/L, while in areas with no known sources, concentrations ranged from <0.1 to 0.2 ng/L for

Table 2 Levels of PFCs in food

Food	PFBS	PFHxA	PFHpA	PFHxS	THPFOS	PFOA	PFNA	PFOS	ΣPFOSA	PFDA	PFUnDA	PFDS	PFDoDA	PFTDA	Ref.
Diet (ng/g)															
Bread	NR	NR	NR	NR	NA	<5	NR	<20	NA	NR	NR	NR	NR	NR	[80]
Miscellaneous cereals	NR	NR	NR	NR	NA	<5	NR	<10	NA	NR	NR	NR	NR	NR	[80]
Carcase meats	NR	NR	NR	NR	NA	<2	NR	<10	NA	NR	NR	NR	NR	NR	[80]
Offal	NR	NR	NR	NR	NA	<2	NR	<20	NA	NR	NR	NR	NR	NR	[80]
Meat products	NR	NR	NR	NR	NA	<2	NR	<10	NA	NR	NR	NR	NR	NR	[80]
Poultry	NR	NR	NR	NR	NA	<2	NR	<10	NA	NR	NR	NR	NR	NR	[80]
Fish	NR	NR	NR	NR	NA	<3	NR	<5	NA	NR	NR	NR	NR	NR	[80]
Oil and fats	NR	NR	NR	NR	NA	<1	NR	<0.5	NA	NR	NR	NR	NR	NR	[80]
Eggs	NR	NR	NR	NR	NA	<1	NR	1	NA	NR	NR	NR	NR	NR	[80]
Sugars & preserves	NR	NR	NR	NR	NA	<1	NR	1	NA	NR	NR	NR	NR	NR	[80]
Green vegetables	NR	NR	NR	NR	NA	<1	NR	<3	NA	NR	NR	NR	NR	NR	[80]
Potatoes	NR	NR	NR	NR	NA	1	NR	10	NA	NR	NR	NR	NR	NR	[80]
Other vegetables	NR	NR	NR	NR	NA	<10	NR	<3	NA	NR	NR	NR	NR	NR	[80]
Canned vegetables	NR	NR	NR	NR	NA	<5	NR	2	NA	NR	NR	NR	NR	NR	[80]
Fresh fruits	NR	NR	NR	NR	NA	<5	NR	<2	NA	NR	NR	NR	NR	NR	[80]
Fruit products	NR	NR	NR	NR	NA	<5	NR	<1	NA	NR	NR	NR	NR	NR	[80]
Beverages	NR	NR	NR	NR	NA	<0.5	NR	<0.5	NA	NR	NR	NR	NR	NR	[80]
Milk	NR	NR	NR	NR	NA	<0.5	NR	<0.5	NA	NR	NR	NR	NR	NR	[80]
Dairy products	NR	NR	NR	NR	NA	<5	NR	<2	NA	NR	NR	NR	NR	NR	[80]
Nuts	NR	NR	NR	NR	NA	<5	NR	<5	NA	NR	NR	NR	NR	NR	[80]
Beef stick	NA	NA	<0.6	NA	NA	<0.5	4.5	2.7	<LOD[a]	<1	<1	NA	<1	<3	[29, 81]
Roast beef	NA	NA	<0.6	NA	NA	2.6	<1	<0.6	<LOD[a]	<2	<2	NA	<1	<3	[29, 81]
Ground beef	NA	NA	<0.5	NA	NA	<0.4	<1	2.1	<LOD[a]	<1	<1	NA	<1	<3	[29, 81]
Luncheon meats, cold cuts	NA	NA	<0.4	NA	NA	<0.4	<1	0.5	<LOD[a]	<1	<LOD	NA	<1	<3	[29, 81]
Fish, marine	NA	NA	<0.4	NA	NA	<0.5	<1	2.6	<LOD[a]	<1	<1	NA	<0.8	<4	[29, 81]
Fish, freshwater	NA	NA	<0.4–1	NA	NA	<0.5–2	<1	1.5–2.0	<LOD[a]	<1–2	<1–2	NA	<0.9–2	<2–5	[29, 81]
Pizza	NA	NA	2.0	NA	NA	0.74	<1	<1	27.3[a]	<1	<1	NA	<1	<1	[29, 81]
Microwave pop corn	NA	NA	1.5	NA	NA	3.6	<1	0.98	15.3–18.9[a]	<1	0.9	NA	<1	<1	[29, 81]
Egg breakfast sandwich	NA	NA	<LOD	NA	NA	<LOD	<LOD	<LOD	11.9[a]	<LOD	<LOD	NA	<LOD	<LOD	[29, 81]
French fries	NA	NA	<LOD	NA	NA	<LOD	<LOD	<LOD	4.11–9.72[a]	<LOD	<LOD	NA	<LOD	<LOD	[29, 81]

(continued)

Table 2 (continued)

Food	PFBS	PFHxA	PFHpA	PFHxS	THPFOS	PFOA	PFNA	PFOS	∑PFOSA	PFDA	PFUnDA	PFDS	PFDoDA	PFTDA	Ref.
Chicken nuggets	NA	NA	<LOD	NA	NA	<LOD	<LOD	<LOD	5.87[a]	<LOD	<LOD	NA	<LOD	<LOD	[29, 81]
Fish burger	NA	NA	<LOD	NA	NA	<LOD	<LOD	<LOD	3.82[a]	<LOD	<LOD	NA	<LOD	<LOD	[29, 81]
Vegetables	<LOD	<LOD	<0.004	<LOD	<LOD	<0.027	<LOD	0.022	<LOD	<LOD	<LOD	<LOD	<LOD	<LOD	[28]
Cereals	<LOD	<LOD	<0.009	<LOD	<LOD	<0.045	<LOD	<0.027	<LOD	<LOD	<LOD	<LOD	<LOD	<LOD	[28]
Pulse	<LOD	<LOD	<0.008	<LOD	<LOD	<0.080	<LOD	<0.069	<LOD	<LOD	<LOD	<LOD	<LOD	<LOD	[28]
White fish	<LOD	<LOD	<0.004	<LOD	<LOD	<0.065	<LOD	0.407	<LOD	<LOD	<LOD	<LOD	<LOD	<LOD	[28]
Sea food	<LOD	<LOD	<0.002	<LOD	<LOD	<0.029	<LOD	0.148	<LOD	<LOD	<LOD	<LOD	<LOD	<LOD	[28]
Tinned fish	<LOD	<LOD	<0.007	<LOD	<LOD	<0.126	<LOD	0.271	<LOD	<LOD	<LOD	<LOD	<LOD	<LOD	[28]
Blue fish	<LOD	<LOD	<0.010	<LOD	<LOD	<0.132	<LOD	0.654	<LOD	<LOD	<LOD	<LOD	<LOD	<LOD	[28]
Pork	<LOD	<LOD	<0.006	<LOD	<LOD	<0.053	<LOD	0.045	<LOD	<LOD	<LOD	<LOD	<LOD	<LOD	[28]
Chicken	<LOD	<LOD	<0.004	<LOD	<LOD	<0.004	<LOD	0.021	<LOD	<LOD	<LOD	<LOD	<LOD	<LOD	[28]
Veal	<LOD	<LOD	<0.003	<LOD	<LOD	<0.003	<LOD	0.028	<LOD	<LOD	<LOD	<LOD	<LOD	<LOD	[28]
Lamb	<LOD	<LOD	<0.012	<LOD	<LOD	<0.012	<LOD	0.040	<LOD	<LOD	<LOD	<LOD	<LOD	<LOD	[28]
Eggs	<LOD	<LOD	<0.005	<LOD	<LOD	<0.005	<LOD	0.082	<LOD	<LOD	<LOD	<LOD	<LOD	<LOD	[28]
Dairy products	<LOD	<LOD	<0.007	<LOD	<LOD	<0.007	<LOD	0.121	<LOD	<LOD	<LOD	<LOD	<LOD	<LOD	[28]
Whole milk	<LOD	<LOD	<0.015	<LOD	<LOD	<0.015	<LOD	<0.014	<LOD	<LOD	<LOD	<LOD	<LOD	<LOD	[28]
Semiskimed milk	<LOD	<LOD	<0.004	<LOD	<LOD	<0.004	<LOD	<0.019	<LOD	<LOD	<LOD	<LOD	<LOD	<LOD	[28]
Fruits	<LOD	<LOD	<0.004	<LOD	<LOD	<0.004	<LOD	<0.017	<LOD	<LOD	<LOD	<LOD	<LOD	<LOD	[28]
Margarine	<LOD	<LOD	<0.014	<LOD	<LOD	<0.014	<LOD	<0.034	<LOD	<LOD	<LOD	<LOD	<LOD	<LOD	[28]
Oil	<LOD	<LOD	<0.035	<LOD	<LOD	<0.035	<LOD	<0.099	<LOD	<LOD	<LOD	<LOD	<LOD	<LOD	[28]
Milk and infant formula															
Infant formula	<0.7	NA	<1.2	<1.34–3.59	NA	<48.3	<2.20	<11.0–11.3	NA	NA	NA	NA	NA	NA	[103]
Dairy milk	<0.7	NA	<1.2	<1.34–3.82	NA	<48.3	<2.20	<11.0	NA	NA	NA	NA	NA	NA	[103]
Drinking water (µg/l)															
Tap water	<2–20	NA	NA	NA	NA	2–4	NA	<1–6	NA	NA	NA	NA	NA	NA	[27]
Tap water	NA	NA	NA	NA	NA	2.3–84	NA	0.16–22	NA	NA	NA	NA	NA	NA	[86]
Tap water	<LOD	<LOD	0.64–3.02	0.28	<LOD	0.32–6.28	0.22–0.52	0.39–0.87	<LOD	<LOD	<LOD	<LOD	<LOD	<LOD	[85]
Bottled water	<LOD	<LOD	0.40	<LOD	<LOD	0.34–0.67	0.13–0.20	<LOD	<LOD	<LOD	<LOD	<LOD	<LOD	<LOD	[85]

<LOQ not detected, NA not analysed, NR not reported, PFBS perfluorobutanesulfonate, PFHxA perfluorohexanoic acid, PFHpA perfluoroheptanoic acid, PFHxS perfluorohexanesulfonate, PFDA perfluorodecanoic acid, PFDoDA perfluorododecanoic acid, PFDS perfluorodecanesulfonate, PFNA perfluorononanoic acid, PFOA perfluorooctanoic acid, PFOS perfluorooctanesulfonate, ∑PFOSA perfluorooctanesulfonamide, PFUnA perfluoroundecanoic acid, THPFOS 1H,1H,2H,2H-perfluorooctanesulfonic acid, PFTDA perfluorotetradecanoic acid, PFUnA perfluoroundecanoic acid, THPFOS 1H,1H,2H,2H-perfluorooctanesulfonic acid

[a]Sum of N-ethyl-PFOSA, PFOSA, N,N-diethyl-PFOSA, N-methyl-PFOSA and N,N'-dimethyl-PFOSA

Table 3 Levels of PFCs in drinking water

Food	PFBS	PFHxA	PFHpA	PFHxS	THPFOS	PFOA	PFNA	PFOS	∑PFOSA	PFDA	PFUnDA	PFDS	PFDoDA	PFTDA	Ref.
Drinking water (μg/l)															
Tap water	<2–20	NA	NA	NA	NA	2–4	NA	<1–6	NA	NA	NA	NA	NA	NA	[27]
Tap water	NA	NA	NA	NA	NA	2.3–84	NA	0.16–22	NA	NA	NA	NA	NA	NA	[86]
Tap water	<LOD	<LOD	0.64–3.02	0.28	<LOD	0.32–6.28	0.22–0.52	0.39–0.87	<LOD	<LOD	<LOD	<LOD	<LOD	<LOD	[85]
Bottled water	<LOD	<LOD	0.40	<LOD	<LOD	0.34–0.67	0.13–0.20	<LOD	<LOD	<LOD	<LOD	<LOD	<LOD	<LOD	[85]
Bottled water	0.00074 <0.00071	ND- <0.00064	<0.00073 <0.00092	<0.00092	NA	<0.0005	NA	0.00075–0.00145	NA	NA	NA	NA	NA	NA	[161]
Tap water	NA	NA	NA	NA	NA	0.00085	NA	0.00039	NA	NA	NA	NA	NA	NA	[162]

<LOQ not detected, NA not analysed, NR not reported, PFBS perfluorobutanesulfonate, PFDA perfluorodecanoic acid, PFDoDA perfluorododecanoic acid, PFDS perfluorodecanesulfonate, PFHpA perfluoroheptanic acid, PFHxA perfluorohexanoic acid, PFHxS perfluorohexanesulfonate PFNA perfluorononanoic acid PFOA perfluorosulfonate acid, PFOS perfluorooctanesulfonate ∑PFOSA perfluorooctanesulfonamide, PFTDA perfluorotetradecanoic acid, PFUnA perfluoroundecanoic acid, THPFOS 1H,1H,2H,2H-perfluorooctanesulfonic acid

aSum of N-ethyl-PFOSA, PFOSA, N,N-diethyl-PFOSA, N-methyl-PFOSA and N,N′-dimethyl-PFOSA

PFOS and from 0.1 to 0.7 ng/L for PFOA. Takagi et al. [86] analysed PFOS and PFOA in raw and tap water samples collected from 14 drinking water treatment plants in winter and summer seasons in Osaka. Concentration ranges of PFOS and PFOA in raw water were 0.26–22 ng/l and 5.2–92 ng/l, respectively. Whereas the concentrations PFOS in raw water from Osaka were similar to those in other areas in Japan, the concentrations of PFOA were higher than in other areas. Concentration ranges of PFOS and PFOA in potable tap water were 0.16–22 ng/l and 2.3–84 ng/l, respectively. There were positive correlations between PFC concentrations in raw water and tap water samples. Therefore, the removal efficiency of PFCs by the present water treatment may be low and their occurrence in waters high. Kunacheva et al. [118] have evaluated tap water in Bangkok and bottled water. The average PFOS and PFOA concentrations in tap water were 0.17 and 3.58 ng/L, respectively. PFOS and PFOA were not similarly distributed in all areas in the city. PFCs concentrations were higher in bottled water than in tap water. But according to the guideline from the New Jersey Department of Environmental Protection, PFOA concentrations in tap water and bottled water found in Bangkok were not expected to cause any health risks. Yim et al. [119] developed a sensitive analytical method for the analysis of 20 PFCs, including several short-chain PFCs, for their quantification in tap water collected in China, Japan, India, the United States and Canada between 2006 and 2008. Of the PFCs measured, PFOS, PFHxS, PFBS, PFPrS, PFEtS, PFOSA, N-EtFOSAA, PFDoDA, PFUnDA, PFDA, PFNA, PFHpA, PFHxA, PFPeA, PFBA and PFPrA were found at detectable concentrations in the tap water samples. The water samples from Shanghai (China) contained the greatest concentrations of total PFCs (arithmetic mean = 130 ng/L), whereas those from Toyama (Japan) contained only 0.62 ng/L. In addition to PFOS and PFOA, short-chain PFCs such as PFHxS, PFBS, PFHxA and PFBA were found to be prevalent in drinking water. However, according to the health-based values (HBVs) and advisory guidelines derived for PFOS, PFOA, PFBA, PFHxS, PFBS, PFHxA and PFPeA by the U.S.EPA and the Minnesota Department of Health, tap water may not pose an immediate health risk to consumers. In a recent study [120], the concentrations of 10 PFCs were investigated in the Hun River, four canals, ten lakes, and influents and effluents from four main municipal wastewater treatment plants (WWTPs) in Shenyang, China. Mass flows of four main PFCs were calculated to elucidate the contribution from different sections of the Hun River. PFO) and perfluorohexanoic acid (PFHxA) were the major PFCs in the river, with ranges of 2.68–9.13 ng/L, and 2.12–11.3 ng/L, respectively, while PFOS was detected at lower levels. The PFC concentrations in the Hun River increased after the river passes through Shenyang and Fushun cities. The concentrations of PFCs in four urban canals were higher than those in the river. Total PFCs in ten lakes from Shenyang were at low levels, with the greatest concentration (56.2 ng/L) detected in a heavily industrialised area. The PFC levels in WWTP effluents were higher than those in surface waters with concentrations ranging from 18.4 to 41.1 ng/L for PFOA, and 1.69–3.85 ng/L for PFOS. Similar PFC profiles between effluents from WWTPs and urban surface waters were found. Finally, it was found that the composition profiles of PFCs in surface waters were similar to those in tap water, but not consistent with those in adult blood from Shenyang. The calculation on total daily intake of PFOS by adults from Shenyang showed that the contribution of drinking water to human exposure was minor [120].

In a recent study in Australia, Thompson et al. [121] assessed the exposure to PFCs via potable water in Australia. Sixty-two samples of potable water were collected from 34 locations across Australia, including capital cities and regional centre. PFOS and PFOA were the most commonly detected compounds, and quantifiable levels were found in 49% and 44% of all samples, respectively. The maximum concentration in any sample was seen for PFOS with a concentration of 16 ng/L, second highest maximums were for PFHxS and PFOA at 13 and 9.7 ng/L. Assuming a daily intake of 1.4 and 0.8 ng/kg b.w. for PFOS and PFOA the average contribution from drinking water was 2–3% with a maximum of 22% and 24%, respectively.

In EEUU, the Bureau of Safe Drinking Water (BSDW) initiated a preliminary occurrence study in July 2006 to determine whether PFOS and PFOA could be found in detectable concentrations in raw and treated public water systems throughout the state of New Jersey (USA). Five out of 23 public water samples showed non-detectable levels of PFOA. Detected and quantifiable levels were found in15 samples with values ranged from 4.5 to 39 ng/L. Additionally, three samples showed levels of PFOA that were detected but not quantified (<4 ng/L). Ten out of 23 water samples collected from public water systems showed non-detectable levels of PFOS. In seven samples, PFOS concentration could be quantified and varied between 4.2 and 19 ng/L. Six samples showed levels of PFOS but could not be quantified (<4 ng/L) (New Jersey Department of Environmental Protection Division of Water Supply 2007). Emmett et al. [122] published their investigations on PFOA contamination of drinking water in the surroundings of a fluoropolymer production site in Washington, West Virginia (USA). The public water supply of Little Hocking, Ohio, draws water from wells across the Ohio River and showed the highest PFOA concentrations reported in drinking water of public water supplies (mean, 3,550 ng/L; range, 1,500–7,200 ng/L).

In Europe, a relevant study was performed by Skutlarek et al. [33]. In this study, drinking water samples were analysed in a contaminated area of the Ruhr and Moehne catchment (Germany). The highest values found were at Neheim, and the PFOA levels found were 519 ng/L, followed by PFHpA (23 ng/L) and PFHxA (22 ng/L). The concentration level measured in the tap water was 767 ng/L for the sum of PFCs analysed. These values were comparable to the levels in water from the Moehneriver that served as a water supply. After detection of PFOA in drinking water at concentrations up to 0.64 μg/l in Arnsberg, Sauerland, Germany, the German Drinking Water Commission (TWK) assessed PFCs in drinking water and set for the first time worldwide in June 2006 a health-based guide value for safe lifelong exposure at 0.3 μg/l (sum of PFOA and PFOS). PFOA and PFOS can be effectively removed from drinking water by percolation over granular activated carbon. Additionally, recent EU-regulations require ***phasing out use of PFOS and ask to voluntarily reduce the one of PFOA. In July 2006, the waterworks of Moehnebogen installed activated charcoal filters, which efficiently decreased PFC concentration in drinking water. Recently, the internal exposure of the residents of this area has been published [123], showing that PFOA levels in blood and serum were 4.5–8.5 times higher than those of the reference population. After installation of activated charcoal filters in the waterworks, PFOA concentrations in Arnsberg were significantly reduced. However, during the study period, filtration performance declined and PFOA concentrations in tap water samples increased from below the LOD to 71 ng/L. Other studies performed in Europe reported values for tap water with values less than 10 ng/L. Loos et al. [96] have investigated PFHpA, PFOA, PFOS, PFNA, PFDA, PFUnA and PFDoA in which tap water that originated from Italy's Lake Maggiore. In this study, all the investigated compounds were detected, with concentrations ranging from 0.1 to 9.7 ng/L. The highest values were 9.7 ng/L for PFOS, 2.9 ng/L for PFOA and 2.8 ng/L for PFDoA. Ericson et al. [93] analysed 14 PFCs in drinking water (tap and bottled) from Spain. The bottled water samples from four commercial companies, whose water spring has different origins, were purchased from a supermarket. Highest values were from Valls sampling site with levels; PFHpA (3.02 ng/L), PFOS (0.44 ng/L) and PFOA (6.28 ng/L). The samples of bottled water contained some PFCs at levels that corresponded to the lowest values observed in tap water.

Although PFCs have been detected in tap water worldwide, very few studies have examined their fate, especially removal in drinking water treatment processes. Takagi et al. [124] studied the concentrations of PFOS and PFOA at every stages of drinking water treatment processes in several water purification plants that employ advanced water treatment technologies. The authors have found that PFC concentrations did not vary considerably in raw water, sand filtered water, settled water and ozonated water. Sand filtration and ozonation did not have an effect on the removal of PFOS and PFOA in drinking water. PFOS and PFOA were removed effectively by activated carbon that had been used for less than one year. However, activated carbon that had been used for a longer period of time (>1 year) was not effective in removing PFOS and PFOA from water. In addition, variations in the removal ratios of PFOS and PFOA by activated carbon were found between summer and winter months.

4.2 Fish Contamination

Among PFC fish contaminants, PFOS is the most crucial and prominent compound. Reports suggest no considerable differences in PFC concentrations among freshwater and marine fish species. PFOA is the second most frequently detected PFCs in fish, but it has been shown that PFOA is detected at much lower concentrations than is PFOS. Quantifiable concentrations of PFOA were detected in lake trout [125, 126], rainbow smelt, and alewife, with concentrations ranging from 0.16 to 6.8 ng/g wet weight (wwt). The difference between the observed PFOS and PFOA concentrations in fish suggests a lower potential of PFOA to bioaccumulate in fish as compared to PFOS. This observation was further confirmed by laboratory experiments, which revealed a 1,000-fold lower bioconcentration factor for PFOA compared to PFOS [114, 127]. A restricted number of studies also reported other PFCs and lower concentrations than PFOS were found. For example, Ye et al. [128] detected PFHxS at a maximum concentration of 1.89 ng/g wwt in a mixture of whole fish in the Missouri River, USA. Concentrations of the other PFCs analysed in this study were found in median concentration of 3.71 (PFHxA), 0.82 (PFDA) and 0.36 ng/g (PFHxS) wwt. Martin et al. [126] detected relatively high concentrations of the longer chain PFCs in fish collected from Lake Ontario, Canada. The highest concentration of these PFCs was 8.3 ng/g wwt for PFUnA. These authors concluded that individual PFCs were generally detected at lower concentrations than were PFOS, and total PFOS equivalents (PFOS and PFOSA) exceeded the sum of all PFCs by a factor of between 1.8 and 12 within each species analysed. Tomy et al. [14] have detected a relatively high mean concentration (92.8 ng/g wwt) of N-ethyl perfluorooctane sulphonamide in Arctic cod, ranging between 9.6 and 144.6 ng/g wwt. Since transformation of N-EtPFOSA to PFOS and PFOSA by rainbow trout microsomes has been reported [129], N-EtPFOSA is an important compound to measure in biota and in human samples. Berger et al. [108] analysed PFCs in muscle tissue from edible fish species caught in the second largest freshwater lake in Sweden, Lake Vättern, and in the brackish water Baltic Sea. Again, PFOS was the predominant PFAS found. PFOS concentrations were higher in Lake Vättern (medians 2.9–12 ng/g fresh weight) than in Baltic Sea fish (medians 1.0–2.5 ng/g fresh weight). Moreover, Lake Vättern fish was more contaminated with several other PFAS than Baltic Sea fish. This may be due to anthropogenic discharges from urban areas around Lake Vättern. The PFAS pattern differed between Lake Vättern and Baltic Sea fish, indicating different sources of contamination for the two study areas. Human exposure to PFOS via fish intake was calculated for three study groups, based on consumption data from literature. The groups consisted of individuals that reported moderate or high consumption of Baltic Sea fish or high consumption of Lake Vättern fish, respectively. The results showed that PFOS intake strongly depended on individual fish consumption as well as the fish catchment area. Median PFOS intakes were estimated to 0.15 and 0.62 ng/kg body weight (bw)/d for the consumers of moderate and high amounts of Baltic Sea fish, respectively. For the group with high consumption of Lake Vättern fish, a median PFOS intake of 2.7 ng/kg bw/d was calculated. Fish consumption varied considerably within the consumer groups, with maximum PFOS intakes of 4.5 (Baltic Sea fish) or 9.6 ng/kg bw/d (Lake Vättern fish). These results suggested that fish from contaminated areas might be a significant source of dietary PFOS exposure. However, some controversial results were obtained by Nania et al. [130] In this study, the objective was to evaluate the contamination levels of PFOS and PFOA in edible fish of the Mediterranean Sea. Twenty-six fish muscles, 17 fish livers, five series of cephalopods (each composed of ten specimens) and thirteen series of bivalves (each composed of about 50 specimens) were used for the investigation. The results showed PFOA and PFOS levels in fishes and molluscs lower than those reported for analogue matrices in different geographic areas. According to their results, no relation can be established between water contamination levels and posterior levels found in sea food. In another work, Llorca et al. [49] analysed eight PFCs in fish samples from Mediterranean Sea. The result of this study showed higher concentrations than those reported by Nania [130]. The results from Nania study also disagree with a recent study carried out under laboratory controlled conditions. To among organisms studied, none of the bivalves accumulated PFCs, and contrarily, insect larvae, followed by fish and crabs contained levels ranging from 0.23 to 144 ng/g wwt of PFOS, from 0.14 to 4.3 ng/g wwt of PFOA, and traces of PFNA and PFHxS.

Haug et al. [62] have studied the possible associations between concentrations of PFCs in serum and consumption of food with particular focus on seafood, and to compare estimated dietary intakes with determined serum PFC concentrations. In this study, the concentrations of 19 PFCs were determined in serum from 175 participants in the Norway. Associations between estimated individual total dietary intakes of PFCs and serum concentrations were also explored. PFC concentrations in serum were significantly associated with the consumption of lean fish, fish liver, shrimps and meat, as well as age, breastfeeding history and area of residence. Although several authorities recommend not eating fish liver because of the risk associated with high intake of persistent organic pollutants (POPs), fish liver (and oil) is still consumed. It should be pointed out that PFC levels in liver are at least two orders of magnitude higher than in muscle tissue [130]. In Japan, concentrations of total PFCs in skipjack tuna livers ranged from <1 to 83 ng/g wwt [131]. PFOS and PFUnA were the prominent compounds.

Similar to fish, PFOS is the dominant PFC found in aquatic invertebrates such as shrimp, mussels, clams and oysters [132, 133]. A few papers report on PFC levels in bivalves. Concentration ranging from 1 to 6.0 ng/g wwt) in oysters were reported from the Ariake Sea [134] and China [133], respectively. Cunha et al. [135] measured high concentrations of PFOS in mussels from several estuaries in the North of Portugal. PFOS was detected in all the samples analysed, and the concentrations were ranging 36.8 to 126.0 ng/g wwt. In a more recent work, Nania et al. [130] found higher PFOA than PFOS in clam but comparable levels were found in mussels, which were attributed to differences in habitat and feeding behaviour.

Nowadays, the bioaccumulation trends of PFCs in aquatic organisms are not clear. In general, concentrations of PFCs are expected to increase with increasing trophic level. This trend has been observed in the Great Lakes food chain [136]. However, higher concentrations of perfluoroalkyl contaminants were reported in lower trophic levels in seafood from China [133] and in invertebrate species from Lake Ontario [126]. However, there are some controversial results. It is clear that different processes are involved at the same time including sorption processes to organic material, metabolic pathways and data continue being inconsistent and the different sorption characteristics of the different types of PFCs should be deeper studied.

Sorption coefficients of PFCs are relatively low for C4–C8-carboxylic acids and increase with increasing chain length [137].

Biomagnification of PFOS in the estuarine food chain of the Western Scheldt estuary was observed by de Vos et al. [116] On the other hand, it is not clear whether there is a difference between the concentrations of PFCs inedible fish from remote versus highly industrialised or urbanised areas or not. However, Gulkowska et al. [133] observed slightly higher PFOS concentrations in fish from the highly urbanised and industrialised areas.

More recently, several authors studied the possible association between fish consumption and levels of PFCs in human blood [138], as well as the evaluation of the risk associated with fish consumption [73, 139]. In recognition of the potential for human exposure to PFCs via fish consumption, the Minnesota Department of Health has issued fish consumption advisories for contaminated sections of the Mississippi River (Minnesota Department of Health 2007). This advisory suggests that people limit their intake of fish to no more than one meal a week if PFOS levels in fillet exceed 38 ng/g.

The provisional tolerable daily intake (TDI) values proposed by the European Food Safety Authority (EFSA 2008) and Health Protection Agency (HPA 2009) amount to 150 ng/kg body weight (bwt)/d and 300 ng/kg bwt/d, for PFOS and PFOA, respectively.

4.3 Foodstuff Studies

Studies that measure PFCs in consumer food are limited. One of the first studies was carried out in EEUU and 3M sponsored it. The study measured PFOA, PFOS and PFOSA in individual food samples including green beans, apples, pork, milk, chicken, eggs, bread, hot dogs, catfish and ground beef [140]. Most samples had levels below the LOD (0.5 ng/g for all chemicals). The highest level of PFOA (2.35 ng/g wwt) was detected in an apple purchased in Decatur, Alabama, the location of a 3M PFOA production plant. The highest level of PFOS (0.85 ng/g wwt) was from milk purchased in Pensacola, Florida. More recently, in another study, PFCs were evaluated among other POPs in composite food samples from Dallas, Texas. The pattern of detection of PFCs varied significantly in this study compared with the previous ones. In previous studies, typically the most commonly detected PFC was PFOS, whereas in the study performed by Schecter [107], PFOS did not exceed the LOD, from 0.01 to 0.5 ng/g wwt, in any samples, which is perhaps not surprising because it has been off the market since 2002. Instead, PFOA was found to exceed the LOD in 17 of 31 samples, with highest levels in butter (1.07 ng/g wwt) and olive oil (1.8 ng/g wwt). The relatively high levels of PFOA detected in the Schecter study might be attributed to the materials used in the processing and packaging of the food. Some food packaging materials contain trace amounts of PFOA, and PFCs have been shown to migrate from packaging materials into food oils [113]. In Canada, a study of chemical contamination of food collected from 1992 to 2004 as part of the Canadian Total Diet Study was conducted. In this Canadian study, the evaluation of PFOS and PFOA was included [27]. PFOA was detected at the highest levels in microwave popcorn (3.6 ng/g wwt) and roast beef (2.6 ng/g wwt), and PFOS was detected at the highest levels in beefsteak (2.7 ng/g wwt) and saltwater fish (2.6 ng/g wwt). PFNA was detected in the beefsteak sample (4.5 ng/g wwt). LODs for PFCs ranged from 0.4 to 5 ng/g wwt. In a later study, Ostertag et al. [141] assessed the dietary exposure to PFCs from the consumption of store-bought and restaurant foods for the Canadian population. PFCs were detected in processed meats, pre-prepared foods, and peppers with a range of concentrations from 0.48 to 5.01 ng/g wwt. 6:2 fluorotelomer unsaturated carboxylate (FTUCA) was detected in cold cuts at a concentration of 1.26 ng/g. Mean daily PFC exposure estimates ranged from 1.5 to 2.5 ng/kg bwt). Perfluorinated carboxylates have been shown to contribute more to PFC exposure than PFOS or FTUCA. Total PFCAs in cakes and cookies, lunch meats and green vegetables were the main contributors to dietary exposure, although these exposure levels were below the provisional toler-able daily intake provided by the German Drinking Water Commission. Dietary exposure to total PFCs has not changed over time, although the contribution of PFOS to total PFC exposure may have increased between 1998 and 2004. Under the light of these results, the authors have concluded that further research on the sources of contamination of processed and prepared foods is required, but the dietary exposure to PFCs among Canadians poses minimal health risks based on current toxicological information.

Daily dietary intake of nine PFCs, including PFOS and PFOA, was assessed in matched daily diet duplicates [142]. Diet samples were collected in year 2004 from 20 women in Osaka and Miyagi, Japan. Only PFOS and PFOA were detected in the diet samples without observing significant difference between cities. After adjusted by water content, diet concentration of PFOA was significantly higher in Osaka. The median daily intake calculated using the measured diet concentrations was 1.47 ng PFOS/kg bw. and 1.28 ng PFOA/kg bw. for Osaka, and 1.08 ng PFOS/kg bw., and 0.72 ng PFOA/kg b.w., for Miyagi.

In Europe, one of the first studies was carried out by the U.K. Food Standards Agency published results of PFC analysis in food collected from the 2004 Total Diet Study [26]. PFOS exceeded the LOD in potatoes, canned vegetables, eggs, sugars and preserves, with highest levels detected in potatoes (10 ng/g wwt), including fresh potatoes as well as potato chips, French fries and hash browns, whereas PFOA was detected only in potatoes (1 ng/g wwt). Fromme et al. [61] have conducted a study in Germany to quantify the dietary intake of PFOS, PFOA, PFHxS, PFHxA and PFOSA using 214 duplicate diet samples and to estimate individual intakes based on blood levels of PFOS and PFOA. The median (90th percentile) daily dietary intake of PFOS and PFOA was 1.4 ng/kg bw (3.8 ng/kg bw) and 2.9 ng/kg bw (8.4 ng/kg bw), respectively. PFHxS and PFHxA were detected only in some samples above detection limit with median (maximum) daily intakes of 2.0 ng/kg b.w. (4.0 ng/kg b.w.) and 4.3 ng/kg b.w. (9.2 ng/kg b. w.), respectively. Because PFOSA could not be detected above, the limit of detection of 0.2 ng/g f.w indicating that this indirect route of exposure is of less significance. Another study examined dietary intake of PFCs has been estimated for various age/gender groups of the population of Tarragona (Catalonia, Spain) [12], during 2006. PFC levels were determined in 36 composite samples of foodstuffs randomly purchased in various locations. PFOS, PFOA and PFHpA were the only detected PFCs in foodstuffs. The most commonly detected PFC was PFOS, in 24 of 36 samples, with the highest levels in an uncooked bluefish composite sample (0.654 ng/g wwt), which included salmon, sardines and tuna. PFOA was found only in whole milk, at relatively low levels (0.055 and 0.058 ng/g wwt). On average, for a standard adult man (70 kg of body weight), the dietary intake of PFOS was estimated to be 62.5 or 74.2 ng/day (assuming $ND = 0$ or $ND = 1/2$ LOD, respectively). Fish, followed by dairy products and meats, were the main contributors to PFOS intake. Haug et al. [110] have studied the levels in food and beverages and daily intake of perfluorinated compounds in Norway. Up to 12 different PFCs were detected in the samples. PFOS and PFOA were found in concentrations similar to or lower than what has been observed in other studies worldwide. Differences in the relative proportion of PFOA and PFOS between samples of animal origin and samples of non-animal origin were observed and support findings that PFOS has a higher bioaccumulation potential in animals than PFOA. Based on these results and consumption data for the general Norwegian population, a rough estimate of the total dietary intake of PFCs was found to be around 100 ng per day, and PFOA and PFOS contributed to about 50% of the total intake. The estimated intakes of PFOS and PFOA in this study were lower than what has been reported in previous studies from Spain, Germany, United Kingdom, Canada and Japan. In a recent study, Noorlander et al. assessed the levels of PFCs in food and dietary intake of PFOS and PFOA in the Netherlands. In this study according to the typical diet in the Netherlands, the main contributors to PFOS and PFOA were identified. For both cases drinking water was a relevant factor. Important contributors of PFOA intake were vegetables/fruit and flour, whereas milk, beef and lean fish were important contributors of PFOS intake.

Several PFCs have been detected in human blood from populations in North and South America, Asia, Australia and Europe [48, 67, 143–146]. Different studies in Europe showed that PFOS is one of the more frequent compound present in human blood [48, 147], and the highest PFOS concentrations were found in Poland, followed by Belgium, being comparable to Sweden, with lowest concentrations in Italy [37]. These results indicate differences in exposure across Europe. However, the sources and pathways of human exposure to PFCs are currently not well understood [27]. The wide variety of industrial and consumer applications leads to numerous possibilities for release of PFCs into the environment and subsequent exposures to humans via environmental routes and media. However, the relative uniform distribution of blood concentrations of PFCs in children and the majority of adult populations points to a common major source, possibly food.

5 Human Bioaccumulation

Bioaccumulation is generally referred to as a process in which the chemical concentration in an organism achieves a level that exceeds that in the respiratory medium (e.g., water for a fish or air for a mammal), the diet, or both. The extent to which chemicals bioaccumulate is expressed by several quantities, including the bioconcentration factor (BCF), bioaccumulation factor (BAF), biomagnification factor (BMF), and trophic or food web magnification factor (TMF) [6]. The ecological, biological and chemical parameters involved in the transfer and accumulation of contaminants in food webs are complex.

Since the toxicological effects of PFCs have been mainly associated with early stages of development, during the last years a great effort has been paid to elucidate human exposure and accumulation and especially infant exposure, because it is a sensitive population. Kärman et al. [9] studied occurrence and levels of PFCs in human milk in relation to maternal serum together with the temporal trend in milk levels between 1996 and 2004 in Sweden. In this study, PFOS and PFHxS were detected in all milk samples at mean concentrations of 0.201 ng/mL and 0.085 ng/mL, respectively. On the other hand, the total PFC concentration in maternal serum was 32 ng/mL, and the corresponding milk concentration was 0.34 ng/mL. The PFOS milk level was on average 1% of the corresponding serum level. Therefore, a strong association was found between increasing serum concentration and increasing milk concentration for PFOS ($R2 = 0.7$) and PFHxS ($R2 = 0.8$). According to these results, the calculated total amount of PFCs transferred by lactation to a breast-fed infant in this study was approximately 200 ng/day. These values were much higher than those reported by Tao et al. [148] in a study performed with breast milk samples collected in 2004 from Massachusetts, USA. In this study, PFOS and PFOA were again the predominant compounds, but mean concentrations for PFOS and PFOA were 131 and 43.8 pg/mL, respectively. In another work, the same group studied the PFCs occurrence and concentrations in several Asian countries. Median concentrations of PFOS in breast milk from Asian countries varied significantly; the lowest concentration of 39.4 pg/mL was found in India, and the highest concentration of 196 pg/mL was found in Japan. The measured concentrations were similar to or less than the concentrations previously reported from Sweden, the United States, and Germany (median, 106–166 pg/mL). PFHxS was found in more than 70% of the samples analysed from Japan, Malaysia, Philippines and Vietnam, at mean concentrations ranging from 6.45 (Malaysia) to 15.8 (Philippines) pg/mL. PFOA was found frequently only in samples from Japan; the mean concentration for that country was 77.7 pg/mL. Völker et al. [149] carried out a pilot study to assess PFCs in breast milk from donors living in Germany and Hungary. PFOS was quantified in all the 70 samples. The concentration in samples from Germany ranged between 28 and 309 ng/l (median: 119 ng/l). Samples from Hungary showed significantly higher PFOS concentrations (median 330 ng/l, range 96–639 ng/l). In only 11 of the 70 samples (16%), PFOA reached the LOQ (200 ng/l); values for PFOA ranged from 201 to 460 ng/l. In another study, Nakata et al. assessed the concentrations of PFCs in Japanese breast milk. Again relatively high levels were detected, which ranged over 0.046–0.098 ng/mL for PFOS and over 0.016–0.270 ng/mL for PFOA. In addition, human milk and maternal plasma collected from the same donor were analysed. As a result, the concentration of PFCs in human milk and maternal plasma showed significant correlations for PFOS. Fromme et al. [121] assessed pre- and postnatal exposure to PFCs in Germany. In this study, PFCs were investigated in maternal blood during pregnancy (at two time points) ($n = 40$ and 38) and 6 months after delivery ($n = 47$), in cord blood ($n = 33$) and in blood of infants six ($n = 40$) and nineteen months ($n = 24$) after birth, and monthly in breast milk samples. Concentrations in maternal serum ranged from 0.5 to 9.4 µg/L for PFOS and 0.7 to 8.7 µg/L for PFOA. In cord serum, the values ranged from 0.3 to 2.8 µg/L and from 0.5 to 4.2 µg/L for PFOS and PFOA, respectively. The median results from serum at six and nineteen months of age were 3.0 and 1.9 µg/L for PFOS and 6.9 and 4.6 µg/L for PFOA, respectively. In breast milk samples, PFOS ranged from <0.03 to 0.11 µg/L (median: 0.04 µg/L), while PFOA was detected only in some samples*** it was were all other PFCs. These results revealed that although the concentrations in breast milk were low, this intake led to a body burden at the age of six months similar to PFOS, or higher than PFOA that found in adults. In Spain, concentration in breast milk reported by Llorca et al. [10] PFOS, and perfluoro-7-methyloctanoic acid (i,p-PFNA) was predominant being present in the 95% of breast milk samples. PFOA was quantified in 8 of the 20 breast milk samples at concentrations in the range of 21–907. ng/L. Recently, Kim et al. studied the distribution of perfluorochemicals between sera and milk from the same mothers and implications for prenatal and postnatal exposures [150]. In this work [151], the levels of six perfluorocarboxylates, four perfluoroalkyl sulfonates and one sulphonamide were measured in paired samples of maternal serum, umbilical cord serum and breast milk. Good correlation was found between the maternal and cord sera for all measured compounds. Nevertheless, there was a significant difference in compound composition profile between the two sera matrices, with a more depletion of the longer chain compounds in cord serum. The transfer efficiency values from maternal to cord serum decreased by 70% with each increasing unit of -CF2 chain within a PFCA group, and for PFOS, by a half compared to PFOA. In contrast to the strong correlation in concentrations between the two sera matrices, the pattern of compounds in breast milk differed considerably with those in sera. Accordingly, compound- and matrix-specific transfer must be considered when assessing prenatal and postnatal exposure.

In conclusion, in spite the levels of PFCs are in general below the maximum limits recommended, a strong contribution can be found during early stages of life. Transplacental exposure to PFCs is not yet well understood; however, most of the authors found a strong correlation between serum samples and breast milk.

In addition to these studies devoted to assess infant and early exposures to PFCs, other human samples have been analysed in order to assess the presence of PFCs in different human fluids and tissues including human liver [8], hair and urine [152].

6 Safety Limits, Exposure and Risk Assessment

Human exposure to PFCs is likely to occur via a number of vectors and routes, for example food, drinking water, the ingestion of non-food materials, dermal contact and inhalation. Circumstantial factors such as place of residence, age, nature of PFCs vector, may also influence exposure. For example, according to Tittlemier et al. [27], food seems to represent the major intake pathway of PFAS in adult Canadians; however, house dust, solution-treated carpeting and treated apparel might contribute a non-negligible 40% to the overall exposure.

In Europe, the first studies of per- and polyfluorinated compounds in air samples were reported in the framework of the Perforce project. The anionic compounds were in general only found in the particulate phase, with PFOA often the predominant analyte. Therefore, it is possible that non-volatile ionic compounds might directly undergo atmospheric transport on particles from source regions. The levels of ionic PFAS at a rural Norwegian site were significantly lower than those found in the UK. Generally, levels of PFAS are reported to be slightly higher in urban areas than rural sites.

Food might become contaminated during production processes and/or cooking due to contact with treated cookware that can release PFCs. However, it must be stressed that the data in general are insufficient to allow for a general evaluation of the contribution of food contact materials to total dietary exposure to PFCs.

Another important route for human exposure to PFCs is drinking water, but it must be noted that the international regulatory organisations (World Health Organization (WHO), European Union (EU)/EFSA, US EPA, etc.) have not established safety limits yet for PFCs in drinking water. However, recently, Schriks et al. [153] derived provisional drinking water guideline values for PFOS and PFOA of 0.5 and 5.3 µg/L, respectively, on the basis of the TDI values proposed by EFSA (2008).

In 2006, EPA and the eight major PFC manufacturing companies in the industry launched the 2010/15 PFOA Stewardship Program, in which companies committed to reduce global facility emissions and product content of PFOA and related chemicals by 95% by 2010 and to work towards eliminating emissions and product content by 2015 (http://www.epa.gov/oppt/pfoa/pubs/stewardship/index.html).

Guidelines have also been developed in Europe. The occurrence of PFCs in surface and drinking waters of the Ruhr and Moehne area [33] caused a high concern, in view of the possible effects on humans and the environment. Immediately after detection of high concentrations of PFOA in drinking water the German Drinking Water Commission (DWC) of the German Ministry of Health at the Federal Environment Agency established guide values for human health protection. In addition, a set of measures were proposed and the local health authorities recommended that residents in parts of Arnsberg to not use the drinking water for preparation of baby food and advised pregnant women to avoid regular intake of such water.

Recently, in New Jersey, the Department of Environmental Protection developed preliminary health-based drinking water guidance for PFOA of 40 ng/L (http://www.defendinscience.org/case_studies/upload/pfoa_dwguidance.pdf).

Most monitoring studies have focused only on PFOS and PFOA, but a few also reported on other PFCs that appear at rather high concentrations in potable water such as PFBS, PFDoA, perfluoropentanoic acid (PFPeA), and PFHxA [33, 96, 154]. Therefore, it is important to increase monitoring efforts with a view to setting more comprehensive safety limits for PFCs in potable water.

The relatively high concentrations of PFCs that have been observed in drinking water samples indicate that the common water treatment steps used do not effectively eliminate perfluorinated compounds. It should be noted that the washing of food samples with tap water may introduce an additional source of PFCs [13].

Several scientific institutions have derived TDIs from toxicological end points by applying an uncertainty factor. The Scientific Panel on Contaminants in the Food Chain (CONTAM) established a TDI for PFOS of 150 ng/kg bwt/d and for PFOA of 1.5 μg/kg bwt/d (EFSA 2008). The UK Committee on Toxicity of Chemicals in Food, Consumer Products and the Environment (COT) proposed a TDI for PFOS and PFOA of, respectively, 300 and 3,000 ng/kg bwt/d (COT 2006a,b). Furthermore, the German Federal Institute for Risk Assessment proposed a TDI of 100 ng/kg bwt/d for both PFOS and PFOA (BfR2006).

The relative importance of metabolic transformation of precursor compounds in exposure to PFOS and PFOA has been scarcely evaluated and, to our knowledge, the only study that afforded the problem by a Scenario-Based Risk Assessment (SceBRA) approach estimated the relative importance of precursor-based doses of PFOS and PFOA of 2–5% and 2–8% in an intermediate scenario and 60–80% and 28–55% in a high-exposure scenario. This indicates that these precursors are of low importance for the general population.

In summary, for the general population, the common routes of exposure to environmental compounds are ingestion, dermal contact and inhalation. Many PFAS are environmentally persistent but not lipophilic; rather they have mixed lipophobic and hydrophobic properties. The exposure scenario is complex as PFAS have a large variety of applications. Oral exposure from materials other than food, inhalation and dermal contact may be important exposure routes for certain segments of the population. Dust inhalation could also be a possible source of exposure. However, the information on concentrations of PFAS in indoor dust is very limited and the bioavailability of the current compounds from dust is unknown.

There are some data on PFOS and PFOA in fish and water from European countries. However, there is a general lack of occurrence data for most foodstuffs. This evaluation, based on food consumption patterns of the EU countries Italy, The Netherlands, Sweden, and the UK, must be regarded as provisional while waiting for the necessary food monitoring results to be gathered.

Fort the risk characterisation of PFOS and PFOA, the currently available information is inadequate to characterise dietary exposure in the different regions in the European Union.

7 Conclusions and Future Trends

In order to assess human exposure and the risk associated with PFCs, there is a strong need to carry out comprehensive food surveys and related studies, such as gastrointestinal uptake studies, which are urgently required for a better understanding of the contribution of food pathway to consumer exposure to PFCs. In addition, is highly required, establishing the levels of PFCs in food and drinking water through the complete European diets. In addition, this information has to include a full range of PFCs, and not limited to PFOS and PFOA. The understanding of exposures to PFCs through the diet is still in its early phase, and only relatively few food samples have been analysed in several countries. Further studies on the correlation between food intake and exposure, as well as food measurements, are needed before reliable conclusions can be made on the source of dietary exposures in humans.

The routes of indirect food contamination, during industrial cooking processes or by packaging migration should be better studied and assessed. In this sense, few studies have been carried out, and the comparison between the results obtained in these studies is in general difficult since the procedures to assess migration during packaging or cooking are not established.

In addition, other routes of human exposure are less studied, and exposure through dust (indoor and outdoor) and inhalation is necessary, and for example, the bioavailability of the current compounds from dust is unknown.

Transformation routes and transport of PFCs in the environment continue being poor understood, and in these sense-specific studies are required.

There is also a well-established record that should be highlighted: the ubiquitous presence and levels of PFOS and PFOA in human milk and blood. However, transplacental studies to elucidate the process involving and the risks associated with early life exposure continue being required.

Acknowledgements This work was supported by the Spanish Ministry of Science and Innovation, through the projects SCARCE Consolider- Ingenio 2010 CSD2009-00065 and CEMAGUA (CGL2007-64551/HID), and by King Saud University grant number (KSU-VPP-105). Thermo Company is acknowledged for the gift of the columns and the training help with the instrument.

References

1. Boulanger B, Vargo JD, Schnoor JL, Hornbuckle KC (2005) Environ Sci Technol 39:5524
2. Clara M, Scheffknecht C, Scharf S, Weiss S, Gans O (2008) In: Water science and technology 58:59
3. Taniyasu S, Kannan K, Man KS, Gulkowska A, Sinclair E, Okazawa T, Yamashita N (2005) J Chromatogr A 1093:89
4. Kim SK, Lee KT, Kang CS, Tao L, Kannan K, Kim KR, Kim CK, Lee JS, Park PS, Yoo YW, Ha JY, Shin YS, Lee JH (2011) Environ Pollut 159:169

5. Nakayama S, Harada K, Inoue K, Sasaki K, Seery B, Saito N, Koizumi A (2005) Environ sci Int J Environ Physiol Toxicol 12:293
6. Houde M, De Silva AO, Muir DCG, Letcher RJ (2011) Environ Sci Technol 45:7962
7. Butt CM, Berger U, Bossi R, Tomy GT (2010) Sci Total Environ 408:2936
8. Kärrman A, Domingo J, Llebaria X, Nadal M, Bigas E, van Bavel B, Lindstrom G (2010) Environmental Science and Pollution Research 17:750
9. Kärrman A, Ericson I, VanBavel B, Ola Darnerud P, Aune M, Glynn A, Ligneli S, Lindström G (2007) Environ Health Perspect 115:226
10. Llorca M, Farré M, Picó Y, Teijón ML, Álvarez JG, Barceló D (2010) Environ Int 36:584
11. Shoeib M, Harner T, Wilford BH, Jones KC, Zhu J (2005) Environ Sci Technol 39:6599
12. Ericson I, Martí-Cid R, Nadal M, Van Bavel B, Lindström G, Domingo JL (2008) J Agric Food Chem 56:1787
13. Ericson I, Nadal M, van Bavel B, Lindström G, Domingo J (2008) Environ Sci Pollut Res 15:614
14. Tomy GT, Budakowski W, Halldorson T, Helm PA, Stern GA, Friesen K, Pepper K, Tittlemier SA, Fisk AT (2004) Environ Sci Technol 38:6475
15. Kelly BC, Ikonomou MG, Blair JD, Surridge B, Hoover D, Grace R, Gobas FAPC (2009) Environ Sci Technol 43:4037
16. Jeon J, Kannan K, Lim HK, Moon HB, Ra JS, Kim SD (2010) Environ Sci Technol 44:2695
17. Bhhatarai B, Gramatica P (2011) Mol Divers 15:467
18. Lau C, Butenhoff JL, Rogers JM (2004) Toxicol Appl Pharmacol 198:231
19. Newsted JL, Jones PD, Coady K, Giesy JP (2005) Environ Sci Technol 39:9357
20. Kudo N, Kawashima Y (2003) J Toxicol Sci 28:49
21. Ji K, Kim Y, Oh S, Ahn B, Jo H, Choi K (2008) Environ Toxicol Chem 27:2159
22. Carmosini N, Lee LS (2008) Environ Sci Technol 42:6559
23. Villagrasa M, López De Alda M, Barcelo D (2006) Anal Bioanal Chem 386:953
24. Suja F, Pramanik BK, Zain SM (2009) Water Sci Technol 60:1533
25. Giesy J, Naile J, Khim J, Jones P, Newsted J (2010) Reviews of environmental contamination and toxicology. Springer, New York, p 1
26. FSA (2006) Fluorinated chemicals - UK dietary intakes. Food Survey Information Sheet No. 11/06. Food Standards Agency, UK
27. Tittlemier SA, Pepper K, Seymour C, Moisey J, Bronson R, Cao XL, Dabeka RW (2007) J Agric Food Chem 55:3203
28. Dietz R, Bossi R, Rigét FF, Sonne C, Born EW (2008) Environ Sci Technol 42:2701
29. BechshÄft T, Sonne C, Dietz R, Born EW, Novak MA, Henchey E, Meyer JS (2011) Sci Total Environ 409:831
30. Ahrens L, Xie Z, Ebinghaus R (2010) Chemosphere 78:1011
31. Houtman CJ (2010) J Integr Environ Sci 7:271
32. Ju X, Jin Y, Sasaki K, Saito N (2008) Environ Sci Technol 42:3538
33. Skutlarek D, Exner M, Färber H (2006) Environ Sci Pollut Res 13:299
34. Martin JW, Kannan K, Berger U, De Voogt P, Field J, Franklin J, Giesy JP, Harner T, Muir DCG, Scott B, Kaiser M, Järnberg U, Jones KC, Mabury SA, Schroeder H, Simcik M, Sottani C, Van Bavel B, Kärrman A, Lindström G, Van Leeuwen S (2004) Environ Sci Technol 38:248A
35. Hansen KJ, Clemen LA, Ellefson ME, Johnson HO (2001) Environ Sci Technol 35:766
36. Holm A, Wilson SR, Molander P, Lundanes E, Greibrokk T (2004) J Sep Sci 27:1071
37. Kärrman A, van Bavel B, Järnberg U, Hardell L, Lindström G (2006) Chemosphere 64:1582
38. Loveless SE, Finlay C, Everds NE, Frame SR, Gillies PJ, O'Connor JC, Powley CR, Kennedy GL (2006) Toxicology 220:203
39. Liu J, Lee LS (2005) Environ Sci Technol 39:7535
40. Fromme H, Tittlemier SA, Völkel W, Wilhelm M, Twardella D (2009) Int J Hyg Environ Health 212:239
41. Dinglasan MJA, Ye Y, Edwards EA, Mabury SA (2004) Environ Sci Technol 38:2857

42. Szostek B, Prickett KB, Buck RC (2006) Rapid Commun Mass Spectrom 20:2837
43. Wang N, Szostek B, Folsom PW, Sulecki LM, Capka V, Buck RC, Berti WR, Gannon JT (2005) Environ Sci Technol 39:531
44. Schultz MM, Barofsky DF, Field JA (2006) Environ Sci Technol 40:289
45. Alzaga R, Bayona JM (2004) J Chromatogr A 1042:155
46. Gonzalez-Barreiro C, Martinez-Carballo E, Sitka A, Scharf S, Gans O (2006) Anal Bioanal Chem 386:2123
47. Moody CA, Field JA (1999) Environ Sci Technol 33:2800
48. Ericson I, Gómez M, Nadal M, van Bavel B, Lindström G, Domingo JL (2007) Environ Int 33:616
49. Llorca M, Farré M, Picó Y, Barceló D (2009) J Chromatogr A 1216:7195
50. Taniyasu S, Kannan K, Horii Y, Hanari N, Yamashita N (2003) Environ Sci Technol 37:2634
51. Saito N, Sasaki K, Nakatome K, Harada K, Yoshinaga T, Koizumi A (2003) Arch Environ Contam Toxicol 45:149
52. Yamashita N, Kannan K, Taniyasu S, Horii Y, Okazawa T, Petrick G, Gamo T (2004) Environ Sci Technol 38:5522
53. Ylinen M, Hahijarvi H, Peura P, Ramo O (1985) Arch Environ Contam Toxicol 14:713
54. Giesy JP, Kannan K (2001) Environ Sci Technol 35:1339
55. Van De Vijver KI, Hoff PT, Van Dongen W, Esmans EL, Blust R, De Coen WM (2003) Environ Toxicol Chem 22:2037
56. Yeung LWY, Taniyasu S, Kannan K, Xu DZY, Guruge KS, Lam PKS, Yamashita N (2009) J Chromatogr A 1216:4950
57. Yeung LWY, Miyake Y, Li P, Taniyasu S, Kannan K, Guruge KS, Lam PKS, Yamashita N (2009) Anal Chim Acta 635:108
58. Haug LS, Thomsen C, Becher G (2009) J Chromatogr A 1216:385
59. Guruge KS, Manage PM, Yamanaka N, Miyazaki S, Taniyasu S, Yamashita N (2008) Chemosphere 73:S210
60. Gosetti F, Chiuminatto U, Zampieri D, Mazzucco E, Robotti E, Calabrese G, Gennaro MC, Marengo E (2010) J Chromatogr A 1217:7864
61. Fromme H, Schlummer M, Møller A, Gruber L, Wolz G, Ungewiss J, Böhmer S, Dekant W, Mayer R, Liebl B, Twardella D (2007) Environ Sci Technol 41:7928
62. Haug LS, Salihovic S, Jogsten IE, Thomsen C, van Bavel B, Lindström G, Becher G (2010) Chemosphere 80:1137
63. Jogsten IE, Perelló G, Llebaria X, Bigas E, Martí-Cid R, Kärrman A, Domingo JL (2009) Food Chem Toxicol 47:1577
64. Kubwabo C, Vais N, Benoit FM (2004) J Environ Monit 6:540
65. Monroy R, Morrison K, Teo K, Atkinson S, Kubwabo C, Stewart B, Foster WG (2008) Environ Res 108:56
66. Szostek B, Prickett KB (2004) J Chromatogr B Analyt Technol Biomed Life Sci 813:313
67. Lien GW, Wen TW, Hsieh WS, Wu KY, Chen CY, Chen PC (2011) J Chromatogr B Analyt Technol Biomed Life Sci 879:641
68. Midasch O, Schettgen T, Angerer J (2006) Int J Hyg Environ Health 209:489
69. Flaherty JM, Connolly PD, Decker ER, Kennedy SM, Ellefson ME, Reagen WK, Szostek B (2005) J Chromatogr B Analyt Technol Biomed Life Sci 819:329
70. Hemat H, Wilhelm M, Völkel W, Mosch C, Fromme H, Wittsiepe J (2010) Sci Total Environ 408:3493
71. Kuklenyik Z, Reich JA, Tully JS, Needham LL, Calafat AM (2004) Environ Sci Technol 38:3698
72. Zhang T, Wu Q, Sun HW, Zhang XZ, Yun SH, Kannan K (2010) Environ Sci Technol 44:4341
73. Zhang W, Lin Z, Hu M, Wang X, Lian Q, Lin K, Dong Q, Huang C (2011) Ecotoxicol Environ Saf 74:1787
74. Kuklenyik Z, Needham LL, Calafat AM (2005) Anal Chem 77:6085

75. Calafat AM, Kuklenyik Z, Caudill SP, Reidy JA, Needham LL (2006) Environ Sci Technol 40:2128
76. Calafat AM, Ye X, Silva MJ, Kuklenyik Z, Needham LL, Kolossa M, Tuomisto J, Astrup Jensen A (2006) Int J Androl 29:166
77. Taniyasu S, Kannan K, Yeung LWY, Kwok KY, Lam PKS, Yamashita N (2008) Anal Chim Acta 619:221
78. Takino M, Daishima S, Nakahara T (2003) Rapid Commun Mass Spectrom 17:383
79. Chu S, Letcher RJ (2008) J Chromatogr A 1215:92
80. Song L, Wellman AD, Yao H, Adcock J (2007) Rapid Commun Mass Spectrom 21:1343
81. Berger U, Haukås M (2005) J Chromatogr A 1081:210
82. Llorca M, Farré M, Picó Y, Barceló D (2010) Anal Bioanal Chem 398:1145
83. D'Eon JC, Mabury SA (2007) Environ Sci Technol 41:4799
84. D'Eon JC, Hurley MD, Wallington TJ, Mabury SA (2006) Environ Sci Technol 40:1862
85. Ellis DA, Martin JW, De Silva AO, Mabury SA, Hurley MD, Sulbaek Andersen MP, Wallington TJ (2004) Environ Sci Technol 38:3316
86. Hagen DF, Belisle J, Johnson JD, Venkateswarlu P (1981) Anal Biochem 118:336
87. Nakayama S, Strynar MJ, Helfant L, Egeghy P, Ye X, Lindstrom AB (2007) Environ Sci Technol 41:5271
88. Young CJ, Mabury SA (2010) Reviews of Environmental Contamination and Toxicology 208:1–110
89. Butt CM, Young CJ, Mabury SA, Hurley MD, Wallington TJ (2009) J Phys Chem A 113:3155
90. Chu S, Letcher RJ (2009) Anal Chem 81:4256
91. Houde M, Czub G, Small JM, Backus S, Wang X, Alaee M, Muir DCG (2008) Environ Sci Technol 42:9397
92. Kissa E (2001) Fluorinated surfactants and repellents, 2nd edn, vol 97. Marcel Dekker, New York
93. Ericson I, Domingo JL, Nadal M, Bigas E, Llebaria X, Van Bavel B, Lindström G (2009) Arch Environ Contam Toxicol 57:631
94. Harada K, Saito N, Sasaki K, Inoue K, Koizumi A (2003) Bull Environ Contam Toxicol 71:31
95. Wilhelm M, Bergmann S, Dieter HH (2010) Int J Hyg Environ Health 213:224
96. Loos R, Wollgast J, Huber T, Hanke G (2007) Anal Bioanal Chem 387:1469
97. Delinsky AD, Strynar MJ, McCann PJ, Varns JL, McMillan L, Nakayama SF, Lindstrom AB (2010) Environ Sci Technol 44:2549
98. Gebbink WA, Hebert CE, Letcher RJ (2009) Environ Sci Technol 43:7443
99. Gebbink WA, Letcher RJ, Burgess NM, Champoux L, Elliott JE, Hebert CE, Martin P, Wayland M, Weseloh DVC, Wilson L (2011) Environ Int 37:1175
100. Pan Y, Shi Y, Wang J, Jin X, Cai Y (2011) Bull Environ Contam Toxicol 87:152
101. Wang M, Park J-S, Petreas M (2011) Environmental Science and Technology Sep 1;45(17):7510–6. Epub 2011 Aug 1
102. Loi EIH, Yeung LWY, Taniyasu S, Lam PKS, Kannan K, Yamashita N (2011) Environ Sci Technol 45:5506
103. Shi Y, Pan Y, Yang R, Wang Y, Cai Y (2010) Environ Int 36:46
104. Van De Vijver KI, Holsbeek L, Das K, Blust R, Joiris C, De Coen W (2007) Environ Sci Technol 41:315
105. Zhao YG, Wan HT, Law AYS, Wei X, Huang YQ, Giesy JP, Wong MH, Wong CKC (2011) Chemosphere 85:277
106. Lu GH, Yang YL, Taniyasu S, Yeung LWY, Pan J, Zhou B, Lam PKS, Yamashita N (2011) Environ Chem 8:407
107. Schecter A, Colacino J, Haffner D, Patel K, Opel M, Päpke O, Birnbaum L (2010) Environ Health Perspect 118:796

108. Berger U, Glynn A, Holmström KE, Berglund M, Ankarberg EH, Törnkvist A (2009) Chemosphere 76:799
109. Ostertag SK, Tague BA, Humphries MM, Tittlemier SA, Chan HM (2009) Chemosphere 75:1165
110. Haug LS, Thomsen C, Brantsæter AL, Kvalem HE, Haugen M, Becher G, Alexander J, Meltzer HM, Knutsen HK (2010) Environ Int 36:772
111. Powley CR, George SW, Ryan TW, Buck RC (2005) Anal Chem 77:6353
112. Del Gobbo L, Tittlemier S, Diamond M, Pepper K, Tague B, Yeudall F, Vanderlinden L (2008) J Agric Food Chem 56:7551
113. Begley TH, White K, Honigfort P, Twaroski ML, Neches R, Walker RA (2005) Food Addit Contam 22:1023
114. Martin JW, Mabury SA, Solomon KR, Muir DCG (2003) Environ Toxicol Chem 22:189
115. Martin JW, Whittle DM, Muir DCG, Mabury SA (2004) Environ Sci Technol 38:5379
116. de Vos MG, Huijbregts MAJ, van den Heuvel-Greve MJ, Vethaak AD, Van de Vijver KI, Leonards PEG, van Leeuwen SPJ, de Voogt P, Hendriks AJ (2008) Chemosphere 70:1766
117. Saito N, Harada K, Inoue K, Sasaki K, Yoshinaga T, Koizumi A (2004) J Occup Health 46:49
118. Kunacheva C, Boontanon SK, Fujii S, Tanaka S, Poothong S, Wongwattana T, Shivakoti BR (2010) Perfluorinated Compounds Contamination in Tap Water and Drinking Water in Bangkok, Thailand. Journal of Water Supply: Research and Technology — AQUA, Vol. 59, No. 5, pp 345–354
119. Yim LM, Taniyasu S, Yeung LWY, Lu G, Jin L, Yang Y, Lam PKS, Kannan K, Yamashita N (2009) Environ Sci Technol 43:4824
120. Sun H, Li F, Zhang T, Zhang X, He N, Song Q, Zhao L, Sun L, Sun T (2011) Water Res 45:4483
121. Fromme H, Mosch C, Morovitz M, Alba-Alejandre I, Boehmer S, Kiranoglu M, Faber F, Hannibal I, Genzel-Boroviczény O, Koletzko B, Völkel W (2010) Environ Sci Technol 44:7123
122. Emmett EA, Shofer FS, Zhang H, Freeman D, Desai C, Shaw LM (2006) J Occup Environ Med 48:759
123. Hölzer J, Midasch O, Rauchfuss K, Kraft M, Reupert R, Angerer J, Kleeschulte P, Marschall N, Wilhelm M (2008) Environ Health Perspect 116:651
124. Takagi S, Adachi F, Miyano K, Koizumi Y, Tanaka H, Watanabe I, Tanabe S, Kannan K (2011) Water Res 45:3925
125. Furdui VI, Stock NL, Ellis DA, Butt CM, Whittle DM, Crozier PW, Reiner EJ, Muir DCG, Mabury SA (2007) Environ Sci Technol 41:1554
126. Martin JW, Mabury SA, Solomon KR, Muir DCG (2003) Environ Toxicol Chem 22:196
127. Gruber L, Schlummer M, Ungewiss J, Wolz G, Moeller A, Weise N, Sengl M, Frey S, Gerst M, Schwaiger J (2007) Organohalogen Compd 69:3
128. Ye X, Strynar MJ, Nakayama SF, Varns J, Helfant L, Lazorchak J, Lindstrom AB (2008) Environ Pollut 156:1227
129. Tomy GT, Tittlemier SA, Palace VP, Budakowski WR, Braekevelt E, Brinkworth L, Friesen K (2004) Environ Sci Technol 38:758
130. Nania V, Pellegrini GE, Fabrizi L, Sesta G, Sanctis PD, Lucchetti D, Pasquale MD, Coni E (2009) Food Chem 115:951
131. Hart K, Kannan K, Tao L, Takahashi S, Tanabe S (2008) Sci Total Environ 403:215
132. Van den Heuvel-Greve M, Leonards P, Vethaak D (2006) Dioxin research Westerschelde: measuring percentages of dioxins, dioxin-like substances and other possible problematic substances in fishery products, sediment and food chains of the Westerschelde [Dioxineonderzoek Westerschelde: meting van gehalten aan dioxinen, dioxineachtige stoffen en andere mogelijke probleemstoffen in visserijproducten, sediment en voedselketens van de Westerschelde] Rapport RIKZ, 2006011 RIKZ: Middelburg, The Netherlands 80 pp
133. Gulkowska A, Jiang Q, So MK, Taniyasu S, Lam PKS, Yamashita N (2006) Environ Sci Technol 40:3736

134. Nakata H, Kannan K, Nasu T, Cho HS, Sinclair E, Takemura A (2006) Environ Sci Technol 40:4916
135. Cunha I, Hoff P, Van De Vijver K, Guilhermino L, Esmans E, De Coen W (2005) Mar Pollut Bull 50:1128
136. Kannan K, Tao L, Sinclair E, Pastva SD, Jude DJ, Giesy JP (2005) Arch Environ Contam Toxicol 48:559
137. Voogt PD, Saez M (2006) TrAC - Trends Anal Chem 25:326
138. Angerer J, Ewers U, Wilhelm M (2007) Int J Hyg Environ Health 210:201
139. Lacina O, Hradkova P, Pulkrabova J, Hajslova J (2011) J Chromatogr A 1218:4312
140. A.o.P. US EPA (Environmental Protection Agency), FOSA, and PFOA from various food matrices using HPLC electrospray/mass spectrometry, 3M study conducted by Centre Analytical Laboratories, Inc (http://www.ewg.org/files/), 200
141. Ostertag SK, Hing Chan MAN, Moisey J, Dabeka R, Tittlemier SA (2009) J Agric Food Chem 57:8534
142. Kärrman A, Harada KH, Inoue K, Takasuga T, Ohi E, Koizumi A (2009) Environ Int 35:712
143. Lau C, Anitole K, Hodes C, Lai D, Pfahles-Hutchens A, Seed J (2007) Toxicol Sci 99:366
144. Guo F, Zhong Y, Wang Y, Li J, Zhang J, Liu J, Zhao Y, Wu Y (2011) Chemosphere 85:156
145. Needham LL, Grandjean P, Heinzow B, Jorgensen PJ, Nielsen F, Sjödin A, Patterson DG Jr, Turner WE, Weihe P (2011) Environ Sci Technol 45:1121
146. Nilsson H, Rotander A, Van Bavel B, Lindström G, Westberg H (2010) Environ Sci Technol 44:7717
147. Olsen GW, Mair DC, Reagen WK, Ellefson ME, Ehresman DJ, Butenhoff JL, Zobel LR (2007) Chemosphere 68:105
148. Tao L, Kannan K, Wong CM, Arcaro KF, Butenhoff JL (2008) Environ Sci Technol 42:3096
149. Völkel W, Genzel-Boroviczény O, Demmelmair H, Gebauer C, Koletzko B, Twardella D, Raab U, Fromme H (2008) Int J Hyg Environ Health 211:440
150. Kim S, Choi K, Ji K, Seo J, Kho Y, Park J, Kim S, Park S, Hwang I, Jeon J, Yang H, Giesy JP (2011) Environ Sci Technol 45:7465
151. Bossi R, Riget FF, Dietz R, Sonne C, Fauser P, Dam M, Vorkamp K (2005) Environ Pollut 136:323
152. Guo F, Wang Y, Li J, Zhang J, Zhao Y, Wu Y (2011) Chinese J Chromatogr (Se Pu) 29:126
153. Schriks M, Heringa MB, van der Kooi MME, de Voogt P, van Wezel AP (2010) Water Res 44:461
154. Brede E, Wilhelm M, Göen T, Müller J, Rauchfuss K, Kraft M, Hölzer J (2010) Int J Hyg Environ Health 213:217
155. Olivero-Verbel J, Tao L, Johnson-Restrepo B, Guette-Fernaĺndez J, Baldiris-Avila R, O'Byrne-Hoyos I, Kannan K (2006) Environ Pollut 142:367
156. Tseng CL, Liu LL, Chen CM, Ding WH (2006) J Chromatogr A 1105:119
157. So MK, Taniyasu S, Lam PKS, Zheng GJ, Giesy JP, Yamashita N (2006) Arch Environ Contam Toxicol 50:240
158. Takagi S, Adachi F, Miyano K, Koizumi Y, Tanaka H, Mimura M, Watanabe I, Tanabe S, Kannan K (2008) Chemosphere 72:1409
159. Liu Y, Wang J, Wei Y, Zhang H, Liu Y, Dai J (2008) Aquat Toxicol 88:183
160. Sinclair E, Mayack DT, Roblee K, Yamashita N, Kannan K (2006) Arch Environ Contam Toxicol 50:398
161. Thompson J, Eaglesham G, Mueller J (2011) Chemosphere 83:1320
162. Yang Y, Lu G, Yeung LW, et al. (2010) Levels and distribution of perfluorinated compounds in water and biological samples from the Shenyang area, China. Acta Scientiae Circumstantiae. 30(10):2097–2107

Fate and Risks of Polar Pesticides in Groundwater Samples of Catalonia

Marianne Köck-Schulmeyer, Antoni Ginebreda, Miren López de Alda, and Damià Barceló

Abstract Contamination of groundwater by pesticides is a subject of growing concern, first, because groundwater is the most sensitive and also the largest body of freshwater in the European Union and, second, because pesticides have been shown to be ubiquitous contaminants in this aquatic compartment. This work presents the results of a monitoring study carried out in Catalonia (NE, Spain) to investigate the occurrence of 22 multiclass polar pesticides in 13 different groundwater bodies where agricultural practice is significant, between 2007 and 2010. Results have shown a pesticide profile dominated by triazines (atrazine, simazine, terbuthylazine), although organophosphates such as dimethoate and phenylureas such as diuron and linuron show also an important contribution. The groundwater quality standards set by Directive 2006/118/EC for both individual and total pesticides levels were surpassed in several cases. The most contaminated groundwater bodies were located in areas with intensive agricultural activity (especially irrigated lands). Temporal trends indicate that the area known to be the most polluted by pesticides in Catalonia (Lleida) is changing over time to better conditions, whereas in others, pesticide pollution remains constant or slightly increases.

Keyword Catalonia, Groundwater, Monitoring, Polar pesticides

M. Köck-Schulmeyer (✉), A. Ginebreda, and M.L. de Alda
Department of Environmental Chemistry, IDAEA-CSIC, c/Jordi Girona 18-26, 08034 Barcelona, Spain
e-mail: mkoqam@cid.csic.es

D. Barceló
Department of Environmental Chemistry, IDAEA-CSIC, c/Jordi Girona 18-26, 08034 Barcelona, Spain

Catalan Institute for Water Research-ICRA, Parc Científic i Tecnològic de la Universitat de Girona, c/Emili Grahit, 101, Edifici H2O, E-17003 Girona, Spain

King Saud University, Box 2454, Riyadh 11451, Saudi Arabia

D. Barceló (ed.), *Emerging Organic Contaminants and Human Health*,
Hdb Env Chem (2012) 20: 375–394, DOI 10.1007/698_2011_132,
© Springer-Verlag Berlin Heidelberg 2012, Published online: 17 January 2012

Contents

1   Introduction

Groundwater, which represents 30% of the total fresh water in the world, is a subject of national importance because it is a major source of water for agricultural and industrial purposes, as well as for drinking water all over Europe.

In Spain, where groundwater was declared a public domain resource in 1985 [1], more than one third of the Spanish territory contains groundwater distributed in around 700 groundwater bodies. These groundwater bodies comprise a set of associated aquifers, which are the major reservoirs of groundwater and are recharged by rain, snowmelt, interchange with surface waters, and/or artificial recharge [2].

In recent years, the growth of industry, agriculture, population and water use has increased the stress upon both our land and water resources. Evaluation of the impact of all these factors is very often performed through the use of indicators. Important groundwater quality indicators are nitrate, pesticides, chloride, alkalinity, pH-value and electrical conductivity [3]. Focusing the attention on the agricultural sources of groundwater contamination, it is possible to identify three main sources: pesticides, fertilisers and animal wastes. Fertilisers and animal wastes are in this order the predominant sources of nitrogen, and by extension of nitrate [2], one of the groundwater quality indicators. But as previously mentioned, many other factors in addition to agriculture are responsible for the pressure put on groundwater resources. Table 1 shows the main identified activities known to contribute to the total chemical pressure observed in the groundwater bodies studied in this chapter. This pressure is irregular and has a relationship with the location, extent and distribution of the groundwater body. For example, body M21 (Detrital Neogene of El Baix Penedès) presents an insignificant agricultural pressure but high impact of wastewater discharges and water catchments (mainly to irrigate other regions) that can contribute to high levels of organic contaminants in the groundwater body. Just to give an idea about the impact of irrigation, current groundwater use in Spain amounts approximately 6,500 Mm^3/year, and around 75% of it is used for irrigation (of about 1 million hectares, which corresponds to 30% of the total irrigated surface of the country) [1]. Groundwater catchments due to tourism also

Table 1 Pressure of different activities in groundwater bodies of Catalonia (NE Spain)

Code	Groundwater bodies	Intensive agriculture	Ornamental and plant nurseries	Cattle manure	Water catchment	Urban and industrial areas	WWTP discharges	Industrial discharges[a]
M6	L'Empordà							
M14	La Selva							
M16	Alluvial of El Vallès							
M18	El Maresme							
M21	Detrital Neogene of El Baix Penedès							
M24	Lower Francolí							
M25	L AltCamp							
M32	Fluviodeltaic of the Fluvia - Muga							
M33	Fluviodeltaic of the Ter							
M35	Alluvial of the lower Tordera and delta							
M46	Alluvial of the middle Segre							
M47	Alluvial of the lower Segre							
M48	Alluvial of Urgell							

Pressure: High Moderate Low Zero

WWTP: wastewater treatment plant

[a]Industrial discharges are usually included within those of WWTP

cause very high pressures on groundwater because of the additional water demand arising during the summer seasons when the groundwater situation may be already rather critical [3]. Furthermore, overexploitation of coastal aquifers has also negative effects on the quality due to salinisation caused by marine intrusion.

This and other human interventions in the hydrological cycle cause profound effects on groundwater quantity and quality, and as a result today, 38.65% of the groundwater bodies in the country are classified "at risk" [1].

1.1 Groundwater and Pesticides

Any pesticide is a mixture of additives and active ingredients, where the latter are the biologically active part of the pesticide that kills or controls the pest. In Europe, around 250 pesticides are approved for use [4] and applied in many different contexts, not just in agriculture. In the private and the public sector, pesticides are mainly used to control insects and to clear outdoor areas of weed vegetation, not only in culture areas but also around railways, roads, car parks and airports, around sports facilities, cemeteries and parks [3]. Specifically, from industrial activities, pesticides may reach groundwater from accidents during production, storage and transport, or through the discharge of industrial effluents or leachate from dumping sites.

For decades, the likelihood of groundwater contamination by pesticides was largely ignored [2], but today, the deterioration of the groundwater quality in connection with the intensive uses of pesticides is clearly evident. Throughout the years, the use of pesticides has changed from persistent organic compounds (mainly organochlorines and organophosphates), due to their obvious bad consequences to human and environmental health, to less persistent but more polar and water soluble compounds, characteristics that make them great candidates for groundwater contamination.

On the other hand, the concentration of pesticides in groundwater depends on many factors such as the nature of the surface to which the pesticide is applied, the weather, the pesticide application (nature and rate), as well as on the physical and chemical characteristics of the pesticide. In this last respect, one of the most relevant parameters is the Groundwater Ubiquity Score (GUS). The GUS index (Gustafson, 1993) takes into account the dissipation half-life ($DT_{1/2}$) – the time that it takes for a pesticide to decrease from its original concentration to half of it – as well as the adsorption coefficient – affinity to be retained in soil – expressed as Koc (mL/g or L/kg), and it is a simple method for assessing pesticides leachability. High $DT_{1/2}$ and low Koc mean high probability for a pesticide to survive its journey through soil and reach the groundwater. Hence, if GUS is >2.8, the pesticide has "high leachability", if GUS is <1.8, the pesticide has "low leachability", and if 1.8 < GUS < 2.8, the pesticide has a "transition state" [5]. This model is an empirical regression. It incorporates only the properties of pesticides, and no information from the soil. Therefore, GUS indicates the intrinsic mobility of pesticides. This index has been used by many researchers, such as Milhome et al. [6], to evaluate the potential for pesticide contamination in surface and ground waters and soils.

1.2   Legislation

Protection of water sources is critical to economic viability as well as to human and environmental health. This has been evident for many decades, but just in the 1980s, the European Community started to adopt legislative measures to evaluate and preserve groundwater and also to implement bans on pesticides which at first concerned mainly persistent organochlorine compounds. Along the years, various European Community Directives have addressed these issues in the European Union, among them:

– Directive 80/68/EEC [7]: concerning the protection of groundwater against pollution caused by certain dangerous substances
– Directive 91/676/EEC [8]: concerning the protection of waters against pollution caused by nitrates from agricultural sources
– Directive 91/414/EEC [9]: concerning the placing of plant protection products on the market
– Directive 98/8/EC [10]: of the European Parliament and of the Council concerning the placing of biocidal products on the market.

After that, the Water Framework Directive 2000/60/EC [11] marked a change in Community water policy establishing a list of 33 priority substances, the third part of which are pesticides. Six years after, the Groundwater Directive 2006/118/EC concerning the protection of groundwater against pollution and deterioration [12] established groundwater quality standards (GQS). For pesticides, the GQS is 0.1 μg/L for individual substances and 0.5 μg/L for the sum of individual pesticides, including their relevant metabolites and degradation and reaction products. A GQS of 50 mg/L was also established for nitrates.

Recently, the Commission has completed the review of existing pesticides that were on the market before 1993. After detailed risk evaluation with respect to their effects on humans and on the environment, only about 250 compounds, out of some 1,000 active substances, have passed the harmonised EU safety assessment [4].

In spite of the legislative measures that have progressively been adopted, many different pesticide substances are detected in Europe's groundwater at levels sometimes greater than the Directive 2006/118/EC maximum allowable concentration, and the pesticides most commonly found in groundwater appear to be atrazine, simazine and lindane [3, 13].

1.3   Analysis of Pesticides in Groundwater

There are many techniques that allow the detection of pesticides in groundwater; however, what is now recognised as a very good technique, is one that:

- Allows detecting pesticides in the low ng/L or pg/L range
- Is compatible for a large number of compounds
- Allows analysis of compounds having a large variety of physicochemical properties
- Is robust for the analysis of pesticides in different environmental matrices.

In this context, the extraction, cleanup and detection methods need to be in accordance with these proposed objectives. Solid-phase extraction (SPE) has been the most widely used technique in the last decade, not just because it allows the extraction of pesticides efficiently but also because it integrates extraction and cleanup in a single step. Other advantages of SPE as compared to other techniques are the small sample and solvent volumes required and the variety of sorbents available. A good example of the multiresidue analysis of pesticides in water is the method described by Marín et al. [14] for the determination of 37 multiclass pesticides in different environmental waters, including groundwater. Other examples of the application of this technique to the analysis of pesticides in groundwater can be found in [15–22]. Another advantage of SPE is the possibility of automation. Online SPE, whose advantages and limitations have been discussed by Rodriguez-Mozaz et al. [23], has also been used by various researchers to detect pesticides in groundwater [24–27].

For the detection, gas chromatography (GC) [15, 18–20, 28] and liquid chromatography (LC) [14–16, 21, 22, 24, 26–29] coupled with mass spectrometry (MS) or tandem mass spectrometry (MS/MS) have been the techniques most frequently used in the determination of pesticides in ground water. Examples of the application of both techniques in the area of study, Catalonia, are the work of Garrido et al. [17], who used GC-MS and GC with electron capture detection (ECD) for the analysis of 44 pesticides in groundwater samples from Catalonia and that of Kampioti et al. [25], who used online SPE-LC-MS/MS to analyse 20 pesticides in river water and

groundwater samples in the area of Barcelona. Nevertheless, nowadays, LC-MS/
MS is considered the technique of choice due to the already mentioned polar
character of most of the currently used pesticides and their metabolites and transformation products [30].

2 Case Study: Fate of Pesticides in Groundwater of Catalonia

2.1 Introduction

Catalonia has 53 groundwater bodies, and they represent approximately 35% of
the total water resources used. The overexploitation of groundwater, especially
in coastal areas, exists side by side with clear situations of surplus. To understand
the gravity of the situation, it may be worth mentioning that all groundwater
bodies identified in Catalonia, except the water mass in the Ebro River Delta,
are catchments for human consumption (10 m^3/day). Each groundwater body
comprises one or more aquifers of different natures (e.g., porous, granular type,
fissured) [31]. The chemical status of these aquifers is questionable, but not all of
them are compromised because of the agricultural activity. Figure 1 shows the
location of the various groundwater bodies, together with the results of the pressure
and impact analysis of intensive agriculture performed by the responsible water
authority, the Catalan Water Agency (ACA).

In order to ascertain the degree of accomplishment of the aforementioned
Directive 2006/118/EC in Catalonia (NE, Spain) and better characterise the nature
of the contamination of these aquifers by pesticides, a monitoring programme was
carried out on various selected hydrological units intended to be among the most
vulnerable and relevant ones because of intensive agricultural activity and use for
human consumption. The number of aquifers analysed depends more or less on the
extension of the groundwater body. For example, bodies M46 and M21 present
extensions of 18 and 72 km^2, respectively, and just one and two aquifers of each
body were analysed, respectively. In contrast, body M18 (Maresme) presents an
extension of more than 400 km^2, and nine different aquifers were analysed.

This chapter presents the results of this monitoring programme where 22
pesticides with relatively high GUS index were investigated in 13 groundwater
bodies. The list of target pesticides included six triazines (atrazine, simazine,
cyanazine, desethylatrazine, terbuthylazine and deisopropylatrazine), four phenylureas (diuron, isoproturon, linuron and chlortoluron), four organophosphates
(diazinon, dimethoate, fenitrothion and malathion), one anilide (propanil), two
chloroacetanilides (alachlor and metolachlor), one thiocarbamate (molinate) and
four acid herbicides (mecoprop, 2,4 D, bentazone and MCPA). Table 2 provides
some chemical information (GUS index), their classification according to the target
organism, and main uses. These pesticides were selected on the basis of extent

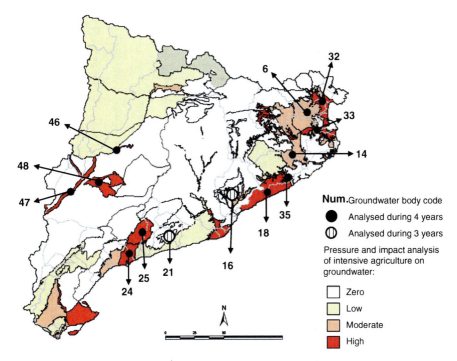

Fig. 1 Map showing the location of the groundwater bodies sampled in 2007, 2008, 2009, and 2010 and results of the pressure and impact analysis of intensive agriculture on the groundwater bodies from Catalonia, as estimated by the Catalan Water Agency (ACA) – Generalitat de Catalunya; http://www.gencat.cat/aca/

of use, legal requirements of the European Unit, information gathered from the water authorities, known use in certain specific crops (i.e., rice), and amenability to LC-MS analysis.

2.2 Sampling and Analysis

A total of 169 groundwater samples were collected from 13 different groundwater bodies (Fig. 1) and 48 different aquifers between 2007 and 2010 (Table 3). Sampled aquifers were mainly unconfined porous aquifers and presented a sandy and gravel composition, with few exceptions. More details on the sampling can be found in Postigo et al. 2010 [27].

About the analysis of pesticides, water samples were collected in amber glass bottles and transported to the laboratory under cooled conditions (4°C). Upon reception, samples were filtered (0.4-µm membrane filters) and then stored at −20°C in the dark until analysis. Analysis was performed by a fully automated multiresidue analytical method based on online SPE-LC-MS/MS [27].

Table 2 Properties of target pesticides

Pesticide	Chemical properties		GUS	Pesticide properties		Description
	Substance group	CAS		Introduction	Pesticide type	
2,4-D	Acid	94-75-7	2.25	~1950	Herbicide	A translocated phenoxy herbicide for use in cereals, grass, and amenity use
Bentazone[b]	Acid	25057-89-0	2.55	1972	Herbicide	A postemergence herbicide used for selective control of broadleaf weeds and sedges (a weed) in beans, rice, corn, peanuts, mint, and others
MCPA	Acid	94-74-6	2.51	1950	Herbicide	A herbicide for control of annual and perennial broad-leaved weeds
Mecoprop[b]	Acid	7085-19-0	2.29	~1956	Herbicide	A postemergence herbicide used to control broad-leaved weeds such as cleavers, chickweed, plantains, and clover
Propanil	Anilide	709-98-8	0.42	1961	Herbicide	A postemergence herbicide used for broad-leaved and annual grass weed control in rice and other crops
Alachlor[a]	Chloroacetanilide	15972-60-8	2.19	1969	Herbicide	A herbicide for preemergence control of annual grasses and broad-leaved weeds
Metolachlor	Chloroacetanilide	51218-45-2	3.32	1976	Herbicide	A preemergence herbicide used to control certain broad-leaf and annual grassy weeds in a variety of situations
Diazinon	Organophosphate	333-41-5	1.14	1953	Insectic.	An insecticide used to control sucking and chewing insects on a wide range of crops including top fruit. It also has livestock applications
Dimethoate	Organophosphate	60-51-5	1.05	~1957	Insectic.	An insecticide and acaricide used to control a wide range of pests including Aphididae, Coleoptera, and Lepidoptera
Fenitrothion	Organophosphate	122-14-5	0.64	1962	Insectic.	An insecticide used to control chewing, sucking, and boring pests on a range of crops
Malathion	Organophosphate	121-75-5	-1.28	1991	Insectic.	An insecticide and acaricide used to control a wide range of pests including Coleoptera, Diptera, Hemiptera, and Lepidoptera
Chlortoluron	Phenylurea	15545-48-9	2.79	1994	Herbicide	A contact and residual urea herbicide used to control broad-leaved weeds and grasses
Diuron[a]	Phenylurea	330-54-1	1.83	~1951	Herbicide	A preemergence residual herbicide for total control of weeds and mosses in noncrop areas and woody crops
Isoproturon[a]	Phenylurea	34123-59-6	2.07	Unknown	Herbicide	A herbicide for use, typically, in cereals to control annual grasses and many broad-leaved weeds
Linuron	Phenylurea	330-55-2	2.03	~1965	Herbicide	A herbicide for the pre- and postemergence control of annual grass and broad-leaved weeds
Molinate	Thiocarbamate	2212-67-1	2.49	1964, USA	Herbicide	A herbicide used to control germinating broad-leaved and grassy weeds
Atrazine[a]	Triazine	1912-24-9	3.75	1957	Herbicide	A triazine herbicide used pre- and postemergence with restricted permitted uses to control broad-leaved weeds and grasses

Cyanacine	Triazine	21725-46-2	2.07	~1967	Herbicide	A preemergence herbicide used for general weed control including grasses and broad-leaved weeds in a range of crops
DIA	Triazine	1007-28-9	3.54	–	Herbicide	Chemical transformation product. Parent: simazine and atrazine
DEA	Triazine	6190-65-4	3.35	–	Herbicide	Chemical transformation product. Parent: atrazine
Simazine[a]	Triazine	122-34-9		~1960	Herbicide	A soil-acting herbicide used to control most germinating annual grasses and broad-leaved weeds
TBA	Triazine	5915-41-3	3.07	1967	Herbicide	A herbicide to control grass and broad-leaved weeds in a variety of situations including forestry and for control slime-forming algae, fungi, and bacteria in nonagricultural situations

DIA deisopropylatrazine, *DEA* desethylatrazine, *TBA* terbuthylazine

[a]Included in the Directive 2008/105/EC as priority substances in the field of water policy

[b]Included in the Directive 2008/105/EC as substances subject to review for possible identification as priority substances or priority hazardous substances

Table 3 Concentration of total pesticides in aquifer samples

Code	Groundwater bodies Name	Aquifer code	Sampling				Total pesticides (ng/L)		Pesticide levels		Pesticide > 100 ng/L (number of samples)
			2007	2008	2009	2010	min	max	TP > 500 ng/L	IP > 100 ng/L	
M6	L'Empordà	117056-0052	x			x	50.1	87.1			
M6	L'Empordà	17221-0016	x			x	1.7	4.0			
M14	La Selva	17020-0026		x	x	x	107.5	272.6		x	2,4-D(1), TBA(1)
M14	La Selva	17148-0060		x	x	x	0.0	66.3			
M14	La Selva	17233-0022	x			x	45.0	225.5		x	Atrazine(1)
M16	Alluvial of El Vallès	08088-0025		x	x	x	31.0	43.0			
M16	Alluvial of El Vallès	08106-0041		x	x	x	7.3	21.4			
M16	Alluvial of El Vallès	08108-0009		x	x	x	19.2	48.4			
M18	El Maresme	08007-0064	x	x	x		1.2	7.4			
M18	El Maresme	08009-0060	x	x	x	x	7.1	9.8			
M18	El Maresme	08029-0029	x	x	x	x	0.9	3.8			
M18	El Maresme	08029-0031	x	x	x	x	15.1	81.3			
M18	El Maresme	08121-0117	x	x	x	x	9.4	28.9			
M18	El Maresme	08197-0033	x	x	x	x	7.5	32.7			
M18	El Maresme	08197-0048	x	x	x	x	8.1	70.8			
M18	El Maresme	08219-0025	x	x	x	x	3.9	7.6			
M18	El Maresme	08261-0025	x	x	x	x	2.7	46.9			
M21	Detrital Neogene of El Baix Penedès	43020-0056		x	x	x	25.1	31.0			
M21	Detrital Neogene of El Baix Penedès	43140-0056		x	x	x	34.9	42.8			
M24	Lower Francolí	43047-0018	x	x	x	x	0.4	6.9			
M24	Lower Francolí	43047-0032	x	x		x	9.0	49.1			
M24	Lower Francolí	43047-0033	x		x	x	175.3	437.8		x	DIA(1), TBA(3)

Code	Site	ID									Compounds
M25	L'Alt Camp	43119-0016	x	x	x	x	41.5	66.4	x		Diuron(1)
M25	L'Alt Camp	43161-0138	x	x	x	x	43.4	196.4	x		DEA(1)
M32	Fluviodeltaic of the Fluvià – Muga	17047-0015		x	x	x	14.7	182.0			
M32	Fluviodeltaic of the Fluvià – Muga	17047-0035	x	x	x	x	9.0	12.4			
M32	Fluviodeltaic of the Fluvià – Muga	17066-0013		x	x	x	128.1	563.3	xx	x	Atrazine(1), DEA(2)
M32	Fluviodeltaic of the Fluvià – Muga	17178-0050	x	x	x		36.4	50.8			
M33	Fluviodeltaic of the Ter	17018-0013		x	x		5.0	10.4			
M33	Fluviodeltaic of the Ter	17067-0001	x	x	x	x	43.3	200.6			
M33	Fluviodeltaic of the Ter	17191-0021		x	x	x	5.4	13.1			
M33	Fluviodeltaic of the Ter	17199-0031	x	x	x	x	5.1	14485.3	x	x	Alachlor(1), atrazine (1), chlortoluron (1), DIA(1), TBA (1), dimethoate(1), linuron(1)
M33	Fluviodeltaic of the Ter	17199-0044		x	x	x	3.1	7.5			
M33	Fluviodeltaic of the Ter	17211-0025	x	x	x	x	1.8	8.9			
M35	Alluvial of the lower Tordera and delta	08110-0025	x	x	x	x	52.9	2293.3	x		Dimethoate(1)
M35	Alluvial of the lower Tordera and delta	08110-0027	x	x	x	x	0.0	17.6			
M35	Alluvial of the lower Tordera and delta	08110-0029	x	x	x	x	37.9	603.0	x		Metolachlor(1)

(continued)

M. Köck-Schulmeyer et al.

Table 3 (continued)

Code	Groundwater bodies Name	Aquifer code	Sampling 2007	2008	2009	2010	Total pesticides (ng/L) min	max	Pesticide levels TP > 500 ng/L	IP > 100 ng/L	Pesticide > 100 ng/L (number of samples)
M35	Alluvial of the lower Tordera and delta	08110-0140	x	x	x	x	1.4	32.6			
M35	Alluvial of the lower Tordera and delta	08110-0185	x	x	x		0.0	7.7			
M35	Alluvial of the lower Tordera and delta	08155-0084	x	x	x	x	20.8	81.4			
M46	Alluvial of the middle Segre	25172-0005	x	x	x	x	35.6	202.2		x	DEA(1)
M47	Alluvial of the lower Segre	25102-0001	x	x	x	x	9.5	23.3			
M47	Alluvial of the lower Segre	25120-0020	x	x	x	x	33.4	483.1		x	Bentazone(1)
M47	Alluvial of the lower Segre	25254-0004	x	x		x	69.3	328.8		x	Atrazine(1), DEA(2)
M48	Alluvial of Urgell	25168-0001	x	x		x	500.0	2155.3	xx	x	Atrazine(2), DEA(3), DIA(1), TBA(1)
M48	Alluvial of Urgell	25168-0003	x	x		x	201.1	1086.2	x	x	Atrazine(1), DEA(1), TBA(1)
M48	Alluvial of Urgell	25168-0004	x	x		x	453.3	1778.5	xx	x	Atrazine(2), DEA(3), DIA(1), TBA(2)
M48	Alluvial of Urgell	25225-0003	x	x	x	x	27.1	559.6	x	x	Atrazine(1), DEA(2)

X means exceedance in one sampling campaign; XX means exceedance in two sampling campaigns
TP total pesticide, *IP* individual pesticide, *DIA* deisopropylatrazine, *DEA* desethylatrazine, *TBA* terbuthylazine

2.3 Results and Discussion

2.3.1 Individual Pesticide Levels

The most ubiquitous pesticide was simazine, present in 80% of the samples, followed by atrazine, diuron, DEA and diazinon, present in more than 50% of the samples (64%, 56%, 56% and 50%, respectively). Cyanazine, molinate, fenitrothion and mecoprop were detected in less than 5% of the samples. The maximum individual concentrations were observed for alachlor (9,950 ng/L in M33, 2008), dimethoate (2,277 ng/L in M35, 2010), DEA (1,370 ng/L in M48, 2007) and linuron (1,010 ng/L in M33, 2008), while many others, such as terbuthylazine, DIA, atrazine and metolachlor, presented levels also higher than 500 ng/L. Results are consistent when evaluated with the GUS index (see Table 2). Mots triazines and metolachlor, i.e., the compounds with GUS index > 3 and therefore with higher leaching potential, were among the most ubiquitous an abundant compounds. In contrast, fenitrothion, which according to its GUS index (0.64) is a nonlixiviable pesticide, was detected at low levels in less than 5% of the samples.

The environmental quality standard of 100 ng/L set for individual pesticides in groundwaters by the Directive 2006/118/EC was surpassed in several cases. Table 3 shows the minimum and maximum total pesticide levels measured in the various aquifer samples as well as the pesticides that exceeded the limit. As previously mentioned, alachlor showed the highest pesticide level (9,950 ng/L in M33, 2008), but this was its only noncompliance with the directive, whereas other compounds such as DEA, atrazine and terbuthylazine exceeded the 100 ng/L limit in 15, 10 and 9 samples, respectively. In general, as it can be seen in Fig. 2a, the pesticide profile was dominated by the presence of triazines, although organophosphates, such as dimethoate, and phenylureas, such as diuron and linuron, also have an important contribution. These overall relative concentrations represent the average of the respective values calculated for each sampling campaign, and in their interpretation, it is important to have in mind that the number of pesticides analysed within each family is variable (e.g., 6 triazines, 1 anilide).

Previous studies carried out in groundwater bodies from Catalonia have also shown the presence of some of these pesticides. Garrido et al. [17] found in groundwater samples from Catalonia collected in 1997 and 1998 significant levels of organophosphates, such as malathion, fenitrothion, diazinon and dimethoate (1,300, 800, 400 and 100 ng/L, respectively), and of triazines, such as atrazine and simazine (1,100 and 150 ng/L, respectively). Quintana et al., Rodriguez-Mozaz et al., Kampioti et al. and Teijon et al. [20, 21, 25, 32] studied the presence of pesticides in groundwater samples collected in the Llobregat area (NE Spain) in 2000, 2002, 2003 and 2007–2008, respectively. Atrazine showed decreasing levels with time: from 25–59 ng/L in 2000 to 7–14 ng/L in 2002, 2.3 ng/L in 2003 and below detection limits in 2007–2008. Simazine presented the same trend with time with levels decreasing from 25–164 ng/L in 2000 to 22–153 ng/L in 2002, 54 ng/L

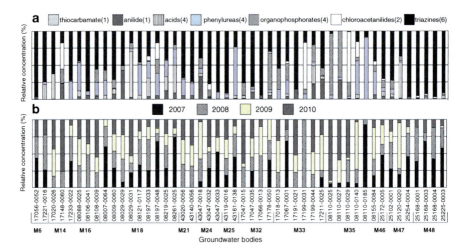

Fig. 2 Relative concentration of (**a**) pesticide classes in each aquifer and (**b**) total pesticides in each sampled year and aquifer

in 2003 and 3 ng/L in 2007–2008. Terbuthylazine was detected in 2000 (30–83 ng/L) and 2003 (5.2 ng/L), but it was not detected in 2007–2008 in the Llobregat zone.

In the Ebro river zone (NE Spain), pesticide concentrations in groundwater were much higher than in the Llobregar river area. Hildebrandt et al. [18, 19] found in groundwater samples collected in 2000–2001 very high levels of metolachlor (10–2000 ng/L) and triazines (2460, 1980, 1270, 790 and 540 ng/L for atrazine, DEA, terbuthylazine, DIA and simazine, respectively). However, 3 years later (2004), triazines concentrations decreased dramatically, whereas metolachlor presented levels even higher (from 2,000 to 5,370 ng/L).

Outside Spain, the profile of pesticide contamination is fairly similar. For instance, Kolpin et al. [28] analysed pesticides in groundwater samples from the United States and found, in samples taken in 2001, triazine concentrations between 50 and 620 ng/L. In groundwater samples collected in Portugal between 2005 and 2008, atrazine and terbuthylazine were also detected [16].

As it can be seen, triazines are very often the main contributors to the pesticide contamination not just in Catalonia but also in Spain and other countries around the world. Indeed, atrazine is very frequently found in groundwater samples even after being banned in the EU due to its carryover (a generally undesirable property for herbicides). The allowed deadline for use of atrazine was September 2005, extended to December 2007 for special application in corn cultivation [16]. In fact, atrazine stands out for being the triazine of greatest concern regarding groundwater contamination because it does not break down readily (within a few weeks) after being applied to soils.

2.3.2 Total Pesticide Levels

Table 3 shows the minimum and maximum total pesticide concentrations found in each aquifer between 2007 and 2010. As it can be seen, the aquifers belonging to the groundwater bodies M48, M35, M47, and M33 showed the highest pesticides levels and M16, M21, M18 and M6, the lowest. According to the Directive 2006/118/EC, the quality standard of 500 ng/L set for total pesticides in groundwaters was surpassed six times in M48, twice in M32 and M35, and once in M33. Moreover, the percentage of aquifer stations in Catalonia with nitrate concentration higher than 50 mg/L was 35%, 30% and 37% in 2007, 2008 and 2009, respectively [33], highlighting the general impact of agriculture in the quality of groundwater in Catalonia.

With the aim to better understand the source of these pesticide levels, each groundwater body land was classified according to their use in (a) urban and industrial, (b) woodland, (c) agricultural land without irrigation (dry) and (d) agricultural land with irrigation. The separation between dry and irrigated agricultural land is critical because irrigation increases considerably the possibility of finding more contaminants in groundwater.

It may be also worth noting that natural recharge happens mainly not only due to irrigation returns but also due to recharge by rainwater infiltration, infiltration of rivers and infiltration of side tributaries. Figure 3 shows the relative distribution of land according to use for each groundwater body [31]. As it can be seen, M47 and M25 present similar profiles of land distribution if both dry and irrigated agricultural lands are considered as a single classification, with more than 80% of the land

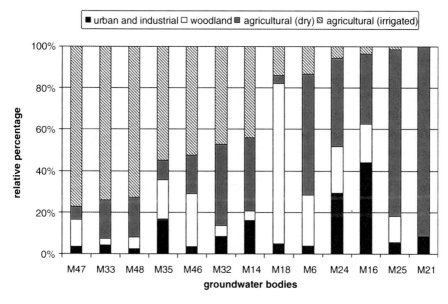

Fig. 3 Relative uses (in percentage) of the groundwater bodies' lands

dedicated to agriculture. Nevertheless, in M47, more than 70% of the land is irrigated constantly, in contrast to M25, which presents 80% of dry land cultivation. These differences are in accordance with results; in general, levels of total pesticides in M47 are higher than in M25. Indeed, most differences between pesticides concentrations can be justified crossing information in Table 3 with information in Fig. 3. M47, M48 (Lleida), M33 (Girona) and M35 (Tordera-Besòs), i.e., the groundwater bodies with the highest percentages of irrigated agricultural land, are exactly the groundwater bodies with higher total pesticides levels. In contrast, M16 (Sabadell-Granollers) and M21 (near Tarragona), the groundwater bodies with the lowest total pesticides levels, are the areas with the less irrigated lands. Only to contextualise, Lleida is well known for its agricultural activities and Granollers for its important industrial activities.

2.3.3 Pesticides Over Time

The previous sections review the geographical distribution of pesticides in the groundwater bodies under study, but the occurrence of pesticide over time is also important to know more about the trends in their use and distribution. Figure 2b shows the relative concentration of pesticides considering time. M47 and M48, the most polluted groundwater bodies, as well as M46, M18 and M25, show generally decreasing levels over time. This means that the Lleida area, known to be the most polluted in Catalonia by pesticides, is changing over time to better conditions. In contrast, M35 shows the opposite profile, with three out of six aquifers presenting total pesticide levels higher over time. In the case of the aquifer #08110-0025, it appears that in 2010, there was a punctual case of pollution by dimethoate that led to levels of this compound above 2,200 ng/L. This groundwater body (M35) has just 26 km^2 of extension and the most contaminated aquifers are from the low part of the groundwater body.

Results have also shown that in some cases, the concentration of pesticides remains fairly constant over time. This is the case of the groundwater body M21 (72 km^2 of extension, near Tarragona), where the pesticide profile is the same over time.

Nevertheless, it is not straightforward to assess the trend of pesticide use over time since there are some other nonnegligible contributing factors that must be also taken into consideration. Among them, the most relevant are related to water quantity, such as the aquifer natural recharge by rainfall and the water abstraction regime.

2.3.4 Risk

The risk of groundwater contamination by pesticides stems mainly from the use made of aquifers for irrigation purposes, way by which pesticides can enter the food chain, and from its use also as resource for the production of drinking water.

According with the most recent and extensive monitoring study carried out in the EU for assessment of the occurrence of pesticides in groundwater [13], triazines and their degradation products (atrazine, desethylatrazine, desethylterbutylazine, simazine, terbutylazine) are the most ubiquitous and abundant polar pesticides found in aquifers. Drinking water treatment plants are not completely effective at removing these compounds from the water phase [21, 25] and, in addition, different transformation products (TPs) can be generated during the treatment process [34]. In general, the amounts found in drinking water of either the active pesticide substances or the transformation products, which can also be toxic, are very low, of just a few ng/L or lower. However, drinking water, alike food, is just one of the various possible ways of human exposure to these compounds, and pesticides (and their TPs) are just one class of chemical contaminants that, together with many others that can also be present in both water and food, can exert pernicious effects on the environment and on humans, and among which synergistic effects can occur. In the past, the main concern with regards to the toxicological effects of pesticides towards humans was associated to their carcinogenic and mutagenic activity, which very often coincided in compounds with also bioaccumulation and biomagnification properties. However, today, this concern goes further due to the increasing weight of evidence on the potential of various pesticides, among them, triazines, to exert endocrine disrupting effects [35–37].

Another example of the potential pernicious effects of pesticides upon human health is the study conducted at the University of Colorado where researchers have found that higher concentrations of four pesticides – atrazine, simazine, alachlor and metolachlor – in groundwater are significantly associated with higher levels of Parkinson disease. For every 10 µg/L increase of pesticide levels in the drinking water, they found that the risk for Parkinson disease increased by 3%; and their water samples had pesticide concentrations ranging from 0.0005 to 20 µg/L [38].

3   Conclusions

The presence of pesticides belonging to different chemical and functional classes in groundwater seems to be ubiquitous especially in those areas subjected to intensive agricultural practices, as it has been illustrated with the case study reviewed corresponding to Catalonia (NE Spain).

Pesticide levels often exceed the requirements posed by the Ground Water Directive (2006/118/EC), thus constituting a serious threat to ground water quality. This becomes especially relevant in those cases in which groundwater are used as human supply source. Of particular concern is the fact that the commercialisation of formulations containing some of the most commonly found pesticides, such as triazines (atrazine, simazine, etc.), lindane have been already banned in Europe.

The above described situation deserves two final comments: First of all, the need of keeping extensive monitoring campaigns for the control of pesticides in groundwater as the only reliable basis to assess quality status, and secondly, it must be once more stressed that protection of groundwater from pesticide pollution will be only possible through a continuous environmental education effort and implementation of good agricultural practices.

References

1. Molinero J, Custodio E (2008) Groundwater in Spain: overview and management practices. Real Academia de Ciencias. http://www.rac.es/ficheros/doc/00587.pdf. Accessed 2011
2. Fait G, Ferrari F, Balderacchi M et al (2008) Herbicide and nitrates groundwater leaching assessment. La Goliardica Pavese, Piacenza
3. Scheidleder A, Grath J, Winkler G et al (1999) Groundwater quality and quantity in Europe. EEA, Copenhagen
4. European Commission (2009) EU action on pesticides: our food has become greener? http://ec.europa.eu/dgs/health_consumer/information_sources/docs/plant/factsheet_pesticides_en.pdf. Accessed 2011
5. AERU (2009) Agriculture & Environment Research Unit – THE PPDB, Pesticide Properties Database. EU-funded FOOTPRINT project. http://sitem.herts.ac.uk/aeru/footprint/en/index.htm. Accessed 2011
6. Milhome MAL, de Sousa DDB, Lima FDF et al (2009) Assessment of surface and groundwater potential contamination by agricultural pesticides applied in the region of Baixo Jaguaribe, CE, Brazil. Engenharia Sanitaria E Ambiental 14:363–372
7. Council of the European Communities (1980) Directive 80/68/EEC on the protection of groundwater against pollution caused by certain dangerous substances. Off J Eur Community L 20:43
8. Council of the European Communities (1991) Directive 91/676/EEC concerning the protection of water against pollution caused by nitrates from agricultural sources. Off J Eur Community L 375:1–8
9. Council of the European Communities (1991) Directive 91/414/EEC concerning the placing of plant protection products on the market. Off J Eur Community L 230:0001
10. Council of the European Communities (1998) Directive 98/8/EC concerning the placing of biocidal products on the market. Off J Eur Community L 123:1
11. Council of the European Communities (2000) Directive 2000/60/EC establishing a framework for Community action in the field of water policy. Off J Eur Community L 327:1
12. Council of the European Communities (2006) Directive 2006/118/EC on the protection of groundwater against pollution and deterioration. Off J Eur Community L 372:19
13. Loos R, Locoro G, Comero S et al (2010) Pan-European survey on the occurrence of selected polar organic persistent pollutants in ground water. Water Res 44:4115–4126
14. Marin JM, Gracia-Lor E, Sancho JV et al (2009) Application of ultra-high-pressure liquid chromatography-tandem mass spectrometry to the determination of multi-class pesticides in environmental and wastewater samples Study of matrix effects. J Chromatogr A 1216:1410–1420
15. Baugros JB, Giroud B, Dessalces G et al (2008) Multiresidue analytical methods for the ultra-trace quantification of 33 priority substances present in the list of REACH in real water samples. Anal Chim Acta 607:191–203
16. Carvalho JJ, Jeronimo PCA, Goncalves C et al (2008) Evaluation of a multiresidue method for measuring fourteen chemical groups of pesticides in water by use of LC-MS-MS. Anal Bioanal Chem 392:955–968
17. Garrido T, Fraile J, Ninerola JM et al (2000) Survey of ground water pesticide pollution in rural areas of Catalonia (Spain). Int J Environ Anal Chem 78:51–65

18. Hildebrandt A, Guillamon M, Lacorte S et al (2008) Impact of pesticides used in agriculture and vineyards to surface and groundwater quality (North Spain). Water Res 42:3315–3326
19. Hildebrandt A, Lacorte S, Barcelo D (2007) Assessment of priority pesticides, degradation products, and pesticide adjuvants in groundwaters and top soils from agricultural areas of the Ebro river basin. Anal Bioanal Chem 387:1459–1468
20. Quintana J, Marti I, Ventura F (2001) Monitoring of pesticides in drinking and related waters in NE Spain with a multiresidue SPE-GC-MS method including an estimation of the uncertainty of the analytical results. J Chromatogr A 938:3–13
21. Rodriguez-Mozaz S, de Alda MJL, Barcelo D (2004) Monitoring of estrogens, pesticides and bisphenol A in natural waters and drinking water treatment plants by solid-phase extraction-liquid chromatography-mass spectrometry. J Chromatogr A 1045:85–92
22. Zhao S, Zhang PF, Crusius J et al (2011) Use of pharmaceuticals and pesticides to constrain nutrient sources in coastal groundwater of northwestern Long Island, New York, USA. J Environ Monit 13:1337–1343
23. Rodriguez-Mozaz S, de Alda MJL, Barcelo D (2007) Advantages and limitations of on-line solid phase extraction coupled to liquid chromatography-mass spectrometry technologies versus biosensors for monitoring of emerging contaminants in water. J Chromatogr A 1152:97–115
24. Ibanez M, Pozo OJ, Sancho JV et al (2005) Residue determination of glyphosate, glufosinate and aminomethylphosphonic acid in water and soil samples by liquid chromatography coupled to electrospray tandem mass spectrometry. J Chromatogr A 1081:145–155
25. Kampioti AA, da Cunha ACB, de Alda ML et al (2005) Fully automated multianalyte determination of different classes of pesticides, at picogram per litre levels in water, by on-line solid-phase extraction-liquid chromatography-electrospray-tandem mass spectrometry. Anal Bioanal Chem 382:1815–1825
26. Kuster M, Diaz-Cruz S, Rosell M et al (2010) Fate of selected pesticides, estrogens, progestogens and volatile organic compounds during artificial aquifer recharge using surface waters. Chemosphere 79:880–886
27. Postigo C, de Alda MJL, Barcelo D et al (2010) Analysis and occurrence of selected medium to highly polar pesticides in groundwater of Catalonia (NE Spain): an approach based on on-line solid phase extraction-liquid chromatography-electrospray-tandem mass spectrometry detection. J Hydrol 383:83–92
28. Kolpin DK, Schnoebelen DJ, Thurman EM (2004) Degradates provide insight to spatial and temporal trends of herbicides in ground water. Ground Water 42:601–608
29. Fenoll J, Hellin P, Martinez CM et al (2011) Determination of 48 pesticides and their main metabolites in water samples by employing sonication and liquid chromatography-tandem mass spectrometry. Talanta 85:975–982
30. Kuster M, de Alda ML, Barcelo D (2009) Liquid chromatography-tandem mass spectrometric analysis and regulatory issues of polar pesticides in natural and treated waters. J Chromatogr A 1216:520–529
31. ACA (Agència Catalana de l'Aigua). Generalitat de Catalunya. http://www.gencat.cat/aca/. Accessed 2011
32. Teijon G, Candela L, Tamoh K et al (2010) Occurrence of emerging contaminants, priority substances (2008/105/CE) and heavy metals in treated wastewater and groundwater at Depurbaix facility (Barcelona, Spain). Sci Total Environ 408:3584–3595
33. MARM (2011) Perfil ambiental de España 2010: informe basado en indicadores. Ministerio de Medio Ambiente y Medio Rural y Marino, Madrid
34. Brix R, Bahi N, Lopez de Alda MJ et al (2009) Identification of disinfection by-products of selected triazines in drinking water by LC-Q-ToF-MS/MS and evaluation of their toxicity. J Mass Spectrom 44:330–337
35. EPA (2010) Atrazine disrupts reproductive development and function across vertebrate classes. http://www.savethefrogs.com/actions/pesticides/images/publications/Hayes-EPA-2010-Submission.pdf. Accessed 2011

36. Fan W, Yanase T, Morinaga H et al (2007) Atrazine-induced aromatase expression is SF-1 dependent: implications for endocrine disruption in wildlife and reproductive cancers in humans. Environ Health Perspect 115:720–727
37. Hayes TB, Falso P, Gallipeau S et al (2010) The cause of global amphibian declines: a developmental endocrinologist's perspective. J Exp Biol 213:921–933
38. Shaw G (2011) New evidence for association of pesticides with Parkinson disease. Neurology Today 11:16–21

Zebrafish as a Vertebrate Model to Assess Sublethal Effects and Health Risks of Emerging Pollutants

Demetrio Raldúa, Carlos Barata, Marta Casado, Melissa Faria, José María Navas, Alba Olivares, Eva Oliveira, Sergi Pelayo, Benedicte Thienpont, and Benjamin Piña

Abstract Zebrafish is developing as a major model for assessing toxicity of pharmaceuticals, drugs, and pollutants. Besides its applications in regulatory toxicity and drug discovery, its characteristics make it a unique system to analyze sublethal toxic effects that only can be studied applying holistic, in toto approaches. Here, we show some of these analyses, in which complex organic systems (neuronal, muscular, sensorial, digestive, thyroid), as well as the embryonic development, show specific effects upon exposure to pharmaceuticals and several environmentally relevant substances, including nanoparticles and other emerging pollutants for which no adequate toxicological profile is still available. These analyses are especially relevant for embryo risk evaluation, given the close similarity of the early stages of the development in all vertebrates, including humans.

Keywords Dioxins, Environmental Pollution, Nanoparticles, Neurotoxicity, Thyroid disrupters

Contents

D. Raldúa, C. Barata, M. Casado, M. Faria, A. Olivares, E. Oliveira, S. Pelayo,
B. Thienpont, and B. Piña (✉)
Institute of Environmental Assessment and Water Research (IDÆA-CSIC), Jordi Girona 18,
1808034 Barcelona, Spain
e-mail: bpcbmc@cid.csic.es

J.M. Navas
Department of Environment, Instituto Nacional de Investigación y Tecnología Agraria y
Alimentaria (INIA), Ctra. de la Coruña Km 7, 28040 Madrid, Spain

D. Barceló (ed.), *Emerging Organic Contaminants and Human Health*,
Hdb Env Chem (2012) 20: 395–414, DOI 10.1007/698_2011_124,
© Springer-Verlag Berlin Heidelberg 2011, Published online: 22 November 2011

1   Zebrafish Model in Environmental Risk Assessment

Zebrafish is rapidly becoming the favorite vertebrate model organism for developmental biology, drug discovery, evaluation of toxicological side effects of the potential drugs, and ecotoxicology [1]. Small and inexpensive to maintain, a single pair of adults breeds once a week, generating 100–200 offspring per brood, and their husbandry costs are 100 and 1,000 times lower than those of mice or other mammals [2]. Their ex utero development and optical clarity during embryogenesis and early larval stages facilitate visual in vivo observation of early developmental processes and organogenesis. These functional and morphological changes may be observed in vivo or in whole-mount fixed specimens by using transgenic lines, vital dyes, fluorescent tracers, antibodies and riboprobes, and fluorescent markers [3]. Zebrafish embryos grow rapidly, with the basic vertebrate body plan laid out within 24 h post-fertilization (hpf). At this stage, embryo length is around 1.9 mm, so several embryos fit easily inside a single well of a 384-well plate [4]. The majority of organs, including the nervous system, cardiovascular system, intestines, liver, and kidneys, can be studied at 5 days post-fertilization (dpf), when the larva is still only 3–4 mm in length, making assays in 96-well plate format entirely possible [2]. These organs and tissues have proved similar to their mammalian counterparts on the anatomical, physiological, and molecular levels [3]. As larvae can live in as little as 50 µl water, only micrograms of compounds are required for assays, representing a major cost saving in screening entire molecule libraries [2]. Consequently, zebrafish represents a unique vertebrate model for high-throughput chemical screening, making them useful for toxicological evaluation [4].

Whereas the main focus of zebrafish research has traditionally been on developmental biology, observations from large-scale genetic screening allowed the identification of mutants phenocopying diseases and developmental pathologies similar to humans. These studies demonstrated the suitability of using zebrafish as a model of human disease, drug discovery, and drug toxicity analyses [5]. In this context, zebrafish is proposed as an intermediate step between single cell-based evaluation and mammalian (and ultimately human) testing. Zebrafish tests contribute to the prioritization of drug candidates [6], and can also be used for the selection of the chemical concentrations to be used in further in vivo assays, thus reducing the number and cost of mammalian studies. It is important to note that zebrafish assays done at the initial endotrophic nutritional period (0–120 hpf) are considered nonanimal-based assays by Directive 2010/63/EU that actualizes the previous directive 86/609/EEC.

The same principles that led to the adoption of zebrafish as a model for discovery and drug toxicology analyses also apply to the analysis of the toxic mechanisms and effects of emerging pollutants. Whereas single-cell bioassays can cover only a limited number of endpoints, the analysis of toxic effects in zebrafish embryos provides a holistic approach, as it includes multiple aspects of the physiology, development, and functionality of complex organic systems. A variety of assays are available for assessing toxicity on the cardiovascular, gastrointestinal, renal, nervous, thyroid, digestive or skeletal systems [2, 5]. In this chapter, we describe some assays for detecting detrimental effects of pollutants on some of these systems, using the whole-animal approach that the zebrafish model allows.

2 Developmental Neurotoxicity Assessment

Zebrafish embryos and larvae are exceptionally well suited for developmental neurotoxicity studies that combine cellular, molecular, and genetic approaches. Since zebrafish embryos and early larvae are transparent, specific neurons and axon tracts can be visualized in vivo using transgenic lines or by injecting reporter dyes [3]. Working at later developmental stages is also possible by using transparent zebrafish, as *casper* mutants. Specific types of neurons can be visualized in fixed intact zebrafish by whole-mount immunohistochemistry (IHC) or in situ hybridization [7, 8]. In addition, the small size of early stage zebrafish permits performance of quantitative whole-animal assays in a 96-well microplate format for neurotoxicity screening. The zebrafish model has been used for assessing the toxic effect of different xenobiotics on specific cell types in the nervous systems, as dopaminergic neurons or the mechanosensory system [7, 8].

2.1 Assessment of Effects on Dopaminergic Neurons

Drug effects on dopaminergic neurons in zebrafish can be assessed by whole-mount immunostaining with an anti-tyrosine hydroxylase antibody [7, 8]. This enzyme catalyzes the first step in the catecholamine biosynthetic pathway and can be used as a biomarker for catecholaminergic neurons, including dopaminergic, noradrenergic, and adrenergic neurons. Although anti-tyrosine hydroxylase antibody is, therefore, not specific for dopaminergic neurons, it has been demonstrated that all tyrosine hydroxylase-positive neurons in zebrafish diencephalon (hypothalamus, posterior tuberculum, ventral thalamus, and pretectum) are DA neurons [7]. By using this approach, it has been demonstrated that DDT, dieldrin, nonylphenol and 6-hydroxydopamine strongly decrease the number of DA neurons in zebrafish ventral diencephalon [8]. Both 1-methyl-4-phenyl-1,2,3,6-tetrahydropyridine

(MPTP) and sodium benzoate induce a strong downregulation in the expression of tyrosine hydroxylase mRNA and in the tyrosine hydroxylase-positive cells in the ventral diencephalon. These chemicals also decreased the expression of the dopamine transporter (DAT), a membrane transport protein involved in dopamine reuptake, that is, a specific marker of dopaminergic neurons [9].

A new biotechnological tool has been developed to assess in vivo the effect of chemicals and drugs on dopaminergic neurons. In the enhancer trap transgenic zebrafish line, ETvmat2:GFP, most monoaminergic neurons are labeled with the green fluorescent protein (GFP) during embryonic development [10]. The GFP expression pattern of this transgenic line is identical to that of the endogenous *vmat2* gene, including the large and pear-shaped tyrosine hydroxylase-positive neurons in the posterior tuberculum of ventral diencephalon (PT). It has been found that these PT neurons were significantly reduced after exposure to MPTP [10]. Thus, this transgenic zebrafish line may facilitate the analysis of environmental pollutants affecting these dopaminergic PT neurons in living transparent embryos [11].

2.2 Assessment of Effects on the Mechanosensory System: Lateral Line

The mechanosensory systems of fish, including the lateral line, are closely related to the mammalian hearing system [12]. Besides possessing the typical vertebrate inner ear, fish possess the lateral line organs that contain sensory hair cells. These analogies are most relevant in toxicology and drug discovery and evaluation, as some of the pharmaceuticals already detected in aquatic ecosystems as emerging pollutants affect the auditory function in humans.

The fish lateral line is composed of rosette-like structures called neuromasts, composed of hair cells and supporting cells. Hair cells are innervated by sensory neurons that are localized in the anterior and posterior lateral line ganglia [13]. The function of the lateral line is mainly to allow the fish to orient relative to a water current, to hold a stationary position in a stream, for school swimming, to detect prey, or to avoid predators [12]. Neuromasts are located on the fish surface, with their peripheral sensory neurons in direct contact with the surrounding water containing pollutants. Hair cells of the neuromast can be easily stained by various fluorescent dyes in vivo or in fixed larvae by using whole-mount immunofluorescence or whole-mount in situ hybridization (WISH) [12, 14]. Figure 1a, c show the stereotyped positions of the neuromasts of the anterior and posterior lateral line on the surface of the head and body, respectively, using an in vivo labeling with DiAsp (Fig. 1a), and immunofluorescence with fixed animals (Fig. 1c). Immunofluorescence allows the labeling of the kinocilia (acetylated alpha-tubulin, red fluorescence in Fig. 1c, d) and the sterocilia (actin filaments stained with phalloidin-Alexa 488, green fluorescence in Fig. 1b, d) of the hair cells of a neuromast. The intensity of hair cell staining can be imaged and quantified using image analysis systems, and the effects can be assessed

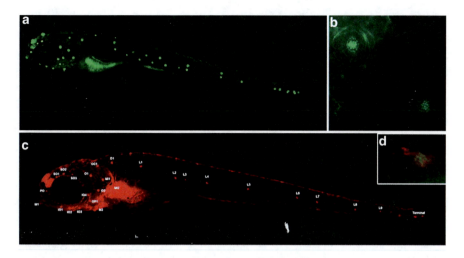

Fig. 1 Lateral line neuromasts in zebrafish larvae. (**a**) Lateral view of live untreated 8 dpf zebrafish larvae (anterior to the left and dorsal up) stained with 5 μM 4-Di-2-Asp. (**b**) A neuromast from a fixed 5 dpf eleutheroembryo containing hair cells labeled for f-actin with phalloidin-Alexa 488. (**c**) Lateral view of a fixed 5 dpf larva containing lateral line neuromast stained with anti-acetylated alpha-tubulin. (**d**) Hair cells in a neuromast, with the stereocilia bundles in green (phalloidin-Alexa 488) and the kinocilia in *red* (alpha-tubulin, 1/1000)

for the screening of potential ototoxic pollutants [12, 15]. In a massive study, 1,040 drugs and bioactives were screened for ototoxic effects in 5 dpf zebrafish larvae by using the vital dye YO-PRO1, and 21 compounds were identified as ototoxic, including aminoglycosides (tobramycin, neomycin, and kanamycin), other antibacterial agents (chlortetracycline, chloramphenicol, and demeclocycline), antiprotozoal agents (pentamidine and mefloquine), an anti-neoplastic agent (cisplatin), an anticholinergic agent (propantheline bromide), and an antihyperlipidemic agent (simvastatin) [15]. Gentamicin and streptomycin are also well known to induce hair cell loss [16, 17], as well as different heavy metals as cadmium, copper, zinc, iron, and silver [12, 14]. Although the mechanism of hair cells disruption for most of these chemicals remains unclear, it appears to involve the formation of free radicals, induction of oxidative stress and activation of the apoptotic pathways [12].

3   Neuromuscular System

3.1   Assessment of Effects on Motor Neurons

Zebrafish skeletal muscle is innervated by primary and secondary motor neurons. There are three primary motor neurons (PMNs) per hemi-segment, one of which innervates the dorsal trunk musculature, whereas the other two innervate mid and ventral regions of the trunk. Secondary motor neurons (SMNs) follow similar paths to those taken by PMNs, although they do not branch as extensively and are more numerous. It has been demonstrated that exposition to some chemicals impairs the motor neuron development. Stereotypic pattern of the axonal projections and specific neurotoxic effects on motor neurons can be assessed by whole-mount IHC with axon-specific antibodies. For instance, by using IHC with znp1 and zn5/zn8, monoclonal antibodies labeling axonal bundles of primary and secondary motor neurons, respectively, an impairment in the stereotypic pattern of the axonal projections of the motor neurons has been demonstrated in zebrafish embryos or larvae exposed to nicotine, caffeine, ethanol, sodium benzoate, cadmium, or metam-sodium [18–21]. Figure 2 summarizes different abnormalities we have found in the stereotypic pattern of PMNs in zebrafish larvae exposed to nicotine, caffeine and sodium benzoate, by using znp-1 (upper panels) or zn8 (lower panels) whole-mount immunofluorescence. This is the first report on the neurotoxic effect of nicotine and caffeine on the development of the PMNs, as well as on the detrimental effect of sodium benzoate on the development of the SMNs.

As a recent development, the use of these transgenic lines in which GFP expression is driven by motor neuron-specific promoters will be useful in screening the effect of libraries of environmental pollutants on the development of the neuromotor system [22, 23].

3.2   Assessment of Effects on Neuromuscular Junctions

Effects of environmental pollutants on the neuromuscular junctions (NMJ) in zebrafish embryos and larvae have been assessed by analyzing different endpoints. Figure 3 shows the effect of the lipid regulator clofibrate on the morphology and the number neuromuscular junctions by using whole-mount acetylcholinesterase staining [24]. On the other hand, fluorophore-conjugated α-bungarotoxin (BTX), which binds specifically to acetylcholine receptors (AChRs), has been used coupled to znp-1/SV2 immunofluorescence in order to assess disruption of the AChR clusters and its effect on the growing motor axons. Using this approach, different defects in the NMJs has been found in zebrafish exposed to sodium benzoate and caffeine [21, 25].

3.3   Assessment of Effects on Muscle Fibers

Zebrafish larvae possess two types of skeletal muscle fibers. Slow (red) muscle fibers, a superficial monolayer on the surface of the myotome, are equipped for oxidative phosphorylation, can generate relatively large stores of energy, and are most resistant to fatigue. Fast (white) muscle fibers, in the deep portion of the myotome, are least resistant to fatigue because they rely on anaerobic glycolysis for

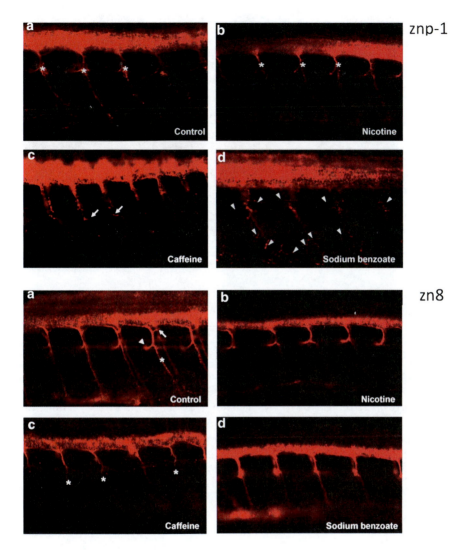

Fig. 2 Some emerging pollutants impair the development of the motor neurons in zebrafish. *Upper panel*: znp1 monoclonal antibody staining of 57 hpf zebrafish eleutheroembryos after exposure in the water (**a**) and water containing 33 μM nicotine (**b**), 0.77 mM caffeine (**c**), or 6.9 mM sodium benzoate (**d**) for 31 h. Embryos treated with nicotine (**b**) exhibited a in some segments shorter common path and decreased axon branching in the caudal primary motor neurons (CaP) *asterisk* labeling the choice point). Embryos treated with caffeine (**c**) exhibited axons of some CaP with abnormal morphology (*arrows*) and decreased branching. Moreover, most of the axons of rostral (RoP) and middle (MiP) PMNs were absent. Embryos treated with sodium benzoate exhibited axonal over-branching of the CaP (*arrowheads*). *Lower panel*: zn8 monoclonal antibody staining of 57 hpf zebrafish larvae after exposure in the water (**a**) or water containing 33 μM nicotine (**b**), 0.77 mM caffeine (**c**), or 6.9 mM sodium benzoate (**d**) for 31 h. Although secondary motor neurons (SMNs) follow similar paths to those taken by PMNs in control embryos (**a**), embryos treated with nicotine (**b**) exhibited severe defects in axogenesis such as early

Fig. 3 Comparison of the
neuromuscular junction
pattern and muscular fiber
organization in control and
clofibrate-treated larvae,
using whole-mount
acetylcholinesterase staining.
(**a**) Control larva at 3 dpf.
(**b**) Clofibrate 0.75 mg/l-
treated larva at 3 dpf. AChE
staining labeling to identify
neuromuscular junctions
(NMJ). Representative larvae
are shown at trunk level in
lateral view with the anterior
part to the left. *mf* muscular
fiber, *mj* myoseptal junctions

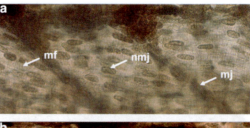
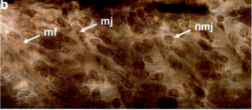

ATP generation. The two types of muscle fibers perform different functions, as fast muscle fibers are inactive during slow swimming episodes and slow muscle fibers are recruited during fast swimming. Fast muscle fibers are innervated by both primary and secondary motor neurons, while slow muscle fibers are likely only innervated by SMNs. Slow and fast muscle fibers can be labeled in zebrafish larvae and subtle changes in the muscle fiber alignments thus easily observed by using whole-mount immunofluorescence with antibodies directed against slow and fast muscle myosin (F59 and F310, respectively), as shown in Fig. 4. Evaluated endpoints included the length, width, and number of the muscle fibers as well as disorganized muscle fiber alignment. This disorganization can be reflected by a lack of segment division, presence of fibers extending over two segments rather than one, altered angles between dorsal and ventral hemi-segments, and smaller muscle fibers. Ethanol, caffeine, sodium benzoate, fipronil, and lovastatin exhibit a myotoxic effect for slow muscle fibers of zebrafish [19, 21, 25–27], whereas ethanol has also been proved to be myotoxic for fast muscle fibers [19].

Fig. 2 (continued) truncation of the axons of those SMNs following the path of the CaP and complete absence of the axons of those SMNs following the axons of MiP. Effect of caffeine (**c**) on SMNs was consistent with the effects on PMNs, with abnormal morphology of the axons (*asterisk*), but also embryos exhibited a complete absence of the axons of those SMNs following the axons of RoP and MiP. Sodium benzoate (**d**) induced early truncation of axons of those SMNs following the path of RoP and a complete absence of the axons of those SMNs following the axons of MiP

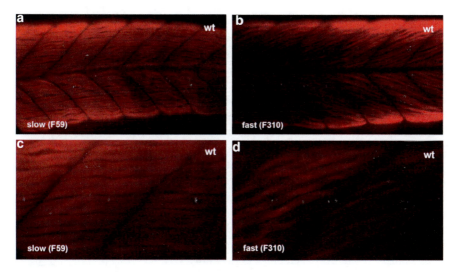

Fig. 4 Wild-type zebrafish eleutheroembryo (57 hpf) stained with antibodies specific for slow muscle fibers (F59; **a**, **c**) or fast muscle fibers (F310; **b**, **d**). Labeling with these antibodies facilitates the analysis of muscle fiber alignments

4   Effects on the Circulatory System

The heart is one of the first organs to develop, and its formation involves a complex series of morphological and morphogenetic events. Because the process occurs through an evolutionary conserved program common to all vertebrates, information about the genes and their mechanisms of action can be extrapolated from small animal models, including zebrafish [28, 29]. As a distinct feature, and because of their small size, zebrafish embryos are not completely dependent on a functional cardiovascular system, carrying on a relatively normal development for several days in the absence of cardiovascular function [29]. Therefore, zebrafish embryos can be analyzed in conditions (genetic, environmental, toxic) leading to serious cardiovascular defects that would be lethal in other systems, thus facilitating the discovery and characterization of toxic compounds severely impairing cardiovascular development.

One of the pollutants known to interfere with cardiovascular development is 2,3,7,8-Tetrachlorodibenzo-p-dioxin (TCDD). TCDD is a persistent, bioaccumulative environmental contaminant, as well as a potent developmental toxicant and human carcinogen [30]. Piscine, avian, and mammalian cardiovascular systems are sensitive to TCDD toxicity, with effects including cardiac enlargement, edema, and several dysfunctions. In zebrafish embryos, these effects include a reduction in cardiomyocyte number at 48 hpf, decreased heart size, altered vascular remodeling, pericardial edema, and decreased ventricular contraction culminating in ventricular standstill [31–34].

An essential step of TCDD toxic effects, including its carcinogenic potential, is its binding to the aryl hydrocarbon receptor (AhR) at the pM range. Deletion of

AhR gene results in mice that are resistant to TCDD and other structurally related pollutants [35–38]. AhR is a transcription factor that regulates expression of many phase I and II metabolic enzymes [35–38]. There are many chemical compounds (sometimes called dioxin-like compounds) able to bind the AhR and to trigger the same kind of transcriptional response, although only a fraction of them are carcinogenic. In fact, the carcinogenic effects associated with dioxin and other dioxin-like compounds are attributed to their potential to stabilize radical species produced by the excess of oxidative enzymes (being CYP1A one of the most important ones) brought upon by the ectopic activation of AhR.

Analysis of the effect of different dioxin-like compounds on the developing zebrafish embryo demonstrated that the cardiopatic effect is independent from the carcinogeneticy. Whereas the knockdown of the AhR signaling pathway in zebrafish by morpholino oligonucleotide (MO) injection [39, 40] severely reduces TCDD toxicity, analogous removal of CYP1A expression (therefore preventing formation of radicals and the consequent carcinogenicity) does not prevent the cardiopatic effect in the developing zebrafish [41, 42].

The dissociation between cardiotoxicity and carcinogenesis can be visualized in zebrafish embryos exposed to known dioxin like carcinogens, like benzo[a]pyrene, or to noncarcinogenic AhR ligands, as β-naphthoflavone (Fig. 5) [41, 43]. The conclusion that cardiotoxic effects in zebrafish (and presumably, other vertebrates') embryos occur by exposition to AhR ligands so-far considered

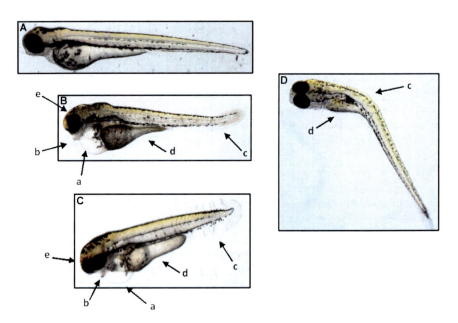

Fig. 5 Typical deformations detected in 96 hpf zebrafish embryos exposed at the indicated concentrations of the carcinogenic polycyclic aromatic hydrocarbons Benzo[a]Pyrene (**b**) and Benazo [k]Fluoranthene (**c**), or to the reportedly nontoxic AhR-ligand β-naphthoflavone (**d**). A nonexposed, normal specimen is shown in (**a**). *Arrows* indicate (*a*) pericardial edema, (*b*) malformation of the lower jaw, (*c*) malformation of the tail, (*d*) color of the yolk, and (*e*) coagulation

innocuous may have many implications on the evaluation of risk of organic compounds for environmental and human health.

5 Disruption of the Thyroid System

The thyroid hormones are central in many physiological processes in vertebrates, including development, growth, and behavior [44]. There is a growing concern about endocrine-disrupting compounds targeting the thyroid system, the so-called thyroid-disrupting chemicals or TDCs. International policies call for the development of high-throughput methods for screening TDCs [45, 46], which cause major concern considering the critical role played by thyroid hormones (TH) during nervous system development [47, 48]. There are many examples of TDCs disrupting key steps in the complex TH regulatory network, including TH synthesis, binding to thyroid transport proteins, metabolism, or transactivation of target genes containing thyroid response elements (Fig. 6) [47–49]. This plurality of possible

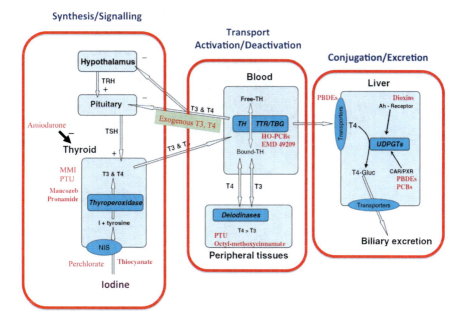

Fig. 6 Thyroid hormone control pathways and sites of disruption by xenobiotic chemicals. Organs, enzymes, and transporters that constitute known or putative targets for thyroid disruption are shown in *blue boxes* and ovals; their corresponding physiological processes are indicated by *red boxes*. Xenobiotics that block, inhibit, or upregulate the different biological functions are shown in *red*. The predicted effect of exogenous T3 or T4 is highlighted in *green*. *MMI* methimazole, *PTU* propylthiouracil, *PCB* polychlorinated biphenyls, *PBDE* polybrominated diphenyl ethers, *T3* triiodothyronine, *T4* thyroxine, *TBG* T4-binding globulin, *TH* T3 + T4, *TRH* thyrotropin-releasing hormone, *TSH* thyrod-stimulating hormone, *TTR* transthyretin, *UDPTG* uridine diphosphate glucuronyltransferase. Modified from [48]

targets makes it challenging to determine from among a very long list of potential thyroid toxicants, primarily tested in rodent and in vitro models, which are relevant to human thyroid function [47]. Very likely, a single assay would never be able to detect all possible TDC activity in the ever-growing number of chemicals continuously released into the environment [49].

The origin and growth of the thyroid gland in zebrafish has been extensively studied and described [50]. Endogenous TH starts to be synthesized and stored in the thyroid follicles at the eleutheroembryo stage [49, 51], but there is no information about its biological availability. Concentrations of the two active forms of TH, T3, and T4 (triiodothyronine and thyroxine, respectively) in 3 dpf eleutheroembryos have been estimated as approximately 0.8 and 2.6 pM, respectively [52]. Despite these low levels of THs, thyroid receptors (TR) and some TH-regulated genes are already expressed and functional during zebrafish early development. Major postembryonic changes (sometimes described as metamorphosis) occur in zebrafish during the third and fourth weeks after fertilization, including

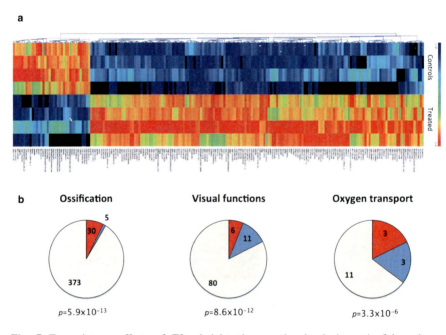

Fig. 7 Transcriptome effects of T3 administration on the developing zebrafish embryo. (**a**) Heatmap of microarray results from 652 probes showing significant differences between T3-treated and control samples. Fold induction values (in log scale) are represented by different shades of color (scale shown in the *far-right bar.*). (**b**) Distribution of overrepresented (*red*), underrepresented (*blue*), and unchanged/undetected (*ivory*) transcripts in T3-treated embryos belonging to the functional categories ossification, visual processes, and oxygen transport. The significance of the observed variations (*p* values) was calculated by the hypergeometric distribution with the Bonferroni correction

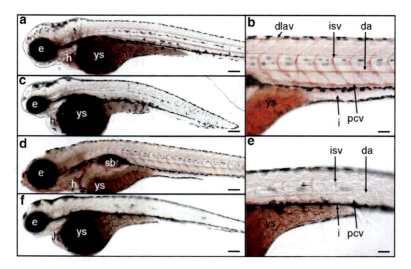

Fig. 8 Clofibrate impairs endotrophic lipid consumption in zebrafish larva. (**a–e**) Whole-mount ORO staining of representative larvae are shown in lateral view with the anterior part to the left. Enlargement at the trunk level is shown in panels **b** and **e**. Control larva at 3 dpf (**a**) and 4 dpf (**d**, **b**). Clofibrate 0.75 mg/l-treated larva at 3 dpf (**c**). Clofibrate 0.5 mg/l-treated larva at 4 dpf (**f**, **e**). *e* eye, *da* dorsal aorta, *dlav* dorsal longitudinal anastomotic vessel, *h* heart, *i* intestine, *isv* intersegmental vessel, *pcv* posterior cardinal vein, *sb* swim bladder, *ys* yolk sac. Reprinted from [24] with permission from Elsevier

resorption of the larval fin fold and development of adult unpaired fins, changes in the gut, peripheral nervous system, and sensory systems (including eyes), alterations in physiology and behavior, substitution of adult hemoglobin for embryonic forms, development of scales, and formation of an adult pigment pattern [51, 53]. Thyroid hormones are supposed to control the time course of these changes. Therefore, any distortion on the timing of the onset of thyroid hormone may have deleterious effects on the developing embryo.

Figure 7 shows the effect of ectopic administration of T3 to the developing zebrafish embryo. At nontoxic concentration (50 nM), only a moderate fraction (less than 5%) of the zebrafish transcriptome shows significant changes. Ossification, visual processes, and the hematopoietic system were the physiological processes most affected by the treatment, in a pattern consistent with an advancement of the development in these particular functions (Fig. 7b). Genes involved in these three processes are known targets for TDCs during metamorphosis in amphibians, teleost fishes, and lampreys [54–60], and constitute molecular counterparts of different endpoints used to test for TDC in amphibians [56, 58]. Therefore, they are excellent candidates for markers of thyroid disruptors in zebrafish at early developmental stages. Chapter 14 provides a more in-deep description of the developmental effects of thyroid disruption in zebrafish embryos.

6   Nutritional Effects

The gastrointestinal system of zebrafish presents clear differences from the human system. The zebrafish does not possess a stomach, the intestine is continuous with the pharynx through a short esophagus, and no sphincters are present [61]. However, zebrafish have most of the cell types observed in the small intestine – absorptive, endocrine, goblet, and interstitial cells of Cajal, although Paneth cells are absent. Gut contractions are under the control of the enteric nervous systems, which respond to different pharmaceuticals in similar way as the mammalian counterpart. For example, zebrafish embryos can be used as predictor of emetic response to pharmaceuticals, one of the most commonly reported clinical adverse effects to be considered in the development of new drugs [61].

In zebrafish, there is an initial endotrophic period, from fertilization to mouth opening, where the embryo and pre-feeding larvae use endogenous yolk nutrients previously accumulated in the oocyte. As the development and survival of fish

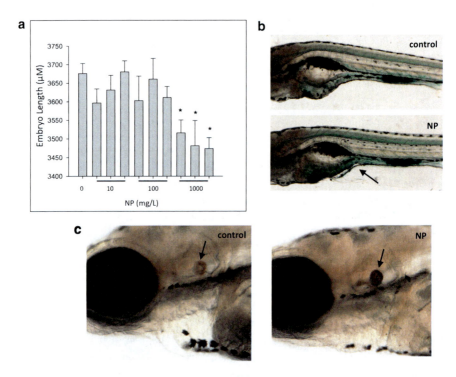

Fig. 9 Effects of TiO$_2$ NP in the developing zebrafish embryos. (**a**) Reduction of embryo length; *asterisks* indicate significant ($P < 0.05$) deviations from the control value following ANOVA and Dunnet Post hoc tests. (**b**) Oxidative stress, as indicated by elevated activity of SA-β-galactosidase in notocorda and gut (*arrow*). (**c**) Alteration of the otoliths to become optically denser than in control (*arrow*). The occurrence of denser otoliths was 10 and 95% out of 60 analyzed larvae in control and NM exposures. *NP* nanoparticle-treated animals

embryos and yolk sac larvae depend, at least partially, on the mobilization of yolk lipid constituents, the presence of blood lipid regulators in surface water may disrupt the endotrophic and endo-exothrophic nutritional phases in fish development. We have recently demonstrated that fibrates, an important group of blood lipid regulators, impair the transfer of neutral lipid between the yolk sac and the embryo. These compounds have recently been reported as pollutants in rivers, due to the high consumption and the low efficiency of the waste water treatment plants to remove them. Figure 8 shows a malabsorption syndrome in zebrafish brought about by the exposure to clofibrate [24]. These results highlight the suitability of the zebrafish model for studying the direct or permissive role of nutrient and hormone deprivation induced by chemicals and human pharmaceuticals in the early stages of vertebrate development.

7 Tissue-Specific Effects of Nanoparticles

Manufactured nanoparticles (NP) are those with at least one dimension less than 100 nm. These particles have unique physicochemical properties, due to their high surface/volume ratio [62]. Titanium dioxide NP (TiO_2 NP) is widely used in various products such as sunscreens, cosmetics, paints, and building materials [63]. Due to such extensive use, TiO_2 NP is found in many water bodies at environmentally relevant concentrations, and the concerns for their putative environmental toxicity are on the rise [62, 64, 65]. TiO_2 NP showed very low acute toxicity in fish and fish embryos [63, 66, 67], although there are reports of several sublethal effects, like some histopathological changes in gills [63, 66] and oxidative stress in different tissues, including brain [63, 66, 68]

Effects on brain are perhaps the most consistent cause of concern about TiO_2 NP exposure [69]. There are reports on the formation of reactive oxygen species (ROS) in brain microglia and induction of damage to neurons in vitro [70, 71], whereas it is known that NP can reach the fish brain in vivo via the nerve endings of the olfactory bulb [62] or through the bloodstream [69, 72]. Different larval swimming parameters, including average and maximum velocity and activity level, are significantly affected by TiO_2 NP, at concentrations far lower than those required for systemic, global toxic effects, like hatchability and survival [73].

The acute toxicity and oxidative effects of nano-scale TiO_2 depend on the size of the nanoparticle (bulk TiO_2 is positively nontoxic) and increase notably through illumination, as this leads to the formation of hydroxyl radicals [74], further indicating oxidative stress as a major candidate for the mechanism of action of NP toxicity. However, a recent microarray analysis of the transcriptome of zebrafish embryos treated with TiO_2 NP showed no major increase of transcripts related to oxidative stress. Instead, significant effects were observed on expression of genes involved in circadian rhythm, kinase activity, vesicular transport, and immune response [75].

Morphological and biochemical analyses of zebrafish embryos exposed to TiO_2 NP show some unexpected effects. Exposed animals exhibited a significant reduction in size (Fig. 9a) and markers of oxidative stress, as revealed by the increase in the SA-β-

Gal activity in the neural tube and the intestine (Fig. 9b). In addition, TiO$_2$ NP exposure induces a clear alteration of the otoliths, which become noticeably denser (Fig. 9c). These effects may well be related to the alterations in behavior and to the slightly less viability observed in TiO$_2$ NP exposed animals, although further research is required to investigate their meaning and relevance in the general toxic effects of NP.

8 Conclusions

Widely used as a toxicological model, zebrafish eleutheroembryo is a valuable biotechnological tool to identify and characterize sublethal effects of emerging pollutants. Pharmaceuticals, NP or new pesticides, while scoring as only moderately toxic in survival tests, have the potential to interact with different complex systems in animal (and human) organisms and to develop new forms of toxicity that only an in toto approach can detect and characterize. This is particularly important when dealing with developing organisms, for little is known about the effects of the accidental administration of many of these substances at the early, most sensitive stages of the development. In this regard, zebrafish opens the unique possibility of following the complete embryo development in a matter of few days, in a system for which abundant and very powerful genetic tools have been developed, and with relatively few requirements in terms of installations and husbandry.

Acknowledgments This work was supported by the Spanish Ministry of Science and Innovation (grants CGL2008-01898/BOS and CTM2011-30471-C02-01), the INIA (grant RTA2009-00074-00-00), the Generalitat de Catalunya (grant 2009 SGR 924) and European Union (FP7 EU Project Managing the Risks of Nanomaterials-MARINA). E.O. acknowledges a predoctoral fellowship from the Portuguese Fundação para a Ciência e Tecnologia (SFRH/BD/48244/2008). B.T. acknowledges a predoctoral fellowship from the Spanish Ministry of Science and Innovation (FPU AP2006-01324).

References

1. Love DR, Pichler FB, Dodd A, Copp BR, Greenwood DR (2004) Technology for high-throughput screens: the present and future using zebrafish. Curr Opin Biotechnol 15:564–571
2. Goldsmith P (2004) Zebrafish as a pharmacological tool: the how, why and when. Curr Opin Pharmacol 4:504–512
3. McGrath P, Li CQ (2008) Zebrafish: a predictive model for assessing drug-induced toxicity. Drug Discov Today 13:394–401
4. Zon L, Peterson R (2005) In vivo drug discovery in the zebrafish. Nat Rev Drug Discov 4:35–44
5. Berghmans S, Butler P, Goldsmith P, Waldron G, Gardner I, Golder Z, Richards FM, Kimber G, Roach A, Alderton W, Fleming A (2008) Zebrafish based assays for the assessment of cardiac, visual and gut function–potential safety screens for early drug discovery. J Pharmacol Toxicol Methods 58:59–68

6. Langheinrich U (2003) Zebrafish: a new model on the pharmaceutical catwalk. Bioessays 25:904–912
7. Parng C, Roy NM, Ton C, Lin Y, McGrath P (2007) Neurotoxicity assessment using zebrafish. J Pharmacol Toxicol Methods 55:103–112
8. Ton C, Lin Y, Willett C (2006) Zebrafish as a model for developmental neurotoxicity testing. Birth Defects Res A Clin Mol Teratol 76:553–567
9. Chen Q, Huang NN, Huang JT, Chen S, Fan J, Li C, Xie FK (2009) Sodium benzoate exposure downregulates the expression of tyrosine hydroxylase and dopamine transporter in dopaminergic neurons in developing zebrafish. Birth Defects Res B Dev Reprod Toxicol 86:85–91
10. Wen L, Wei W, Gu W, Huang P, Ren X, Zhang Z, Zhu Z, Lin S, Zhang B (2008) Visualization of monoaminergic neurons and neurotoxicity of MPTP in live transgenic zebrafish. Dev Biol 314:84–92
11. Rubinstein AL (2006) Zebrafish assays for drug toxicity screening. Expert Opin Drug Metab Toxicol 2:231–240
12. Froehlicher M, Liedtke A, Groh KJ, Neuhauss SC, Segner H, Eggen RI (2009) Zebrafish (Danio rerio) neuromast: promising biological endpoint linking developmental and toxicological studies. Aquat Toxicol 95:307–319
13. Bricaud O, Chaar V, Dambly-Chaudiere C, Ghysen A (2001) Early efferent innervation of the zebrafish lateral line. J Comp Neurol 434:253–261
14. Yang L, Kemadjou J, Zinsmeister C, Bauer M, Legradi J, Muller F, Pankratz M, Jakel J, Strahle U (2007) Transcriptional profiling reveals barcode-like toxicogenomic responses in the zebrafish embryo. Genome Biol 8:R227
15. Chiu LL, Cunningham LL, Raible DW, Rubel EW, Ou HC (2008) Using the zebrafish lateral line to screen for ototoxicity. J Assoc Res Otolaryngol 9:178–190
16. Seiler C, Nicolson T (1999) Defective calmodulin-dependent rapid apical endocytosis in zebrafish sensory hair cell mutants. J Neurobiol 41:424–434
17. Ton C, Parng C (2005) The use of zebrafish for assessing ototoxic and otoprotective agents. Hear Res 208:79–88
18. Chow ESH, Cheng SH (2003) Cadmium affects muscle type development and axon growth in zebrafish embryonic somitogenesis. Toxicol Sci 73:149–159
19. Sylvain NJ, Brewster DL, Ali DW (2010) Zebrafish embryos exposed to alcohol undergo abnormal development of motor neurons and muscle fibers. Neurotoxicol Teratol 32:472–480
20. Tilton F, Tanguay RL (2008) Exposure to sodium metam during zebrafish somitogenesis results in early transcriptional indicators of the ensuing neuronal and muscular dysfunction. Toxicol Sci 106:103–112
21. Tsay HJ, Wang YH, Chen WL, Huang MY, Chen YH (2007) Treatment with sodium benzoate leads to malformation of zebrafish larvae. Neurotoxicol Teratol 29:562–569
22. Bruses JL (2011) N-cadherin regulates primary motor axon growth and branching during zebrafish embryonic development. J Comp Neurol 519:1797–1815
23. Kanungo J, Zheng YL, Amin ND, Kaur S, Ramchandran R, Pant HC (2009) Specific inhibition of cyclin-dependent kinase 5 activity induces motor neuron development in vivo. Biochem Biophys Res Commun 386:263–267
24. Raldua D, Andre M, Babin PJ (2008) Clofibrate and gemfibrozil induce an embryonic malabsorption syndrome in zebrafish. Toxicol Appl Pharmacol 228:301–314
25. Chen YH, Huang YH, Wen CC, Wang YH, Chen WL, Chen LC, Tsay HJ (2008) Movement disorder and neuromuscular change in zebrafish embryos after exposure to caffeine. Neurotoxicol Teratol 30:440–447
26. Cao P, Hanai J, Tanksale P, Imamura S, Sukhatme VP, Lecker SH (2009) Statin-induced muscle damage and atrogin-1 induction is the result of a geranylgeranylation defect. FASEB J 23:2844–2854
27. Stehr CM, Linbo TL, Incardona JP, Scholz NL (2006) The developmental neurotoxicity of fipronil: notochord degeneration and locomotor defects in zebrafish embryos and larvae. Toxicol Sci 92:270–278

28. Lambrechts D, Carmeliet P (2004) Genetics in zebrafish, mice, and humans to dissect congenital heart disease: insights in the role of VEGF. Curr. Topics in Develop. Biol. 62:189–224

29. Stainier DYR (2001) Zebrafish genetics and vertebrate heart formation. Nat Rev Genet 2:39–48

30. Cole P, Trichopoulos D, Pastides H, Starr T, Mandel JS (2003) Dioxin and cancer: a critical review. Regul Toxicol Pharmacol 38:378–388

31. Antkiewicz DS, Burns CG, Carney SA, Peterson RE, Heideman W (2005) Heart malformation is an early response to TCDD in embryonic zebrafish. Toxicol Sci 84:368–377

32. Antkiewicz DS, Peterson RE, Heideman W (2006) Blocking expression of AHR2 and ARNT1 in zebrafish larvae protects against cardiac toxicity of 2,3,7,8-tetrachlorodibenzo-p-dioxin. Toxicol Sci 94:175–182

33. Belair CD, Peterson RE, Heideman W (2001) Disruption of erythropoiesis by dioxin in the zebrafish. Dev Dyn 222:581–594

34. Bello SM, Heideman W, Peterson RE (2004) 2,3,7,8-Tetrachlorodibenzo-p-dioxin inhibits regression of the common cardinal vein in developing zebrafish. Toxicol Sci 78:258–266

35. Gonzalez FJ, Fernández-Salguero P (1998) The aryl hydrocarbon receptor: studies using the AHR-null mice. Drug Metab Dispos 26:1194–1198

36. Hankinson O (1995) The aryl hydrocarbon receptor complex. Annu Rev Pharmacol Toxicol 35:307–340

37. Nebert DW, Puga A, Vasiliou V (1993) Role of the Ah receptor and the dioxin-inducible [Ah] gene battery in toxicity, cancer, and signal transduction. Ann N Y Acad Sci 685:624–640

38. Shimizu Y, Nakatsuru Y, Ichinose M, Takahashi Y, Kume H, Mimura J, Fujii-Kuriyama Y, Ishikawa T (2000) Benzo[a]pyrene carcinogenicity is lost in mice lacking the aryl hydrocarbon receptor. Proc Natl Acad Sci USA 97:779–782

39. Prasch AL, Tanguay RL, Mehta V, Heideman W, Peterson RE (2006) Identification of zebrafish ARNT1 homologs: 2,3,7,8-tetrachlorodibenzo-p-dioxin toxicity in the developing zebrafish requires ARNT1. Mol Pharmacol 69:776–787

40. Prasch AL, Teraoka H, Carney SA, Dong W, Hiraga T, Stegeman JJ, Heideman W, Peterson RE (2003) Aryl hydrocarbon receptor 2 mediates 2,3,7,8-tetrachlorodibenzo-p-dioxin developmental toxicity in zebrafish. Toxicol Sci 76:138–150

41. Carney S, Prasch A, Heideman W, Peterson R (2006) Understanding dioxin developmental toxicity using the zebrafish model. Birth Defects Res A Clin Mol Teratol 76:7–18

42. Prasch AL, Heideman W, Peterson RE (2004) ARNT2 is not required for TCDD developmental toxicity in zebrafish. Toxicol Sci 82:250–258

43. Scott JA, Incardona JP, Pelkki K, Shepardson S, Hodson PV (2011) AhR2-mediated, CYP1A-independent cardiovascular toxicity in zebrafish (Danio rerio) embryos exposed to retene. Aquat Toxicol 101:165–174

44. Scholz S, Mayer I (2008) Molecular biomarkers of endocrine disruption in small model fish. Mol Cell Endocrinol 293:57–70

45. DeVito M, Biegel L, Brouwer A, Brown S, Brucker-Davis F, Cheek AO, Christensen R, Colborn T, Cooke P, Crissman J, Crofton K, Doerge D, Gray E, Hauser P, Hurley P, Kohn M, Lazar J, McMaster S, McClain M, McConnell E, Meier C, Miller R, Tietge J, Tyl R (1999) Screening methods for thyroid hormone disruptors. Environ Health Perspect 107:407–415

46. OECD (2006) Detailed review paper on thyroid hormone disruption assays. OECD series on testing and assessment. OECD, Paris

47. Brent GA, Braverman LE, Zoeller RT (2007) Thyroid health and the environment. Thyroid 17:807–809

48. Crofton KM (2008) Thyroid disrupting chemicals: mechanisms and mixtures. Int J Androl 31:209–223

49. Raldúa D, Babin PJ (2009) Simple, rapid zebrafish larva bioassay for assessing the potential of chemical pollutants and drugs to disrupt thyroid gland function. Environ Sci Technol 43:6844–6850

50. Alt B, Reibe S, Feitosa NM, Elsalini OA, Wendl T, Rohr KB (2006) Analysis of origin and growth of the thyroid gland in zebrafish. Dev Dyn 235:1872–1883

51. Brown DD (1997) The role of thyroid hormone in zebrafish and axolot development. Proc Natl Acad Sci USA 94:13011–13016

52. Walpita CN, Crawford AD, Janssens EDR, Van der Geyten S, Darras VM (2009) Type 2 iodothyronine deiodinase Is essential for thyroid hormone-dependent embryonic development and pigmentation in zebrafish. Endocrinology 150:530–539

53. Parichy DM, Tumer JM (2003) Zebrafish puma mutant decouples pigment pattern and somatic metamorphosis. Dev Biol 256:242–257

54. Allison WT, Dann SG, Veldhoen KM, Hawryshyn CW (2006) Degeneration and regeneration of ultraviolet cone photoreceptors during development in rainbow trout. J Comp Neurol 499:702–715

55. Benbassat J (1974) The transition from tadpole to frog haemoglobin during natural amphibian metamorphosis. I. Protein synthesis by peripheral blood cells in vitro. J Cell Sci 15:347–357

56. Opitz R, Braunbeck T, Bogi C, Pickford DB, Nentwig G, Oehlmann J, Tooi O, Lutz I, Kloas W (2005) Description and initial evaluation of a xenopus metamorphosis assay for detection of thyroid system-disrupting activities of environmental compounds. Environ Toxicol Chem 24:653–664

57. Solbakken JS, Norberg B, Watanabe K, Pittman K (1999) Thyroxine as a mediator of metamorphosis of Atlantic halibut, *Hippoglossus hippoglossus*. Environ Biol Fishes 56:53–65

58. Tata JR (2006) Amphibian metamorphosis as a model for the developmental actions of thyroid hormone. Mol Cell Endocrinol 246:10–20

59. Temple SE, Ramsden SD, Haimberger TJ, Veldhoen KM, Veldhoen NJ, Carter NL, Roth WM, Hawryshyn CW (2008) Effects of exogenous thyroid hormones on visual pigment composition in coho salmon (*Oncorhynchus kisutch*). J Exp Biol 211:2134–2143

60. Youson JH, Manzon RG, Peck BJ, Holmes JA (1997) Effects of exogenous thyroxine (T4) and triiodothyronine (T3) on spontaneous metamorphosis and serum T4 and T3 levels in immediately premetamorphic sea lampreys, *Petromyzon marinus*. J Exp Zool 279:145–155

61. Barros TP, Alderton WK, Reynolds HM, Roach AG, Berghmans S (2008) Zebrafish: an emerging technology for in vivo pharmacological assessment to identify potential safety liabilities in early drug discovery. Br J Pharmacol 154:1400–1413

62. Oberdörster G, Oberdörster E, Oberdörster J (2005) Nanotoxicology: an emerging discipline evolving from studies of ultrafine particles. Environ Health Perspect 113:823–839

63. Federici G, Shaw BJ, Handy RD (2007) Toxicity of titanium dioxide nanoparticles to rainbow trout (Oncorhynchus mykiss): Gill injury, oxidative stress, and other physiological effects. Aquat Toxicol 84:415–430

64. Johnson AC, Bowes MJ, Crossley A, Jarvie HP, Jurkschat K, Jurgens MD, Lawlor AJ, Park B, Rowland P, Spurgeon D, Svendsen C, Thompson IP, Barnes RJ, Williams RJ, Xu N (2011) An assessment of the fate, behaviour and environmental risk associated with sunscreen TiO(2) nanoparticles in UK field scenarios. Sci Total Environ 409:2503–2510

65. Perez S, Farre M, Barcelo D (2009) Analysis, behavior and ecotoxicity of carbon-based nanomaterials in the aquatic environment. Trends Analyt Chem 28:820–832

66. Hao LH, Wang ZY, Xing BS (2009) Effect of sub-acute exposure to TiO(2) nanoparticles on oxidative stress and histopathological changes in Juvenile Carp (Cyprinus carpio). J Environ Sci (China) 21:1459–1466

67. Zhu XS, Zhu L, Duan ZH, Qi RQ, Li Y, Lang YP (2008) Comparative toxicity of several metal oxide nanoparticle aqueous suspensions to zebrafish (Danio rerio) early developmental stage. J Environ Sci Health A Tox Hazard Subst Environ Eng 43:278–284

68. Ramsden CS, Smith TJ, Shaw BJ, Handy RD (2009) Dietary exposure to titanium dioxide nanoparticles in rainbow trout, (Oncorhynchus mykiss): no effect on growth, but subtle biochemical disturbances in the brain. Ecotoxicology 18:939–951

69. Hu YL, Gao JQ (2010) Potential neurotoxicity of nanoparticles. Int J Pharm 394:115–121

70. Long TC, Saleh N, Tilton RD, Lowry GV, Veronesi B (2006) Titanium dioxide (P25) produces reactive oxygen species in immortalized brain microglia (BV2): implications for nanoparticle neurotoxicity. Environ Sci Technol 40:4346–4352

71. Long TC, Tajuba J, Sama P, Saleh N, Swartz C, Parker J, Hester S, Lowry GV, Veronesi B (2007) Nanosize titanium dioxide stimulates reactive oxygen species in brain microglia and damages neurons in vitro. Environ Health Perspect 115:1631–1637

72. Handy RD, Henry TB, Scown TM, Johnston BD, Tyler CR (2008) Manufactured nanoparticles: their uptake and effects on fish-a mechanistic analysis. Ecotoxicology 17:396–409

73. Chen TH, Lin CY, Tseng MC (2011) Behavioral effects of titanium dioxide nanoparticles on larval zebrafish (Danio rerio). Mar Pollut Bull 63:303–308

74. Xiong DW, Fang T, Yu LP, Sima XF, Zhu WT (2011) Effects of nano-scale TiO(2), ZnO and their bulk counterparts on zebrafish: acute toxicity, oxidative stress and oxidative damage. Sci Total Environ 409:1444–1452

75. Jovanovic B, Ji TM, Palic D (2011) Gene expression of zebrafish embryos exposed to titanium dioxide nanoparticles and hydroxylated fullerenes. Ecotoxicol Environ Saf 74:1518–1525

Disrupting Effects of Single and Combined Emerging Pollutants on Thyroid Gland Function

Demetrio Raldúa, Patrick J. Babin, Carlos Barata, and Benedicte Thienpont

Abstract Inadequate thyroid hormone (TH) production in mothers during the first months of pregnancy can produce irreversible neurological effects in the offspring. Even though the main cause of insufficient synthesis of TH is the lack of iodine in the diet, TH insufficiency can also be caused by the presence of some naturally occurring and synthetic chemicals disrupting the thyroid gland function. The identification of emerging pollutants that may interfere with mammalian thyroid gland function is still in progress. The goal of this chapter is to review the potential of the zebrafish eleutheroembryos, a vertebrate model used in toxicology and by pharmaceutical companies in drug discovery, as a predictive model for screening emerging pollutants and drugs having a direct effect on the mammalian thyroid gland function.

Keywords Goitrogens, Thyroid follicle, Thyroid gland, Thyroid gland function disruptor, Thyroid hormone, Zebrafish

Contents

D. Raldúa (✉), C. Barata and B. Thienpont
Institute of Environmental Assessment and Water Research (IDÆA-CSIC), Jordi Girona 18, 08034 Barcelona, Spain
e-mail: drpqam@cid.csic.es; cbmqam@cid.csic.es; bthqam@cid.csic.es

P.J. Babin
University of Bordeaux, Maladies Rares: Génétique et Métabolisme (MRGM) EA4576, 33400 Talence, France
e-mail: p.babin@gpp.u-bordeaux1.fr

D. Barceló (ed.), *Emerging Organic Contaminants and Human Health*,
Hdb Env Chem (2012) 20: 415–434, DOI 10.1007/698_2011_123,
© Springer-Verlag Berlin Heidelberg 2011, Published online: 22 November 2011

Abbreviations

API	Average pixel intensity
BP2	Benzophenone 2
CA	Concentration-addition
CNS	Central nervous system
DIT	Diiodotyrosine
EGL	External granule layer
ETU	Ethylene thiourea
IA	Independent-action
ID	Iodine deficiency
IT4C	Intrafollicular T4 content
MIT	Monoiodotyrosine
MMI	Methimazole
MoA	Mode of action
NIS	Sodium-iodide symporter
NOEC	No observed effect concentration
PTUracil	6-Propyl-2-thiouracil
T3	$3,5,3'$-Triiodothyronine
T4	Thyroxine
TG	Thyroglobulin
TGFD	Thyroid gland function disruptor
TH	Thyroid hormones
TIQDT	Thyroxine immunofluorescence quantitative disruption test
TPO	Thyroperoxidase
TSH	Thyroid-stimulating hormone (thyrotropin)

1   Thyroid Hormone and Brain Development

The expression of several thyroid hormones (TH)-regulated genes in a strictly regulated spatiotemporal arrangement is essential during the development of the central nervous system (CNS). Changes in the expression pattern of these genes during early development may impair CNS maturation and organization, thereby affecting its function. Thus, THs play essential roles in neural proliferation, differentiation, migration, synaptogenesis, and myelination during human nervous system development [1]. The extent of impact of TH deficiency on the future neuropsychological development of the child depends on its timing and the brain area concerned. For example, in the developing cerebellum, TH is important in granule cell proliferation, differentiation, migration, and survival [2]. In hypothyroidal rats, granule cells in the external granule layer (EGL) remain in the proliferative phase longer than in controls, resulting in decreased cell differentiation [3], a delay in granule cell migration, and the persistence of the EGL. Moreover, there is also a drastic increase in the apoptosis of granule cells reaching the internal granule layer [4]. Hypothy-roidism during the first weeks of postnatal development impairs also the develop-ment of the Purkinje cells, inducing a reduction in the dendritic arborization and synaptogenesis between Purkinje cell dendritic spines and the parallel fibers of the granule cells in the molecular layer [3]. Similarly, maternal hypothyroidism or hypothyroxinemia may produce drastic alterations in the fetal cerebral cortex [1]. Since most of the neurogenesis and migration take place before the onset of the fetal thyroid gland function [5], maternal TH is essential for normal brain development of the fetus.

The role of the mother's TH status on the future neuropsychological develop-ment of the child has been emphasized in several studies in regions with a high prevalence of severe iodine deficiency (ID). It has been clearly demonstrated in these studies that maternal hypothyroxinemia early in pregnancy is not only the cause of neurological cretinism, but also of less severe mental deficits affecting a large proportion of the apparently "normal" population in those areas. These epidemiological studies clearly indicate these deficits and the cretin births, as irreversible consequences of IDs and can be prevented with an adequate supply of iodine during the first months of gestation [6]. On the other hand, several studies carried out in areas without severe ID have suggested that maternal thyroxine (T4) plays an important role in the neuropsychological development of the progeny [6].

2   Thyroid Gland Function and Disruption

2.1   Thyroid Gland Function

In humans, the thyroid gland is butterfly shaped endocrine gland located at the front of the neck. The thyroid gland is responsible for producing TH in all vertebrates. TH production takes place in the thyroid follicles, the functional units of the thyroid gland (Fig. 1). Each follicle consists of a single layer of thyroid epithelial cells adjacent to the follicular lumen. In the human thyroid gland, thyroid follicles are supported by connective tissue that forms a framework for the entire gland. The follicular lumen of the thyroid follicles is filled with a colloid, mainly consisting of thyroglobulin (TG), a large glycoprotein essential for TH synthesis [7]. In a first step, there is iodide uptake across the basolateral membrane of the thyroid follicular cells by the sodium-iodide symporter (NIS). This is driven by the sodium gradient created by sodium-potassium ATPase. The iodide is exported across the apical

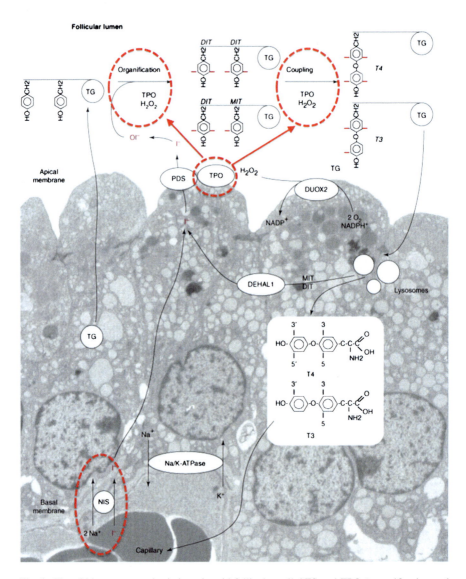

Fig. 1 Thyroid hormone synthesis in a thyroid follicular cell. NIS and TPO (organification and coupling reaction) have been marked in *red dashed line* as the two main targets for direct thyroid gland function disruptors. *DEHAL1* iodotyrosine dehalogenase 1, *DIT* diiodotyrosine, *DUOX2* dual oxidase 2, *MIT* monoiodotyrosine, *Na/K-ATPase* sodium-potassium ATPase, *NIS* sodium-iodide symporter, *PSD* pendrin, *TG* thyroglobulin, *TPO* thyroperoxidase. Reprinted from [7] with permission from Elsevier

membrane through partial mediation of pendrin [8]. Thyroperoxidase (TPO) is a transmembrane enzyme located at the apical membrane of the thyroid follicular cells. TPO, in the presence of hydrogen peroxide (H_2O_2), oxidizes iodide at the cell–colloid interface. H_2O_2 is generated by a calcium-dependent flavoprotein enzyme system that includes DUOX2, an NADPH oxidase [7]. TG acts as a matrix for the synthesis of T4 and 3,5,3'-triiodothyronine (T3) and is secreted into the follicular lumen. TG iodination is a highly regulated process, primarily controlled by H_2O_2 generation and iodide supply. First, TPO iodinates certain tyrosyl residues (iodination reaction) to form mono- and di-iodotyrosines (MIT, DIT) [9]. In a second step, two iodotyrosines are coupled to form T4 or T3 (coupling reaction), a reaction also catalyzed by TPO. TG typically consists of 0.5% iodine in individuals with adequate iodine supplies, distributed in five MIT, five DIT, 2.5 T4, and 0.7 T3 residues [10]. More iodine increases the DIT/MIT and T4/T3 ratios, while iodine insufficiency reduces them [10]. To liberate THs, TG is engulfed by pinocytosis, digested in lysosomes by cathepsin B, and then secreted into the bloodstream at the basolateral membrane.

2.2   Thyroid Gland Function Disruption

2.2.1   Endemic Goiter and Goitrogens

Endemic goiter is characterized by enlargement of the thyroid gland in a significantly large fraction of a population group, and is generally considered to be due to insufficient iodine in the daily diet. The thyroid glands are often diffusely enlarged in childhood, but are almost always nodular in adults [11]. Endemic goiter is the result of different adaptative processes of the thyroid gland aiming to achieve an adequate secretion of THs when iodine intake is abnormally low. These adaptive processes are triggered and maintained by increased secretion of thyroid-stimulating hormone (TSH). The morphological consequence of prolonged thyrotropic stimulation is thyroid hyperplasia. However, large colloid goiters in endemic ID represent maladaptation instead of adaptation to iodine deficiency. In these goiters, extremely distended vesicles filled with colloid with a flattened epithelium occupy the major part of the gland, and although the mechanism is not fully understood, it does not appear to be TSH hyperstimulation [11].

In 1990, it was estimated that 655 million people worldwide had endemic goiter [12]. Although a high percentage of the people suffering endemic goiter live in developing countries where ID is prevalent, there are also endemic goiter in some areas of highly developed countries with iodine prophylaxis [13]. Thus, although the main cause of severe thyroid gland dysfunction throughout the world is most certainly ID, a number of natural-occurring and synthetic chemicals also inhibit TH synthesis, even in iodine-sufficient regions [5, 14, 15]. These thyroid gland function disruptors (TGFD) can have a direct effect on thyroid follicles, by impairing iodide uptake at the NIS (e.g., perchlorate and thiocyanate), by inhibiting iodide organification by TPO (e.g., methimazole (MMI), 6-propyl-2-thiouracil (PTUracil), and flavonoids), or by inducing selective cytotoxic effects on the thyroid follicular cells (amiodarone and pyrazole) [16]. Other TGFDs, including dioxin-like compounds and polybrominated diphenyl ethers, impair the thyroid function via indirect mechanisms, essentially by boosting the T4 metabolism and activating the HPT axis in reaction to the induced hypothyroidal state [17].

2.2.2   Sources of Goitrogenic Compounds

Many common foods, such as cassava, sweet potatoes, corn, apricot, almonds, cauliflower, cabbage, broccoli and Brussels sprouts have very high levels of thiocyanate. Raw and cooked cauliflower can contain as much as 9,000 and 250 mg/kg thiocyanate, respectively [18]. Thiocyanate and goitrin are also present in milk from dairy cattle that graze on cruciferous plants. Endemic cretinism provides the best evidence of the risk for human brain development of a chemical that inhibits TH production. This condition, involving severely hindered physical and mental development, occurs in iodide-deficient areas where cassava represents a major food source and has been linked with the inhibitory effect of thiocyanate on iodide uptake [6]. Fruit and vegetables, such as soy, onion, apple, legumes, and grapes are rich in flavonoids. The daily intake of flavonoids in the Western world has been estimated at between 20 and 1,000 mg/d [19]. Moreover, flavonoids are not only present in the regular diet, but they are also ingested as dietary supplements. Some reports mention goiters induced in infants fed with soy formula, which were reversed by changing to cows' milk [20], and raised doubts concerning the impact of isoflavones on human thyroid function. Finally, residues of some fungicides and herbicides, such as mancozeb and amitrole, also hinder TH synthesis by inhibiting TPO [17]. In resumed, all these fresh foods are rich in thiocyanate and flavonoids and provoke the exposure of women of childbearing age daily to a wide range of potential thyroid disrupting compounds [21, 22].

There are also a large variety of TGFDs potentially present in drinking water. Thus, coal is a source of TGFDs as phenol, dihydrophenols (resorcinol), substituted dihydrobenzenes, thiocyanate, disulfides, phatalic acids, pyridines, and halogenated and polycyclic aromatic hydrocarbons [13]. Most of these compounds have been identified in drinking water from iodine-sufficient goitrous areas. Moreover, the antithyroid and goitrogenic effect of coal-water extracts form iodine-sufficient goiter areas have been demonstrated [15]. Perchlorate and humic substances (e.g., resorcinol and phloroglucinol) are also potent thyroid disruptors, potentially present in drinking water supply [13].

Cigarette smoking is the most ubiquitous source of exposure to thiocyanate [23], and it has been demonstrated that smoking during the period of breastfeeding increases the risk of ID-induced brain damage in the child [24]. Moreover, TG and thiocyanate concentrations at birth and at 1 year of age in infants of smoking parents are greater than in infants with non-smoking parents [25].

Finally, some drugs widely used including lithium, ethionamide, sulfonamides carbutamide, p-aminobenzoic acid, p-aminosalicylic acid, ketoconazole, or phenylbutazone have adverse side effects on the thyroid [26].

2.2.3 Combined Effect of Goitrogenic Compounds

Most TGFDs are present in the environment at concentrations far below their individuals EC50, possibly also below their individual no observed effect concentration (NOEC), yet still they may contribute to substantial effects. The relevance of joint action of this group of chemicals in foodstuff and drinking water has long been recognized [13, 15]. Estimating the potential hazards associated with exposures to mixtures of TGFDs in general, and direct-TGFDs in particular, is facilitated by empirical data derived from single chemical exposure and mathematical models. Concentration-addition (CA) and independent-action (IA) models are commonly used to predict the behavior of mixtures of chemicals with similar and dissimilar mode of actions (MoAs), respectively [27]. However, the number of studies testing the hypothesis of CA or IA with mixtures of TGFDs is still very scarce, and most of them have been focus on the indirect-TGFDs. Thus, exposition of rodents to a complex mixture of TGFDs affected production and clearance of T4 additively [28]. The contribution of TH synthesis inhibitors and TH clearance enhancers resulted in predictions that were closer to the empirical data than to the IA model alone [28]. When rats were exposed to indirect-TGFDs with dissimilar mechanisms of action, such as dioxins and polychlorinated biphenyls, the observed effect could be predicted by the CA model, although at the highest doses, there was a greater-than-additive effect [29].

3 Zebrafish: A Vertebrate Model for Screening Thyroid Gland Function Disruptors

3.1 Zebrafish Eleutheroembryo: Vertebrate Model in Biomedical and Toxicological Research

Zebrafish (*Danio rerio*) is a small tropical fish species native to the rivers of India and South Asia. It is a member of the minnow family (Cyprinidae). It has become one of the most popular model organisms in developmental genetics and (eco) toxicology [30, 31]. George Streisinger at the University of Oregon established its utility in the early 1970s. Zebrafish is now widely used as a model organism of vertebrate development and is receiving increasing attention as a toxicological model due to several innate advantages [32, 33].

There is a lot of research performed on zebrafish early development and large databases are available about genetics of the zebrafish. Zebrafish are small and inexpensive to maintain. A single pair of adults generates 100–200 offspring per brood, up to once a week. They develop ex utero and are semi-transparent during embryogenesis and early larval stages, facilitating in vivo observation of early developmental processes and organogenesis. Functional and morphological changes may be highlighted in vivo or in whole-mount fixed specimens by using transgenic lines, vital dyes, fluorescent tracers, antibodies and riboprobes, and fluorescent markers [33]. Zebrafish embryos grow rapidly, with the basic body plan laid out within 24 hours post-fertilization (hpf). At this stage, embryo length is around 1.9 mm, so several embryos fit easily inside a single well of a 96-well plate [34]. As larvae can live in as little as 50 µL water, so that only insignificant amounts of chemicals are needed for assays. This represents a major cost saving in screening entire molecule libraries [34]. The majority of organs, including the nervous system, cardiovascular system, intestines, liver, and kidneys can be studied at 5 days post-fertilization (dpf). Consequently, zebrafish represent a unique vertebrate model for high-throughput chemical screening, making them useful for toxicological evaluation and studying contamination at a toxicodynamic level [35].

Zebrafish are more and more used for assessing drug toxicity and safety and numerous studies have established that mammalian and zebrafish toxicity profiles are very similar [36]. Although in vitro assays are widely used in drug discovery and safety pharmacology, the results cannot always be extrapolated to in vivo effects. Thus, zebrafish are increasingly used in drug-screening assays as an intermediate step after cell-based evaluation to select drug candidates for conventional animal testing, hereby reducing the number and cost of mammalian studies [33].

Based on the view that the transition to external feeding is a critical step to independence, and that the stages before this step constitute an extension of embryonic development after hatching (the eleutheroembryo) [37], zebrafish assays during the initial endotrophic nutritional period (0–120 hpf) are considered non-animal based assays by Directive 86/609/EEC and 2010/63/EU. In the principle of 3Rs (reduce, refine, replace), it is important to have an animal model that meets these requirements to a maximum level and the zebrafish eleutheroembryo is for all the aforementioned reasons an ideal candidate to study toxicology.

3.2 Zebrafish Thyroid Gland

Klaus B. Rohr's lab has studied extensively the development and expansion of the zebrafish thyroid gland, and a recent review is available [38]. Therefore, we decide to only briefly summarize thyroid gland development and function in the zebrafish model, emphasizing its similarities with the higher vertebrates. While the morphology of thyroid gland in humans and zebrafish seems apparently very different, the morphology of the thyroid follicles is strictly similar. In zebrafish, thyroid follicles are loosely spread along the ventral midline of the pharyngeal mesenchyme, rather than encapsulated in connective tissue, as in higher vertebrates. In addition, some studies have suggested that the development of zebrafish thyroids is generally comparable to humans' thyroid gland [39]. The expression of key transcription factors, such as hhex, nk2.1a, pax2a, and pax8, defines the presumptive region where the primordium develops at about 24–28 hpf. These four transcription factors play similar late roles in the differentiation and growth of thyroid follicular cells. The zebrafish thyroid gland develops from precursor cells positioned in the endoderm just previous to pharynx formation. Two morphogenetic phases take place in thyroid relocalization during both zebrafish and mouse development [39]. After induction and evagination from the pharyngeal epithelium at around 32 hpf, the thyroid primordium adopts a position close to the cardiac outflow tract in zebrafish, with the first differentiated thyroid follicle appearing in this position at around 2.5 dpf. In the second phase of relocalization, dependent on the ventral aorta [40], the thyroid grows along the ventral pharyngeal midline. By 5 dpf, the second phase of thyroid relocalisation has produced follicles oriented longitudinally along the ventral aorta, from the second pharyngeal arch element, the ceratohyal, to ceratobranchial III.

The expression of genes encoding TG (*tg*) and the NIS (*slc5a5*) have been used as specific marker of differentiated thyroid follicular cells during the zebrafish thyroid gland organogenesis [39]. Initially, *tg* expression is only observed in the thyroid primordium, at around 32 hpf, and the TG protein is first detected at the end of the first relocalisation phase, in a single small follicle, at around 55 hpf. At around 40 hpf, *slc5a5* expression is initiated in the thyroid primordium. Recently, the expression pattern of other benchmark differentiation markers of thyroid follicular cells, such as genes encoding the TSH receptor (*tshr*), has been determined in zebrafish eleutheroembryos [41]. Thus, cloning of a *tshr* cDNA in zebrafish has shown conservation of primary structure and functional properties between zebrafish and mammalian TSHR. A faint *tshr* expression in thyroid primordium is detected from 40 hpf on, with also a weak extrathyroidal expression in lens and brain. Between 40 and 42 hpf, the first weak expression of *tpo* is detected in the thyroid primordium, and stronger staining is observed from 46 to 100 hpf. This expression is confined to developing thyroid [41].

Although it is well known that, in humans, *TG*, *SLC5A5*, and *TPO* expression are regulated by specific combinations of NKX2-1 (TTF-1), FOXE1 (TTF-2), and PAX8 [42], no data are currently available on this mechanism in zebrafish. Although knockdown of zebrafish *tshr* function by morpholino microinjection does not affect early thyroid morphogenesis, the functional differentiation is impaired, as indicated by reduction in number and size of functional follicles, downregulation of differentiation markers (*tg*, *tpo*, *slc5a5*), as well as reduced thyroid transcription factor expression (*nkx2.1a*, *pax2a*, *pax8* and *foxe1*) [41].

The first T4-producing thyroid follicle is located in an area corresponding to the first relocalization phase and is detected at around 72 hpf. From that time onward, a growing number of T4-positive thyroid follicles are detectable in the pharyngeal area, spread along the ventral aorta corresponding to the second relocalisation phase. It was recently demonstrated that genetic backgrounds or chemicals that hinder the second phase of relocalization did not affect the functionality of the first thyroid follicle [40, 43]. The yolk contains maternal TH, and this TH is required for many developmental processes [44]. The larval thyroid follicles only begins to produce T4 as the yolk sac diminishes in the free-swimming larvae and the supply of maternal TH is used up [45].

There is no information existing about the exact stage when NIS and TPO, the two main targets for direct TGFD, become functional in zebrafish follicular cells. The presence of T4 in the first thyroid follicle at about 72 hpf suggests that both mechanisms are likely to be functional by that time. Moreover, exposing 48 hpf larvae to prototypic inhibitors of both NIS, e.g., perchlorate, and TPO e.g., MMI and PTUracil, for 3 days, completely abolishes T4 immunofluorescence in the thyroid follicles [46]. All these data indicate that zebrafish eleutheroembryos provide a powerful vertebrate model for analyzing thyroid gland function and disruption.

3.3   Thyroxine Immunofluorescence Quantitative Disruption Test: A Predictive and Alternative System for Screening Chemicals Disrupting the Mammalian Thyroid Gland Function

The function of the thyroid gland is the synthesis of THs. Recently, the Organization for Economic Co-operation and Development proposed intra-thyroidal T4 content as a sensitive endpoint to be included, after development and validation, in screening programs for thyroid disrupting chemicals. The endpoint that we have selected to assess thyroid function on zebrafish eleutheroembryos is based on measuring intrafollicular T4-content (IT4C) by whole-mount T4-immunofluorescence [46, 47]. The recently developed "T4 immunofluorescence quantitative disruption test" (TIQDT) is a simple and fast test on zebrafish eleutheroembryos for assessing the potential of chemical pollutants and drugs to disrupt thyroid gland function [46], where the endpoint evaluated is IT4C. The 3-day exposure window protocol of TIQDT, from 2 to 5 dpf, avoids any potential side effects on thyroid gland morphogenesis. Moreover, because TIQDT use zebrafish eleutheroembryos, it is considered a non-animal based assay by Directive 86/609/EEC and 2010/63/EU.

The first step during the development of TIQDT was to check whether the design of this short-duration screening assay, produced detectable changes in the T4-immunofluorescence signal in thyroid follicles after exposure to TGFDs with different MoA: methimazole (MMI), 6-n-propyl-2-thiouracil (PTuracil), potassium perchlorate ($KClO_4$), amiodarone, and exogenous T3. MMI and PTuracil are goitrogens, which reduce TH synthesis by inhibiting TPO function. $KClO_4$, another goitrogen, inhibits iodide uptake through the follicular NIS. Amiodarone, an iodine-rich benzofuranic derivative, widely used to treat tachyarrhythmias, has cytotoxic effects on thyrocytes [48, 49]. Finally, administration of exogenous T3 induces thyroid toxicity via a secondary mechanism, the inhibition of $TSH\beta$ mRNA expression [50, 51]. All these chemicals induced a clear decrease in the T4 immunofluorescence (Fig. 2), although quantitative analysis was not performed at this step.

The next step in the development of the assay was to design a system for the quantitative analysis of whole-mount T4-immunofluorescence signals at the thyroid follicles (Fig. 3). High resolution TIFF images showing a ventral view of the heads of T4-immunostained 5 dpf larvae were analyzed using ImageJ (NIH) software. In the first step, background fluorescence was estimated by analyzing average pixel intensity (API) in image areas that did not contain any immunolabeled objects, i.e., the background threshold, and this background was calculated individually for each image.

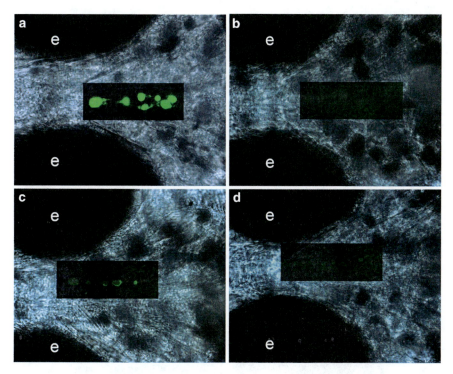

Fig. 2 Whole-mount T4 immunofluorescence staining superposed on brightfield illumination of thyroid follicles in (a) 5 dpf control (0.1% DMSO used as a vehicle control), (b) 1.5 mM MMI-treated larvae, (c) 1 μM amiodarone-treated larvae, and (d) 50 nM T3-treated larvae. Heads of representative larvae are shown in ventral view with the anterior part on the left. Abbreviation: *e* eye. Reprinted with permission from [46], Copyright 2009 American Chemical Society.

In order to calculate the background levels, the images were first converted to 8-bit grayscales, inverted, and thresholded. Thyroid follicles were then selected, using the "wand tool" function of the computer program. Selected areas were transferred to the original 4 three-channel image: red, green, and blue (RGB), the inverse selection was created and the API for the background was calculated. This background threshold was then removed by subtracting the background API from each pixel in the image. Selected follicle areas were transferred to the subtracted image and finally, the API of the selected areas was calculated.

After these initial steps, TIQDT was tested for its capacity to screen environmentally relevant compounds. 2,4-Dichlorophenoxyacetic acid (2,4-D), 1,1-bis(4-chlorophenyl)-2,2,2-trichloroethane (DDT), and 4-nonylphenol (4-NP) were selected as they are usually considered to be TGFDs [52–55]. Reports concerning the effect of methylmercury (MeHg) on thyroid function are contradictory [56–58]. Two environmentally relevant compounds with no reported effects on the thyroid system, fenoxicarb and atrazine [59, 60], were included as negative controls to assess the specificity of the assay. Our data strongly suggest that TIQDT may be

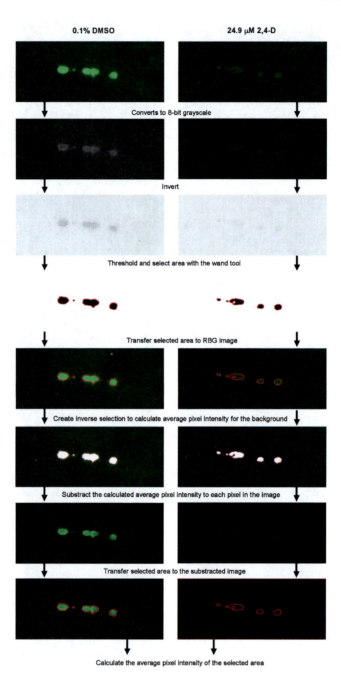

Fig. 3 Quantitative analysis of whole-mount T4-immunofluorescence signals for the Thyroxine Immunofluorescence Quantitative Disruption Test (TIQDT). The intrafollicular T4-content was calculated in the initial development of the TIQDT by measuring the average pixel intensity of the thyroid follicles. Reprinted with permission from [46],Copyright 2009 American Chemical Society.

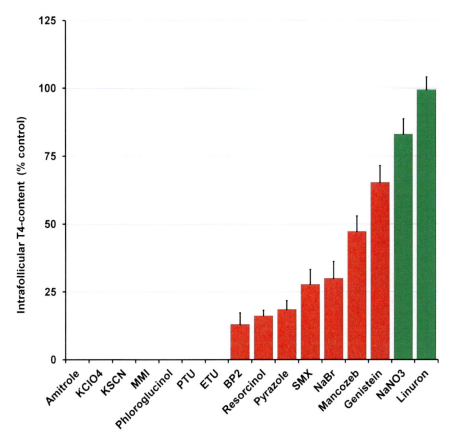

Fig. 4 Screening for drugs, environmental pollutants, and naturally occurring substances with a direct effect on the thyroid gland function, using intrafollicular T4-content (IT4C) in zebrafish eleutheroembryos as the endpoint. *Red bars* indicate "thyroid gland function disruptors" (TGFDs), with IT4C values significantly lower than control ($p < 0.05$); *Green bars* indicate compounds that were not considered "TGFD," as their IT4C values were not significantly different from control ($p > 0.05$). Data represent mean $\pm$ SE from at least two independent experiments ($n = 18$–24)

used to assess exposure to environmental pollutants and as an in vivo bioassay for screening chemicals and drugs [46]. These initial results encourage us to optimize the protocol and to start the pre-validation of the assay.

In spite of the promising results obtained by using TIQDT, we found also some problems to be fixed. The main detected problems were (1) the variability in the size and number of thyroid follicles in each animal, (2) the fact that some TGFDs could impair the size or the number of thyroid follicles, but not the concentration/API of T4 signal inside of the quantified follicles, (3) some autofluorescence may be found with the set of immunofluorescence filters used, and (4) clear signs of systemic toxicity were found in eleutheroembryos exposed to some chemicals at the maximum tolerated concentration. For an optimized TIQDT protocol, we increased the

Fig. 5 Heat-map
summarizing mammalian
versus zebrafish
eleutheroembryo responses
using 16 known or suspected
direct-TGFDs with different
modes of action. *Red*
indicates thyroid gland
function disrupting activity;
Green indicates that the
chemical is not a TGFD

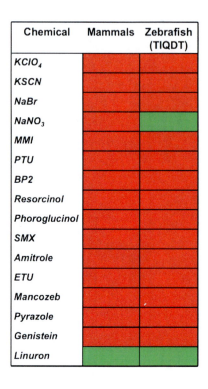

Chemical	Mammals	Zebrafish (TIQDT)
KClO₄		
KSCN		
NaBr		
NaNO₃		
MMI		
PTU		
BP2		
Resorcinol		
Phoroglucinol		
SMX		
Amitrole		
ETU		
Mancozeb		
Pyrazole		
Genistein		
Linuron		

number of individual of the different experimental groups, analyzing at least 18 eleutheroembryos per group, in order to decrease the intra-group variability. Moreover, IT4C was calculated from the "integrated density," i.e., the product of "area" and "API" instead of API alone, of the fluorescence in the thyroid follicles. A decrease in the autofluorescence of the embryos was found using a secondary antibody conjugates to Alexa 555 instead of Alexa 488. Finally, systemic toxicity, evaluated at the phenotypic and behavioral levels, was evaluated in all the chemicals inducing a significant decrease in IT4C. Only those compounds positive in TIQDT at concentrations below systemic toxicity were considered TGFDs.

The next step was to analyze the concordance of zebrafish versus mammalian assays to assess the potential utility of the optimized TIQDT as a screening assay for thyroid disruptors in an endocrine disruptor screening and testing program [47]. IT4C in eleutheroembryos exposed to substances with a direct effect on the thyroid gland function, e.g., amitrole, KClO₄, KSCN, MMI, phloroglucinol, PTUracil, ethylene thiourea (ETU), benzophenone 2 (BP2), resorcinol, pyrazole, sulfamethoxazole, pyrazole, sodium bromide, mancozeb, and genistein, was significantly lower than in the controls ($p < 0.05$); therefore, those chemicals were classified as direct-TGFDs on zebrafish eleutheroembryos (Fig. 4). IT4 levels in eleutheroembryos exposed to nitrate and linuron were similar to controls ($p > 0.05$), so these compounds were classified as "non-TGFDs" (Fig. 4). The concordance analysis (Fig. 5) showed a high degree of concordance with the mammalian data (15/16, 93.75%).

We have also analyzed the suitability of TIQDT for assessing the thyroid disrupting potency and hazard of selected direct-TGFDs (BP2, resorcinol, PTUracil, MMI and phloroglucinol). While thyroid disrupting potency describes the concentration range over which thyroid gland function is impaired, thyroid disrupting "hazard" is our term for describing the relationship between effective concentrations and systemic toxicity. To obtain accurate concentration-response curves, 5–8 different concentrations were used for each test compound. EC10 and EC50 were the parameters selected to describe thyroid disrupting potency and the thyroid disrupting index (TDI: LC50/EC50) was used as a descriptor of thyroid disrupting hazard. Our results show BP2 exhibited the highest thyroid disrupting potency but also the lowest thyroid disrupting hazard, because its thyroid disrupting activity was present only at concentrations very close to lethality.

Finally, we recently found that TIQDT is also suited for estimating the potential hazards associated with mixtures of TGFDs [Thienpont et al., manuscript in preparation]. We have found that mixtures of eight compounds with both similar MoA act according to the concentration-addition model (Fig. 6). Moreover, a mixture of these eight compounds at their NOEC inhibited almost completely

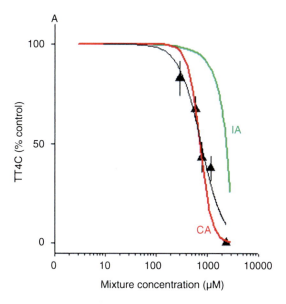

Fig. 6 Mixture toxicity analyses of eight putative thyroid peroxidase inhibitors (methimazole, 6-propyl-2-thiouracil, benzophenone 2, resorcinol, amitrole, phloroglucinol, ethylenethiourea, and sulfamethoxazol) in zebrafish prefeeding larvae using the TIQDT assay. Chemical mixtures were dosed using a fixed ratio design proportional to the EC10 of mixture constituents. Symbols and error bars corresponds to the mean and SE values ($n = 18$). Observed (*black lines*) and predicted joint toxicity effects according to the concentration-addition (*CA, red lines*) and independent-action (*IA, green lines*) concepts are indicated

IT4C, which emphasizes the risk for the thyroid gland function of the mixture of chemicals with thyroid disrupting activity.

In summary, our data provide evidence for the suitability of zebrafish eleutheroembryos as a predictive vertebrate model for evaluating the effect of individual chemicals and mixtures on thyroid gland function. TIQDT performed on zebrafish eleutheroembryos is an alternative whole-organism screening assay that provides relevant information for environmental and human risk assessments.

Acknowledgments This work has been supported by the Spanish Ministry of Science and Innovation (CTM2011-30471-C02-01 and PA1002979) and the Generalitat de Catalunya (2010 BE1 00623) to D.R. and C.B. B.T. was supported by a fellowship of the Spanish Government (AP2006-01324). This work was supported by a Conseil Régional d'Aquitaine grant (200881301031/TOD project) to P.J.B.

References

1. Berbel P, Bernal J (2010) Hypthyroxinemia: a subclinical condition affecting neurodevelopment. Exp Rev Endocrinol Metab 5:563–575
2. Zoeller RT, Tyl RW, Tan SW (2007) Current and potential rodent screens and tests for thyroid toxicants. Crit Rev Toxicol 37:55–95
3. Nicholson JL, Altman J (1972) The effects of early hypo- and hyperthyroidism on the development of the rat cerebellar cortex. II. Synaptogenesis in the molecular layer. Brain Res 44:25–36
4. Xiao Q, Nikodem VM (1998) Apoptosis in the developing cerebellum of the thyroid hormone deficient rat. Front Biosci 3:A52–A57
5. Brown V (2003) Disrupting a delicate balance: environmental effects on the thyroid. Environ Health Perspect 111:A642–A649
6. de Escobar G, Obregon M, del Rey F (2000) Is neuropsychological development related to maternal hypothyroidism or to maternal hypothyroxinemia? J Clin Endocrinol Metab 85:3975–3987
7. Kopp P, Pesce L, Solis-S JC (2008) Pendred syndrome and iodide transport in the thyroid. Trends Endocrinol Metab 19:260–268
8. Bizhanova A, Kopp P (2009) The sodium-iodide symporter NIS and pendrin in iodide homeostasis of the thyroid. Endocrinology 150:1084–1090
9. Dedieu A, Gaillard JC, Pourcher T, Darrouzet E, Armengaud J (2011) Revisiting iodination sites in thyroglobulin with an organ-oriented shotgun strategy. J Biol Chem 286:259–269
10. Alvino CG, Acquaviva AM, Memoli Catanzano AM, Tassi V (1995) Evidence that thyroglobulin has an associated protein kinase activity correlated with the presence of an adenosine triphosphate binding site. Endocrinology 136:3179–3185
11. Eastman CJ, Zimmermann MB (2009) The iodine deficiency disorders. Thyroidmanager.org. http://www.thyroidmanager.org/Chapter20/20-frame.htm
12. WHO/UNICEF/ICCIDD (1991) Global prevalence of iodine deficiency disorders. Micronutrient Deficiency Information System Working Paper No. 1. Geneva, WHO
13. Gaitan E (1990) Goitrogens in food and water. Annu Rev Nutr 10:21–39
14. Brucker-Davis F (1993) Effects of environmental synthetic chemicals on thyroid function. Thyroid 8:827–856

15. Gaitan E, Cooksey RC, Legan J, Cruse JM, Lindsay RH, Hill J (1993) Antithyroid and goitrogenic effects of coal-water extracts from iodine-sufficient goiter areas. Thyroid 3:49–53
16. Cunha GCP (2005) Evaluation of mechanisms inducing thyroid toxicity and the ability of the enhanced OECD Test Guideline 407 to detect these changes. Arch Toxicol 79:390–405
17. Mastorakos G, Karoutsou EI, Mizamtsidi M, Creatsas G (2007) The menace of endocrine disruptors on thyroid hormone physiology and their impact on intrauterine development. Endocrine 31:219–237
18. De Groef B, Decallonne BR, Van der Geyten S, Darras VM, Bouillon R (2006) Perchlorate versus other environmental sodium/iodide symporter inhibitors: potential thyroid-related health effects. Eur J Endocrinol 155:17–25
19. Giuliani C, Noguchi Y, Harii N, Napolitano G, Tatone D, Bucci I et al (2008) The flavonoid quercetin regulates growth and gene expression in rat FRTL-5 thyroid cells. Endocrinology 149:84–92
20. Doerge DR, Sheehan DM (2002) Goitrogenic and estrogenic activity of soy isoflavones. Environ Health Perspect 110:349–353
21. Steinmaus C, Miller M, Howd R (2007) Impact of smoking and thiocyanate on perchlorate and thyroid hormone associations in the 2001–2002 national health and nutrition examination survey. Environ Health Perspect 115:1333–1338
22. Vanderver GB (2007) Cigarette smoking and iodine as hypothyroxinemic stressors in US women of childbearing age: a NHANES III analysis. Thyroid 17:741–746
23. Zimmermann MB (2009) Iodine deficiency. Endocr Rev 30:376–408
24. Laurberg P, Nøhr SB, Pedersen KM, Fuglsang E (2004) Iodine nutrition in breast-fed infants is impaired by maternal smoking. J Clin Endocrinol Metab 89:181–187
25. Gasparoni A, Autelli M, Ravagni-Probizer MF, Bartoli A, Regazzi-Bonora M, Chirico G, Rondini G (1998) Effect of passive smoking on thyroid function in infants. Eur J Endocrinol 138:379–382
26. Sarne D (2010) Effects of the environment, chemicals and drugs on thyroid function. Thyroidmanager.org. http://www.thyroidmanager.org/Chapter5/5a-frame.htm
27. Crofton K (2008) Thyroid disrupting chemicals: mechanisms and mixtures. Int J Androl 31:209–223
28. Flippin JL, Hedge JM, DeVito MJ, LeBlanc GA, Crofton KM (2009) Predictive modeling of a mixture of thyroid hormone disrupting chemicals that affect production and clearance of thyroxine. Int J Toxicol 28:368–381
29. Crofton KM, Craft ES, Hedge JM, Gennings C, Simmons JE, Carchman RA, Carter WH Jr, DeVito MJ (2005) Thyroid-hormone-disrupting chemicals: evidence for dose-dependent additivity or synergism. Environ Health Perspect 113:1549–1554
30. Parng C, Seng WL, Semino C, McGrath P (2002) Zebrafish: a preclinical model for drug screening. Assay Drug Dev Technol 1:41–48
31. Ton C, Lin Y, Willett C (2006) Zebrafish as a model for developmental neurotoxicity testing. Birth Defects Res A Clin Mol Teratol 76:553–567
32. Love DR, Pichler FB, Dodd A, Copp BR, Greenwood DR (2004) Technology for high-throughput screens: the present and future using zebrafish. Curr Opin Biotechnol 15:564–571
33. McGrath P, Li CQ (2008) Zebrafish: a predictive model for assessing drug-induced toxicity. Drug Discov Today 13:394–401
34. Goldsmith P (2004) Zebrafish as a pharmacological tool: the how, why and when. Curr Opin Pharmacol 4:504–512
35. Zon LI, Peterson RT (2005) In vivo drug discovery in the zebrafish. Nat Rev Drug Discov 4:35–44
36. Berghmans S, Butler P, Goldsmith P, Waldron G, Gardner I, Golder Z et al (2008) Zebrafish based assays for the assessment of cardiac, visual and gut function-potential safety screens for early drug discovery. J Pharmacol Toxicol Methods 58:59–68
37. Strähle U, Schloz S, Geisler R, Greiner P, Hollert H, Rastegar S et al (2011) Zebrafish embryos as an alternative to animal experiments – a commentary on the definition of the onset of protected life stages in animal welfare regulations. Reprod Toxicol. doi:10.1016/j.reprotox.2011.06.121

38. Porazzi P, Calebiro D, Benato F, Tiso N, Persani L (2009) Thyroid gland development and function in the zebrafish model. Mol Cell Endocrinol 312:14–23

39. Alt B, Reibe S, Feitosa N, Elsalani O, Wendl T, Rohr KB (2006) Analysis of origin and growth of the thyroid gland in zebrafish. Dev Dyn 235:1872–1883

40. Alt B, Elsalani O, Schrumpf P, Haufs N, Lawson N, Schwabe G et al (2006) Arteries define the position of the thyroid gland during its developmental relocalisation. Development 133:3797–3804

41. Opitz R, Maquet E, Zoenen M, Dadhich R, Costagliola S (2011) TSH receptor function is required for normal thyroid differentiation in zebrafish. Mol Endocrinol 25:1579–1599

42. Dohan O, De la Vieja A, Carrasco N (2000) Molecular study of the sodium-iodide symporter (NIS): a new field in thyroidology. Trends Endocrinol Metab 11:99–105

43. Raldúa D, Andrè M, Babin PJ (2008) Clofibrate and gemfibrozil induce an embryonic malabsorption syndrome in zebrafish. Toxicol Appl Pharmacol 228:301–314

44. Walpita CN, Van der Geyten S, Rurangwa E, Darras VM (2007) The effect of 3,5,3'-triiodothyronine supplementation on zebrafish (Danio rerio) embryonic development and expression of iodothyronine deiodinases and thyroid hormone receptors. Gen Comp Endocrinol 152:206–214

45. Wendl T, Lun K, Mione M, Favor J, Brand M, Wilson SW et al (2002) Pax2.1 is required for development of thyroid follicles in zebrafish. Development 129:3751–3760

46. Raldúa D, Babin PJ (2009) Simple, rapid zebrafish larva bioassay for assessing the potential of chemical pollutants and drugs to disrupt thyroid gland function. Environ Sci Technol 43:6844–6850

47. Thienpont B, Tingaud-Sequeira A, Prats E, Barata C, Babin PJ, Raldúa D (2011) Zebrafish eleutheroembryos provide a suitable vertebrate model for screening chemicals that impair thyroid hormone synthesis. Environ Sci Technol 45:7525–7532

48. Pitsiavas V, Semerdely P, Li M, Boyages SC (1997) Amiodarone induces a different pattern of ultrastructural change in the thyroid to iodine excess alone in both BB/W rat and the Wistar rat. Eur J Endocrinol 137:89–98

49. Martino E, Bartalena L, Bogazzi F, Braverman LE (2001) The effects of amiodarone on the thyroid. Endocr Rev 22:240–254

50. Cunha GC, van Ravenzwaay B (2005) Evaluation of mechanisms inducing thyroid toxicity and the ability of the enhanced OECD Test Guideline 407 to detect these changes. Arch Toxicol 79:390–405

51. MacKenzie DS, Jones RA, Miller TC (2009) Thyrotropin in teleost fish. Gen Comp Endocrinol 161:83–89

52. Charles JM, Cunny HC, Wilson RD, Bus JS (1996) Comparative studies on 2,4- dichlorophenoxyacetic acid, amine and ester in rats. Fundam Appl Toxicol 33:161–165

53. Santini F, Vitti P, Mammoli C, Rosellini V, Pelosini C, Marsili A et al (2003) In vitro assay of thyroid disruptors affecting TSH-stimulated adenylate cyclase activity. J Endocrinol Invest 26:950–955

54. Rossi M, Dimida A, Dell'anno MT, Trincavelli ML, Agretti P, Giorgi F et al (2007) The thyroid disruptor 1,1,1-trichloro-2,2-bis(p-chloropheyl)ethane appears to be an uncompetitive inverse agonist for the thyrotropin receptor. J Pharmacol Exp Ther 320:465–474

55. Pandey AK, George KC, Mohamed MP (1995) Effect of DDT on the thyroid gland of the mullet Liza parsia (Hamilton-Buchanan). J Mar Biol Assoc India 37:287–290

56. Bleau H, Daniel C, Chevalier G, van Tra H, Hontela A (1996) Effects of acute exposure to mercuric chloride and methylmercury on plasma cortisol, T3, T4, glucose, and liver glycogen in rainbow trout (Oncorhynchus mykiss). Aquat Toxicol 34:221–235

57. Kirubagaran R, Joy KP (1994) Effects of short-term exposure to methylmercury chloride and its withdrawal on serum levels of thyroid hormones in the catfish Clarias batrachus (L.). Bull Environ Contam Toxicol 63:166–170

58. Nishida M, Muraoka K, Nishikawa K, Takagi T, Kawada J (1989) Differential effects of methylmercuric chloride and mercuric chloride on the histochemistry of rat thyroid peroxidase and thyroid peroxidase activity of isolated pig thyroid cells. J Histochem Cytochem 37:723–727
59. Stoker TE, Guidici DL, Cooper RL (2002) The effects of atrazine metabolites on puberty and thyroid function in male Wistar rat. Toxicol Sci 67:198–206
60. Flatt T, Moroz LL, Tatar M, Heyland A (2006) Comparing thyroid and insect hormone signalling. Integr Comp Biol 46:777–794

Psychoactive Substances in Airborne Particles in the Urban Environment

M. Viana, C. Postigo, C. Balducci, A. Cecinato, M. J. López de Alda, D. Barceló, B. Artíñano, P. López-Mahía, A. Alastuey, and X. Querol

Abstract The Earth's atmosphere is affected by the presence of psychotropic chemicals, both licit and illicit substances, not only in major city centres but also in suburban and rural regions. Dedicated analytical procedures, most of them based on gas or liquid chromatography coupled to mass spectrometry, have been optimised for the detection of these substances. Nicotine and caffeine (licit substances), are widespread in the world at concentrations sometimes reaching 100 ng/m^3. Conversely, drugs of abuse (namely cocaine, cannabinoids, heroin and amphetamines, which are in most countries illicit) rarely exceed 1 ng/m^3 each. However, their presence in airborne particles is virtually ubiquitous in agreement with what was observed in the past for surface and waste waters. The spatial and temporal variability of psychotropic substances in the atmosphere has been an object of study in different types of urban areas, whereas data are scarcer for rural areas. In the current ambient concentrations, personal exposure to airborne drugs of abuse may be considered negligible, posing no harm to human health. The possibility of drawing abuse prevalence indicators from the drug contents in the air merits, however, to be explored.

M. Viana (✉), C. Postigo, MJ. López de Alda, D. Barceló, A. Alastuey, and X. Querol
IDAEA-CSIC, Institute for Environmental Assessment and Water Research, C/Jordi Girona 18, 08034 Barcelona, Spain
e-mail: mar.viana@idaea.csic.es

C. Balducci and A. Cecinato
Istituto Inquinamento Atmosferico CNR, Via Salaria km 29.3, I-00015 Monterotondo Stazione RM, Italy

B. Artíñano
CIEMAT, Centre for Energy, Environment and Technology Research, Av. Complutense 40, 28040 Madrid, Spain

P. López-Mahía
Department of Analytical Chemistry, University of A Coruña, Campus A Zapateira, 15071 A Coruña, Spain

D. Barceló (ed.), *Emerging Organic Contaminants and Human Health*,
Hdb Env Chem (2012) 20: 435–460, DOI 10.1007/698_2011_135,
© Springer-Verlag Berlin Heidelberg 2012, Published online: 2 March 2012

Keywords Air quality, Drugs of abuse, Human health, Psychotropic substances

Contents

1   Introduction

The occurrence of psychotropic substances in the air of cities is nowadays ascertained. Indeed, not only nicotine, caffeine, and the corresponding by-products, whose existence could be foreseen due to the wide use by population, but also cocaine, cannabinoids, heroin, and the so-called smart drugs have been found to affect our environment.

As illicit psychotropic substances were detected in the atmosphere, their real presence and environmental relevance were immediately questioned. Are cocaine and cannabis markers really ubiquitous? Are the amounts detected representative of total contents of drugs affecting the environment? Are their corresponding burdens enough to justify any health effect on population? And besides, what is/are the predominant source(s) of atmospheric drugs? How long might they survive in the environment? What is, if any, the relationship between the drug abuse prevalence and the corresponding concentration in the air? What is, if any, the relationship between the illicit drug contents in the air and in surface/waste waters?

To find answers to these crucial questions and to establish if the occurrence of psychotropic substances in the atmospheric aerosols is just a curiosity or rather a potential problem for the community, a series of investigations have been carried out both in the laboratory and in the field. Dedicated procedures have been optimised, for instance, for cocaine and cannabinoids (see sections below), and the chemical stability of cocaine in airborne particulates and its partition between gas and aerosol phases were estimated, as well as its accumulation in fine rather than coarse particles. Furthermore, cocaine and cannabinoids concentrations have been measured in several cities over the world through field studies. After the first detection of cocaine in ambient air by Hannigan et al. [1] in Los Angeles, measurements were performed more extensively in Italy (for instance over 10 consecutive months in downtown Rome, or in 38 Italian localities) and Spain (where analysis covered also other drugs such as opioids and amphetaminic compounds, and assessed their spatial–temporal variability on a city scale). In conclusion, relatively detailed investigations have been conducted in a limited number of countries (mainly Italy and Spain), but by contrast a lack of information remains in other regions of the world, since scarce measurements have only been carried out in Algiers, Brazil, Chile, Mexico, Portugal, Serbia, and the USA.

2 Psychotropic Substances: Definition

A number of different terms may be found in the literature to refer to these kinds of substances: psychoactive drugs, drugs of abuse, illicit drugs, illicit substances, recreational drugs, or simply, drugs. By definition, psychoactive substances are chemicals which cross the blood–brain barrier and act primarily upon the central nervous system, where they affect brain function, resulting in changes in perception, mood, consciousness, cognition, and behaviour [2]. These substances may be used recreationally, to purposefully alter one's consciousness, for ritual and/or spiritual purposes, as a tool for studying or augmenting the mind, or therapeutically as medication. Thus, some substances may have uncontrolled or illegal uses while others may have medicinal use. The absence of a single term to define these substances stems mainly from the ambiguity linked to their definition, given that psychotropic substances refer to both legal and illegal substances. Legal substances and legal uses may be social drinking or sleep aids, as well as nicotine and caffeine. Caffeine is the world's most widely consumed psychoactive substance, being legal and unregulated in nearly all jurisdictions.

According to national statistics [3, 4], tobacco smokers account for ca. 23% of the Italian population and consume 13 cigarettes per day. Percentages show small variations along the peninsula and over the last 5 years. These data indicate that both policy [5–8] and general knowledge of dangerous effects of tobacco smoking on health have drastically reduced the amount of consumers with regards to previous decades. Nevertheless, a hard core of smokers remains; in particular, both (very) young and female people are attracted by tobacco, heedless of risk. By consequence, whilst other potential causes of morbidity and death are declining, the clinical statistics associated with smoking are gaining importance. Nicotine was discovered in 1826 as a tobacco constituent [9]. The psychoactive effects of tobacco were soon associated with nicotine [10, 11]. Nevertheless, formerly this substance was not regarded as addictive, as it happened only at the end of the 1900s [12–15]. Nicotine was found poisonous and toxic [16] but not directly responsible for lung cancer promotion, so it was still admitted as surrogate of tobacco smoking (nicotine gums, patches and inhalers). Most researches focussed on nicotine belong to physiology and pharmacology [17]. Relatively few articles deal with nicotine occurrence in the air; in those articles, the composition of secondary tobacco smoke [18], the partition between gas and particulate phases [19] depending on soot acidity, the concentrations detected indoors [20–26] and the by-products of nicotine [27, 28] are discussed.

Illicit drugs, on the other hand, are defined as substances that not at all or rarely find medical applications and are consumed in the absence of medical supervision. Their use is thus mainly recreational, for their mood and perception altering effects. The term "illegal" is in itself not fully objective, given that it depends on the legal boundaries set by each country or State (see the case of The Netherlands). The legality of psychoactive drugs has been controversial through most of recent history; the Second Opium War (when opium trade was legalised) and the Prohibition era in the United States (where alcohol was made illegal for 13 years) are two historical examples of legal controversy surrounding psychoactive drugs. However, in recent years, the most important document regarding the legality of psychoactive drugs is the Single Convention on Narcotic Drugs, an international treaty signed in 1961 as an Act of the United Nations, signed by 73 nations including the United States, the USSR, India, and the United Kingdom. All countries that signed the treaty passed laws to implement these rules within their borders [29]. However, the chemical composition or the active ingredients are not the only factors defining an illicit substance but also the way in which they are manufactured, labelled, distributed, acquired or even used [30]. This means that the definition of illicit drugs may include legal pharmaceuticals, which are illegally manufactured, distributed or acquired. Adulterated legal pharmaceuticals are also considered illicit drugs.

Finally, aside from strictly illegal substances and their metabolites, psychoactive substances or drugs of abuse may also include active ingredients from pharmaceutical products with legal therapeutic uses such as morphine.

In order to limit this ambiguity, the terms psychoactive substances (referring to licit and illicit substances, e.g., caffeine and nicotine but also cannabis and heroin) and drugs of abuse (referring only to recreational drugs, e.g., cocaine, cannabis, etc.) will be used throughout the present work.

3   Monitoring Methods of Psychoactive Substances in Ambient Air

Nicotine is by far the most extensively studied among psychotropic substances. Although its occurrence in the environment is not a novelty, nonetheless, investigations are overall restricted to indoor or artificial environments. Thanks to its basic properties, nicotine is usually collected from air by using nonvolatile acid reagents (e.g., sulphuric or benzenesulphonic acid) loaded on the surfaces of denuder devices or filters for the gaseous and particulate fractions, respectively; alternatively, vapours are trapped into adsorbent resins (XAD-4) [19, 31, 32]. Due to the recognised dangerous effects of tobacco exhausts, which hit even non-smokers, detailed research has been carried out to abate soot, nitrogen dioxide and tar, to control the nicotine release into the environment, to identify the soot components including polycyclic aromatic hydrocarbons (PAHs) and to understand the phase distribution of emissions as a function of smoke concentration, temperature and cigarette pH [33, 34]. Besides that, both educational campaigns and legislation have been implemented to reduce the health impact of tobacco smoking, which continues to be among the principal causes of tumours and deaths, reaching 5.4 million people in 2005 [8, 35, 36]. On the other hand, it is worthwhile to remark that very few measurements have been made till now in the open air, despite the fact that particulate nicotine can exceed 100 ng/m^3 in the open air (aerosol phase), and 100 µg/m^3 in public indoor places (gas phase) [37–41].

Whilst nicotine, caffeine, amphetamine and methamphetamine are expected to partition between gas and particulate phases, most drugs of abuse (i.e., cocaine, cannabinoids, heroin and their respective by-products) are high boiling. Whenever an estimate for nicotine and caffeine is provided limited to the particulate fraction, results seem to be affected by higher uncertainty than those drawn for illicit compounds [37]. Due to their physicochemical properties (low vapour pressure, high and medium polarity, weak alkalinity and molecular weight range between 135 and 360 g/mol), these substances are associated primarily with particulates in the atmosphere [42–44]. As for the grain size distribution of drugs of abuse in particles, different grain size fractions were targeted by different studies available in the literature, mainly PM_{10}, $PM_{2.5}$ and $PM_{2.5-10}$. A comparison of the cocaine and benzoylecgonine contents in simultaneous PM_{10} and $PM_{2.5}$ samples [45] evidenced that the grain size distribution of these substances is predominantly fine (>99% in $PM_{2.5}$, within the uncertainty of the determinations). In a study by Cecinato et al. [46], the following percentages of psychoactive substances in fine ($PM_{2.5}$) airborne particulates with respect to PM_{10} were found: 90–95% for cocaine and cannabinol, 75–80% for THC, and 80–100% for cannabidiol. Thus, similarly to PAHs, psychotropic substances accumulate in fine particles. Thus, after PM_{10} or $PM_{2.5}$ samples are collected in urban air quality monitoring networks, psychoactive substance determination may be carried out concurrently to that of PAHs. In this way, not only costs are reduced but also links between drugs and soot may be assessed, as well as between drugs and other anthropogenic pollution sources (see sections below).

Sampling of these substances has been carried out following three approaches: liquid absorbents [47], solid-phase microextraction (SPME) fibres [43] and filter substrates (mostly quartz fibre filters but also PTFE membranes [1, 42, 48, 49]). When filter substrates are used, atmospheric particles are collected over 24-h periods using high-volume (dichotomous or single-filter instruments [1, 48]), medium-volume or low-volume samplers (operated to ensure collection of sufficient aerosol mass [37, 50]). Samples were always stored at low temperatures (refrigerated or frozen) to ensure sample preservation.

After sample collection, drugs of abuse concentrations may be determined chemically by means of different methodologies, which target at times common but mostly different groups of substances. A brief description of the methodologies is provided below:

– Liquid chromatography coupled with electrochemical detection: this technique, in combination with a high-throughput liquid-absorption pre-concentrator (HTLAP) that sampled air and collected analytes from vapours or aerosols into a small volume of liquid absorbent, was used by Zaromb et al. [47] for the detection and analysis of cocaine and heroin in air.

– Gas chromatography coupled to mass spectrometry (GC-MS [1]): particulate matter filters were Soxhlet extracted with dichloromethane (DCM), fractionated by single size exclusion chromatography (SEC) followed by normal-phase liquid chromatography (NP-LC) with a cyanopropyl (CN) column, and reduced in volume under nitrogen. Analysis of the aromatic polar sub-fractions (where cocaine was present) was then performed by GC-MS in full scan mode.

– High-resolution gas chromatography coupled to mass spectrometry (HRGC-MS [42]): analysis of cocaine in ambient aerosols required a multi-step procedure comprising Soxhlet extraction of the organic matter with a mixture of dichloromethane:acetone (80:20), solvent reduction through Kuderna-Danish distillation, redissolution of the extract in isooctane:dichloromethane (9:1), fractionation by column chromatography through basic alumina, solvent evaporation under nitrogen and redissolution in toluene. Analysis of the high-polarity fraction (where cocaine in addition to nicotine, caffeine and cannabinol would be present) was performed by HRGC-MS, first in full scan mode and then in the selected ion monitoring (SIM) mode.

– Ion mobility spectrometry (IMS [43]). Solid phase microextraction (SPME) using a 100 µm polydimethylsiloxane (PDMS) SPME fibre was used for headspace sampling and preconcentration of volatile markers of cocaine, MDMA and marijuana (methyl benzoate, piperonal and terpenes, respectively) in cargo containers. Analysis was then performed by IMS after thermal desorption of the drug markers from the fibre into the IMS analyser.

– Liquid chromatography-electrospray-tandem mass spectrometry (LC-ESI-MS/MS [44]): this technique was the basis of a methodology developed for the multianalyte determination of 17 different drugs of abuse and metabolites (three cocainics, five amphetamine-like compounds, three opiates, three cannabinoids, and lysergic acid and two of its metabolites) in atmospheric airborne particles. Extraction of the target analytes from the filters was performed by pressurised liquid extraction (PLE) using methanol followed by a mixture of methanol-acetone. Further analysis of the extracts, once reduced in volume and reconstituted in methanol, was performed by LC-MS/MS in the selected reaction monitoring (SRM) mode, following a previously optimized protocol developed for the determination of these substances in sewage waters [51].

– Gas chromatography coupled to tandem mass spectrometry (GC-MS/MS [49]): prior to analysis of the three primary active components of cannabis (THC, cannabidiol and cannabinol) by this technique, airborne particulates were extracted in an ultrasonic bath with chloroform. The extracts were combined and concentrated under a stream of nitrogen. The residue was filtered through pre-washed disposable PTFE membranes, dried and finally reconstituted with cyclohexane. After separation and solvent partitioning, the extracts were analysed by GC-MS/MS.

4 Concentrations and Sources of Airborne Psychoactive Substances Around the World

4.1 Major Drugs of Abuse: Cocaine and Cannabinoids

Cocaine is clearly the drug of abuse most extensively studied around the world, when compared to other substances such as cannabinoids, heroin or amphetamine-like compounds.

The ambient concentrations of cocaine detected at urban and rural locations around the world are summarised in Table 1 and Fig. 1 (with a detailed compilation of results in Table 2). Cocaine concentrations in airborne particulates have been measured at 71 locations between 1998 and 2008, in a total of 9 different countries. On average, the amounts of cocaine measured in urban atmospheres rarely exceeded 1,000 pg/m^3 (1 ng/m^3), with usual values ranging between 50 and 200 pg/m^3. Maximum concentrations were registered in Santiago de Chile (3,550 pg/m^3) and followed closely by Mexico City (1,404 pg/m^3), although it must be noted that these were point measurements both in time and space. Studies with a larger spatial and temporal coverage in Brazil, Portugal, Spain and USA estimated mean ambient cocaine concentrations in the range 150–200 pg/m^3 (Table 2), reaching up to 500–600 pg/m^3 as maximum daily values in Spain and Brazil, respectively. For countries with >1 monitoring locations an average value was computed for all sites, disregarding the type (urban, rural, etc.). Maximum daily levels of 267 pg/m^3 were detected in Italy, with a mean concentration for all sampling locations of around 60 pg/m^3. Finally, cocaine concentrations were below the method's detection limit (2 pg/m^3) in Algiers and Serbia, respectively. As shown in Table 1, the largest standard deviation was reported for Mexico (387 pg/m^3), followed by Brazil (192 pg/m^3) and Spain (153 pg/m^3). The high standard deviation reflects a large variability of levels across the different types of monitoring sites, be it urban vs. rural sites as in the case of Brazil or residential vs. nightlife areas in Spain.

Table 1 Airborne concentrations of cocaine detected around the world

	No. of locations	Mean (pg/m^3)	Cocaine Stdev (pg/m^3)	Min (pg/m^3)	Max (pg/m^3)	Range (pg/m^3)
Algiers	5	ND	ND	ND	ND	ND
Brazil	7	179	192	47	590	47–590
Chile	1	3,550	1,098	2,440	4,635	2,440–4,635
Italy	39	59	65	5	267	<5–267
Mexico	1	1,404	468	1,130	1,678	1,130–1,678
Portugal	1	148	$n = 1$	148	148	$n = 1$
Serbia	5	ND	ND	ND	ND	ND
Spain	11	172	153	11	480	11–480
USA	1	200	ND	200	200	$n = 1$

ND not detected

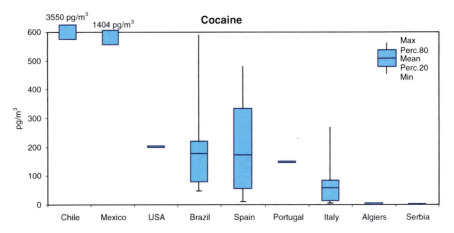

Fig. 1 Mean, maximum and minimum airborne concentrations of cocaine at urban and rural locations around the world

As shown in Table 2, cocaine concentrations are detected in almost all the studies targeting psychoactive substances or drugs of abuse in the literature, with the exception of Algiers and Serbia, where cannabinoids were found whilst cocaine was not (Tables 3 and 4). In the literature, cannabinoid concentrations are either expressed as Δ^9-tetrahydrocannbinol (THC) or as the sum of cannabinol, cannabidiol and THC (known as CBs).

Cocaine and cannabinoids are also detected in suburban areas and only seldom in rural sites. For instance, in two Italian rural localities (Monte Monaco, 1100 ma.s.l., and San Lucido, on the coast) drugs of abuse were only detected in winter. As an example, cocaine and CBs levels in the metropolitan region of Rome and in three major Spanish cities, are shown in Fig. 2. In Italy, three sites (namely Belloni Street, Cipro Street and Francia Boulevard) were located downtown, while Ciampino, Colleferro, Civitavecchia and Guidonia were four towns located 20–80 km from the capital; Montelibretti was a semi-rural area. The highest values of cocaine were reached at Belloni Street. Excluding Montelibretti, the average "suburban" and "urban" concentrations were similar in all the three periods investigated. In the Spanish cities, cocaine levels were higher in Barcelona and Madrid when compared to A Coruña (NW Spain, 250,000 inhabitants), but conversely THC was detected more frequently in A Coruña than in the two major Spanish cities. The unexpectedly high cocaine levels detected in the residential area in Barcelona (Fig. 2) might indicate consumption of this psychoactive substance in an area in which low levels were expected.

The ambient concentrations of THC and CBs detected at urban and rural locations in Algiers, Italy, Mexico and Spain are summarised in Table 3 and Fig. 3 (complete dataset available in Table 4).

On average, CBs levels in Italy were 126 pg/m³, ranging between 30 and 537 pg/m³, much higher than those measured in Mexico city (11 pg/m³, only one location available). When looking at THC only, mean levels were significantly higher

Table 2 Detailed airborne concentrations of cocaine at urban and rural locations around the world

		Cocaine		
	Location	Date	Conc. (pg/m^3)	References
Algeria				
Algiers (downtown)	Urban	2003–2006	ND	A
Algiers (Suburban)	Suburban	2003–2006	ND	A
Algiers (Tafoura)	Downtown	Summer 2007	ND	C
Algiers (Roubia)	Residential	Summer 2007	ND	C
Algiers (Reghaia)	Industrial	Summer 2007	ND	C
Brazil				
Sao Paulo	Urban	Winter 2003	590	C
Sao Paulo	Urban	Autumn 2003	250	C
Piracicaba	Rural	Winter 2003	76	C
Piracicaba	Rural	Winter 2003	89	C
Araraquara	Rural	Winter 2003	101	C
Ouro Preto	Industrial	Winter 02 to autumn 03	47	C
Ouro Preto	Industrial	Winter 02 to autumn 03	101	C
Chile				
Santiago de Chile	Urban	June 2007	2,800	C
Italy				
Rome (University district)	Urban	2003–2006	98 ± 13	A
Rome (Business quarter)	Urban	2003–2006	12 ± 10	A
Rome (Canyon street)	Urban	2003–2006	15 ± 17	A
Rome (Dwelling zone)	Urban	2003–2006	21 ± 20	A
Rome (City garden)	Urban	2003–2006	70 ± 47	A
Rome	Suburban	2003–2006	10 ± 10	A
Rome	Suburban	2003–2006	11 ± 6	A
Rome (downtown)	Urban	Sept 2007	52 ± 26	B
Rome (Fermi)	Traffic	Winter 2006	69	C
Rome (Bissolati)	Urban	Winter 2006	36	C
Rome (Belloni)	Urban garden	Winter 2006	70	C
Rome (Villa Ada)	Urban	2006–2007	63	C
Rome (Villa Ada)	Urban	Winter 2006	83	C
Rome (Villa Ada)	Urban	Summer 2007	30	C
Rome (Malagrotta)	Suburban-industrial	Spring 2007	47	C
Rome city	Urban	Winter 2009	95 ± 38	D
Rome city	Urban	Summer 2009	60 ± 19	D
Montelibretti	Semi-rural (EMEP)	Sept 2007	15 ± 8	B
Montelibretti	Extra-urban	2006–2007	24	C
Taranto (downtown)	Urban	2003–2006	<10	A
Taranto (residential)	Urban	2003–2006	10 ± 10	A
Taranto (rural)	Rural	2003–2006	<5	A
Bari	Traffic	Winter	13	C
Milano (Torre Sarca)	Urban background	2006–2007	208	C
Milano (Torre Sarca)	Urban background	Winter 2006	267	C

(continued)

Table 2 (continued)

	Location	Date	Conc. (pg/m^3)	References
		Cocaine		
Milano (Torre Sarca)	Urban background	Summer 2007	85	C
Milan	Urban	Summer 2006	129	D
Milan	Urban	Winter 2007	267	D
Northern Italy	Urban	Winter 2009	116 ± 118	D
Northern Italy	Urban	Summer 2009	40 ± 28	D
Emilia Romagna region	Urban	Winter 2009	88 ± 39	D
Emilia Romagna region	Urban	Summer 2009	40 ± 13	D
Marche region	Urban	Winter 2009	20 ± 15	D
Marche region	Urban	Summer 2009	9 ± 10	D
S. Italy and Islands	Urban	Winter 2009	24 ± 6	D
S. Italy and Islands	Urban	Summer 2009	19 ± 12	D
Rome province	Urban	Winter 2009	35 ± 17	D
Rome province	Urban	Summer 2009	20 ± 12	D
Central Italy (exc. Rome)	Urban	Summer 2009	12 ± 10	D
Mexico				
Mexico City	Urban	Feb, Apr, Aug 2008	1404 ± 387	H
Portugal				
Oporto (Ermesinde)	Urban background	July 2006	148	C
Serbia				
Pancevo (PHI)	Urban	Summer 06 to spring 07	ND	C
Pancevo (FB)	Urban	Summer 06 to spring 07	ND	C
Pancevo (UR)	Urban	Summer 06 to spring 07	ND	C
Pancevo (FF)	Urban	Summer 06 to spring 07	ND	C
Pancevo (BZ)	Urban	Summer 06 to spring 07	ND	C
Spain				
Barcelona (urban)	Urban	Winter 2007	204 ± 172	E
Madrid (urban)	Urban	Winter 2007	480 ± 276	E
Barcelona (residential)	Urban	Winter 2008	336 ± 173	F
Madrid (residential)	Urban	Winter 2008	28 ± 21	F
Coruña (residential)	Urban	Winter 2008	11 ± 6	F
Barcelona (university)	Urban	Winter 2008	98 ± 66	F
Madrid (university)	Urban	Winter 2008	63 ± 12	F
Coruña (suburban)	Urban	Winter 2008	57 ± 32	F
Barcelona (nightlife)	Urban	Winter 2008	194 ± 141	F
Madrid (nightlife)	Urban	Winter 2008	334 ± 151	F
Coruña (nightlife)	Urban	Winter 2008	85 ± 28	F
USA				
Los Angeles	Urban	Annual 1993	200	G

ND not detected

References: (A) Cecinato and Balducci [42]; (B) Cecinato et al. [37]; (C) Cecinato et al. [50]; (D) Cecinato et al. [46]; (E) Viana et al. [48]; (F) Viana et al. [45]; (G) Hannigan et al. [1]; (H) Cecinato (2011) (see *Acknowledgments*)

Table 3 Airborne concentrations of THC and cannabinoids (CBs) detected around the world

		THC/CBs				
	No. of locations	Mean (pg/m^3)	Stdev (pg/m^3)	Min (pg/m^3)	Max (pg/m^3)	Range (pg/m^3)
Algeria (CNB)	1	<1	ND	<1	<1	<1
Italy (CBs)	20	126	126	30	537	30–537
Mexico (CBs)	1	11	5	11	11	11
Spain (THC)	6	30	8	23	44	23–44

CBs sum of cannabinol, cannabidiol and THC, *CNB* cannabinol

Table 4 Airborne concentrations of THC and cannabinoids (CBs) at urban and rural locations around the world

		Cannabinoids			
Location		Date	Conc. (pg/m^3)	Compound	References
Algeria					
Algiers	Urban	October 2007	<1	THC	C
Italy					
Northern Italy	Urban	Winter 2009	107 ± 88	CBs	D
Emilia Romagna region	Urban	Winter 2009	82 ± 53	CBs	D
Marche region	Urban	Winter 2009	154 ± 95	CBs	D
Southern Italy and Islands	Urban	Winter 2009	168 ± 170	CBs	D
Rome city	Urban	Winter 2009	537 ± 420	CBs	D
Rome province	Urban	Winter 2009	370 ± 213	CBs	D
Northern Italy	Urban	Summer 2009	37 ± 22	CBs	D
Emilia Romagna region	Urban	Summer 2009	41 ± 24	CBs	D
Central Italy (except Rome)	Urban	Summer 2009	107 ± 108	CBs	D
Marche region	Urban	Summer 2009	42 ± 42	CBs	D
Southern Italy and Islands	Urban	Summer 2009	125 ± 126	CBs	D
Rome city	Urban	Summer 2009	78 ± 48	CBs	D
Rome province	Urban	Summer 2009	63 ± 28	CBs	D
Milan	Urban	Summer 2006	113	CBs	D
Milan	Urban	Winter 2007	224	CBs	D
Rome	Urban	September 2008	44 ± 5	THC	C
Rome	Urban	October 2008	104 ± 61	THC	C
Montelibretti	EMEP	September 2008	30 ± 5	THC	C
Montelibretti	EMEP	October 2008	58 ± 15	THC	C
Bari	Urban	March 2007	39 ± 7	THC	C
Mexico					
Mexico City	Urban	Feb, Apr, Sep 2008	11 ± 5	CBs	G
Spain					
Barcelona	Urban	Winter 2007	27 ± 42	THC	E
Madrid	Urban	Winter 2007	44 ± 35	THC	E
Barcelona (residential)	Urban	Winter 2008	26 ± 10	THC	F
Coruña (suburban)	Urban	Winter 2008	26 ± 8	THC	F
Madrid (nightlife)	Urban	Winter 2008	23 ± 19	THC	F
Coruña (nightlife)	Urban	Winter 2008	34 ± 23	THC	F

CBs sum of cannabinol, cannabidiol and THC
References: (C) Cecinato et al. [37]; (D) Cecinato et al. [46]; (E) Viana et al. [48]; (F) Viana et al. [45]. (G) Cecinato (2011) (see *Acknowledgments*)

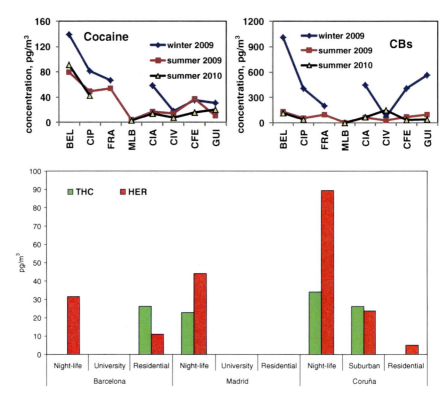

Fig. 2 *Top*: Cocaine and CBs levels in the metropolitan region of Rome, detected during field campaigns in 2009–2010. *COC* cocaine, *CBs* cannabinol + cannabidiol + Δ^9-tetrahydrocannabinol). *BEL* Belloni Street, *CIP* Cipro Street, *FRA* Francia Boulevard, *MLB* Montelibretti, *CIA* Ciampino, *CIV* Civitavecchia, *CFE* Colleferro, *GUI* Guidonia. *Bottom*: Heroin and THC levels in three Spanish cities, detected during field campaigns in 2008

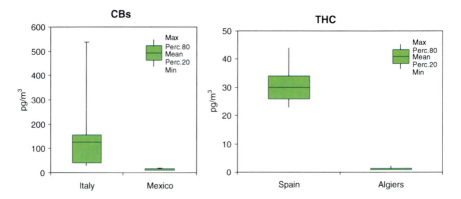

Fig. 3 Mean, maximum and minimum airborne concentrations of THC and cannabinoids (CBs) at urban and rural locations around the world. *CBs* sum of cannabinol, cannabidiol and THC

in Spain (30 pg/m^3, with a maximum of 44 pg/m^3) than in Algiers (<1 pg/m^3). In Italian cities, cannabinol accounted for more than 90% of the total CBs in ~90% of cases but in specific samples (the richest ones, which were recorded always during winter) THC predominated. Moreover, it was surprising to observe that the sum of the three cannabinoids (CBs) was usually lower than the ambient concentration of cocaine, even though the abuse of cannabis/marijuana smoking is generally estimated to be higher.

4.2 Minor Drugs of Abuse: Heroin and Amphetamine-Like Compounds

Heroin and amphetamine-like compounds (amphetamine, methamphetamine, MDMA or ecstasy, (R,R)($-$)-pseudoephedrine (PS-EPH), and (1S,2R)(+)-ephedrine hydrochloride (EPH-HCl), the last two measured together as total ephedrine) have so far only been detected in airborne particulates in Spain (Tables 5, 6 and 7). Mean heroin concentrations ranged between 10 and 50 pg/m^3 (Table 5), with maximum daily levels reaching 80–90 pg/m^3. The maximum concentrations were detected in Madrid and A Coruña, and seemed to be independent of population size. As for amphetamine-like compounds, airborne levels were always below 15 pg/m^3

Table 5 Airborne concentrations of heroin, amphetamine, ephedrine, metamphetamine and MDMA (ecstasy) detected in Spain

Spain	No. of locations	Mean (pg/m^3)	Stdev (pg/m^3)	Min (pg/m^3)	Max (pg/m^3)	Range (pg/m^3)
Heroin	7	41	34	5	90	5–90
Amphetamine	4	8	7	1.4	15	1.4–15
Ephedrine	3	3.9	0.7	3.1	4.4	3.1–4.4
Metamphetamine	3	6.3	3.6	2.8	10	2.8–10
MDMA (ecstasy)	2	4.0	1.4	3.0	5	3–5

Table 6 Airborne concentrations of heroin at urban locations in Spain

Heroin	Location	Date	Conc. (pg/m^3)	References
Spain				
Madrid (urban)	Urban	Winter 2007	84 ± 53	E
Barcelona (residential)	Urban	Winter 2008	11 ± 4	F
Coruña (residential)	Urban	Winter 2008	5	F
Coruña (suburban)	Urban	Winter 2008	24 ± 14	F
Barcelona (nightlife)	Urban	Winter 2008	26	F
Madrid (nightlife)	Urban	Winter 2008	44 ± 21	F
Coruña (nightlife)	Urban	Winter 2008	90 ± 52	F

References: (E) Viana et al. [48]; (F) Viana et al. [45]. It is worth noting that Cecinato et al. [37, 50] reported heroin levels below the method's quantification limit (70 pg/m^3)

Table 7 Airborne concentrations of amphetamine-like compounds at urban locations in Spain

Spain	Amphetamine			
	Location	Date	Conc. (pg/m3)	References
Barcelona	Urban	Winter 2007	2.3 ± 1.2	E
Madrid	Urban	Winter 2007	1.4 ± 0.9	E
Barcelona (nightlife)	Urban	Winter 2008	13 ± 0.5	F
Barcelona (residential)	Urban	Winter 2008	15	F
Ephedrine				
Barcelona (nightlife)	Urban	Winter 2008	3.1	F
Barcelona (university)	Urban	Winter 2008	4.4	F
Barcelona (residential)	Urban	Winter 2008	4.3 ± 1.5	F
Metamphetamine				
Barcelona (nightlife)	Urban	Winter 2008	6	F
Barcelona (residential)	Urban	Winter 2008	2.8	F
Madrid (nightlife)	Urban	Winter 2008	10 ± 10	F
MDMA (ecstasy)				
Madrid (nightlife)	Urban	Winter 2008	3	F
Coruña (nightlife)	Urban	Winter 2008	5	F

References: (E) Viana et al. [48]; (F) Viana et al. [45]

(Table 5), with amphetamine as the highest contributor to ambient levels (mean of 8 pg/m^3).

These mean levels are markedly lower than those described above for cocaine (ranging between 50 and 200 pg/m^3 but occasionally exceeding 1,000 pg/m^3) and CBs (126 pg/m^3 in Italy) in urban environments. Conversely, heroin and THC were detected at similar airborne concentrations in urban areas in Spain (THC 23–44 pg/m^3, heroin 5–90 pg/m^3), and at a similar number of study locations (six sites for THC, seven for heroin). The different consumption patterns of each drug of abuse should be assessed in order to interpret the ambient concentrations found: cocaine is generally consumed as powder, whereas cannabis (THC or CBs) is mostly smoked, amphetamines are handled as pills and heroin may be injected and/or smoked. Thus, the emission of cocaine, cannabis and heroin into the atmosphere is direct through consumption and handling, whereas amphetamine traces are emitted mainly through handling of the pills. This probably explains the lower concentrations of amphetamines (always <15 pg/m^3) found. In addition, the low molar mass of amphetamines (135 g/mol) when compared to that of other drugs (e.g., heroin, 369 g/mol) and their high vapour pressure could also be the causes of the lower levels of these compounds detected in particulate form.

4.3 Nicotine and Caffeine

Till now nicotine and caffeine have been monitored outdoors only to chemically characterise the particulates containing illicit substances, with the purpose of pointing out quantitative links among the respective airborne concentrations [37, 46]. Indeed, no toxic properties have been ascribed to the two pure compounds at the amounts expected in the atmosphere. On the other hand, looking to the worldwide consumption of nicotine and caffeine, very little is known about their time and space modulations. A short example is provided in Fig. 4 where the average concentrations of nicotine and caffeine recorded in Italy during 2009 are shown. The measurements were made in January–February and in June.

The highest airborne concentrations of nicotine and caffeine have been recorded in the Northern cities (see Fig. 4). The gross average of nicotine reached 36 ± 8 ng/m^3 in winter and 14 ± 3 ng/m^3 in summer across Italy, and that of caffeine 15 ± 8 ng/ m^3 and 1.2 ± 0.6 ng/m^3, respectively. In the Rome metropolitan region, nicotine was 28 ± 8 ng/m^3 and 11 ± 2 ng/m^3 in the two periods, and caffeine 1.5 ± 0.5 ng/m^3 and 0.3 ± 0.1 ng/m^3.

According to the data available in the literature, both nicotine and caffeine reach much higher concentrations in winter than in summer. The ambient temperature probably influences the net amounts of compounds in the particulate phase, through promoting volatilization in the warm season; for instance, nicotine was ≈ 2.2 times more abundant in winter than in summer. Nonetheless, atmospheric oxidants seem to play a complementary role; in fact, with regard to nicotine this hypothesis is in

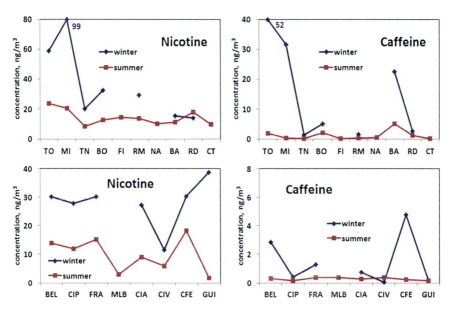

Fig. 4 Average atmospheric concentrations of nicotine and caffeine in Italy (2009). *Top*: Cities lying in Northern (N), Central (C) and Southern (S) Italy: TO: Turin (N); MI: Milan (N); Bo: Bologna (N); FI: Florence (C); RM: Rome (C); NA: Naples (S); BA: Bari (S); RD: Rende (S); CT: Catania (Sicily). *Bottom*: Rome metropolitan region, *BEL* Belloni Street, *CIP* Cipro Street, *FRA* Francia Boulevard, *MLB* Montelibretti, *CIA* Ciampino, *CIV* Civitavecchia, *CFE* Colleferro, *GUI* Guidonia

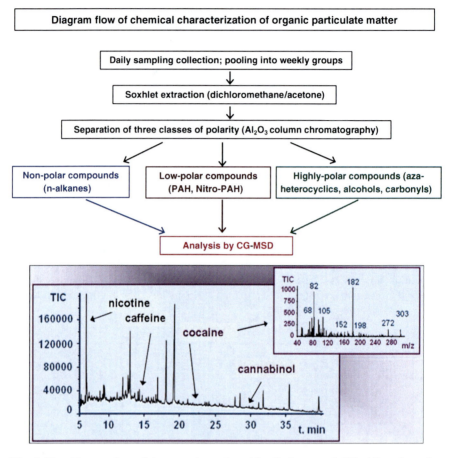

Fig. 5 *Top*: Diagram flow of the procedure adopted by Cecinato et al. [37, 46] to determine psychotropic substances. *Bottom*: Drug detection through GC-MSD (EI-SIM) analysis of an airborne particulate organic extract, highly polar fraction. (Rome, 2006)

agreement with the frequent detection of cotinine in summer samples; moreover, the winter/summer concentration ratio of caffeine reached ≈10 (Fig. 5).

5 Spatial and Temporal Variability of Airborne Psychoactive Substances in Cities

The atmospheric and chemical processes controlling the spatial and temporal variability of psychoactive substances in urban atmospheres are largely uncertain, mostly due to the fact that the atmospheric residence time of these compounds is so far unclear. The transport, transformation and deposition/atmospheric removal mechanisms of psychoactive substances are a current subject of study. One incognita is the prevalence of these compounds in the atmosphere, i.e., whether they remain in suspension and propagate unaltered after their release, or whether chemical transformations occur while the drugs are airborne. In addition, the social, geographical and meteo-climatic contexts are factors to be taken into account. High airborne contents of cocaine were detected in areas of high drug prevalence, which are usually considered large metropolis. It is worth noting, in this regard, that densely inhabited areas are probably the more prone both to atmospheric pollution and drug of abuse consumption. However, drug levels did not always correlate with population size, as seen in Spain where cocaine levels were highest in Madrid and Barcelona (the country's two major cities, see Table 2), whereas heroin levels were highest in A Coruña (Table 6).

Studies carried out in Italy evidenced that airborne drug levels did not always follow atmospheric pollution indicators such as particulate PAH or PM_{10}, and that low concentrations were concurrent with high levels of ozone and/or air temperature. In order to be able to correctly interpret temporal and spatial variations of drugs of abuse in ambient air, the concentrations of these substances should be normalised by means of usual pollution indicators such as $PM_{2.5}$ or soot. This normalisation allows for the direct comparison of drug concentrations between different localities and study periods. This approach is of course advisable for legal psychoactive substances (caffeine and nicotine) as well. For instance, the cocaine/nicotine, cocaine/caffeine and cocaine/PAH concentration ratios (data not reported here) at Belloni Street (Italy) were 1.5–2 times higher than in the other downtown sites and 2–4 times higher than in the towns, roughly proportional to the local densities of population. In Spain, cocaine levels were potentially linked to population size (not population density), whereas the levels of heroin and cannabinoids in the three cities studied (Madrid, Barcelona and A Coruña) seemed to be independent of it. To this respect, it is important to consider that cocaine and cannabis are mainly perceived as recreational drugs, whereas heroin abuse is a more marginal conduct.

Other than population size, studies in Spain have shown a correlation between airborne concentrations of drugs of abuse and social environments with regard to drug abuse: nightlife, university and residential areas. Independently of population size, higher levels were detected in the nightlife areas, pointing towards consumption and trafficking as major emission sources, and possibly ruling out drug manufacture. With only one exception in Barcelona, a clear decreasing trend was detected in the levels of cocaine, heroin and cannabinoids from the nightlife areas, to the university campus and the residential areas (see Fig. 6 for one example). These results suggest that the concentrations of drugs of abuse in ambient air are a good indicator of social behaviour with respect to these substances (consumption and/or small-scale trafficking). The determination of the levels of psychotropic substances in ambient particulates may even be considered a useful tool to monitor drug abuse or trafficking in urban environments, and to detect it even in areas where it might not be expected.

The temporal variability of these substances has been analysed as a function of annual and weekly trends. Weekly concentrations monitored for almost 2 years in

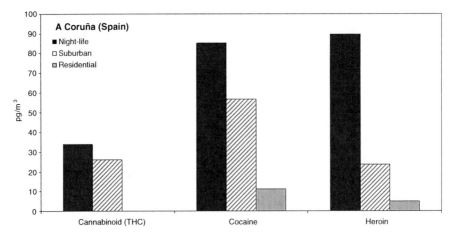

Fig. 6 Airborne concentrations of cocaine, heroin and cannabinoid (THC) in three areas in the city of A Coruña (Spain): recreational nightlife, suburban and residential

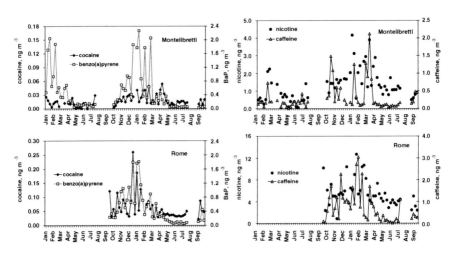

Fig. 7 Weekly modulation of cocaine, benzo(a)pyrene, nicotine and caffeine in the air of downtown Rome and Montelibretti from January 2006 to September 2007

Rome and Montelibretti (Italy) evidenced slightly higher cocaine concentrations during the summer, which in Rome correlated with the levels of benzo(a)pyrene (but not at Montelibretti, Fig. 7). This finding should be ascribed to the different features of locations. In fact, the residential zones were preferably examined in Rome, whilst Montelibretti was affected by duty vehicle traffic, that probably increased the PAH release into the atmosphere. In Spain, the weekly evolution of drugs of abuse in atmospheric particles (normalised by PM, to remove the influence of meteorology) evidenced a clear weekly trend for ambient concentrations of all substances (cocaine, heroin, cannabinoids and ephedrine, Fig. 8), with higher levels

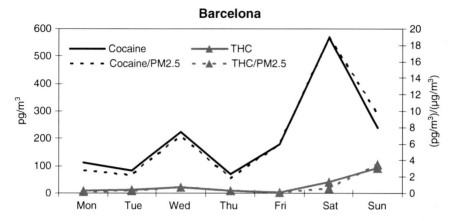

Fig. 8 Example of the weekly trend of ambient concentrations of cocaine and THC in Barcelona (Spain)

during weekends (including Fridays) and with weekend/week day ratios of up to 20 for cocaine and 48 for THC. This trend, expected for nightlife areas due to the increased consumption or trafficking, was however unexpected at university and residential areas. In the latter, either: (a) levels were expected to be higher on weekdays due to the presence of students (university), or (b) levels were expected to remain constant due to the absence of consumption (residential area). Conversely, higher levels were always detected during weekends. This result suggests a long residence time of these substances in the atmosphere, which seem to remain in suspension and propagate unaltered after their release. Thus, we may hypothesize that drug emissions into the atmosphere take place mostly in the nightlife areas and on weekends, and that drugs are then transported on the city-scale and detected in residential and university areas. Because of dilution processes during transport of the drugs, it is not possible with the current data to quantitatively estimate the atmospheric residence time of the drugs analysed.

One issue which remains unclear in the literature is if the occurrence of drugs of abuse in the atmosphere is related to trafficking, manufacture or abuse [46]. The temporal or spatial patterns of drug manufacture in cities are uncertain. However, the concentrations detected in Spain peaked on weekends and in the nightlife areas, thus suggesting consumption and/or small-scale trafficking (dealing) as the main emission source.

6 Consumer Groups and Consumption Patterns

According to the most recent statistics, tobacco smokers exceed one billion in number and about 15 billion cigarettes are consumed every day [8]. In particular, more than 60% of 14÷65y males are smokers in China and Russia, ≈50% in

Eastern Europe, 30% in Western Europe and Brazil, 19% in USA, Canada and Mexico and 14% in India. Female smokers are lesser (25% in Eastern and Western Europe, 20% in Brazil, 17% in USA) or much lesser (10% in Russia, China and India; 2.5% in Mexico), with the exception of France, Germany and Argentina (35%). Eleven to fifteen cigarettes per capita are smoked every day on the average. Alternatives to cigarettes (cigars, electronic devices, beedis, pipes and vaporizers) are appreciated by small fractions of population.

Caffeine-containing beverages are consumed by a large fraction of world populations. Coffee is drunk preferably by European and American people (11.9 kg/year per capita in Finland, 6.50 kg/y p.c. in Germany, 5.8 kg/y p.c. in Italy, 5.7 kg/y p.c. in Brazil), while consumption is much lower in Asia and Africa (0.9 kg/y p.c. in Ivory Coast, 0.1 kg/y p.c. in Benin, 0.1 kg/y p.c. in India) [52]. The caffeine contents range from 3 mg/cup for decaffeinate, to 65 for instant and to 115 mg for roasted and ground coffee. On the other hand, non-coffee consumers very often drink tea, which contains $0.16 \div 0.40$ mg/mL of caffeine, and caffeine is present also in chocolate ($0.4 \div 0.9$ mg/g) and in cola beverages (~120 mg/L). For instance, when the world tea crop production reaches 3.95×10^{12} g per year, China and India account for 1.38 and 0.80×10^{12} g, respectively [53].

The consumption of drugs of abuse may follow patterns which define different types of consumer groups [54]. This means that certain drugs may be preferred by certain types of consumers, and thus be consumed at the same location at a given point in time (by single or different individuals). This pattern is consequently reflected in the airborne concentrations of these substances. By means of statistical analyses, high degrees of correlation ($r^2 = 0.98$) were found between cocaine and amphetamine in Spain, whereas weak relationships were found between cocaine and heroin ($r^2 = 0.26$). The consumption of heroin and cocaine was not expected to be linked in space and time, given the different profiles of consumers of both substances [55, 56]. THC was rarely consumed in combination with any of the other substances, showing low correlation coefficients.

In addition, the ratios between weekend and weekday levels for the different substances in Spain also revealed characteristic consumption patterns: cocaine levels were higher on weekends than on weekdays by a factor of 1–3, cannabinoids were higher by a factor of 1–2, and heroin was higher by a factor of only 1.1–1.7. As described above, cocaine and cannabis are perceived as recreational drugs, whereas heroin abuse is a more marginal conduct, and therefore cocaine and cannabinoid levels show a much stronger increase during weekends than those of heroin. Conversely, heroin consumption is slightly more constant throughout the week, according to airborne concentrations of these substances.

The correlation of airborne drugs of abuse and other anthropogenic atmospheric pollutants might provide insights into potential emission sources of psychotropic substances. In Spain, the atmospheric pollutants used for this analysis were $PM_{2.5}$ chemical components. At one urban background location in Barcelona and Madrid, cocaine and THC levels were correlated with tracers of vehicular traffic: mineral matter, copper (Cu) and antimony (Sb) in Barcelona (the monitoring station was located in the vicinity of an unpaved parking lot, thus the high mineral matter emissions), and total carbon in Madrid. Total carbon, copper and antimony are well-known tracers of vehicular emissions (Adachi and Tainosho 2004) [59]. Higher vehicle traffic indicates higher potential consumer traffic. As expected, no significant correlations were found between drugs of abuse and sulphate (SO_4^{2-}), a well-known tracer of long-range air mass transport (Seinfeld and Pandis 1998) [60]. Despite these findings, the determination of the local or regional origin of the drugs of abuse in atmospheric particles is a very complex issue involving large uncertainties regarding source–receptor relationships at such scales.

7 Exposure to Airborne Psychotropic Substances: Potential Health Effects

With regard to the potential health hazards related to the inhalation of these substances by non-consumers, an estimation was made based on data provided by the Spanish Observatory for Drugs [54]. According to these data, average doses for private consumption are 197 mg of cocaine, 106 mg of heroin and 2–5 g of cannabis (THC). As a result, and based on the maximum daily ambient concentrations detected in Spain, it was estimated that it would require breathing over a period of 32,000–47,000 years (based on average human breathing of 20 m^3/day) to inhale the equivalent of 1 dose of cocaine. In the case of THC, it would take $> 4 \times 10^6$ years to inhale the equivalent to 1 dose, and >100,000 years for heroin. As a conclusion, personal exposure to the levels of drugs detected in ambient particulates, in the levels reported in Viana et al. [45, 48] may be considered negligible.

As for ambient toxicity associated with drugs, no studies have been conducted till now and probably will not be made in the near future. The knowledge about their occurrence in the atmosphere and the potential medium- and long-term effects of exposition to the low concentrations normally detected in air is too poor. However, cocaine levels have been detected which exceeded benzo(a)pyrene or nicotine, reaching hundreds of nanograms per cubic metre in open air. Thus, both licit and illicit substances should be reasonably included among emerging contaminants. Combining progress, scientific knowledge and legislation, the community is clearing the atmosphere from "old" pollutants; but the lack of sensibility and awareness is introducing new dangerous species into the environment.

8 Discussion: Drawing Abuse Prevalence Indicators from Drug Contents in the Air

Unfortunately, till now measurements were unavailable from drug production regions as well as from sites suspected of drug refinery or trafficking. By consequence, the possibility of identifying point-type sources like refineries and drug shops is still unexplored. Due to the variety of factors determining the concentration of psychotropic substances in the atmosphere, whatever algorithm capable of linking abuse prevalence in local communities (inhabitants of city districts, urban areas or metropolitan regions) to the airborne contents of drugs must be complex. Thus, the route to transform the airborne concentration of cocaine, cannabinoids or chemicals into prevalence indicators (similar to what is happening for the drug residues contaminating waste waters [57]) bristles with difficulties and uncertainties.

To reach this purpose, two ways seem practicable. The first would require to estimate all parameters determining the actual content of cocaine in the air, including the emission population density, the emission factor (i.e., the net drug amount released per dose consumed), the percentage existing in the particulate phase and the average height of the boundary layer along with other meteorological parameters controlling atmospheric dilution. Moreover, the persistence of cocaine in the atmosphere should be presumed. The second way would consist of selecting one co-pollutant presenting a similar environmental behaviour, but whose release into the atmosphere is quite independent of land-characterized population. Whereas measurable, this co-pollutant would act as "normalisation" factor accounting for all environmental variables except for the prevalence of abuse. Both approaches proposed should be verified and would require to be "tuned" on the basis of historical archives of drug consumption.

The determination of the levels of drugs of abuse in atmospheric particulates collected at regular air quality monitoring sites could constitute a useful public health tool with a range of potential applications

- Long-term monitoring of consumption patterns: the determination of the levels of drugs of abuse on a broader temporal scale (e.g., years) would allow for the identification of changes in consumption patterns, linked to the decrease in the consumption of certain substances and the increase of new emerging drugs.
- Mapping of consumption and/or drug dealing areas within cities: drug monitoring with a high spatial resolution would allow detecting and identifying potential hotspots in any given city, which would indicate high consumption or drug dealing areas.
- Modelling of levels and atmospheric transport of drugs of abuse in the urban environment: results from ambient levels of drugs of abuse within the city could be introduced in dispersion models for simulate atmospheric transport of these substances in urban environments. This methodology can be combined with health population data and other tools such as GIS-based systems in order to generate health-risk maps.
- This tool, based on regular monitoring sites from existing networks presents a number of major advantages. One of the main ones is that chemical determinations of psychoactive substances may be performed on particulates collected on quartz-fibre filters, which are the EU reference collection substrates EU 2008 [58], and are therefore currently available throughout Europe and the United States in local air quality networks. This minimises the cost of these determinations with no new sampling instrumentation needed. Other advantages over the usual indicators are semi real-time information (12–24 h sampling

duration), anonymity and cost. In addition, the determination of drugs in airborne particulates would allow for the detection of activities such as drug handling and dealing, which are related to consumption but which are not detectable by indicators such as population health statistics.

Acknowledgements This work was partially funded by the Spanish Ministry of Science and Innovation through the Ramón y Cajal programme and the project NANO-TROJAN (CTM2011-24051). Cristina Postigo acknowledges the European Social Fund and AGAUR (Generalitat de Catalunya, Spain) for their economical support through the FI pre-doctoral grant. The authors would also like to thank Prof. Amador Muñoz and his staff (Atmospheric Science Center, Universidad Nacional Autónoma de México), who kindly provided particulate samples from Mexico City.

References

1. Hannigan MP, Cass GR, Penman BW, Crespi CL, Busby JWF, Lafleur AL, Thilly WG, Simoneit BRT (1998) Bioassay-directed chemical analysis of Los Angeles airborne particulate matter using a human cell mutagenicity assay. Environ Sci Technol 32:3502–3514
2. ADCA (2007) Alcohol and other drugs: chapter 1. In: Laycock A (ed) The public health bush book: a resource for working in community settings in the Northern Territory. The Alcohol and Other Drugs Council of Australia, Dept. of Health and Community Services, Australia
3. Italian National Institute of Statistics ISTAT (2010) The 2010 Annual Report (Annuario 2010). ISTAT Publ., Rome, 858 pp
4. Italian Superior Institute of Health ISS (2010) Rapporto Nazionale PASSI 2009: Guadagnare salute. Rome. http://www.epicentro.iss.it/passi/default.asp
5. Food and Drug Administration (FDA) (1996) Regulations restricting the sale and distribution of cigarettes and smokeless tobacco to protect children and adolescents; final rule. Fed Regist 61:44396–45318
6. Gazzetta Ufficiale della Repubblica Italiana (GURI). Italian Parliament. Disposizioni ordinamentali in materia di pubblica amministrazione. Released on January the 16th, 2003. In: GURI, January the 20th, 2003, 15, Suppl. 5
7. U.K. Parliament (2006) Health Act 2006, Chapter 28. The stationery office limited, pp 1–91
8. World Health Organization (WHO) (2011) Report on the global tobacco epidemic, 2011: warning about the dangers of tobacco. WHO Press, Geneva, CH, pp 1–164
9. Posselt W, Reimann L (1828) Chemischeuntersuchungen des tabaksund darstellung des eigenthumlichenwirksamenprincipsdieserpflanze. Geigers Magazin der Pharmazie 24:138–161
10. Russell MAH (1971) Cigarette smoking: natural history of a dependence disorder. Br J Med Psychol 44:1–16
11. Russell MAH (1979) Tobacco dependence: is nicotine rewarding or aversive? In: Krasnegor NA (ed) Cigarette smoking as a dependence process. NIDA Research Monograph 23. Public Health Service, US Department of Health Education & Welfare, Washington, DC, pp 100–122
12. Henningfield JE, London ED, Jaffe JH (1987) Nicotine reward: studies of abuse liability and physical dependence potential. In: Orleand L, Engel J (eds) Brain reward systems and abuse. Raven Press, New York, pp 147–164
13. Kessler DA (1995) Statement of David Kessler, Commissioner of Food and Drugs, Food and Drug Administration, Accompanied by Jack E. Henningfield, Chief, Clinical Pharmacology Branch, National Institute on Drug Abuse. Hearings before the Subcommittee on Health and

the Environment of the Committee on Energy and Commerce, House of Representatives, 103rdCongress, 2nd Session. U.S.Government Printing Office, Washington DC, 1995, 28–43

14. Food and Drug Administration (FDA) (1995) FDA regulations restricting thesale and distribution of cigarettes and smokeless tobaccoproducts to protect children and adolescents; proposed ruleanalysis regarding FDA's jurisdiction over nicotine-containingcigarettes and smokeless tobacco products; notice. Fed Regist 60:41314–41792

15. Van De Ven MOM, Greenwood PA, Engels RCME, Olsson CA, Patton GC (2010) Patterns of adolescent smoking and later nicotine dependence in young adults: a 10-year prospective study. Public Health 124:65–70

16. Yildiz D, Liu Y-S, Ercal N, Armstrong DW (1999) Comparison of pure nicotine- and smokeless tobacco extract-induced toxicities and oxidative stress. Arch Environ Contam Toxicol 37:434–439

17. Henningfield JE, Zeller M (2006) Nicotine psychopharmacology research contributionsto United States and global tobacco regulation: a look back and a look forward. Psychopharmacology 184:286–291

18. Eatough DJ, Benner CL, Bayona JM, Richards G, Lamb JD, Lee ML, Lewis EA, Hansen LD (1969) Chemical compositlon of environmental tobacco smoke. 1 Gas-phase acids and bases. Environ Sci Technol 23:679–687

19. Häger B, Niessner R (1997) On the distribution of nicotine between the gas and particle phase and its measurement. Aerosol Sci Technol 26:163–174

20. Lewis DA, Colbeck I, Mariner DC (1995) Dilution of mainstream tobacco smoke and its effects upon the evaporation and diffusion of nicotine. J Aerosol Sci 26:841–846

21. Lewis DA, Colbeck I (1996) Modelling of nicotine diffusion from mainstream tobacco smoke within denuder tubes. J Aerosol Sci 27:S319–S320

22. Michael CM, Demetriou E, Kosmas V, Krashia A, Akkelidou D (1996) Nicotine levels in indoor athletic centres. Sci Total Environ 192:213–216

23. Schorp MK, Leyden DE (2002) Distribution analysis of airborne nicotine concentrations in hospitality facilities. Environ Int 27:567–578

24. Hengel MJ, Hung BK, Engebretson JA, Shibamoto T (2005) Analysis of nicotine in California air samples from XAD-4 resin. Bull Environ Contam Toxicol 74:445–455

25. Kim S, Aung T, Berkeley E, Diette GB, Breysse PN (2008) Measurement of nicotine in household dust. Environ Res 108:289–293

26. Sureda X, Fu M, José López M, Martínez-Sánchez JM et al (2009) Second-hand smoke in hospitals in Catalonia (2009): a cross-sectional study measuring PM2.5 and vapor-phase nicotine. Environ Res 110:750–755

27. Sleiman M, Destaillats H, Smith JD, Liu C-L, Ahmed M, Wilson KR et al (2010) Secondary organic aerosol formation from ozone-initiated reactions with nicotine and second hand tobacco smoke. Atmos Environ 44:4191–4198

28. Petrick L, Svidoski A, Dubovski Y (2011) Third-hand smoke: heterogeneous oxidation of nicotine and secondary aerosol formation in the indoor environment. Environ Sci Technol 45:328–333

29. UN (1972) United Nations Single Convention on Narcotic Drugs, United Nations, New York, 25 March 1972, p 44

30. Daughton CG (2011) Illicit drugs: contaminants in the environment and utility in forensic epidemiology. In: Reviews of environmental contamination and toxicology, USEPA Ed. Las Vegas, NV, pp 59–110

31. Leaderer BP, Hammond SK (1991) Evaluation of vapor-phase nicotine and respirable suspended particle mass as markers for environmental tobacco smoke. Environ Sci Technol 25:770–777

32. Ongwandee M, Sawanyapanich P (2011) Influence of relative humidity and gaseous ammonia on the nicotine sorption to indoor materials. Indoor Air 2011, wileyonlinelibrary.com/journal/ina, doi:10.1111/j.1600-0668.2011.00737.x

33. Liang C, Pankow JF (1996) Gas/particle partitioning of organic compounds to environmental tobacco smoke: partition coefficient measurements by desorption and comparison to urban particulate material. Environ Sci Technol 30:2800–2805
34. Pankow J, Mader B, Isabelle LM, Luo W, Pavlik A, Liang C (1997) Conversion of nicotine in tobacco smoke to its volatile and available free-base form through the action of gaseous ammonia. Environ Sci Technol 31:2428–2433
35. World Health Organization (WHO) (2008) Report on the global tobacco epidemic, 2008: the MPOWER package. WHO Press, Geneva, CH, pp 1–342
36. World Health Organization (WHO) (2009) Report on the global tobacco epidemic, 2009: implementing smoke-free environments. WHO Press, Geneva, CH, pp 1–568
37. Cecinato A, Balducci C, Nervegna G (2009) Occurrence of cocaine in the air of the World's cities an emerging problem? a new tool to investigate the social incidence of drugs? Sci Total Environ 407:1683–1690
38. Hammond SK (1999) Exposure of U.S. workers to environmental tobacco smoke. Environ Health Perspect 107:329–340
39. Nebot M, López MJ, Gorini G, Neuberger M, Axelsson S, Pilali M, Fonseca C, Abdennbi K, Hackshaw A, Moshammer H, Laurent AM, Salles J, Georgouli M, Fondelli MC, Serrahima E, Centrich F, Hammond SK (2005) Environmental tobacco smoke exposure in public places of European cities. Tobac Contr 14:60–63
40. Stillman F, Navas-Acien A, Ma J, Ma S, Avila-Tang E, Breysse P, Yang G, Samet J (2007) Second-hand tobacco smoke in public places in urban and rural China. Tobac Contr 16:229–234
41. Selvavinayagam TS (2010) Air nicotine monitoring study at Chennai, Tamil Nadu to assess the level of exposure to second hand smoke in public places. Indian J Commun Med 35:186–188
42. Cecinato A, Balducci C (2007) Detection of cocaine in the airborne particles of the Italian cities Rome and Taranto. J Sep Sci 30:1930–1935
43. Lai H, Corbin I, Almirall JR (2008) Headspace sampling and detection of cocaine, MDMA, and marijuana via volatile markers in the presence of potential interferences by solid phase microextraction-ion mobility spectrometry (SPME-IMS). Anal Bioanal Chem 392:105–113
44. Postigo C, Alda MJLd, Viana M, Querol X, Alastuey A, Artiñano B, Barcelo D (2009) Determination of drugs of abuse in airborne particles by pressurized liquid extraction and liquid chromatography-electrospray-tandem mass spectrometry. Anal Chem 81(11):4382–4388
45. Viana M, Postigo C, Querol X, Alastuey A, Alda MJLd, Barceló D, Artiñano B, López-Mahia P, Gacio DG, Cots N (2011) Cocaine and other illicit drugs in airborne particulates in urban environments: a reflection of social conduct and population size. Environ Pollut 159 (5):1241–1247
46. Cecinato A, Balducci C, Budetta V, Pasini A (2010) Illicit psychotropic substance contents in the air of Italy. Atmos Environ 44:2358–2363
47. Zaromb S, Alcaraz J, Lawson D, Woo CS (1993) Detection of airborne cocaine and heroin by high-throughput liquid-absorption preconcentration and liquid chromatography-electrochemical detection. J Chromatogr 643:107–115
48. Viana M, Querol X, Alastuey A, Postigo C, López de Alda MJ, Barceló D, Artiñano B (2010) Drugs of abuse in airborne particulates in urban environments. Environ Int 36:527–534
49. Balducci C, Nervegna G, Cecinato A (2009) Evaluation of principal cannabinoids in airborne particulates. Anal Chim Acta 641:89–94
50. Cecinato A, Balducci C, Nervegna G, Tagliacozzo G, Allegrini I (2009) Ambient air quality and drug aftermaths of the Notte Bianca (White Night) holidays in Rome. J Environ Monit 11:200–204
51. Postigo C, Alda MJLd, Barcelo D (2008) Fully automated determination in the low nanogram per liter level of different classes of drugs of abuse in sewage water by on-line solid-phase extraction-liquid chromatography-electrospray-tandem mass spectrometry. Anal Chem 80(9):3123–3134
52. International Coffee Organization ICO (2011) Country datasheets 2009. London. http://www.ico.org/profiles_e.asp

53. Food and Agriculture Organization (FAO) (2011) FAOSTAT. Rome. http://faostat.fao.org/site/567/DesktopDefault.aspx?PageID=567#ancor
54. OED (2007) Annual report 2007: Situación y tendencias de los problemas de drogas en España; MINISTERIO DE SANIDAD Y CONSUMO: Madrid, Spain; 199 pp
55. United Nations Office Drug & Crime (UNODC) (2006) 2006 World Drug Report, vol 1–2. UN Publications, Geneva
56. United Nation Office on Drug And Crime (2007) 2007 World Drug Report. United Nations Publications, Geneva
57. Postigo C, Alda MJLd, Barcelo D (2008) Analysis of drugs of abuse and their human metabolites in water by LC-MS(2): a non-intrusive tool for drug abuse estimation at the community level. TRAC-Trends Anal Chem 27(11):1053–1069
58. European Union (2008) European Normative EN 15549 - Air Quality - Standard method for the measurement of the concentration of benzo[a]pyrene in ambient air. EU Publ., Brussels
59. Adachi K, Tainosho Y (2004) Characterization of heavy metal particles embedded in tire dust. Environment International 30:1009–1017
60. Seinfeld JH, Pandis SN (1998) Atmospheric Chemistry and Physics: From air pollution to climate change. John Wiley & Sons, Inc.: pp 1323

Index

Barceló (ed.), *Emerging Organic Contaminants and Human Health,*   461
Hdb Env Chem (2012) 20: 461–466, DOI 10.1007/698_2011,
© Springer-Verlag Berlin Heidelberg 2011

Printed by Publishers' Graphics LLC